Cancer Nanotechnology
Principles and Applications
in Radiation Oncology

IMAGING IN MEDICAL DIAGNOSIS AND THERAPY

William R. Hendee, Series Editor

Forthcoming titles in the series

Cancer Nanotechnology
Principles and Applications
in Radiation Oncology

Edited by

Sang Hyun Cho, PhD
Sunil Krishnan, MD

CRC Press
Taylor & Francis Group
Boca Raton London New York

CRC Press is an imprint of the
Taylor & Francis Group, an **informa** business

A TAYLOR & FRANCIS BOOK

CRC Press
Taylor & Francis Group
6000 Broken Sound Parkway NW, Suite 300
Boca Raton, FL 33487-2742

First issued in paperback 2020

© 2013 by Taylor & Francis Group, LLC
CRC Press is an imprint of Taylor & Francis Group, an Informa business

No claim to original U.S. Government works

Version Date: 20120703

ISBN 13 : 978-0-367-57656-1 (pbk)
ISBN 13 : 978-1-4398-7875-0 (hbk)

Library of Congress Cataloging-in-Publication Data

Cancer nanotechnology : principles and applications in radiation oncology / editors, Sang Hyun Cho and Sunil Krishnan.
    p. ; cm. -- (Imaging in medical diagnosis and therapy)
  Includes bibliographical references and index.
  ISBN 978-1-4398-7875-0 (alk. paper)
  I. Cho, Sang Hyun. II. Krishnan, Sunil. III. Series: Imaging in medical diagnosis and therapy.
  [DNLM: 1. Nanostructures--therapeutic use. 2. Neoplasms--radiotherapy. 3. Nanostructures--diagnostic use. 4.
Nanotechnology--methods. 5. Neoplasms--diagnosis. 6. Neoplasms--therapy. QZ 269]

  616.99'407572--dc23

2012025201

Visit the Taylor & Francis Web site at
http://www.taylorandfrancis.com

and the CRC Press Web site at
http://www.crcpress.com

To my family

-S.H.C.

To my wife and parents

For their love, support and encouragement

-S.K.

Contents

SECTION I Basic Principles of Radiation Oncology and Radiosensitization

SECTION II Synthesis, Safety, and Imaging of Nanomaterials for Cancer Applications

SECTION III Nanomaterials for Radiation Therapy

SECTION IV Nanomaterials for Hyperthermia and Thermal Therapy

SECTION V Future Outlook

Series Preface

Advances in the science and technology of medical imaging and radiation therapy are more profound and rapid than ever before, since their inception over a century ago. Furthermore, the disciplines are increasingly cross-linked as imaging methods become more widely used to plan, guide, monitor, and assess treatments in radiation therapy. Today, the technologies of medical imaging and radiation therapy are so complex and so computer-driven that it is difficult for the persons (physicians and technologists) responsible for their clinical use to know exactly what is happening at the point of care, when a patient is being examined or treated. The persons best equipped to understand the technologies and their applications are medical physicists, and these individuals are assuming greater responsibilities in the clinical arena to ensure that what is intended for the patient is actually delivered in a safe and effective manner.

The growing responsibilities of medical physicists in the clinical arenas of medical imaging and radiation therapy are not without their challenges, however. Most medical physicists are knowledgeable in either radiation therapy or medical imaging, and expert in one or a small number of areas within their discipline. They sustain their expertise in these areas by reading scientific articles and attending scientific talks at meetings. In contrast, their responsibilities increasingly extend beyond their specific areas of expertise. To meet these responsibilities, medical physicists periodically must refresh their knowledge of advances in medical imaging or radiation therapy, and they must be prepared to function at the intersection of these two fields. How to accomplish these objectives is a challenge.

At the 2007 annual meeting of the American Association of Physicists in Medicine in Minneapolis, this challenge was the topic of conversation during a lunch hosted by Taylor & Francis Publishers and involving a group of senior medical physicists (Arthur L. Boyer, Joseph O. Deasy, C.-M. Charlie Ma, Todd A. Pawlicki, Ervin B. Podgorsak, Elke Reitzel, Anthony B. Wolbarst, and Ellen D. Yorke). The conclusion of this discussion was that a book series should be launched under the Taylor & Francis banner, with each volume in the series addressing a rapidly advancing area of medical imaging or radiation therapy of importance to medical physicists. The aim would be for each volume to provide medical physicists with the information needed to understand technologies driving a rapid advance and their applications to safe and effective delivery of patient care.

Each volume in the series is edited by one or more individuals with recognized expertise in the technological area encompassed by the book. The editors are responsible for selecting the authors of individual chapters and ensuring that the chapters are comprehensive and intelligible to someone without such expertise. The enthusiasm of volume editors and chapter authors has been gratifying and reinforces the conclusion of the Minneapolis luncheon that this series of books addresses a major need of medical physicists.

Imaging in Medical Diagnosis and Therapy would not have been possible without the encouragement and support of the series manager, Luna Han of Taylor & Francis Publishers. The editors and authors, and most of all I, are indebted to her steady guidance of the entire project.

William Hendee
Series Editor
Rochester, Minnesota

Preface

Nanotechnology, the study and manipulation of matter and phenomena at the nanoscale, involves sensing, imaging, measuring, modeling, and manipulating matter in a size regime of about 1 to 100 nm. At these dimensions, matter exhibits unique physical and chemical properties by virtue of its nanoscale proportions that distinguish it from individual molecules or bulk matter composed of the same material. Rapid advances in nanotechnology have enabled the fabrication of nanoparticles from various materials with different shapes, sizes, and properties, and efforts are ongoing to exploit these materials for practical applications. Quite naturally, this promise of functional utility has fueled the quest for clinically meaningful applications. Within the National Institutes of Health, these inquiries have coalesced under the auspices of the National Nanotechnology Initiative. In the field of oncology, an early finding that made nanotechnology highly relevant was the recognition that leaky immature and chaotic vasculature of tumors, a hallmark of unrestrained growth, also results in passive accumulation of nanoparticles preferentially within tumors. This unique feature (often referred to as the enhanced permeability and retention effect) of tumors in conjunction with the unique physical/chemical properties of nanoparticles offers many novel strategies for cancer diagnosis and treatment. Notable approaches in the radiation oncology setting include the use of nanoparticles for radiation response modulation of tumors or normal tissues, and thermal ablation or hyperthermia treatment of tumors. Indeed, many of these techniques hold promise for clinical deployment, and some of them are rapidly advancing from preclinical validation in animal models to early clinical evaluation. Furthermore, various computational and experimental techniques have been employed to explain and predict the biophysical consequences of nanoparticles interacting with radiation to enhance antitumor effects.

Clearly, these research endeavors lie at the intersection of disciplines as diverse as clinical radiation oncology, radiation physics, nanotechnology, material science, biomedical engineering, pharmacology, chemistry, tumor biology, and radiation biology; disciplines that do not often overlap in terms of researcher cross-training, journal(s) and professional meeting(s) that bring research(ers) together under one umbrella, or even similar vocabularies and terminologies. The literature that spans these diverse topics is dispersed across specialty journals with a readership often restricted to researchers with a specific area of interest. To our knowledge, there is no single compilation of extant research in the arena of nanoparticles and radiation oncology that provides a comprehensive survey of basic principles, research techniques, and outcomes with a view toward eventual clinical translation of research findings. As this book covers most of these aspects in one venue, we envision it serving as a valuable reference for a wide spectrum of readers such as physicists, clinicians, engineers, chemists, and biologists in industry and academia, who have an interest in nanotechnology applications in radiation oncology. It may also be used as a textbook or key reference for a graduate level special topic course in medical physics or biomedical engineering or any other disciplines dealing with nanotechnology applications in cancer therapy.

This book has five distinct sections and 19 individual chapters under these sections. Contents are grouped under five major categories: basic principles of radiation oncology, synthesis and imaging of nanomaterials, nanotechnology applications for radiation therapy, nanotechnology applications for hyperthermia and thermal therapy, and future outlook. The first two sections cover the basics of radiation oncology as well as a general introduction to imaging, fabrication, and preferential tumor targeting of nanoparticles. The next two sections cover specific applications of nanomaterials in the realms of radiation therapy, hyperthermia, thermal therapy, and normal tissue protection from radiation exposure. The last section presents an outlook for future research and clinical translation including regulatory issues for ultimate use of nanomaterials in humans. The last chapter provides an overall summary and outlook of the topics covered in the preceding chapters as well as other important topics omitted in this book for various reasons but worth noting for their potential applications in radiation oncology.

It has been exciting to assemble this team of scientists and researchers as we embarked on this journey and gratifying to see it through to completion. We hope this book will encourage further research in this rapidly expanding field and inspire a new generation of multidisciplinary researchers to improve extant paradigms for treatment of cancer with radiation therapy.

Acknowledgments

We would like to thank all of our friends and colleagues who have graciously contributed their time, energy, and expertise in compiling individual chapters of this book. We are also grateful to Luna Han at Taylor & Francis Group for encouraging us to put this book together.

S.H.C. acknowledges the following people who have contributed to his research efforts over the years at Georgia Tech: Drs. S.-K. Cheong, B. L. Jones, Y. Yang; Mr. A. Siddiqi, N. Manohar, F. Reynoso; Ms. F. Liu, K. Dextraze. S.H.C. would also like to acknowledge Drs. P. Diagaradjane, J. Stafford, O. Vassiliev, S. Jang, C. Wang, E. Elder, J. Oshinski, and A. Karellas.

S. K. is grateful to his colleagues Parmeswaran Diagaradjane, Amit Deorukhkar, Dev Kumar Chatterjee, Edward Agyare, and Shanta Bhattarai for their invaluable contributions to research efforts in his laboratory and the numerous enlightening discussions that have provoked, inspired, sharpened, and focused these efforts. He thanks the laboratory's students and visitors whose perceptive and inquisitive minds helped fashion individual research endeavors and forced him to stay abreast of this rapidly advancing field. He is also grateful for the opportunity to collaborate with many of the authors of individual chapters of this book.

About the Editors

 **Sang Hyun Cho, PhD,** is currently an associate professor of medical physics at Georgia Tech, and previously served as associate professor at the University of Texas M. D. Anderson Cancer Center in Houston, Texas. He earned his PhD from Texas A&M University, College Station, Texas. Dr. Cho is a certified medical physicist in therapeutic medical physics by the American Board of Radiology and a licensed medical physicist in the same specialty by the State of Texas. He has been involved with numerous grant-funded activities as a principal or coinvestigator, has been a member of various national committees, has served as a referee for the top scientific journals in medical physics and nanotechnology, and has been a reviewer for major granting agencies. Dr. Cho's research program covers a wide spectrum of research topics ranging from traditional medical physics to nanotechnology, and combines both experimental and computational approaches. His recent research efforts have been devoted to the use of nanoparticles, especially gold nanoparticles, for cancer detection and therapy.

 **Sunil Krishnan, MD, FACP,** is an associate professor in the Department of Radiation Oncology at the University of Texas M. D. Anderson Cancer Center. He serves as the director of translational research for the gastrointestinal cancer service. As a clinician, his research focus is on developing novel therapies for patients with gastrointestinal tumors—he is the principal investigator of several clinical trials focusing on radiation sensitization strategies for liver, pancreatic, and rectal cancers. His basic research laboratory's primary motivation is the development and advancement of nanoparticles for tumor-specific targeting, image guidance, and radiosensitization. His laboratory has utilized conjugated nanoparticles for noninvasive real-time quantitative and repetitive imaging of cancers, thermally activatable nanoparticles for minimally invasive image-guided interventions that complement standard cancer therapies, activatable nanoparticles for spatially and temporally synchronized drug delivery within tumors, and conjugated nanoparticles to optimize radiation dose enhancement. His laboratory also has a radiobiology focus on identifying and targeting inducible radioresistance pathways in gastrointestinal cancers. These research endeavors are funded by the National Institutes of Health, the Department of Defense, and other funding agencies. Dr. Krishnan is the recipient of various fellowships and awards, and has served in leadership positions at the American Society for Radiation Oncology and Radiation Therapy Oncology Group.

Contributors

Peter Alexander
Department of Cancer Biology
Wake Forest School of Medicine
Wake Forest University
Winston Salem, North Carolina

Jacob M. Berlin
Department of Molecular Medicine
City of Hope
Beckman Research Institute
Duarte, California

Vikas Bhardwaj
Department of Radiation Oncology
M. D. Anderson Cancer Center
Houston, Texas

Andrew R. Burke
Department of Cancer Biology
Wake Forest School of Medicine
Wake Forest University
Winston Salem, North Carolina

Dev K. Chatterjee
Department of Radiation Oncology
M. D. Anderson Cancer Center
Houston, Texas

Devika B. Chithrani
Department of Physics
Ryerson University
Toronto, Ontario, Canada

Sang Hyun Cho
School of Mechanical Engineering
Georgia Institute of Technology
Atlanta, Georgia

M. Angelica Cortez
Department of Radiation Oncology
M. D. Anderson Cancer Center
Houston, Texas

Laurence Court
Department of Radiation Physics
M. D. Anderson Cancer Center
Houston, Texas

Cathy Cutler
Missouri University Research Reactor
University of Missouri–Columbia
Columbia, Missouri

Mohamad Fakhreddine
Baylor College of Medicine
Houston, Texas

Glenn Goodrich
Nanospectra Biosciences
Houston, Texas

John D. Hazle
Department of Imaging Physics
M. D. Anderson Cancer Center
Houston, Texas

Mark Hurwitz
Department of Radiation Oncology
Harvard Medical School
Fall River, Massachusetts

Bernard L. Jones
Department of Radiation Oncology
University of Colorado School of
    Medicine
Aurora, Colorado

Amit Joshi
Department of Radiology
Baylor College of Medicine
Houston, Texas

Raghuraman Kannan
Departments of Radiology
University of Missouri–Columbia
Columbia, Missouri

Kattesh V. Katti
Departments of Radiology and Physics and
    Missouri University Research Reactor
University of Missouri–Columbia
Columbia, Missouri

Sunil Krishnan
Department of Radiation Oncology
M. D. Anderson Cancer Center
Houston, Texas

Nana Liang
School of Chemistry and Chemical
    Engineering
Shandong University
Jinan, China

Aijuan Liu
School of Chemistry and Chemical
    Engineering
Shandong University
Jinan, China

Yanyan Liu
School of Chemistry and Chemical
    Engineering
Shandong University
Jinan, China

Kathy A. Mason
Department of Experimental Radiation
    Oncology
M. D. Anderson Cancer Center
Houston, Texas

Radhe Mohan
Department of Radiation Physics
M. D. Anderson Cancer Center
Houston, Texas

Qingxin Mu
Department of Chemical Biology and
    Therapeutics
St. Jude Children's Research Hospital
Memphis, Tennessee

Yan Mu
School of Chemistry and Chemical
    Engineering
Shandong University
Jinan, China

Varun P. Pattani
Department of Biomedical Engineering
The University of Texas at Austin
Austin, Texas

J. Donald Payne
Oncolix
Houston, Texas

Francisco Reynoso
School of Mechanical Engineering
Georgia Institute of Technology
Atlanta, Georgia

Ke Sheng
Department of Radiation Oncology
UCLA School of Medicine
Los Angeles, California

Ravi N. Singh
Department of Cancer Biology
Wake Forest School of Medicine
Wake Forest University
Winston Salem, North Carolina

R. Jason Stafford
Department of Imaging Physics
M. D. Anderson Cancer Center
Houston, Texas

Gaoxing Su
Department of Chemical Biology and
    Therapeutics
St. Jude Children's Research Hospital
Memphis, Tennessee

Frank M. Torti
Departments of Medicine
University of Connecticut School of
    Medicine
Farmington, Connecticut

Suzy V. Torti
Departments of Molecular, Microbial
    and Structural Biology
University of Connecticut School of
    Medicine
Farmington, Connecticut

James W. Tunnell
Department of Biomedical Engineering
The University of Texas at Austin
Austin, Texas

James Welsh
Department of Radiation Oncology
M. D. Anderson Cancer Center
Houston, Texas

Bing Yan
Department of Chemical Biology and
    Therapeutics
St. Jude Children's Research Hospital
Memphis, Tennessee

and

School of Chemistry and Chemical
    Engineering
Shandong University
Jinan, China

Yi Zhang
Department of Chemical Biology and
    Therapeutics
St. Jude Children's Research Hospital
Memphis, Tennessee

I

Basic Principles of Radiation Oncology and Radiosensitization

Basic Principles of Radiation Therapy of Cancers

Laurence Court
M. D. Anderson Cancer Center

Radhe Mohan
M. D. Anderson Cancer Center

1.1 Introduction and Historical Background

Approximately 41% of men and women born in the United States today will be diagnosed with cancer at some point in their lifetime. In the past year, this amounted to 1,500,000 people in the United States being diagnosed with cancer. Unfortunately, although the science and technology of cancer treatments is continually advancing, we still have a long way to go. Currently, the overall 5-year survival of cancer patients, compared with the general population, is only 65%, although this does vary widely depending on the type of cancer. In 2010, almost 600,000 people in the United States died of cancer. There is, therefore, clearly a huge need for dramatic improvements in cancer prevention and cure. The focus of this volume is to describe the different applications of nanotechnology in radiation oncology, each with the ultimate goal of improving patient outcomes. The focus of this first chapter is to introduce the reader to the basic principles of radiation therapy of cancers.

X-rays have been used for cancer treatments since the very end of the nineteenth century; the first reports of medical uses of radiation were only a year or two after Wilhelm Roentgen discovered x-rays (1895) and Becquerel discovered radioactivity (1897). There were also early reports of the toxic effects of radiation—Becquerel himself described the skin erythema and ulceration that occurred when he left a container of radium in his pocket. Skin toxicity was thus an issue from the very start. We will demonstrate below how modern treatment techniques aim to deliver radiation dose to deep tumors while minimizing skin toxicity.

Although initial cathode-ray tubes were unreliable and produced radiation at a very low intensity, it was not long before William Coolidge developed the "hot" cathode tube (1912–1913). This was a much more reliable device that could produce x-ray spectra with peak energies of 200–250 kV, and associated higher penetrating power, allowing the treatment of relatively deep-seated tumors. It was the direct ancestor of today's x-ray tube. The first treatment unit using spectra with peak x-ray energy exceeding 1 MV appeared in 1937, soon to be followed by the use of Cobalt-60 gamma sources for external beam treatments in 1951 and medical linear accelerators (LINACs) with megavoltage (MV) beams in 1952. This progressive development of high energy treatment units was important not only because of the increase in penetrating power, but because these beams deposit their maximum dose some distance below the skin surface, facilitating the treatment of deep tumors with reduced skin reactions. Multileaf collimators (MLCs), which are a set of small, individually motorized collimator blades attached to the treatment unit gantry, were commercialized in 1984. Intensity-modulated radiation therapy (IMRT) where the MLCs move across the radiation field to modulate its intensity was introduced commercially in the 1990s. This treatment technique, which allowed treatment planners to sculpt the dose distribution, delivering high dose to the targets while minimizing dose to adjacent tissues, has become the standard-of-care for many patients. The complex fields treated in IMRT can mean longer treatment times—this has now been addressed with the introduction of volume-modulated arc therapy (VMAT), where the gantry rotates around the patient while the MLCs move, reducing beam-on times from 5–10 to 1–2 min. Other forms of radiation used in radiotherapy include electrons and protons, both of which are described below.

Imaging has played an important role in radiation therapy from the early days. Radiation therapy simulators, kilovoltage (kV) x-ray units with the same geometry as the treatment machine, were introduced in the late 1960s and early 1970s. Physicians acquired images using the same perspective as the treatment beams, showing which organs would be irradiated. This allowed them to use standard or custom radio-opaque blocks to shield these organs. Around this time, Hounsfield introduced computed tomography, but it was not until the 1990s that computers and associated networks had sufficiently advanced to allow routine CT-based treatment planning. The use of other imaging modalities, such as positron emission tomography (PET) and magnetic resonance imaging, to aid in defining tumor volumes (which may not show on CT images) is also common.

Radiographic film was used for many years to take beams-eye-view images of the treatment field, using the MV x-rays from the treatment unit. Film has now been mostly replaced by electronic portal imaging devices. Furthermore, modern LINACs include kV x-ray tubes and detectors attached to the gantry, allowing high-quality images to ensure patient setup. This arrangement is also capable of taking CT images, and in many clinics it is now routine practice to take CT images daily to ensure accurate patient positioning.

The use of external radiation beams is not the only way to use radiation to treat a patient. Another option is the placement of sealed radioactive sources in or on the tumor directly—a technique known as brachytherapy from the Greek word *brachys*, meaning short distance. Initially, most brachytherapy used radium, but the discovery of artificial radioactivity in 1934 eventually allowed the use of many other materials, including Ir-192, Au-198, and I-125. In this work, we focus on the use of external beam radiotherapy, as this is the predominant modality is use today—however, it is likely that nanotechnology may also serve to improve the efficacy of brachytherapy treatments.

1.2 Therapeutic Ratio

In radiation therapy, we rely on being able to destroy the tumor cells without causing intolerable damage to nearby normal tissues. If the dose response of the normal tissue is similar or close to that of the tumor, then it will be very difficult to destroy the tumor without excessive complications, as illustrated in Figure 1.1. Some tumors are particularly sensitive to radiation

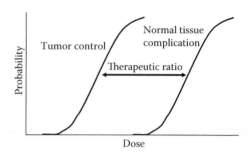

FIGURE 1.1 The therapeutic ratio describes the relationship between tumor control probability and the probability of normal tissue complication.

(e.g., lymphoma), but in many cases these two curves are very close. That is, the dose required to control the tumor is close to the tolerance levels of surrounding tissues. One of the goals of radiation therapy is to manipulate the treatment design to maximize the therapeutic ratio. One simple way is to irradiate the tumor with beams coming in from multiple angles, catching it in their crossfire. We can build on this approach by selecting beam angles to avoid tissues that are either more sensitive, or for which we can less tolerate any damage (e.g., the spinal cord). Therapeutic ratio enhancing techniques routinely used in the clinic are described below. One of the major challenges for future radiation treatment techniques is to maximize the therapeutic ratio and improve tumor control. A real-life example is given in Figure 1.2a, which shows the 2-year disease-free survival of patients with lung cancer from 16 different clinical trials. Figure 1.2b shows the relationship between the probability of pneumonitis and the mean dose to the lung. Low-grade pneumonitis is often considered an acceptable result of radiation therapy for lung cancer, but higher grades can require hospitalization and, in extreme cases, can result in the patient's death. It can be seen, from this figure, that although increases in tumor dose may increase tumor control, this can also lead to an increase in normal tissue (i.e., lung) toxicity. Other critical tissues that must be considered when treating lung tumors include the esophagus and the heart. It has been reported that 15–25% of patients receiving either concurrent chemoradiotherapy or hyperfractionation can experience severe acute esophagitis. This can necessitate surgical intervention, hospitalization, or breaks in radiation therapy, which can lower local tumor control. It can also be a dose-limiting factor that prevents dose escalation to the tumor. Finally, acute esophagitis predicts long-term esophageal sequelae that undermine patients' ongoing quality of life. Similar issues exist for most radiation therapy treatments, particularly when combined with chemotherapy. For prostate treatments, we must minimize dose to the rectum and bladder; for treatments of tumors in the head and neck region, the concern is dose to the spinal cord, the brainstem, optical structures, etc. As we introduce the different aspects of radiation therapy, we will illustrate their importance in terms of the therapeutic ratio.

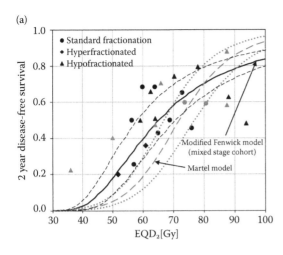
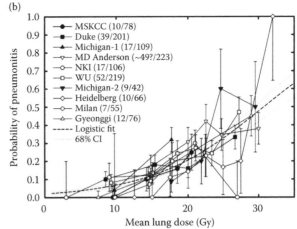

FIGURE 1.2 (a) Relationship between 2-year disease-free survival and delivered dose from different clinical trials. Also shown are two tumor control models. (From Partridge, M. et al., *Radiotherapy and Oncology*, 99(1), 6–11, 2011. Reproduced with permission.) EQD₂ is the dose corrected for different fractionation schemes. (b) Relationship between the probability of pneumonitis and the mean lung dose from 11 clinical trials. (From Marks, L.B. et al., *International Journal of Radiation Oncology*Biology*Biophysics*, 76(3, Supplement), S70–S76, 2010. Reproduced with permission.)

1.3 Physical Basis of Radiation Therapy

Radiation therapy involves the use of x-rays, gamma rays, or charged particles, to deposit dose. First, we describe the physical processes by which the beams are attenuated and dose is deposited. Later, we describe the biological effect of radiation, and then how the characteristics of these different modalities can help maximize the therapeutic ratio when treating tumors in different locations.

1.3.1 X-Ray Beams

When x-rays pass through a material, they may be transmitted through the material without interaction, or may be attenuated by absorption or scattering processes (Johns and Cunningham 1982; Hendee and Ibbott 2005; Khan 2007; Metcalfe et al. 2007; Podgorsak 2005). The fractional number of photons attenuated by an infinitesimally thin slab of material is proportional to its thickness and a parameter known as its attenuation coefficient, μ. This coefficient includes the individual coefficients for various absorption and scattering effects. We describe them briefly in the following paragraphs since they are of relevance to interactions of radiation with nanoparticles.

Coherent scatter. Coherent scatter is the process by which the x-rays are scattered without losing energy. First, the electromagnetic wave sets the electrons in the atom into vibration, which then emit radiation of the same wavelength. The emitted waves combine to form the scattered x-ray. The likelihood of this interaction is extremely low for high-energy photons interacting in soft tissue, and there is no deposition of energy in the medium, so it plays no role in radiation therapy. It does play a role in diagnostic imaging, where it results in increasing scatter in the patient, thus impacting image quality.

Photoelectric effect. A photoelectric interaction is an interaction between a photon and an inner electron of the medium. This interaction results in the ejection of an electron from the atom with kinetic energy equal to the difference between that of the incident photon and the binding energy of the ejected electron. The mass attenuation coefficient for photoelectric absorption generally varies as Z^3/E^3, where Z is the atomic number of the medium and E is the photon energy. For soft tissue, this is the dominant interactive process for low-energy incident photons (<0.03 MeV), but rarely occurs at the megavoltage energies used in external beam radiation therapy. The Z dependence of the photoelectric effect explains why bones (which have relatively high Z) have high contrast in diagnostic x-ray images. An example of how this dependency can be used to improve the therapeutic effect of radiation is the use of gold nanoparticles, described in later chapters. The atomic number of gold is more than 10 times higher than that of soft tissue, so the introduction of gold nanoparticles results in an increase in the absoption of x-rays and release of electrons by the photoelectric effect.

Compton interactions. Whereas the photoelectric effect dominates x-ray interactions with soft tissue for low x-ray energies, Compton interactions dominate for higher energies. In these interactions the x-ray photon interacts with relatively loosely bound electrons. In each interaction, some energy is scattered, and some is transferred to the electron, which is then emitted from the atom. The likelihood of Compton interaction decreases slowly with increasing photon energy. Because these interactions are with loosely bound electrons, the interaction probability is almost independent of atomic number, and depends primarily on electron density. This means that the presence of bone in the path of the radiation beam does not dramatically alter the dose to the tissues downstream from the bone.

Pair production. Pair production is an x-ray absorption event. It occurs when the x-ray photon passes close to the nucleus of an atom, and undergoes conversion into mass in the form of a positron–electron pair. The mass equivalent of this pair is 1.02 MeV (2 × 0.51 MeV), so the incident x-ray must have at least this threshold energy for this interaction to occur. Energy in excess of this threshold is distributed between the two particles

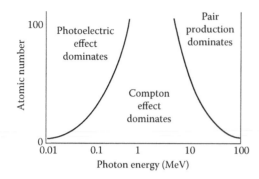

FIGURE 1.3 Relative importance of photoelectric, Compton, and pair production for different atomic number media and photon energies. (Modified from Hendee, W.R. et al., *Radiation Therapy Physics*, Wiley, Hoboken, NJ, 2005.)

as kinetic energy. The likelihood of this interaction occurring increases with increasing E and Z.

1.3.1.1 Relative Importance of Different Types of Interactions

The relative importance of photoelectric, Compton, and pair production for different atomic number media and photon energies are illustrated in Figure 1.3. Of particular note is that we may be able to modulate these interactions by introducing nanoparticles of different atomic numbers.

1.3.1.2 Characteristics of Therapeutic Photon Beams

Variation of dose with depth. Although the attenuation of x-ray beams can be described as a combination of exponential attenuation (described above) and $1/r^2$ fall-off as the distance from the source increases, the variation of deposited dose with depth in a media is quite a different shape, as can be seen in Figure 1.4. This figure shows the percentage depth dose (PDD) curve for different incident x-ray energies. Photon interactions in the patient's surface tissues result in the ejection of high-energy electrons, which deposit their energy some distance downstream. At progressively greater depths, more and more electrons become involved,

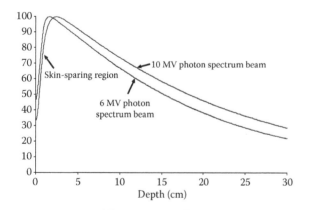

FIGURE 1.4 Percentage depth dose for 6- and 10-MV x-ray beams.

giving an initial increase in dose in the buildup region, until a maximum is reached. This phenomenon is known as the *skin-sparing effect* (the skin, which is relatively sensitive, is "spared" because it receives a lesser dose than slightly deeper tissues). The photon fluence is decreasing with depth because of attenuation and $1/r^2$ fall-off, so the density of dose-depositing electrons falls off with depth, leading to an eventual *fall-off* in dose with depth.

Dose distributions. The shape of the dose distribution is described using isodose curves, which are lines connecting the points that receive the same particular dose. The dose is usually very uniform across the central region of a broad beam. As will be described below, the use of various beam modifiers can change this, and some modern LINACs have very nonuniform dose distributions that require modifiers if a uniform distribution is needed. The shape of the dose distribution at the edges of and outside the beam, however, depends on the geometric penumbra (resulting from the finite size of the LINAC's focal spot), collimation, and beam energy (i.e., to what extent photons are scattered in the forward direction).

1.3.2 Therapeutic Electron Beams

As described above, when photons interact with the media via photoelectric, Compton, or pair production events, electrons in the media are set in motion. It is the electrons that deposit radiation dose, not photons. Electrons may also come directly from the LINAC—modern LINACS typically produce a range of electron energies from 4 to 20 MeV or higher. Whatever the source of the electrons, they travel through the medium and gradually lose energy until they are slow enough to be captured by atoms. The main processes by which electrons lose energy are through Coulombic (electric) interactions with either atomic electrons or atomic nuclei, as discussed below.

Inelastic electron collisions with atomic electrons. The rate of energy loss through the excitation or ionization of atomic electron clouds depends on the electron density, and is generally lower for higher-Z materials, whose electrons tend to be more tightly bound. For high-energy electrons ($E > 1$ MeV), the rate of energy loss in passage through water (or soft tissue) is fairly constant at 2 MeV/cm. This is important, because it determines the maximum depth to which electrons will penetrate; for example, if an 8-MeV electron beam is used to irradiate a neck node, the spinal cord will receive a very low dose if it is at least 4 cm deep.

Inelastic collisions with atomic nuclei. Bremsstrahlung x-ray energy loss occurs when an electron passes near a nucleus and is deflected and decelerated by its Coulombic field. The probability of occurrence of a bremsstrahlung interaction increases with electron kinetic energy and with Z. It is this interaction that is responsible for the creation of high-energy photon beams when high-energy electrons collide with the targets of diagnostic x-ray tubes and LINACs. High-energy photo- and Compton electrons follow tortuous paths through tissue because of multiple Coulomb scatterings. The scattering cross section is approximately proportional to Z^2 and inversely proportional to the electron energy.

1.3.2.1 Protons

As is the case for electrons, protons and other charged particles interact with the media in which they are traveling by interactions between their electric field and the electric field in the media. Unlike electrons, the mass of proton is large compared to the atomic electron, and there is very little scatter. Toward the end of their range, where they have little energy left, they experience a large increase in stopping power, giving a peak (known as the *Bragg peak*) in absorbed dose. The same thing happens also with electron beams, but the multiple scattering that electrons experience causes their Bragg peaks to be washed out completely.

1.3.2.2 Characteristics of Therapeutic Electron and Proton Beams

1.3.2.2.1 Electron Depth Dose Curves

As with photons, electron beams can be characterized with PDD curves. Figure 1.5 shows the PDD for three different electron beams. They are characterized by a small skin-sparing region, a fairly flat region of high dose (depending on the energy), and a rapid fall-off. The use of electrons in radiation therapy takes advantage of this shape to optimize the therapeutic ratio.

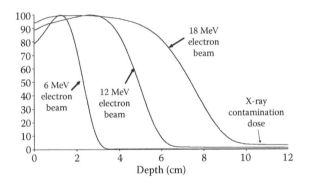

FIGURE 1.5 Percentage depth dose curves for 6-, 12- and 18-MeV electron beams.

Specifically, we take advantage of the fact that, in contrast to photon beams, the electrons are actually stopped in the patient. Tissues distal to the treatment volume receive minimal dose (just a small dose due to x-ray contamination of the beam). This is useful, for example, when treating the neck while avoiding the spinal cord, or the breast while avoiding the lung and heart.

1.3.2.2.2 Proton Depth Dose Curves

Figure 1.6a shows a proton PDD. Like electrons, and unlike x-rays, there is virtually no dose beyond the practical range (ICRU 1998). Because the Bragg peak is very narrow, and certainly much narrower than most tumors, for actual patient treatment this is spread out (spreadout Bragg peak) by using multiple incident energies, as illustrated in Figure 1.6b. As with electrons, we use protons to optimize the therapeutic ratio by taking advantage of the fact that there is virtually no dose distal to the target.

1.3.3 Use of Radioactivity in Cancer Therapy

1.3.3.1 Introduction to Radioactivity

Radioactivity refers to the emission of particles and gamma rays from an unstable nucleus as it transitions (decays) toward a more stable configuration of neutrons and protons. The rate of decay is characterized by the half-life of the sample. This is simply the time required for half the atoms in the sample to decay. It can be short (Au-198, e.g., has a half-life of 3 days) or very long (Ra-226 has a half-life of 1622 years). There are several modes in which a nucleus decays; these are briefly summarized here.

1.3.3.2 Alpha Decay

Alpha decay is a process that results in the emission of a single alpha particle from the nucleus, increasing the nuclear stability. This particle, which is a helium nucleus, contains two protons and two neutrons. This type of decay only happens for large nuclei. It was used in therapy for many years in the form of a decay from Ra-226 to Rn-222, with a half-life of 1622 years.

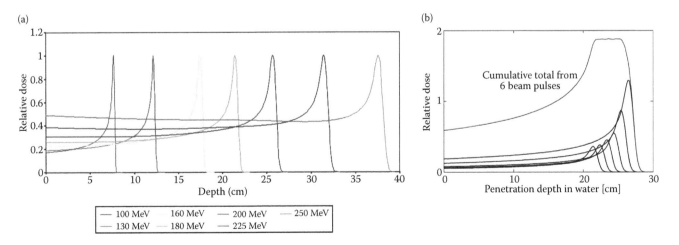

FIGURE 1.6 (a) Percentage depth dose for proton beams of different energies. (b) Generation of spread-out Bragg peak.

The radon itself eventually decays to stable lead, with numerous gamma rays being emitted in the process. It is these photons that were used to deliver dose in cancer treatments.

1.3.3.3 Beta Decay

Beta decay involves the emission of a positive or negative electron (beta particle) from the nucleus. If the nucleus has an unstably high neutron/proton ratio, a neutron is changed into a proton, giving an increase in the atomic number of the nucleus. An electron is simultaneously emitted. The beta particles are emitted as a continuous spectrum of energies up to an energy characteristic of the decay.

If the nucleus has a neutron/proton ratio lower than needed for stability, it may decay by converting a proton to a neutron, resulting in emission of a positive electron (positron). PET involves attaching a positron emitting isotope such as F-18 to a metabolically active compound. The isotope emits positrons that quickly interact with electrons in the tissue, resulting in the annihilation of the positron and electron, and produce two 511-keV photons that travel in opposite directions. Detection of these photons allows the distribution of the metabolically active compound to be reconstructed.

1.3.3.4 Gamma Emission and Internal Conversion

Gamma emission can occur during transitions between isomeric states of the nucleus. That is, when an atom goes from an excited energy state to a more stable one without a change in Z, N, or A. Internal conversion is a competing process in which energy is transferred from the nucleus to an inner electron, which is then ejected. This is followed by emission of x-rays and auger electrons as the atom returns to a stable structure.

1.3.3.5 Characteristics of Brachytherapy Dose Distributions

In brachytherapy the source is positioned generally in, but sometimes on, the patient. Because of this proximity, the dose distributions are dominated by the extremely rapid $1/r^2$ fall-off, meaning that dose to adjacent normal tissues can be very low. The shape of the dose distribution very close to the source is determined by details of the construction of the source; further away, the distribution is more or less spherical. Volumes of varying complexity can be treated by using inserting multiple sources (e.g., as in prostate brachytherapy) or by moving a single source to different positions in a catheter (e.g., as in high dose-rate brachytherapy). Brachytherapy, because of the lower energies of radiation particles, may have greater clinical effectiveness than x-rays used for external beam therapy when combined with nanoparticles

1.4 Radiobiological Basis of Radiation Therapy

1.4.1 What Is Cancer?

Tumors occur when a single cell suffers a disruption in its mechanisms for proliferation and self-elimination. This disruption is caused by a genetic mutation that might be a result of random events during normal cellular replication, but can also be caused by carcinogens such as radiation or cigarette smoke. The malignant cells may be immortal or divide many more times than normal cells, and often grow much more rapidly than normal cells. Additionally, they can exhibit abnormal interactions with other cells, allowing them to metastasize and grow in places they would not normally be found.

1.4.2 Role of Radiation Therapy in Cancer Treatments

The primary goal of radiation therapy is to kill the cancerous cells. This is achieved by damaging biologically important molecules, particularly deoxyribonucleic acid (DNA). Interestingly, only about one-third of biological damage is caused by the radiation interacting directly with these molecules. The majority of damage is achieved when radiation interacts with water which, upon excitation or ionization, transforms into highly reactive chemical species (free radicals) that themselves damage biological molecules. This latter phenomenon is known as indirect action and accounts for around two-thirds of the biological damage caused by x-rays. Of particular importance is the fact that indirect action is open to modification by chemical sensitizers or protectors. The sensitivity of cells to radiation is described by four biological processes (known as the four R's) (Hall 2000):

1.4.2.1 Repair

Radiation-induced damage to the DNA can be categorized as DNA protein crosslink, base alterations, single-strand breaks, and double-strand breaks. The most important of these is double-strand breaks, since they are repaired very inefficiently. The others are either infrequent (DNA protein cross-links) or efficiently repaired (base alterations and single-strand breaks). Note that all four mechanisms are capable of playing a role in the separate but related process of radiation carcinogenesis.

1.4.2.2 Repopulation

Given time, some types of undamaged cells will divide and repopulate, replacing those that were killed by the irradiation. The ideal scenario is that this occurs more rapidly in healthy tissues, as with repair.

1.4.2.3 Redistribution

Different phases of the cells cycle are more resistant to radiation than others, so after irradiation, more cells are left in radiation-resistant phases than in sensitive phases. This means that an immediate subsequent irradiation would be less successful in killing tumor cells. It would be advantageous if we could time subsequent treatments such that the cohort tumor cells had returned to a sensitive phase, whereas the healthy cells had not.

1.4.2.4 Reoxygenation

Oxygen-starved (hypoxic) cells are particularly resistant to radiation damage. The ratio of dose needed under hypoxic to aerated conditions to achieve the same biological effect is known as the oxygen enhancement ratio (OER). It is dependent on the radiation type (see below), and also on the phase of the cell cycle. Because of the limited diffusion distance of oxygen in tissues, cells at the center of a pocket of tumor may be hypoxic, and therefore more difficult to kill, than those surrounding it. By spreading the irradiation over many fractions, so that outer portions of the tumor are killed, previously hypoxic regions may become oxygenated, and therefore more sensitive to irradiation.

The relative importance and effectiveness of these processes can be significantly different for different tissues. For example, rapidly dividing cells, such as cells of skin or the lining of the gut, are more sensitive to irradiation than nondividing cells, such as neurons.

1.4.3 Cell-Survival Curves

The radiation sensitivity of cells can be expressed with cell survival curves, such as those shown in Figure 1.7, where one curve applies to a tumor or early responding tissue and the other to late-responding tissues (e.g., lung or kidney). At low doses, the late-reacting normal tissues are better at repairing themselves. This is a motivation for fractionating radiation therapy treatments (i.e., splitting the dose delivery into multiple smaller treatments). The shapes of these curves mean that by splitting the delivery of a large prescription dose into multiple smaller fractions, we can achieve significantly better survival of normal tissues but still slowly kill the tumor cells.

1.4.4 Linear Energy Transfer and Relative Biological Effectiveness

The effect of radiation on tissue is not dependent solely on the amount of absorbed energy, but is dependent on details of how the energy is deposited at the microscopic level. This can be understood in terms of the average energy transfer per unit length of the track—linear energy transfer (LET). Note that microscopically the energy transfer varies widely, and the use of the average is not without its controversies (see Hall 2000). The

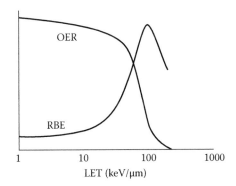

FIGURE 1.8 Effect of linear energy transfer (LET) on relative biological effectiveness (RBE) and oxygen enhancement ratio (OER). (Modified from Hall, E.J., Giaccia, A.J., *Radiobiology for the Radiologist*, Lippincott Williams & Wilkins, Philadelphia, PA, 2006.)

LET is around 0.2 keV/μm for Co-60 gamma rays and increases with decreasing energy to ~2.0 for 250-kV x-rays; it is about 0.5 for 150 MeV protons, again increasing with decrease in energy, to about 4.7 for 10 MeV protons. Thus, the ionization densities will vary widely between these different radiation sources. The relative biological effect of these is described by comparing the dose needed to reach some endpoint, such as death of half the cells in a sample) for the test radiation with the dose needed to reach the same endpoint for a standard radiation. The effect of the LET on relative biological effectiveness (RBE) (mammalian cells) is shown in Figure 1.8. There is little variation in RBE for LET <10 keV/μm, but it rises quickly after that, reaching a peak for an LET of about 150 keV/μm, after which the RBE falls again. This peak in RBE occurs when the average separation between ionization events is about the same as the diameter of the DNA double helix (2 nm), and the passing of a single particle has the highest probability of resulting in a double-strand break. The fall-off in the curve is seen because there are more ionization events than needed, so the biological effect per dose is reduced.

LET also has an impact on the OER described above. This is also illustrated in Figure 1.8. For low LET radiation the OER is about 3, but it falls with increasing LET, and is 1 at about 200 keV/μm. This means that reoxygenation may play less of a role when irradiating with high LET radiation than when irradiating with low LET radiation.

It should be noted that determining RBE is very complicated, because it strongly depends on the cell and tissue type, the endpoint used, and the dose per fraction.

1.4.5 Toxicity to Normal Tissues

We have already pointed out that the major limitation to how much dose we can deliver to the tumor is often limited by the risk of toxicity to normal tissues. Toxicity can be categorized as either acute or late. Acute effects are mostly found in tissues with rapidly dividing cells. This is because of radiation-induced cell death during mitosis, so cells that are dividing rapidly show the most rapid cell loss. Examples include the skin or mucosal

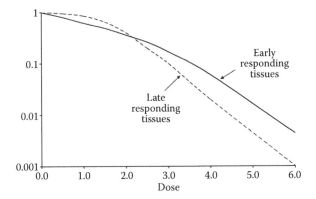

FIGURE 1.7 Cell survival curves for early and late responding tissues.

surfaces such as the oropharynx, esophagus, and rectum. When treating head and neck cancer, mucositis is experienced early on, and is typically at its worst 3–4 weeks into treatment. In many cases, it then stabilizes as proliferation of normal mucosal cells increases in response to the cell loss. Sometimes this reaction is so severe that it forces a break in treatment, although clinical staff try to avoid this because of the negative impact of extending overall treatment time on the tumor control. Mucositis is a short-term effect and usually resolves within a couple of weeks of treatment completion.

Late effects are seen between 6 months and many years after irradiation. They include lung fibrosis, esophageal stricture, and other organ damage. In some cases (e.g., esophagitis), acute symptoms are directly predictive of future late effects; in other cases (e.g., heart damage), there are no apparent acute symptoms.

Radiation therapy can also cause secondary cancers. Unlike the toxic effects described above, and for which the severity increases with dose, cancer induction is considered a stochastic effect. That is, although the probability of occurrence increases with dose, the severity is independent of dose. Examples of radiation-induced secondary cancers include breast cancers years after treatment for childhood Hodgkin's lymphoma, and osteogenic sarcoma after radiation of childhood retinoblastoma. The relative risks of the current cancer versus a possible future cancer are such that clinicians are very unlikely to not treat a patient because of concerns about secondary cancers. However, the risk of secondary cancers is often discussed when comparing treatment techniques with apparently very similar tumor control outcomes, for example, when comparing IMRT with proton treatments. For this reason, there is active research into the development and improvement of treatment techniques to reduce dose to tissues outside the target.

1.4.6 Fractionation

Fractionation refers to the dose delivered per fraction, the number of fractions per day, the total number of fractions, and the total amount of time required to complete the course of radiation. Most patients receive what is called standard fractionation, or 1.8–2.25 Gy/day. One of the justifications for this fraction size is the shape of the cell survival curve described above (Figure 1.7)—by splitting the dose delivery into smaller fractions, we give the normal tissues a better chance of survival. Over time, this fractionation scheme has become standard, and clinicians have a reasonable understanding of the expected outcomes (tumor control and normal tissue toxicity) when patients are treated with standard fractionation. Reasons to diverge from standard fractionation include the potential of improved outcomes, or to allow patients receiving palliative care to be treated in a shorter time.

Accelerated fractionation refers to a treatment schedule that allows the treatment to be completed in a time shorter that would be achieved with standard fractionation. The motivation for this came from head and neck treatments, where it was found that increasing the tumor dose resulted in an increase in tumor

control. However, approximately 2 weeks into treatment the tumors began to repopulate at an accelerated rate, so the gain in tumor control with increase in dose was not as high as expected. The idea of accelerated fractionation is to complete the treatment in a shorter time, so the accelerated proliferation is not an issue. One way this can be achieved is to treat a standard fraction in the morning and a boost treatment in the afternoon.

A second altered fraction approach is called hyperfractiom. This refers to the treatment of more than one fraction per day, with a dose per fraction lower than with standard fractionation. The idea is that acute toxicity for rapidly dividing normal tissues and also for tumors is the same as for standard fractionation, but there should be fewer late complications.

A third scheme is hypofractionation, which is the use of a smaller number of fractions than standard fractionation. Historically, because of concerns about increases in late toxicity, hypofractionation was used mainly for palliative cases, where the goal was symptom relief. More recently, however, the use of advanced imaging techniques in the treatment room have allowed margin reduction and the irradiation of less normal tissue (essentially separating the curves of the therapeutic curve), and hypofractionation is also being used for potentially curative treatments.

1.4.7 Modulating Radiobiological Damage

In some cases, it may be possible to increase the therapeutic ratio by modulating the relative radiation sensitivities of the tumor cells and the cells of the surrounding normal tissues. This could involve making the tumor cells more sensitive to radiation than the surrounding normal tissues, or by making the normal tissues more resistant to radiation damage.

1.4.8 Radiation Sensitizers

One way to increase the effect of radiation is to administer a chemical or pharmacologic agent that acts as a radiation sensitizer. For a sensitizer to be useful, it must increase the sensitivity of tumor cells to radiation more than it does for normal tissues. Most radiation sensitizers do not do this, and so do not have a useful role in cancer therapy. One type of radiation sensitizer that does impact tumors more than normal tissues is hypoxic-cell sensitizers. We have described how tumors can contain hypoxic cells that are relatively resistant to radiation. Normal tissues do not contain hypoxic cells, so any agent that increases the radiation sensitivity of hypoxic cells will differentially act on tumor cells. A second type of radiation sensitizer that differentially impacts tumor cells is halogenated pyrimides, which sensitize cells dependent on how much of the drug is incorporated. This assumes that the tumor cells are cycling faster than normal tissues. These are only examples. Cells can also be sensitized to radiation using heat—various roles of nanotechnology in hyperthermia (application of heat for therapeutic uses) are described later in this volume. Nanoparticles may sensitize tumor cells through other mechanisms as well.

1.4.9 Radiation Protectors

Sulfhydryl compounds can act as radiation protectors by scavenging free radicals (from low LET radiation) before they damage the DNA and/or by donating hydrogen atoms to facilitate DNA repair. This is not fully understood, however, and these compounds also act as radioprotectors with high LET radiation. In any case, these compounds are extremely toxic, and so are not useful in cancer therapy. Amifostine is a drug that came about from many years of military research; it has the same protective function, but without the severe toxicity. Animal studies indicate that it is absorbed quickly into normal tissues, but slowly into tumors—and should therefore differentially protect the normal tissues provided the radiation is given shortly after administration of the drug. It is approved by the Food and Drug Administration for use in radiation therapy, but its use is not widespread, mainly because of concerns about its potential protection of the tumor itself.

1.5 Treatment Planning and Delivery

The goal of the treatment planning process is to develop a treatment plan, in terms of number of beams, beam angles, etc., that maximizes the therapeutic ratio of the treatment. There are many counterbalancing considerations including, of course, the dose that we need to deliver to the tumor and the maximum dose that we can safely deliver to adjacent normal tissues. Uncertainties in the treatment must also be considered, including the fact that the patient position and shape will not be the same every time they lie on the treatment couch. Respiratory motion and other issues must also be considered.

1.5.1 Planning Volumes

In order to maximize the therapeutic ratio, radiation treatments are planned to shape the dose distribution to a specific target volume, and avoiding normal tissues. It is useful to understand how the volume to be treated is determined. The volume of the known disease (e.g., visible on a CT image) is called the gross tumor volume (GTV) (ICRU 1999). In addition to this volume, regions of suspected subclinical, microscopic malignant disease may exist, but may not be visible or palpable. These may be in tissues immediately surrounding the visible tumor, or may be along pathways where the tumor cells are known to travel (e.g., lymph nodes). This region, which should also be treated if local failure is to be avoided, is known as the clinical target volume (CTV). The GTV and CTV comprise the primary targets in radiation therapy. However, the location of these volumes is not stationary in space. For example, targets in the thorax or abdomen are subject to respiratory motion. Most lung tumors move less than 1 cm, but those closer to the diaphragm can move up to around 2 cm. The liver and kidneys can move even more. There are also numerous geometric uncertainties in the exact location of the target on a day-to-day basis. The prostate, for example, may move relative to the bones, based on rectal and bladder

filling. All of these uncertainties must be accounted for if we are to avoid missing parts of the target—therefore, we expand the GTV and/or CTV to give what is known as a planning target volume (PTV). This can be a substantial increase in target volume. Depending on the complexity of the shape of the PTV, the treatment technique, and how well we can conform the dose distribution, the volume to which the prescription dose is delivered may be even larger than the PTV.

As an illustration, when we treat the prostate, the GTV is the prostate itself; the CTV is also the prostate if there is no subclinical spread of disease outside the prostate. The GTV is then expanded by 0.5–1.0 cm to account for daily variations in the prostate position. This results in an increase in the treatment volume by a factor of 3 to 4. This is why much effort is taken to reduce or mitigate these uncertainties. Examples include immobilization techniques, such as head masks, and image-guided radiation therapy procedures.

1.5.2 Dose Boosting Techniques

Regions that are suspected of containing occult disease may be treated to doses between 54 and 65 Gy, depending on the likelihood that disease is present. The gross tumor is typically treated to a higher dose of 70 Gy or more. Traditionally, this is achieved by treating the entire treatment volume to the lower dose, and then reducing the treatment portals and continuing the treatment to give the additional boost dose. This technique is known as the shrinking field technique. All the treatment volume receives the same daily dose, but the gross tumor is treated more times, to a higher total dose. In the past 10 years, an alternative technique, known as simultaneous boost, has become popular. In this technique, all regions are treated simultaneously, with each region getting a different daily dose. For example, the gross tumor might be treated in 2-Gy fractions and the CTV treated in 1.7-Gy fractions.

1.5.3 Beam Arrangements

As shown in Figure 1.4, a single-photon field will give a dose distribution with maximum dose 1.5–3.5 cm below the surface, depending on the x-ray energy. The dose fall-off past the depth of maximum dose means that the use of a single field will give a relatively inhomogeneous dose distribution. For example, if a single 6-MV beam was used to treat a tumor with a distal edge 10 cm below the skin, then the maximum dose would be 50% higher than the minimum dose to the tumor. This issue can be partly overcome by using parallel-opposed fields. That is, radiation fields coming from opposite sides of the patient. In this case, the dose is deposited fairly uniformly throughout the patient.

However, for thicker patients (or low photon energies), the subcutaneous dose can be relatively high. If the tumor is centrally located (e.g., the prostate), then four fields, arranged as two pairs of opposed beams, can be used. This arrangement will give a box of fairly uniform high dose. The dose along the fields outside this central box will be around 50% of the prescribed dose. Using the example of the prostate, this arrangement is useful in reducing the dose to adjacent tissue, particularly the rectum and the bladder. If we needed to avoid a structure immediately below (posterior) the target, we could have angled the vertical fields, avoiding this structure, but still having a high dose volume in the center of the patient. Thus, by careful use of different beam arrangements, we have essentially managed to manipulate the dose distribution to maximize the therapeutic ratio (or reduce the dose to critical tissues). For each beam, the MLC is used to create an aperture that blocks the radiation outside of the shape of the PTV visible from the beam's eye view. The actual aperture is somewhat larger than the PTV to allow for the lateral penumbra of the beam.

1.5.4 Intensity Modulation

For many treatment scenarios, the beam arrangements discussed above are sufficient to give a reasonably uniform high dose to the tumor target, while controlling the dose to normal tissues. The external shape of the patient, the shape of the target, the location of critical structures, or any combination of these, can mean that additional planning tricks are needed. For example, target shapes when treating head and neck tumors are often concave, and partially surround the spinal cord. One such trick is to modify the x-ray intensity. This can be achieved in a variety of ways. The simplest is to insert a wedge-shaped metal attenuator into the path of the beam. Because more x-rays pass through the thin part of the wedge than through the thick part, the resulting isodose curve is titled. The same effect can be achieved by moving one of the collimators across the field while the beam is on. In this case, if the inferior collimator is moved into the field, less x-rays reach the inferior part of the field—much like the thick part of the wedge. Even more complex x-ray intensity distributions can be achieved by moving the leaves of MLCs across the radiation field when the beam is on. This process is called IMRT (Ezzell et al. 2003; Hartford et al. 2009). A recent extension of this technique is to rotate the LINAC gantry at varying speeds at the same time as moving the MLCs (again, while the beam is on); this technique is called VMAT (Otto 2008).

1.5.5 Inverse and Forward Planning

In traditional radiation treatment planning, the treatment planner (a dosimetrist or physicist) determines the shape of the treatment portals, the entrance angles, and beam energies using treatment planning software, and then calculates the dose. The planner may then adjust the relative weights of different beams. This process is known as "forward planning." In IMRT and VMAT, there are too many variables for this process to be possible. Instead, the planner determines the required dose distribution and then uses automatic algorithms to try to find the appropriate fields that can achieve the optimum approximation of this distribution. Generally, the planner will decide the number of beams and the beam angles, and then the optimization software will determine the motion of the MLCs and (for VMAT) gantry speeds. There are many published algorithms that also include optimization of the gantry angles.

1.5.6 Image-Guided Radiation Therapy

Traditionally, patients were positioned for radiation therapy by aligning tattoos on their skin with lasers in the treatment room. The skin, of course, is flexible, so daily uncertainties in patient positioning can be quite large. As treatment delivery became increasingly accurate, there has, therefore, been an associated requirement to position the patient in a more accurate and consistent manner. The use of x-ray imaging in the treatment room to position the patient has been standard-of-practice for many years. It started with the use of electronic portal imaging, where the radiation from the LINAC is used to create images. Because these images are created using high energy photons (MV) where Compton interactions dominate, they have less inherent contrast than is possible with low energy x-rays. In many cases, however, they are good enough for patient alignment. Modern LINAC units overcome this issue by attaching kV x-ray tubes and detectors on arms to the side of the LINAC gantry. Thus, we can now take high-quality x-ray images with the patient in the treatment position. The therapists can compare daily x-ray images with images taken from the treatment plan, and then make adjustments to the couch position before treatment commences. The advantage of this imaging approach is the excellent image quality. The disadvantage is that it is still planar imaging, which is good for imaging bone, but it is often not possible to visualize the tumor and soft tissues. This means we are aligning the patient based on bone, which is a step up from using skin marks, and is a reasonable approach if the tumor is attached to bone; however, there will still be some uncertainties if the tumor can move relative to the boney landmarks. This is the case for prostate and other tumors in the abdomen or thorax. This issue is overcome by either implanting radio-opaque markers into the tumor or by rotating the kV x-ray tube and detector around the patient to take a CT image. This latter approach, known as cone-beam CT because of the geometry, gives CT images which, although not of diagnostic image quality, are often sufficient for visualizing and localizing soft tissue targets.

1.5.7 Brachytherapy

Brachytherapy is the placement of sealed radioactive sources in or on the tumor directly (Thomadsen 2005; Khan 2003). It is characterized by a high dose to close to the source, which falls off with distance quickly ($1/r^2$), giving low doses to adjacent or distant normal tissues. In this way, brachytherapy is maximizing the therapeutic ratio. Because it is difficult to use brachytherapy for large fields, it is often used in combination with external beam radiation therapy, with brachytherapy fulfilling the role of the boost field described above.

1.5.7.1 Brachytherapy Application Techniques

There are several standard ways in which brachytherapy sources can be used to treat tumors; the choice depends on the size and location of the tumor.

- *Interstitial brachytherapy.* This is the insertion of radioactive sources directly into the tissue. The sources may be permanently implanted, such as when I-125 seeds are placed in the prostate, or may be temporary, and removed after the required dose has been delivered. The advantage of a permanent implant is that it involves a one-time procedure. A temporary implant, however, may allow better control and adaptation of the source distribution and resultant dose distribution. With a typical temporary implant, one or several catheters are first inserted into the tissues. Dummy sources are then inserted into the catheters, and x-ray images are taken and used to localize the sources and calculate the dose distribution. The real radioactive sources are then inserted, and removed after the required dose has been delivered. For some treatments, the sources are inserted remotely using a computerized afterloading system.

- *Intracavitary brachytherapy.* This is the insertion of radioactive sources into a cavity in the body. The most common example is the treatment of uterine cancers. It is also possible to insert sources directly into the cavity created by a lumpectomy procedure for breast cancer. Intracavitary brachytherapy is always temporary, and, as with interstitial brachytherapy, the radioactive sources are handled either manually or remotely, depending on the strength of the sources.

- *External applicators.* When the tumor is close to the skin surface, radioactive sources can be inserted into tubes in specially fabricated molds 0.5 to 1.0 cm away from the skin surface. This may be preferable over external beam techniques for complicated irregular external surfaces.

1.6 Conclusions

Radiation has been used in the treatment of cancer for more than 100 years. Current treatment approaches are extremely complex, and there are many tools available to sculpt the radiation dose such that it maximizes dose to the tumor while minimizing dose to surrounding normal tissues. However, in spite of the huge technological strides that we have made in the past decades, there are still some cancers for which we are struggling to improve patient survival. There is, therefore, a significant need for additional developments—the use of nanotechnology is one of these. The following chapters describe different approaches to the use of nanotechnology to improve the therapeutic ratio of radiotherapy treatments, with the ultimate goal of improving patient outcomes.

References

American Association of Physicists in Medicine, D. W. O. Rogers, and J. E. Cygler. 2009. Clinical dosimetry measurements in radiotherapy. *Proceedings of the American Association of Physicists in Medicine Summer School*, Colorado College, Colorado Springs, CO, June 21–25, 2009. Madison, WI: Medical Physics Pub.

American Association of Physicists in Medicine, B. H. Curran, J. E. Balter et al. 2006. Integrating new technologies into the clinic: Monte Carlo and image-guided radiation therapy. *Proceedings of the American Association of Physicists in Medicine Summer School*, University of Windsor, Windsor, Ontario, Canada, June 18–22, 2006. Madison, WI: Medical Physics Pub.

Bentzen, S. M., L. S. Constine, J. O. Deasy et al. 2010. Quantitative Analyses of Normal Tissue Effects in the Clinic (QUANTEC): An introduction to the scientific issues. *International Journal of Radiation Oncology Biology Physics* 76:S3–S9.

Cody, D., and O. Mawlawi. 2008. *The Physics and Applications of PET/CT Imaging.* Madison, WI: Medical Physics Publ. Corp.

Cox, J. D., and K. K. Ang. 2003. *Radiation Oncology: Rationale, Technique, Results.* 8th ed. St. Louis, MO: Mosby.

Ezzell, G. A., J. M. Galvin, D. Low et al. 2003. Guidance document on delivery, treatment planning, and clinical implementation of IMRT: Report of the IMRT subcommittee of the AAPM radiation therapy committee. *Medical Physics* 30:2089–2115.

Hall, E. J. 2000. *Radiobiology for the Radiologist.* 5th ed. Philadelphia, PA: Lippincott Williams & Wilkins.

Hall, E. J., and A. J. Giaccia. 2006. *Radiobiology for the Radiologist.* 6th ed. Philadelphia, PA: Lippincott Williams & Wilkins.

Hartford, A. C., M. G. Palisca, T. J. Eichler et al. 2009. American Society for Therapeutic Radiology and Oncology (ASTRO) and American College of Radiology (ACR) Practice Guidelines for intensity-modulated radiation therapy (IMRT). *International Journal of Radiation Oncology Biology Physics* 73:9–14.

Hendee, W. R., G. S. Ibbott, and H. G. Hendee. 2005. *Radiation Therapy Physics.* 3rd ed. Hoboken, NJ: Wiley.

International Commission on Radiation Units and Measurements. 1997. *Dose and Volume Specification for Reporting Interstitial Therapy.* Bethesda, MD: International Commission on Radiation Units and Measurements.

International Commission on Radiation Units and Measurements. 1998. *Clinical Proton Dosimetry.* Bethesda, MD: International Commission on Radiation Units and Measurements.

International Commission on Radiation Units and Measurements. 1999. *Prescribing, Recording, and Reporting Photon Beam Therapy.* Bethesda, MD: International Commission on Radiation Units and Measurements.

Johns, H. E., and J. R. Cunningham. 1983. *The Physics of Radiology.* 4th ed. Springfield, IL: Charles C. Thomas.

Karzmark, C. J., and R. J. Morton. 1998. *A Primer on Theory and Operation of Linear Accelerators in Radiation Therapy.* 2nd ed. Madison, WI: Medical Physics Pub.

Khan, F. M., AAPM Radiation Therapy Committee. Task Group No. 25. 1991. *Clinical Electron-Beam Dosimetry: Report of Task Group No. 25, Radiation Therapy Committee, AAPM.* New York: Published for the American Association of Physicists in Medicine by the American Institute of Physics.

Khan, F. M. 2003. *The Physics of Radiation Therapy.* 3rd ed. Philadelphia, PA: Lippincott Williams & Wilkins.

Khan, F. M. 2007. *Treatment Planning in Radiation Oncology.* 2nd ed. Philadelphia, PA: Lippincott Williams & Wilkins.

Marks, L. B., S. M. Bentzen, J. O. Deasy, F.-M. Kong, J. D. Bradley, I. S. Vogelius et al. 2010. Radiation dose–volume effects in the lung. *International Journal of Radiation Oncology*Biology*Physics* 76(3, Supplement):S70–S76.

Metcalfe, P., T. Kron, and P. Hoban. 2007. *The Physics of Radiotherapy X-rays and Electrons.* Madison, WI: Medical Physics Pub.

Otto, K. 2008. Volumetric modulated arc therapy: IMRT in a single gantry arc. *Medical Physics* 35:310–317.

Partridge, M., Mi. Ramos, A. Sardaro, and M. Brada. 2011. Dose escalation for non-small cell lung cancer: Analysis and modelling of published literature. *Radiotherapy and Oncology* 99(1):6–11.

Podgoršak, E. B., International Atomic Energy Agency. 2005. *Radiation Oncology Physics: A Handbook for Teachers and Students.* Vienna: International Atomic Energy Agency.

Thomadsen, B., M. J. Rivard, W. M. Butler et al. 2005. *Brachytherapy Physics.* 2nd ed. Madison, WI: Published for American Association of Physicists in Medicine by Medical Physics Pub.

Washington, C. M., and D. T. Leaver. 2004. *Principles and Practice of Radiation Therapy.* 3rd ed. St. Louis, MO: Mosby Elsevier.

2

Biological Principles of Radiosensitization

Vikas Bhardwaj
M. D. Anderson Cancer Center

M. Angelica Cortez
M. D. Anderson Cancer Center

Mohamad Fakhreddine
Baylor College of Medicine

James Welsh
M. D. Anderson Cancer Center

2.1 Historical Perspective

In 1895, W. C. Roentgen reported the discovery of invisible rays that were capable of passing through cardboard, paper, and other substances and could cast shadows of solid objects on film. Roentgen also found that these rays (called x-rays, with "x" denoting an unknown quantity) could pass through human tissues, casting shadows of bones and metal objects. Soon after the discovery of x-rays, radiation biologists began to experiment with the use of radiation to treat a variety of nonmalignant and malignant conditions. In 1900, Thor Stenbeck used radiation to treat a patient with skin cancer. Since that time, radiation has been used extensively for the treatment of cancer. The medical field concerned with various aspects of using radiation to treat cancers is known as *radiation oncology*. Radiology, by contrast, has come to be used to describe the use of radiation for imaging rather than for treatment.

2.2 Radiation-Induced Cell Death

The radiation used for cancer therapy is ionizing, that is, comprising particles that can liberate an electron from an atom or molecule. Radiation thus interacts with cells by liberating electrons from the component atoms, rendering those atoms positively charged. The liberated electrons then interact with water molecules, leading to the formation of free radicals such as hydroxyl ions (OH⁻), which interact with the deoxyribonucleic acid (DNA) in cells to produce various types of lesions such as single-strand or double-strand breaks within the DNA, cross-links between two parts of the DNA or between DNA and proteins, and other damage to the bases constituting the DNA (Hutchinson 1961; Johansen and Howard-Flanders 1965; Nygaard et al. 1975). The cell death induced by radiation was classically thought to result mainly from double-strand breaks; however, current evidence suggests that radiation can cause damage by means of other non-DNA–centric effects, such as the bystander response, adaptive responses, and low-dose hypersensitivity. The two major mechanisms by which irradiation of normal or cancerous cells can kill those cells are by halting reproduction and by causing apoptosis. Most often, ionizing radiation prevents the multiplication of cells after one or two divisions, with the number of divisions depending on the size of the radiation dose (Sanchez et al. 1994). This mode of cell death is known as mitotic death. Those cells that do not lose the ability to divide after radiation continue to multiply and form *clones*, which represents the basis of the clonogenic assay (Figure 2.1) (discussed in the following section). Apoptosis or programmed cell death, on the other hand, is a process induced to maintain homeostasis or in response to stimuli such as radiation. The apoptotic process includes stepwise morphologic changes in the cells and fragmentation of DNA.

15

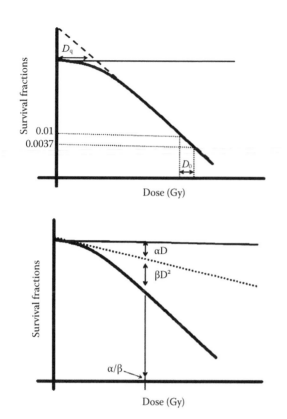

FIGURE 2.1 Proposed models for nonexponential cell killing by radiation. Top: the single- versus double-hit model; bottom: the linear-quadratic model.

2.3 Survival Curve

Various assays have been developed to gain further insight into how cells respond to radiation. As noted in the previous paragraph, those cells that do not lose the ability to divide after being exposed to radiation multiply to form colonies that eventually become visible to the naked eye. The capability of these cells to survive and reproduce forms the basis of evaluating the effect of radiation on different cell types. The assay technique involves preparing single-cell suspensions of the cell types of interest, which are then irradiated, often in combination with an investigational agent. The cells are next seeded in culture dishes and allowed to grow for several weeks until visible colonies are produced. Care must be taken in preparing these suspensions, because each colony is considered to arise from a single cell (i.e., "clonogenic," derived from the word "clone"). Before undertaking a clonogenic assay, one must first establish the *plating efficiency* of the cells, which represents the percentage of seeded cells that form colonies. For example, if only 70 of 100 plated cells go on to form colonies, then the plating efficiency is considered to be 70%.

Analysis of the effect of radiation on the survival and proliferation of cells requires the cells to be seeded in parallel with untreated samples and irradiated at specific doses. The cells that retain the ability to replicate form visible colonies, and these colonies are counted and compared with the numbers of colonies of the unirradiated cells. The *survival fraction* for a particular

radiation dose is thus defined as (number of colonies in irradiated samples × 100)/(number of cells seeded × the plating efficiency of the unirradiated cells).

2.4 Shape of Survival Curve

The first survival curve for mammalian cells was published by Puck and Marcus (1956), who presented their findings in graphs comparing survival fraction versus radiation dose, with survival fraction plotted on a logarithmic scale and dose on a linear scale. Survival curves for mammalian cells exposed to sparsely ionizing radiation such as gamma rays are usually exponential at lower doses (i.e., a straight line on the logarithmic scale), followed by a shoulder [extending over a few units of absorbed dose, expressed as Gray (Gy)] and then again exponential at higher doses. However, the survival curves for densely ionizing radiation such as alpha particles tend to be exponential along the entire dose range. The presence of the shoulder reflects the extent of radiation resistance, which is a common characteristic of some types of cancer cells such as melanoma.

One proposed explanation for the presence of shoulders in survival curves relies on whether the radiation occurs as single or double "hits" on the cell. At lower doses, a single lethal dose of radiation that causes the cells to lose their ability to replicate is represented as D_1. During this "single-hit" event, electrons or free radicals liberated through the interaction of radiation with the DNA kills the cell. However, as the radiation dose increases, the cells activate a repair process, thus reducing the dose-dependent reduction in the clonogenic survival. At increasingly higher doses, the multiple-hit model comes into play, in which the occurrence of two irreparable events before repair signaling can be activated leads to cellular lethality. At high doses such as these, the accumulation of ionization injury leads to the formation of "locally multiple damaged sites" (Ward 1988). High levels of injury can also reduce the ability of the cell to repair the damage, with the slope of this event denoted as D_0. The width of the shoulder, n, is calculated as

$$\log_e n = D_q/D_0$$

where D_q represents the quasi-threshold dose (i.e., the radiation dose below which no effect is produced).

Another model used to describe the survival curve is the linear quadratic model (Read 1952; Lea 1955). According to this model, radiation-induced cell killing depends on two components, one of which is proportional to the dose and the other proportional to the square root of the dose. According to this theory, the survival fraction (S) for the cells at a dose D is represented as

$$S = e^{\alpha D - \beta D2}$$

where S is the survival fraction (fraction of cells surviving a dose), α describes the initial slope of the survival curve, and β describes the quadratic component of cell killing. The higher the α/β ratio, the more linear the response and the less sensitive to dose fractionation.

2.5 Fractionation

Delivery of the first radiation treatments soon led to observations of radiation-induced complications. By 1900, five cases of radiation-induced leukemia and malignant skin changes had been reported. In 1922, lung fibrosis was observed after the treatment of breast cancer. In 1927, Claude Regaud and others observed that exposing the scrotums of rams to small daily fractions of radiation caused minimal skin reactions as compared with one single large dose (Regaud 1927). By the 1930s, fractionated treatment was becoming preferred over the use of single large doses. In 1934, Coutard established that the reaction of normal tissues to radiation therapy depends on the dose, treatment time, and number of treatment sessions. Later, Elkind hypothesized that the maximum dose tolerated by a particular normal tissue is related to both the number of fractions and the period over which the fractions are administered. Elkind and Sutton (1959) found that sublethal DNA damage could be repaired within a few hours. In 1965, Elkind et al. explained that the increase in patient survival associated with the fractionation of ionizing radiation is determined by the repair of sublethal damage by normal tissues. Indeed, the response of normal tissues and tumors to radiation reflects the four "R's" of radiation therapy and biology: repair, redistribution, reoxygenation, and repopulation (Withers 1975; Kallman 1972). For normal tissues, fractionation of a radiation dose is beneficial because such tissues divide slowly and may be able to repair the radiation-induced damage to the DNA. "Redistribution" refers to the relative proportions of cells at different points in the cell cycle. Cells in M (mitotic) phase are the most sensitive to radiation, whereas cells in S phase are the most resistant. Thus, ideally the timing for the radiation fractions should account for the point at which the tumor cells are most sensitive to radiation to enhance the efficiency of fractionation. Reoxygenation of typically hypoxic tumor tissues also makes tumors more sensitive to radiation; however, normal tissues are not sensitized, and this difference in sensitivity is often referred to as the oxygen enhancement ratio. The goal of fractionation is to achieve cumulative doses that result in tumor sterilization without severely affecting normal tissues. This leads to the concept of therapeutic ratio, discussed in the sections below.

2.6 Therapeutic Ratio and Probability of Tumor Control

In any radiation therapy application, the radiation dose to be delivered must be balanced in such a way as to be maximally effective but have minimal toxicity. The *therapeutic ratio* is the ratio of the maximum tolerated radiation dose to a minimum effective dose. Figure 2.2 illustrates a graph of a positive (beneficial) therapeutic ratio, in which the treatment dose produces more damage to the tumor than to the normal tissues.

Traditionally, definitive radiation therapy involved the daily administration of 1.5- to 3-Gy fractions (or about 9–10 Gy/

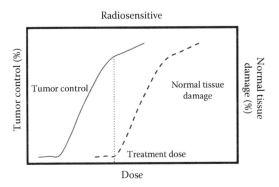

FIGURE 2.2 Representation of a positive therapeutic ratio with minimal normal tissue damage observed at the treatment dose.

week) to total doses of about 60–70 Gy. However, higher radiation doses can be achieved without causing deleterious effects to normal tissues if the dose per fraction is smaller (hyperfractionation); such an approach is often used when tumors are near critical normal-tissue structures, such as a lung tumor near the brachial plexus. On the other hand, larger doses per fractions (hypofractionation) can be used when the tumor is relatively far from a critical structure. Hypofractionated approaches such as stereotactic body radiation therapy have become increasingly feasible with recent advances in techniques for target delineation, motion management, and conformal inverse treatment planning (Potters et al. 2004; Nguyen et al. 2008). However, whereas hypofractionated radiation often results in obliteration of tumor tissues, it can also damage proximal normal tissues, and therefore the success of this approach depends on two factors: if the region receiving a high dose represents a small portion of the affected organ and if the organ can sustain its normal function despite damage to that region; an example would be small, circumscribed lung tumors located where obliteration of a small area would have a negligible influence on the overall function of the organ.

Since the early days of radiation treatment, investigators have attempted to characterize tumors in terms of their sensitivity to radiation. Paterson (1936) grouped tumors into three types: radiosensitive (e.g., germ-cell tumors), intermediate (e.g., adenocarcinomas), and radioresistant (e.g., melanomas). However, these assessments can vary substantially because of the presence of mixed cell populations, with different proliferation rates, within tumors. Because no single characteristic has been found that can reliably predict the response of tumor cells to radiation, a group of several such characteristics are used to deduce tumor control probability (TCP). Fletcher (1980) noted that a tumor cell population must be homogenous to produce a meaningful radiation dose–response curve and that any given dose will kill a specific fraction of cells. Therefore, the number of cells that remain (and remain reproductively active) after radiation will depend on the number of cells irradiated. As noted earlier in this section, the goal of radiation therapy is to reduce tumor size without affecting normal tissues. Therefore, dose–response curves for tumor cells (estimated by using TCP) and normal tissues are plotted to assess the clinical favorability of radiation therapy.

2.7 Radiation Therapy and Normal Tissue Effects

Irradiating tumors almost inevitably involves irradiating some portion of the surrounding tissues as well. The reaction of a normal tissue to radiation can be transient or prolonged. Because some effects of radiation are not apparent immediately, limits on the deliverable dose can be difficult to determine. In general, the development of radiation-induced injury in normal tissues depends on the turnover time of the affected tissue. Thus, the responses of a normal tissue can be categorized as acute, subacute, or late effects. Moreover, because organs often consist of different cell types, the same organ can manifest acute and late responses, thus complicating the simple grouping of organs as either acute-responding or late-responding.

2.7.1 Acute Effects

Acute effects are typically observed in tissues consisting of rapidly proliferating cells, such as skin, the gastrointestinal mucosa, and bone marrow (Leach et al. 2001, 2002; Lee and Bernstein 1993). Such tissues include small numbers of highly proliferative progenitor or stem cells that give rise to mature functional cells. These stem or progenitor cells are the most affected by irradiation; any stem cells that remain are by definition more resistant to radiation. Proliferation of these cells is triggered to compensate for the loss, and eventually the tissue recovers. Some reactions arising from acute responses of tissues to irradiation include edema, inflammation, vascular injury, and activation of the coagulation cascade.

2.7.2 Subacute Effects

Subacute effects or responses are observed within weeks to months after radiation in tissues that have longer turnover periods. These effects are observed during the remodeling phase of the irradiated tissue; an example is Lhermitte's syndrome after spinal cord irradiation.

2.7.3 Late Effects

Late effects develop months to years after radiation and are observed in tissues with slow turnover rates such as the brain, muscles, kidneys, and fatty tissues. The mechanisms underlying late radiation effects is not well understood but may involve the inability of tissue-based stem cells to repopulate and the consequent effects on interactions between various cells types. An example of a reaction arising from late effects is late demethylation after brain irradiation caused by loss of oligodendrocytes and neurons (Chehab et al. 2000).

2.8 Types of DNA Damage

Mammalian DNA is constantly subjected to stimuli, both external and internal, that produce thousands of lesions, which if left unrepaired could block DNA replication and lead to mutations, that is, causing genomic instability (Lindahl and Barnes 2000). Lesions can be caused by oxidative damage (Lindahl 1993), alkylating damage (Sedgwick et al. 2007), and deamination events (Kavli et al. 2007) that can lead to loss of the bases constituting the DNA, single-strand or double-strand breaks in the DNA, and impaired base pairing. Cells have evolved multiple repair mechanisms to counter the damage to DNA depending on the type of damage. A fundamental protein often considered to be a master regulator of this process is p53, which responds to low levels of damage by activating cell cycle arrest and DNA repair and responds to high levels of damage by inducing cell death via apoptosis. Unfortunately, mutations in p53, present in up to 50% of human cancers, can contribute to radiation resistance (Budanov 2011).

Ionizing radiation usually causes double-strand breaks in the DNA but can also cause single-strand breaks at lower doses. DNA damage can be spontaneous or induced by a particular cellular product or chemical compound. Endogenous DNA damage generated by metabolic processes occurs at a rate of 1000 to 1,000,000 molecular lesions cell^{-1} day^{-1} and most often involves chemical modification of nitrogenous bases and disruption of the helical structure of the DNA (Lodish 2003). The damage is caused mainly by reactive oxygen species such as oxygen ions and peroxides produced by the oxidative phosphorylation of mitochondria (Muller 2000; Han et al. 2001). Reactive oxygen species can cause point mutations in nuclear and mitochondrial DNA and large-scale genomic rearrangements (Huang et al. 2003; Hartman et al. 2004). About 100 types of oxidative DNA lesions have been described, including base modifications (such as 8-oxo-2′deoxyguanosine, thymidine glycol, and 8-hydroxycytosine), single- and double-strand DNA breaks, and interstrand cross-links (Cadet et al. 1997). DNA base alkylation, such as formation of 7-methylguanine, and hydrolysis (deamination, depurination and depyrimidination) are other sources of endogenous DNA damage. Mismatch of bases can also occur as the result of copying errors introduced by DNA polymerases during replication (Lodish 2008).

In addition to genetic insults caused by metabolic processes inside the cell, innumerable exogenous agents and compounds can also cause DNA damage. Industrial and environmental compounds such as vinyl chloride and the polycyclic hydrocarbons present in smoke can induce several types of injury to DNA. Nonetheless, the major exogenous sources of DNA damage are ultraviolet and ionizing radiation, either of which can damage DNA components. Ultraviolet radiation can produce free radicals and pyrimidine dimers formed by adjacent cytosine and thymine bases (Goodsell 2001). Ionizing radiation such as that created by radioactive decay or in cosmic rays causes breaks in DNA strands that can induce replicational and transcriptional errors leading to premature aging and cancer (Lodish 2003). In this context, it is interesting that the induction of DNA damage by radiation has been used as a therapeutic strategy to combat cancer.

The damage to DNA induced by radiation, therapeutic or otherwise, is caused by energy deposition along the track of the charged particle (electrons for x-rays and protons and alpha particles for neutrons). Energy is not deposited uniformly along the track; deposition patterns can be broadly categorized as spur (4-nm, 2× DNA diameter; 100 eV energy deposited; about three ion pairs) or blob (7-nm diameter; 100–500 eV energy deposited; about 12 ion pairs). X-rays deposit 95% of their energy via spurs. Neutrons and alpha particles deposit much of their energy via blobs, which can cause more severe DNA damage than spurs. Damage from spur and blob energy depositions can lead to multiple close sites of DNA damage called locally multiple damaged sites. Statistically, about 37% of cells in a population exposed to a dose that induces an average of one lethal event per cell (D_0) will survive, and other cells will accumulate one or multiple lethal events. For most mammalian cells, the D_0 for photons is about 1–2 Gy (Hanai et al. 1998).

2.9 DNA Repair Processes

The consequences of DNA damage are essentially twofold. After misrepair or replication of the damaged template, surviving cells may be subject to permanent changes in the genetic code in the form of mutations or chromosomal aberrations, both of which increase the risk of cancer. Alternatively, damage may interfere with transcription or induce replication arrest, which in turn can trigger cell death or cellular senescence. Damage-induced cell death protects the body from cancer (Hoeijmakers 2009). As a consequence, the DNA repair process is constantly active as it responds to damage in the DNA structure. To prevent the harmful consequences of DNA damage and recover the lost information, a variety of strategies and a complex network of complementary DNA-repair mechanisms have evolved that depend on the type of damage inflicted. The DNA damage response pathways can activate cell cycle checkpoints (which can involve p53) to arrest the cell either transiently or permanently (senescence) or they can activate specific DNA repair pathways in response to certain types of DNA damage (Altieri et al. 2008). As noted above, failure to repair DNA lesions can result in blockages of transcription and replication as well as mutagenesis and cytotoxicity (Friedberg 2006). In humans, DNA damage has been shown to be involved in a variety of genetically inherited disorders, in aging (Finkel and Holbrook 2000), and in carcinogenesis (Hoeijmakers 2001). The complex cellular network that collectively forms the DNA damage response machinery encompasses a plethora of dynamic, hierarchically ordered, and mutually coordinated pathways capable of detecting the lesions and signaling their presence to many DNA damage response proteins and protein complexes that promptly repair the damaged DNA (Hoeijmakers 2001; Bartek and Lukas 2007). When a cell encounters damage that is more difficult to repair, the DNA damage response machinery delays

cell-cycle progression (the so-called "cell cycle checkpoints") to provide more time for repair of the lesions (Lukas and Bartek 2004). Details of the processes by which cells repair single-strand and double-strand breaks in DNA are described in the following sections.

2.9.1 Single-Strand DNA Damage Response

Single-strand DNA breaks are usually not lethal, because most cells have efficient means of repairing them. However, if these breaks remained unrepaired, their accumulation could lead to double-strand breaks, which ultimately can cause cell death. The three mechanisms by which cells repair single-strand DNA breaks—base-excision repair, nucleotide-excision repair, and mismatch repair—are described in the following paragraphs.

2.9.1.1 Base Excision Repair

The base-excision repair pathway was discovered in the 1970s during a search for an enzymatic activity that could remove mutagenic uracil from DNA bases in *Escherichia coli* (Lindahl 1974, 1980). This enzyme was subsequently found to be conserved in other organisms as well (Friedberg et al. 1975; Olsen et al. 1989). Base-excision repair is evoked by DNA damage due to alkylation, deamination, and oxidation, and consists of the damaged DNA base being detected, removed, and replaced with the correct base. The recognitions of damaged base is done by DNA glycosylase, which catalyzes the cleavage of an *N*-glycosidic bond, creating an apurine/apyrimidine site (Laval 1977). Then, DNA endonuclease or DNA lyase creates a single-strand "nick" in the DNA that is processed by the DNA endonuclease, forming a single-nucleotide gap (O'Connor and Laval 1989; Robson and Hickson 1991; Boiteux et al. 1987) that is then filled by DNA polymerase with the correct nucleotide (Matsumoto and Kim 1995).

The base-excision repair pathway in humans was initially considered to involve only four enzymes—uracil-DNA glycosylase, an apurinic/apyrimidinic endonuclease (also called APEX nuclease 1 or APEX1), DNA polymerase beta (POLB), and LIG3 or LIG1 (DNA ligase III or I) (Kubota et al. 1996)—but others involved in another form of repair (long-patch base-excision repair) were subsequently described (Frosina et al. 1996). The initial step in long-patch base-excision repair involves APEX1 catalyzing 5′ nick formation. This catalysis recruits DNA polymerase beta or delta, proliferating cell nuclear antigen (PCNA), flap structure–specific endonuclease 1 (FEN1), and LIG1. The DNA polymerase then polymerizes and displaces DNA by more than one base in a PCNA-dependent manner. This strand displacement produces a "flapped" substrate that resists ligation, and thus FEN1 catalyzes the removal of the flapped substrate for ligation to take place. How the decision is made to proceed with long-patch or short-patch base-excision repair is not well understood, although the concentration of ATP near the apurinic/apyrimidinic site could be a determinant (Klungland and Lindahl 1997; Petermann et al. 2003).

2.9.1.2 Nucleotide Excision Repair

The glycosylase enzyme that recognizes and initiates the base-excision repair process does not recognize bulky and complex lesions; such lesions are recognized and repaired via the nucleotide-excision repair process (Leadon 1996). This process can eliminate lesions formed by ultraviolet light, cisplatin, psoralen, and polycyclic carcinogens such as acetylaminofluorene (Hansson et al. 1989, 1991; Svoboda et al. 1993; Mu et al. 1994). The process is not required for viability, as individuals with xeroderma pigmentosa are completely deficient in the nucleotide-excision repair process; however, such individuals are highly susceptible to skin cancer (Kraemer et al. 1994). Nucleotide-excision repair is initiated by structure-specific endonucleases that recognize structural distortions in the DNA and incise the affected DNA at the 3′ and 5′ ends (Scherly et al. 1993; O'Donovan et al. 1994). After the damaged DNA is removed, the process resembles that of base-excision repair in that DNA polymerase fills in the base and ligase seals the gap. The main difference between the two types of repair is that base-excision repair involves the removal and repair of a base, whereas nucleotide-excision repair involves the repair of longer segment of DNA.

2.9.1.3 Mismatch Repair

As its name suggests, the mismatch repair pathway is responsible for removing base–base mismatches; deficiencies in this process increase the susceptibility to cancer. The proteins involved in mismatch repair are the MutS homologues (MSH)-2, MSH3, and MSH6, MutL homologue 1 (MLH1), and postmeiotic segregation 2 (PMS2) (Lipkin et al. 2000). Mismatch repair eliminates errors such as base–base mismatches and insertion/deletion loops in DNA strands newly synthesized by DNA polymerase. Errors are recognized by MutSα, a heterodimer of MSH2 and MSH6. Another heterodimer, MutSβ, is formed by MSH2 and MSH3 but cannot detect base–base mismatches (Jiricny 1998). After recognition, MLH1 and PMS2 bind to each other and the MutSα heterodimer, which produces nicks in the DNA resulting in a 100- to 1000-nucleotide gap filled by DNA polymerase and PCNA (Jiricny 1998).

2.9.2 Double-Strand DNA Damage Response

Breaks in both strands of DNA can result from ionizing radiation, replication errors, oxidizing agents, or other metabolites. If they are to survive, cells must quickly repair these breaks before they lead to fragmentation of chromosomes and loss of genes. Cells repair double-strand breaks by one of two mechanisms—nonhomologous end-joining or homologous recombination. The first of these mechanisms, nonhomologous end-joining, can be conceptualized as a "quick fix" that is used for the overwhelming majority of DNA double-strand breaks; however, this process is more prone to errors compared to homologous recombination, a more complex process that can repair double-strand breaks with perfect fidelity (Alberts 2008).

As its name implies, nonhomologous end joining does not require a homologous template to join the two broken ends; rather, the two DNA strands that resulted from a break of a single strand are brought together at their ends. This process takes place in three steps: tethering, by which the two strands are brought together to form a scaffolding for other repair proteins; end processing, by which damaged or mismatched nucleotides are resynthesized; and ligation, or rejoining of the two ends. Tethering begins with the recognition of double-strand breaks by the MRN protein complex, which acts to bring together the DNA ends at areas of microhomology consisting of 1–6 bp (Zha et al. 2009; Barlow et al. 1992). The protein heterodimer Ku 70/80 also recognizes the double-strand break and recruits other proteins involved in repair, such as DNA-dependent protein kinase catalytic subunits (DNA-PKcs) (DeFazio et al. 2002). The next step, end processing, involves the repair of the damaged or mismatched nucleotides by the MRN complex, which has nuclease activity, and the DNA-PKcs, which are thought to induce conformational changes that allow other end-processing enzymes access to the double-strand break (Meek et al. 2008; Williams et al. 2007). The third, completing step in nonhomologous end-joining is ligation. In this step, Ku 70/80 interacts with DNA ligase IV (Palmbos et al. 2008), the catalytic unit of the XRCC4-ligase IV complex, to complete the ligation (Callebaut et al. 2006).

The importance of homologous recombination, the other major type of double-strand break repair, is underscored by its strong conservation in most cells (Alberts 2008). In addition to its function in repairing double-strand DNA breaks induced by ionizing radiation or metabolites, homologous recombination also has a central role in promoting genetic diversity by promoting "crossing-over" events during meiosis. Because this process relies on a homologous chromosome to provide a complementary template for repair of the damaged DNA strand, homologous recombination takes place only during and shortly after the S and G_2 phases, when sister chromatids are available as templates (Alberts 2008).

The homologous recombination process begins, as does nonhomologous end joining, with recognition of the double-strand break by the MRN protein complex (Mimitou and Symington 2009). This complex, with the assistance of other recruited factors such as the SAE2 protein, creates a 3′ overhang by removing nucleotides from the 5′ end of the double-strand break (Mimitou and Symington 2009). This 3′ strand is then coated with replication protein A to form a nucleoprotein filament (Wold 1997), and other proteins (e.g., RAD51, RAD52, BRCA1, and BRCA2) are recruited. This nucleoprotein filament then undergoes *strand invasion*, in which the filament interacts with a sister chromatid as it searches for a complementary sequence. Strand invasion results in a heteroduplex DNA structure, which is a DNA double helix consisting of two strands that were initially part of two different DNA molecules. After strand invasion and recognition of a complementary sequence, a DNA polymerase uses the sister chromatid sequence as a template for extending the 3′ end of the invading strand. This extension turns the heteroduplex DNA structure into an intermediate four-strand DNA structure called a "Holliday junction" shared between two DNA helices (Sung and Klein 2006). Further DNA synthesis restores the strand on the homologous chromosome, on one of the original 3′ overhangs that was displaced by the invading strand (Sung and Klein 2006). Homologous recombination is completed by resolution of the Holliday complexes into two independent DNA strands by nick endonucleases (McMahill et al. 2007).

Another form of double-stranded DNA repair that can be considered a type of homologous recombination is *single-stranded annealing*, which specifically repairs double-strand breaks between two repeat sequences. The process is relatively simple. The two strands that result from a break between the repeated sequence each have the same repeated sequence near their terminus. A single-stranded 3′ overhang is first created by trimming the 5′ terminus of each strand. The 3′ overhangs are prevented from annealing to each other by replication protein A, which allows the protein Rad52 to bind to and align the repeat sequence on either side of the break, which allows annealing at the repeated sequence (Lyndaker and Alani 2009). Because this process results in loss of the DNA repeats as well as the DNA sequence between the repeats, the single-stranded annealing process is considered potentially mutagenic (Helleday et al. 2007).

2.10 Conclusion

A cell's response to radiation is characterized by various factors including ionizing property of radiation and repair capacity of the cell. The process is initiated by generation of reactive oxygen species followed by DNA damage in the form of single-strand or double-strand breaks. The cells have developed DNA-repair mechanisms for these insults, which, if unrepaired lead to cell death. Proteins such as p53, MRN complex, and RAD51 play an important role in sensing DNA damage and initiate repair mechanism (if the damage is reversible) or cell death (lethal/ irreversible damage). The understanding of DNA-damage and DNA-repair mechanisms would help us understand the radiosensitization effects of nanoparticles discussed later in this book.

References

Alberts, B. 2008. *Molecular Biology of the Cell*. 5th ed., vol. 1. New York: Garland Science.

Altieri, F., C. Grillo, M. Maceroni, and S. Chichiarelli. 2008. DNA damage and repair: From molecular mechanisms to health implications. *Antioxidants & Redox Signalling* 10(5):891–937.

Barlow, J. W., L. E. Raggatt, C. F. Lim, D. J. Topliss, and J. R. Stockigt. 1992. Characterization of cytoplasmic T3 binding sites by adsorption to hydroxyapatite: Effects of drug inhibitors of T3 and relationship to glutathione-*S*-transferases. *Thyroid* 2(1):39–44.

Bartek, J., and J. Lukas. 2007. DNA damage checkpoints: From initiation to recovery or adaptation. *Current Opinion in Cell Biology* 19(2):238–245.

Boiteux, S., T. R. O'Connor, and J. Laval. 1987. Formamido-pyrimidine-DNA glycosylase of *Escherichia coli*: Cloning and sequencing of the fpg structural gene and overproduction of the protein. *EMBO Journal* 6(10):3177–3183.

Budanov, A. V. 2011. Stress-responsive sestrins link p53 with redox regulation and mammalian target of rapamycin signaling. *Antioxidants & Redox Signalling* 15(6):1679–1690.

Cadet, J., M. Berger, T. Douki, and J. L. Ravanat. 1997. Oxidative damage to DNA: Formation, measurement, and biological significance. *Reviews of Physiology, Biochemistry & Pharmacology* 131:1–87.

Callebaut, I., L. Malivert, A. Fischer, J. P. Mornon, P. Revy, and J. P. de Villartay. 2006. Cernunnos interacts with the XRCC4 × DNA–ligase IV complex and is homologous to the yeast nonhomologous end-joining factor Nej1. *Journal of Biological Chemistry* 281(20):13857–13860.

Chehab, N. H., A. Malikzay, M. Appel, and T. D. Halazonetis. 2000. Chk2/hCds1 functions as a DNA damage checkpoint in G(1) by stabilizing p53. *Genes & Development* 14(3):278–288.

DeFazio, L. G., R. M. Stansel, J. D. Griffith, and G. Chu. 2002. Synapsis of DNA ends by DNA-dependent protein kinase. *EMBO Journal* 21(12):3192–3200.

Elkind, M. M., and H. Sutton. 1959. X-ray damage and recovery in mammalian cells in culture. *Nature* 184:1293–1295.

Elkind, M. M., H. Sutton-Gilbert, W. B. Moses, T. Alescio, and R. W. Swain. 1965. Radiation response of mammalian cells grown in culture: V. Temperature dependence of the repair of X-ray damage in surviving cells (aerobic and hypoxic). *Radiation Research* 25:359–376.

Finkel, T., and N. J. Holbrook. 2000. Oxidants, oxidative stress and the biology of ageing. *Nature* 408(6809):239–247.

Fletcher, G. H. 1980. *Textbook of Radiotherapy*. 3rd ed. Philadelphia, PA: Lea & Febiger.

Friedberg, E. C. 2006. *DNA Repair and Mutagenesis*. 2nd ed. Washington, D.C.: ASM Press.

Friedberg, E. C., A. K. Ganesan, and K. Minton. 1975. *N*-Glycosidase activity in extracts of *Bacillus subtilis* and its inhibition after infection with bacteriophage PBS2. *Journal of Virology* 16(2):315–321.

Frosina, G., P. Fortini, O. Rossi, F. Carrozzino, G. Raspaglio, L. S. Cox, D. P. Lane, A. Abbondandolo, and E. Dogliotti. 1996. Two pathways for base excision repair in mammalian cells. *Journal of Biological Chemistry* 271(16):9573–9578.

Goodsell, D. S. 2001. The molecular perspective: Ultraviolet light and pyrimidine dimers. *Oncologist* (3):298–299.

Han, D., E. Williams, and E. Cadenas. 2001. Mitochondrial respiratory chain-dependent generation of superoxide anion and its release into the intermembrane space. *Biochemical Journal* 353(Pt 2):411–416.

Hanai, R., M. Yazu, and K. Hieda. 1998. On the experimental distinction between ssbs and dsbs in circular DNA. *International Journal of Radiation Biology* 73(5):475–479.

Hansson, J., M. Munn, W. D. Rupp, R. Kahn, and R. D. Wood. 1989. Localization of DNA repair synthesis by human cell extracts to a short region at the site of a lesion. *Journal of Biological Chemistry* 264(36):21788–21792.

Hansson, J., S. M. Keyse, T. Lindahl, and R. D. Wood. 1991. DNA excision repair in cell extracts from human cell lines exhibiting hypersensitivity to DNA-damaging agents. *Cancer Research* 51(13):3384–3390.

Hartman, P., R. Ponder, H. H. Lo, and N. Ishii. 2004. Mitochondrial oxidative stress can lead to nuclear hypermutability. *Mechanisms of Ageing and Development* 125(6):417–420.

Helleday, T., J. Lo, D. C. van Gent, and B. P. Engelward. 2007. DNA double-strand break repair: From mechanistic understanding to cancer treatment. *DNA Repair (Amsterdam)* 6(7):923–935.

Hoeijmakers, J. H. 2001. Genome maintenance mechanisms for preventing cancer. *Nature* 411(6835):366–374.

Hoeijmakers, J. H. 2009. DNA damage, aging, and cancer. *New England Journal of Medicine* 361(15):1475–1485.

Huang, M. E., A. G. Rio, A. Nicolas, and R. D. Kolodner. 2003. A genomewide screen in Saccharomyces cerevisiae for genes that suppress the accumulation of mutations. *Proceedings of the National Academy of Sciences of the United States of America* 100(20):11529–11534.

Hutchinson, F. 1961. Molecular basis for action of ionizing radiations. *Science* 134:533–538.

Jiricny, J. 1998. Replication errors: Cha(lle)nging the genome. *EMBO Journal* 17(22):6427–6436.

Johansen, I., and P. Howard-Flanders. 1965. Macromolecular repair and free radical scavenging in the protection of bacteria against X-rays. *Radiation Research* 24:184–200.

Kallman, R. F. 1972. The phenomenon of reoxygenation and its implications for fractionated radiotherapy. *Radiology* 105(1):135–142.

Kavli, B., M. Otterlei, G. Slupphaug, and H. E. Krokan. 2007. Uracil in DNA—general mutagen, but normal intermediate in acquired immunity. *DNA Repair (Amsterdam)* 6(4):505–516.

Klungland, A., and T. Lindahl. 1997. Second pathway for completion of human DNA base excision-repair: Reconstitution with purified proteins and requirement for DNase IV (FEN1). *EMBO Journal* 16(11):3341–3348.

Kraemer, K. H., M. M. Lee, A. D. Andrews, and W. C. Lambert. 1994. The role of sunlight and DNA repair in melanoma and nonmelanoma skin cancer. The xeroderma pigmentosum paradigm. *Archives of Dermatology* 130(8):1018–1021.

Kubota, Y., R. A. Nash, A. Klungland, P. Schar, D. E. Barnes, and T. Lindahl. 1996. Reconstitution of DNA base excision-repair with purified human proteins: Interaction between DNA polymerase beta and the XRCC1 protein. *EMBO Journal* 15(23):6662–6670.

Laval, J. 1977. Two enzymes are required from strand incision in repair of alkylated DNA. *Nature* 269(5631):829–832.

Lea, D. E. 1955. *Actions of Radiations on Living Cells.* 2nd ed. Cambridge, UK: University Press.

Leach, J. K., G. Van Tuyle, P. S. Lin, R. Schmidt-Ullrich, and R. B. Mikkelsen. 2001. Ionizing radiation-induced, mitochondria-dependent generation of reactive oxygen/nitrogen. *Cancer Research* 61(10):3894–3901.

Leach, J. K., S. M. Black, R. K. Schmidt-Ullrich, and R. B. Mikkelsen. 2002. Activation of constitutive nitric-oxide synthase activity is an early signaling event induced by ionizing radiation. *Journal of Biological Chemistry* 277(18):15400–15406.

Leadon, S. A. 1996. Repair of DNA damage produced by ionizing radiation: A minireview. *Seminars in Radiation Oncology* 6(4):295–305.

Lee, J. M., and A. Bernstein. 1993. p53 mutations increase resistance to ionizing radiation. *Proceedings of the National Academy of Sciences of the United States of America* 90(12):5742–5746.

Lindahl, T. 1974. An *N*-glycosidase from *Escherichia coli* that releases free uracil from DNA containing deaminated cytosine residues. *Proceedings of the National Academy of Sciences of the United States of America* 71(9):3649–3653.

Lindahl, T. 1980. Uracil-DNA glycosylase from *Escherichia coli*. *Methods in Enzymology* 65(1):284–290.

Lindahl, T. 1993. Instability and decay of the primary structure of DNA. *Nature* 362(6422):709–715.

Lindahl, T., and D. E. Barnes. 2000. Repair of endogenous DNA damage. *Cold Spring Harbor Symposia on Quantitative Biology* 65:127–133.

Lipkin, S. M., V. Wang, R. Jacoby, S. Banerjee-Basu, A. D. Baxevanis, H. T. Lynch, R. M. Elliott, and F. S. Collins. 2000. MLH3: A DNA mismatch repair gene associated with mammalian microsatellite instability. *Nature Genetics* 24(1):27–35.

Lodish, H. F. 2003. *Molecular Cell Biology.* 5th ed. New York: W. H. Freeman.

Lodish, H. F. 2008. *Molecular Cell Biology.* 6th ed., 1 vol. New York: W. H. Freeman.

Lukas, J., and J. Bartek. 2004. Watching the DNA repair ensemble dance. *Cell* 118(6):666–668.

Lyndaker, A. M., and E. Alani. 2009. A tale of tails: Insights into the coordination of 3′ end processing during homologous recombination. *Bioessays* 31(3):315–321.

Matsumoto, Y., and K. Kim. 1995. Excision of deoxyribose phosphate residues by DNA polymerase beta during DNA repair. *Science* 269(5224):699–702.

McMahill, M. S., C. W. Sham, and D. K. Bishop. 2007. Synthesis-dependent strand annealing in meiosis. *PLoS Biology* 5(11):e299.

Meek, K., V. Dang, and S. P. Lees-Miller. 2008. DNA-PK: The means to justify the ends? *Advances in Immunology* 99:33–58.

Mimitou, E. P., and L. S. Symington. 2009. Nucleases and helicases take center stage in homologous recombination. *Trends in Biochemical Sciences* 34(5):264–272.

Mu, D., E. Bertrand-Burggraf, J. C. Huang, R. P. Fuchs, A. Sancar, and B. P. Fuchs. 1994. Human and *E. coli* excinucleases are affected differently by the sequence context of acetyl-aminofluorene–guanine adduct. *Nucleic Acids Research* 22(23):4869–4871.

Muller, F. 2000. The nature and mechanism of superoxide production by the electron transport chain: Its relevance to aging. *Journal of the American Aging Association* 23(4):227–253.

Nguyen, N. P., L. Garland, J. Welsh, R. Hamilton, D. Cohen, and V. Vinh-Hung. 2008. Can stereotactic fractionated radiation therapy become the standard of care for early stage non-small cell lung carcinoma. *Cancer Treatment Reviews* 34(8):719–727.

Nygaard, O. F., H. I. Adler, W. K. Sinclair, and International Association for Radiation Research, and Radiation Research Society. 1975. *Radiation Research: Biomedical, Chemical, and Physical Perspectives: Proceedings of the Fifth International Congress of Radiation Research*, Seattle, WA, U.S.A., July 14–20, 1974. New York: Academic Press.

O'Connor, T. R., and J. Laval. 1989. Physical association of the 2,6-diamino-4-hydroxy-5*N*-formamidopyrimidine-DNA glycosylase of *Escherichia coli* and an activity nicking DNA at apurinic/apyrimidinic sites. *Proceedings of the National Academy of Sciences of the United States of America* 86(14):5222–5226.

O'Donovan, A., A. A. Davies, J. G. Moggs, S. C. West, and R. D. Wood. 1994. XPG endonuclease makes the 3′ incision in human DNA nucleotide excision repair. *Nature* 371(6496):432–435.

Olsen, L. C., R. Aasland, C. U. Wittwer, H. E. Krokan, and D. E. Helland. 1989. Molecular cloning of human uracil-DNA glycosylase, a highly conserved DNA repair enzyme. *EMBO Journal* 8(10):3121–3125.

Palmbos, P. L., D. Wu, J. M. Daley, and T. E. Wilson. 2008. Recruitment of *Saccharomyces cerevisiae* Dnl4–Lif1 complex to a double-strand break requires interactions with Yku80 and the Xrs2 FHA domain. *Genetics* 180(4):1809–1819.

Paterson, R. 1936. The radical x-ray treatment of the carcinomata. *British Journal of Radiology* 9:671–679.

Petermann, E., M. Ziegler, and S. L. Oei. 2003. ATP-dependent selection between single nucleotide and long patch base excision repair. *DNA Repair (Amsterdam)* 2(10):1101–1114.

Potters, L., M. Steinberg, C. Rose, R. Timmerman, S. Ryu, J. M. Hevezi, J. Welsh, M. Mehta, D. A. Larson, and N. A. Janjan. 2004. American Society for Therapeutic Radiology and Oncology and American College of Radiology practice guideline for the performance of stereotactic body radiation therapy. *International Journal of Radiation Oncology Biology Physics* 60(4):1026–1032.

Puck, T. T., and P. I. Marcus. 1956. Action of x-rays on mammalian cells. *Journal of Experimental Medicine* 103(5):653–666.

Read, J. 1952. The effect of ionizing radiations on the broad bean root: XI. The dependence of the alpha-ray sensitivity on dissolved oxygen. *British Journal of Radiology* 25(300):651–661.

Regaud, C. 1927. Paris Université Institut du radium. [from old catalog], and Fondation Curie Paris. [from old catalog]. *Radiophysiologie et radiothérapie*. Paris: Les Presses universitaires de France.

Robson, C. N., and I. D. Hickson. 1991. Isolation of cDNA clones encoding a human apurinic/apyrimidinic endonuclease that corrects DNA repair and mutagenesis defects in *E. coli* xth (exonuclease III) mutants. *Nucleic Acids Research* 19(20):5519–5523.

Sanchez, I., R. T. Hughes, B. J. Mayer, K. Yee, J. R. Woodgett, J. Avruch, J. M. Kyriakis, and L. I. Zon. 1994. Role of SAPK/ERK kinase-1 in the stress-activated pathway regulating transcription factor c-Jun. *Nature* 372(6508):794–798.

Scherly, D., T. Nouspikel, J. Corlet, C. Ucla, A. Bairoch, and S. G. Clarkson. 1993. Complementation of the DNA repair defect in xeroderma pigmentosum group G cells by a human cDNA related to yeast RAD2. *Nature* 363(6425):182–185.

Sedgwick, B., P. A. Bates, J. Paik, S. C. Jacobs, and T. Lindahl. 2007. Repair of alkylated DNA: Recent advances. *DNA Repair (Amsterdam)* 6(4):429–442.

Sung, P., and H. Klein. 2006. Mechanism of homologous recombination: Mediators and helicases take on regulatory functions. *Nature Reviews Molecular Cell Biology* 7(10):739–750.

Svoboda, D. L., J. S. Taylor, J. E. Hearst, and A. Sancar. 1993. DNA repair by eukaryotic nucleotide excision nuclease. Removal of thymine dimer and psoralen monoadduct by HeLa cell-free extract and of thymine dimer by *Xenopus laevis* oocytes. *Journal of Biological Chemistry* 268(3):1931–1936.

Ward, J. F. 1988. DNA damage produced by ionizing radiation in mammalian cells: Identities, mechanisms of formation, and reparability. *Progress in Nucleic Acid Research and Molecular Biology* 35:95–125.

Williams, R. S., J. S. Williams, and J. A. Tainer. 2007. Mre11–Rad50–Nbs1 is a keystone complex connecting DNA repair machinery, double-strand break signaling, and the chromatin template. *Biochemistry and Cell Biology* 85(4):509–520.

Withers, H. R. 1975. Cell cycle redistribution as a factor in multifraction irradiation. *Radiology* 114(1):199–202.

Wold, M. S. 1997. Replication protein A: A heterotrimeric, single-stranded DNA-binding protein required for eukaryotic DNA metabolism. *Annual Review in Biochemistry* 66:61–92.

Zha, S., C. Boboila, and F. W. Alt. 2009. Mre11: Roles in DNA repair beyond homologous recombination. *Nature Structural & Molecular Biology* 16(8):798–800.

Principles and Application of Hyperthermia Combined with Radiation

Mark Hurwitz
Harvard Medical School

3.1 Introduction

Hyperthermia is the application of heat in a therapeutic setting. When cells are heated beyond their normal temperature, they can become sensitized to therapeutic agents including radiation and chemotherapy. If cells are heated to still higher temperatures, the heat will cause irreparable damage resulting in cell death, a process referred to as thermal ablation.

3.1.1 History

The value of heat as a therapeutic modality has been recognized for thousands of years. The Egyptians documented use of hyperthermia as a medical treatment more than 5000 years ago (Smith, E. Egyptian surgical papyrus dated around 3000 BC, cited by van der Zee 2002). In ancient Greece, Hippocrates described the first use of hyperthermia in cancer therapy in treatment of breast cancer (Seegenschmiedt and Vernon 1995). He noted: "That which drugs fail to cure, the scalpel can cure. That which the scalpel fails to cure, heat can cure. If heat cannot cure, it must be deemed incurable."

In the nineteenth century, Busch (1866) and Coley (1893) successfully utilized infection and toxins, respectively, in cancer treatment. In 1893, Coley reviewed 38 patients with advanced cancer who developed high fevers secondary to either accidental or deliberate infection with erysipelas. Twelve patients experienced complete regression of their tumors, and 19 others experienced a partial response (Coley 1893). The first documentation of benefit with combined hyperthermia and radiation dates to a 1910 publication of a phase II trial performed in Germany. One hundred patients with histologically confirmed advanced cancer were treated with diathermy and radiation. Thirty-two patients experienced complete regression, whereas 32 experienced a rapid but temporary improvement (Muller 1910). In 1935, Warren reported on combining induced fever with roentgen therapy that resulted in significant improvement and palliation in 29 of 32 oncology patients whose condition was described as "hopeless."

3.1.2 Clinical Trials

In the 1970s and 1980s, new understandings of thermal biology and its potential to complement radiation or directly radiosensitize led to a wave of enthusiasm for new clinical trials. Several phase III trials were initiated in the 1980s. These trials, however, were hampered by several key limitations including the inability of early thermal therapy systems to effectively heat many tumors, lack of appreciation for thermal dosimetry and therefore meaningful treatment goals, and lack of quality assurance guidelines.

The Radiation Therapy Oncology Group (RTOG), a large National Cancer Institute–sponsored cooperative cancer research group, conducted trials for both superficial and deep heating for a varied group of malignancies in the 1980s. RTOG 81-04 compared radiation with or without hyperthermia for a range of malignancies including breast, head and neck, trunk, and extremity tumors. There was no difference in complete response between groups; however, a difference was noted for tumors less than 3 cm, which were more likely to be effectively heated with the technology available at the time (Perez et al. 1991). RTOG 84-19 was a phase III study of the use of radiation with or without interstitial hyperthermia for persistent or recurrent tumors after previous radiation or surgery. A total of 184 patients were enrolled. There was no difference in any of the study endpoints; however, when the quality of hyperthermia was assessed, only one of 173 evaluable patients met the minimum accepted criteria for adequate hyperthermia (Emami et al. 1996). Given the difficulties with hyperthermia delivery with the available technology and lack of widely applicable quality assurance guidelines, enthusiasm for hyperthermia waned by the late 1980s. Despite the challenges with hyperthermia delivery in the early 1990s, a number of investigators persevered and phase III trials have since been completed showing benefit, including survival benefit in several instances, with addition of hyperthermia to radiation or chemotherapy. Head and neck malignancies, cervical cancer, and glioblastoma multiforme are among the malignancies for which hyperthermia combined with radiation has been shown to have an overall survival advantage in phase III randomized trials (Valdagni and Amichetti 1994; van der Zee et al. 2000; Sneed 1998). Ongoing advances in treatment technology, planning, monitoring, and application have brought new attention to the field of thermal medicine. The application of nanomedicine to thermal therapy is one exciting example of the convergence of multiple fields of applied science for the betterment of cancer patients.

3.2 Thermal Biology and Physiology

The clinical field of hyperthermia emerged in the 1970s on a foundation built on compelling biologic evidence that hyperthermia was an ideal complementary treatment to radiation and certain chemotherapeutic agents (Dewey et al. 1977; Westra and Dewey 1971; Kim et al. 1976; Gerweck et al. 1979; Hahn and Shiu 1986; Henle and Leeper 1976; Kano 1985). Cancer cells that are the most radio-resistant are precisely those most sensitive to heat. Cells resistant to radiation include those that are hypoxic, at low pH, nutritionally deprived, or in S phase of the cell cycle. All of these characteristics are associated with sensitivity to heat. In regard to physiology, hyperthermia also has been shown to increase perfusion resulting in improved tumor oxygenation for subsequent radiation treatments (Eddy 1980; Song 1984).

Low pH, particularly when associated with acute acidification, results in sensitization of cells to heat as has been demonstrated in preclinical models and has been studied in humans as well (Engin et al. 1994; Gerweck 1977; Gerweck et al. 1982; Leeper et al. 1994; Mueller-Klieser et al. 1996; Song et al. 1994; Wahl et al. 1997). This enhanced cytotoxicity is attributable to the limited reserve of cells in a low pH environment to further upregulate proton pumping (Wahl et al. 1997). There is evidence, however, that this enhanced thermal sensitivity does not occur in cells conditioned to a low pH environment (Hahn and Shiu 1986). Furthermore, low pH has been associated with inhibition of thermal tolerance and repair of thermal damage (Lin et al. 1992). Methods to induce acidification have included induction of hyperglycemia (Ward et al. 1991; Leeper et al. 1994; Snyder et al. 2001) and use of the respiratory inhibitor metaiodobenzylguanidine (Biaglow et al. 1998; Zhou et al. 2000). The study of application of acute acidification in humans is limited, and the clinical practicality and efficacy of this strategy to enhance thermal cytotoxicity remains to be defined.

Poor nutritional status and low energy stores have also been found to result in increased thermal sensitivity (Kim et al. 1992; Koutcher et al. 1990). Hyperthermia has been linked with cellular energy depletion. Decreases in adenosine triphosphate (ATP) and phosphocreatinine with increases in inorganic phosphate have been noted with hyperthermia (Vaupel and Kelleher 2010). Hydrolysis of ATP results in accumulation of purine catabolites and proton formation with resultant acidification. Acidification is due in large part to an increase in lactic acid resulting from impairment of the oxidative pathway with hyperthermia (Streffer 1985). Hyperthermic sensitization may also result via production of free radicals leading to DNA damage (Vaupel and Kelleher 2010). It is noteworthy that most studies of metabolism and energy status have been done in preclinical systems with relatively high temperatures in the range of 44°C. The impact of nutritional status in clinical studies is yet to be fully defined (Sostman et al. 1994).

Hypoxia has long been recognized for its association with radioresistance given the role of free radicals in DNA damage. Whereas hypoxia increases radioresistance by up to threefold, the oxygen status of cells, however, has no impact on sensitivity to hyperthermia. Blood perfusion also increases within minutes of initiation of heating. Although increased perfusion will result in decreased temperatures, if measures such as increased power distribution are not applied, this enhanced blood flow may result in improved tumor oxygenation. Enhanced perfusion has been noted to last in some instances for more than 24 h (Song 1984; Lüdemann et al. 2009; Brizel et al. 1996). In addition to increased perfusion, improved tumor oxygenation may occur by reducing tissue oxygen consumption such as through impairment of respiration in sublethally heated cells (Griffin and Corry 2009). When radiation is administered after thermal therapy, these sensitizing effects may augment the effects of at least two standard fractionated daily radiation treatments. Along with increase in perfusion, hyperthermia results in increased vascular permeability, which can enhance delivery of drugs or other agents including radioisotopes to tumor.

Hyperthermia and radiation are complementary in regard to cell cycle sensitivities. Cells are most susceptible to radiation during mitosis and are most resistant to radiation in late S phase. This resistance of cells to radiation in S-phase has been postulated to be attributable to increased time for sublethal damage repair to occur. The opposite is true with hyperthermia for which greatest sensitivity occurs during S phase (Westra and Dewey 1971; Kim et al. 1976). Prior research with cell cultures has explored synchronization of the cell cycle as a strategy to enhance tumor kill with combined radiation and hyperthermia (Figure 3.1). Application of such strategy in the clinical setting is of course a greater challenge; nevertheless, this complementary cell kill contributes to the effects of hyperthermia.

The ultimate ways in which hyperthermia augments the effects of radiation are a complex interplay of the biologic and physiologic factors discussed here. The interdependency of perfusion, oxygenation, pH, nutritional status, and cell cycle effects are readily apparent. In the clinical setting, the impact of any one of these factors in isolation is difficult to discern and certainly varies in relation to the heating profile and both the macro- and microenvironments to which heat is applied. The benefits of hyperthermia seen in combination with radiation in numerous clinical trials speak to the veracity of the underlying precepts of thermal biology.

3.3 Mechanisms of Cell Kill with Hyperthermia

Recognition of the principles of thermal biology provides compelling rationale for combined hyperthermia and radiation. Appreciation of the underlying mechanisms for how heat results in cell death has furthered our understanding of how best to apply this therapeutic modality. Further research into the many ways heat may lead to cancer cell death, and how to augment these pathways remains an active area of investigation.

Heating of cells to high temperature as is done with thermal ablation results in rapid cell death. The mechanisms for cell death with mild to moderate temperature elevation between 39°C and 44°C, referred to subsequently in this chapter as MTH, in combination with radiation or chemotherapy is, however, much more complex. Although the clinical focus of thermal ablation strategies has been on identification of tissue to be heated and directly destroyed, there is increased awareness that attention to the heated but non-ablated rim of tumor tissue surrounding the ablated region will lead to more effective thermal therapies (Horkan et al. 2005; Jernberg et al. 2001). An understanding of the biology, physiology, and immunologic effects of MTH are therefore of ever-increasing importance in the field of thermal ablation (Figure 3.2).

Temperatures above 44–45°C sustained for a sufficient time will lead to protein denaturization and cell death typically via necrosis. High temperatures typically associated with thermal ablation, generally in the range of 55–85°C, result in cell death within seconds. At moderate temperature elevations, hyperthermic cell kill occurs in a log-linear manner with an initial shoulder region followed by a steeper decline in cell survival correlated with increasing temperatures up to 45°C. With temperature elevations in the range of 42–45°C, hyperthermia can result

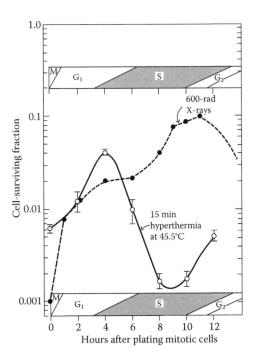

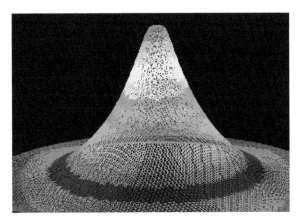

FIGURE 3.1 Comparison of the fraction of cells surviving heat or x-irradiation delivered at various phases of the cell cycle. The heat treatment consisted of 15 min at 45.5°C and the x-ray dose was 600 cGy. (Reproduced with permission from Hall, E.J., *Radiobiology for the Radiologist*, Lippincott and Company, Philadelphia, PA, 1994; from Westra, A., Dewey, W.C., *Int. J. Radiat. Biol.*, 19, 467–477, 1971, with permission.)

FIGURE 3.2 **(See color insert.)** Schematic of thermal ablation temperature map. The heated but not ablated region below approximately 50°C is shown in blue and purple. Tumor kill in this region may be enhanced by combination of radiation with thermal ablation greatly expanding the effective "kill zone" while minimizing risk of thermal damage to adjacent tissues and organs. (Reproduced with permission from Ahmed, M., Goldberg, S.N., *Int. J. Hyperthermia*, 20(7), 781–802, 2004.)

in cell death with sufficiently long exposure but is not a practical or particularly effective therapeutic strategy as monotherapy. It is in the range between 39°C and 44°C that hyperthermia has been extensively studied for its combined and sensitizing effects with radiation. In addition to complementary biologic effects, hyperthermia results in radiosensitizion by inhibition of sublethal and potentially lethal damage repair through inactivation of DNA repair pathways. (Kampinga 2006; Mivechi and Dewey 1985; Raaphorst et al. 1994, 1999).

The mechanisms by which MTH results in tumor kill are complex and highly dependent on the heating profile. Mechanisms may include protein denaturization, induction of apoptosis, senescence, mitotic catastrophe, inhibition of sublethal and potentially lethal damage repair through inactivation of DNA repair pathways, and necrosis. (Gabai et al. 1995, 1998; Kampinga and Dikomey 2001; Mivechi and Dewey 1985; Raaphorst et al. 1994, 1999; Vidair et al. 1995; Westra and Dewey 1971). Early research revealed that protein denaturization was a key biologic effect of hyperthermia with MTH (Dewey 1994). The activation energies for protein denaturization and heat-induced cell death were noted to be within the same range. Further research suggested that nuclear proteins are most sensitive (Lepock et al. 1993, 2001; Lepock 2004), and a high degree of correlation of nuclear protein aggregation and heat-induced cell kill has been noted. Nuclear protein aggregation appears to inhibit the DNA repair process (Lepock 2004; Kampinga et al. 2001). Research to date has made it clear, however, that other mechanisms including apoptosis and senescence also have important roles in hyperthermic cell kill. Although research continues to define precise mechanisms for how heat results in cell death, the clinical utility of this therapeutic modality has become increasingly clear.

3.4 Thermal Dosimetry

The definition of thermal dosimetry and validation of clinically meaningful thermal dosimetric parameters has proven challenging. In contrast to radiation, a modality defined strictly by physical criteria with well-established concepts of dose, thermal dose is dependent on both physics and physiology. Given the impact of perfusion on tissue heating, and the variable thermal conductivity of different tissues further impacted by different thermal delivery methods, temperatures across the target region are heterogeneous. As such, the ability to prescribe thermal dose is a tenuous concept. Science and medicine seek reproducible and therefore quantifiable parameters for treatment. Therefore, ever since the onset of MTH in clinical practice, there has been appreciable interest in defining thermal dosimetric parameters, despite these challenges, that reliably define how the biologic principles of hyperthermia translate into clinical results.

Clinically relevant parameters have been defined despite challenges including the initial need for invasive thermometry, limited temperature measurements taken across the treatment area, as well as accessibility of target areas for invasive monitoring particularly with deep-seated lesions. With the advent of noninvasive thermometry, most notably via the method of proton resonance shift with magnetic resonance, new and more complete understanding of relevant thermal dose parameters should emerge given the capability of this technology to monitor thousands of temperature points in real time.

3.4.1 Specific Absorption Rate

The specific absorption rate (SAR) is one strategy applied to define thermal dose and plan treatment. SAR is a measure of the rate at which energy is absorbed by the body or a portion thereof when exposed to an energy source such as radiofrequency or ultrasound. It is defined as the power absorbed per mass of tissue and has units of watts per kilogram (W/kg). SAR has been shown to correlate with clinical outcome including, in one large study, local control for recurrent breast cancers treated with hyperthermia and radiation (Lee et al. 1998).

3.4.2 Thermal Dose Parameters

The relationship of the rate of cell killing to temperature with MTH can be described with Arrhenius plots. Arrhenius plots have biphasic curves with a breakpoint at which the slopes of the curves diverge. Early thermobiologic research consistently demonstrated a breakpoint in the range of 43°C. For every 1°C above 43°C, the time needed for an isoeffect at 43°C is halved, whereas for each 1°C below 43°C cell killing declines by a factor of 4. Sapareto and Dewey defined this relationship between temperature and cell kill as follows:

$$\text{CEM43°C} = tR^{(43°C-T)}$$

CEM43°C represents cumulative equivalent minutes at 43°C, t is time, R is a constant, and T denoted the average temperature achieved over the heating period. For temperatures above 43°C $R = 0.5$, and for temperatures below 43°C, $R = 0.25$ (Sapareto and Dewey 1984).

In clinical practice, the importance of minimal temperatures achieved within the tumor has been widely recognized (Jones et al. 2005; Oleson et al. 1993; Dewhirst and Sim 1984; Hand et al. 1997; Sherar et al. 1997; Seegenschmiedt et al. 1994; Issels and Schlemmer 2002; Kapp and Cox 1995). Therefore, thermal goals and results are often reported as CEM43°CT90 describing the equivalent minutes at 43°C achieved by 90% of the measured temperature points.

3.5 Thermal Tolerance

3.5.1 Definition

Early researchers in the field of thermal biology quickly came to the realization that the lethal effects of heat decreased in time, and subsequent heating repeated within a few days of the prior treatment was found to be less effective than the initial treatment. Thermal cell survival curves demonstrated not only an initial shoulder but also, particularly at lower temperatures, a

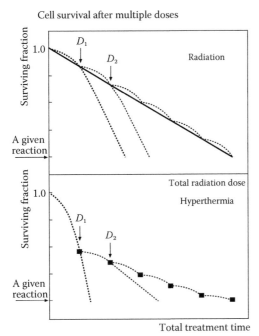

Cell survival after multiple doses

FIGURE 3.3 Thermal tolerance. Top: x-rays. Each dose in a fractionated regiment has about the same effect (i.e., kills the same proportion of cells). The shoulder of the curve must be reexpressed for each dose fraction. Bottom: Hyperthermia. The first heat treatment results in substantial biologic effect but also triggers the development of thermal tolerance, which may take several days to decay. When thermal tolerance is induced, subsequent daily heat treatments may be relatively ineffective because of the acquired thermoresistance of the cells. (Reproduced with permission from Hall, E.J., *Radiobiology for the Radiologist*, Lippincott and Company, Philadelphia, PA, 1994; from Urano, M., *Cancer Res.*, 46, 474–482, 1986, with permission.)

tail to the curve where further cell kill was abrogated. This phenomenon is known as thermal tolerance (Figure 3.3).

3.5.2 Heat Shock Proteins and Thermal Tolerance

Chanel and Maury first described the effects of thermal shock on blood proteins in fish in 1961 (Chanel and Maury 1961). In 1962, Ritossa reported that heat and dinitrophenol induced a characteristic pattern of puffing in the chromosomes of *Drosophila*. This discovery eventually led to the identification of the heat shock proteins (HSPs), whose expression these puffs represented. Over time, HSPs have come to be recognized as having a central role in the cellular stress response for which heat shock is just one of many potential cellular stresses. HSPs serve as chaperones, intracellularly playing a key role in stabilizing damaged proteins through a "holding and folding" mechanism allowing for repair. The induction of thermal tolerance was found to be associated with HSP up-regulation (Li et al. 1982; Berger and Woodward 1983).

3.5.3 Step Down Heating

Study of thermal cell kill led to the recognition that heating for a brief period above the breakpoint in the Arrhenius plot followed by reduction to mild hyperthermic temperatures results in inhibition of thermal tolerance (Dewhirst 1995). The phenomenon is known as step down heating. The ability to utilize step down heating for MTH has been limited by the inability in most cases to rapidly heat to sufficiently high temperatures because of equipment limitations and patient tolerance. Thermal ablation provides an opportunity to assess the role of step down heating in thermal therapy. The elevation in temperatures necessary to inhibit thermal tolerance can be easily achieved, and patient comfort measures such as conscious sedation or anesthesia are typically applied. The thoughtful application of step down heating to thermal ablative therapies as a way to enhance the effects on heated but non-ablative tissue is but one example of how the application of thermal biology may enhance the effects of thermal ablation.

3.6 Heat Shock Proteins

The central importance of HSPs to cellular stress response and therefore life itself is apparent when one recognizes that there is significant conservation of HSPs across all kingdoms of organisms (Lowe et al. 1983; Bardwell and Craig 1984; Chen et al. 2006). HSPs also have a central role in the cellular immune response.

The importance of HSPs is now widely recognized to extend well beyond thermal therapy to general cellular stress response and immune regulation. HSPs are a ubiquitous class of proteins that, although first identified in relationship to hyperthermia (McKenzie et al. 1975; Lindquist and Craig 1988), are now widely recognized as playing a vital role in general cellular stress response as well as immune regulation. HSPs are downstream effectors of multiple signal transduction pathways. Intracellularly, HSPs protect cells from proteotoxic stress through "holding and folding" pathways that prevent denaturation and progression of lethal pathways (Beckman et al. 1990; Gething and Sambrook 1992; Jones et al. 2004; Liu et al. 1992; Netzer and Hartl 1998; Xiao et al. 2006).

In the extracellular environment, HSPs have chaperokine effects with a central role in immune system response coupled with the ability to stimulate proinflammatory cytokine production (Asea et al. 2000). The chaperone function of HSPs relates to their role in chaperoning immunogenic peptides onto major histocompatibility complexes for presentation to T cells. The central importance of HSPs to this process is clear in that induction of immunity requires HSP-PC complexes with functional CD8+ cells and macrophages (Srivastava and Maki 1991). Furthermore, several investigators have shown that HSP-PC complexes can be tumor-specific and thereby induce tumor-specific immunity (Udono and Srivastava 1993, 1994; Udono et al. 1994).

An intriguing area of ongoing investigation is how extracellular HSP release in response to cancer treatment may stimulate an immune response including a tumor-specific immune response. Although controversy exists as to the impact of radiation on HSP response, a recent pilot study demonstrated a consistent increase in serum HSP70 over the duration of a standardly fractionated course of radiation therapy for prostate cancer patients. This increase in HSP70 was associated as expected with increase in proinflammatory cytokines and components of the cellular immune response including CD8+ and natural killer cells. Cell and animal modeling confirmed a tumor-specific response (Hurwitz et al. 2010). Given the recognized ability of hyperthermia to elicit tumor-specific responses (Udono and Srivastava 1993, 1994; Udono et al. 1994), the combination of radiation and hyperthermia might lead to an augmented effect that remains to be clinically investigated. It is intriguing to note the occasional abscopal effects reported with hyperthermia and radiation that are theorized to be secondary to immune modulation, and how such effects could be enhanced through use of hyperthermia and radiation (Dickson and Shah 1982; Szmigielski et al. 1991; Stawarz et al. 1993).

3.7 Combining Hyperthermia and Radiation: Clinical Considerations

3.7.1 Thermal Enhancement Ratio and Therapeutic Gain Factor

Thermal enhancement ratio (TER) is the ratio of radiation doses required to produce a given level of biological damage with versus without heat. Therapeutic gain factor is the ratio of TER in tumor versus normal tissues. Most studies to date have indicated that the TER of tumor is greater than that in normal tissues. Enhanced tumor TER appears due in large part to the tumor microenvironment including atypical vasculature leading to diminished effects of increases in perfusion.

3.7.2 Timing of Hyperthermia and Radiation

It is widely recognized that the greatest benefit to combined radiation and hyperthermia occurs with simultaneous administration of these modalities. In the case of MTH, there is great diminution in benefit if radiation and hyperthermia are separated by more than 1–2 h (Overgaard 1989; Law et al. 1977; Mittal et al. 1984) (Figure 3.4). In the clinical setting, simultaneous administration has not proven practical; however, systems are in development to facilitate this goal (Peñagarícano et al. 2008). The timing of radiation and hyperthermia has varied across clinical trials, and the benefit accrued likely varies with the strategy used. As an example, hyperthermia administered before radiation may enhance perfusion and thereby tumor, oxygenation leading to increased radiosensitization. Increased perfusion has been shown to last for more than 24 h and thereby may result in sensitization of an additional fraction or two of standardly administered daily radiation therapy (Song 1984; Lüdemann et al. 2009; Brizel et al. 1996). However, if temperatures are sufficiently high, vascular damage may occur thus decreasing tumor perfusion (Vaupel and Kelleher 2010). Conversely, a theoretical concern of application of hyperthermia before radiation therapy is up-regulation of HSPs with their protective effects,

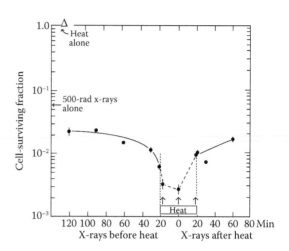

FIGURE 3.4 Timing of radiation and hyperthermia. Survival of Chinese hamster cells irradiated with 500 cGy of x-rays before, during, or after a heat treatment of 40 min at 42.5°C. No killing was observed with this heat treatment alone (open triangle at top left-hand corner). The effect of 500 cGy of x-rays alone is shown by the arrow (at left). There is a clear interaction between heat and x-rays, with the maximum effect being produced if the x-rays are delivered midway through the heat treatment. (Reproduced with permission from Hall, E.J., *Radiobiology for the Radiologist*, Lippincott and Company, Philadelphia, PA, 1994.)

although—as DNA is the target for radiation—the clinical significance of such up-regulation remains to be determined.

3.7.3 Duration of Heating and Temperature Goals

Clinical studies, until recently, have typically prescribed hyperthermia to be administered for a set amount of time in the range of 30–60 min to temperatures of 42–43°C. These treatment goals, initially derived in the 1980s, were based on preclinical models. It is clear from accumulated clinical experience that meaningful time and temperature parameters are complex. As CEM43°CT90 has been shown to be of clinical significance, treatment durations of 1 h are often desirable both in consideration of maximizing equivalent minutes at 43°C as well as practical considerations in regard to patient tolerance. Temperature goals based on preclinical models calling for higher temperatures may not always be necessary as significant clinical benefits have been seen with modest temperature elevations in the lower range of the MTH spectrum. The fact that clinical benefit has been witnessed across the range of 39–44°C speaks to the many potential ways in which hyperthermia and radiation may interact.

The ability to prescribe thermal dose in an analogous manner to radiation is a considerable, but not insurmountable, challenge. In a seminal study addressing this issue, researchers at Duke designed a trial to test whether a thermal dose of more than 10 CEM43°CT90 resulted in improved complete response rate and duration of local control compared with a thermal dose of ≤1 CEM43°CT90. A total of 122 patients with superficial tumors ≤3 cm in depth received a test hyperthermia treatment, of which 113 were successfully "heatable." These patients were randomized to no further hyperthermia or ten total hyperthermia fractions given in combination with radiation therapy. Median CEM43°CT90 for the low and high dose groups was 0.74 and 14.3, respectively. Overall complete response rate was significantly higher for patients in the high dose hyperthermia group, with the greatest benefit noted for patients who had received prior radiation and therefore could only receive limited radiation dose with retreatment.

3.7.4 Radiation and Hyperthermia Fractionation

MTH has typically been administered only one or two times per week, whereas radiation is most often administered daily 5 days per week. Limitation of hyperthermia to a weekly or biweekly schedule is in deference to concerns about development of thermal tolerance and the time needed for resolution of this heat shock response. The clinical significance of thermal tolerance has not been clearly elucidated. The optimal schedule for hyperthermia remains to be definitively defined and likely varies based on the particulars of a given treatment scenario. As hyperthermia results in increased tissue oxygenation, daily administration may prove advantageous, concerns with thermal tolerance aside. Another appealing strategy to consider is increasing the radiation dose administered on days when hyperthermia is given. A current trend in radiation oncology is for hypofractionated treatment. This treatment schedule is an attractive complement to use of hyperthermia.

3.8 Future Directions

3.8.1 Nanotechnology, Radiation, and Thermal Therapy

Nanotechnology has much to offer in furthering the field of thermal medicine. There are already a myriad variety of therapeutic strategies under investigation, including some that have advanced into clinical trials. In additional to strictly therapeutic strategies, nanotechnology is being actively applied to theranostics, the combination of therapy and imaging including applications with paramagnetic nanoparticles and gold nanoparticles for use with hyperthermia and radiation (Kelkar and Reineke 2011).

Paramagnetic nanoparticles are furthest along in clinical development including their use with radiotherapy. The initial report of use of iron oxide particles for magnetically induced hyperthermia in an alternating magnetic field dates to 1957 (Gilchrist et al. 1957). More recently, magnetic nanoparticles have been actively studied as a thermal therapeutic strategy in oncology. (Gazeau et al. 2008; Zhang et al. 2008; Jordan et al. 2006). Two phase I studies for prostate cancer both as standalone treatment and in combination with permanent interstitial prostate brachytherapy have now been completed with intraprostatic injection of magnetic nanofluid (Johannsen et al. 2007a, 2007b). To date, this approach appears both safe and feasible (Figure 3.5).

Use of gold in nanomedicine, an inert metal with an extensive history of medical use, is also under active investigation. A range of platforms including spheres and rods incorporating gold have been used to generate heat that potentially can be applied in oncology (Krishnan et al. 2010; Hirsch et al. 2003; O'Neal et al. 2004; Huang et al. 2006; Hu et al. 2006; Huff et al. 2007; von Maltzahn et al. 2009; Dickerson et al. 2008; Huang et al. 2008a; Skrabalak et al. 2007; Wu et al. 2010; Schwartz et al. 2009; Cheng et al. 2009; Kawano et al. 2009; Huang et al. 2008b; Liu et al. 2010; Bernardi et al. 2008; Ma et al. 2009; Sokolov et al. 2003; Kumar et al. 2008; Diagaradjane et al. 2008). One strategy employed has been the use of gold nanoshells, consisting of a dielectric core surrounded by a thin gold shell, which have resonances about 800 nm that can be stimulated by near-infrared light, thus generating heat sufficient for ablation and vasculature disruption (Diagaradjane et al. 2008) (Figure 3.6). Gold nanoparticles may likewise be used for MTH resulting in radiosensitization.

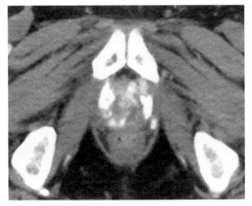 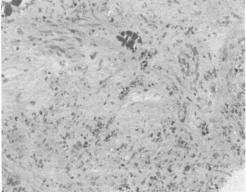

FIGURE 3.5 Example of an unenhanced computed tomography scan obtained 1 year after a single injection of magnetic fluid into the prostate, followed by six thermotherapy sessions (left). Hyperdense nanoparticle deposits in the prostate are still clearly visible. Histology image obtained by prostate biopsy of the same patient 1 year after treatment (right). Iron-oxide nanoparticles (dark gray) are still present in the prostate tissue (hematoxylin–eosin staining, ×200). (Reproduced with permission from Johannsen, M., Thiesen, B., Wust, P., Jordan, A., *Int. J. Hyperthermia*, 26(8), 790–795, 2010.)

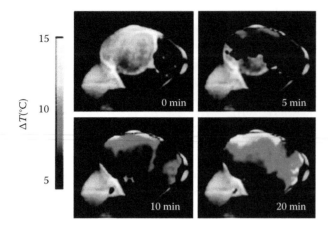

FIGURE 3.6 Magnetic resonance thermal imaging demonstrating the spatial and temporal profile of temperature (°C rise above baseline) distribution across a subcutaneous tumor noninvasively illuminated with an infrared laser following accumulation of intravenously administered gold nanoshells. (Reproduced with permission from Diagaradjane, P. et al., *Nano Lett.*, 8(5), 1492–1500, 2008.)

A range of other nanoplatforms have been developed including carbon nanotubes and temperature-sensitive peptides. Carbon nanotubes are highly efficient in photon to thermal energy conversion with a high absorption cross section in the near-infrared region. Preclinical studies have been completed showing promise (Kam et al. 2005; O'Connell et al. 2002; Chakravarty et al. 2008; Wang et al. 2009; Torti et al. 2007; Biris et al. 2009; Mahmood et al. 2009; Gannon et al. 2007; Moon et al. 2009; Burke et al. 2009). Questions remain, however, regarding *in vivo* toxicity and clinical utility. Temperature-sensitive peptides undergo inverse phase transitions when heated, converting them from a solid state to aggregates (MacKay and Chilkoti 2008). Although there is considerable focus on enhancing drug delivery and local effect, these platforms may also have applicability to radiation including radionuclide therapy in a multimodality setting. Development of these and other platforms as applied to radiation and thermal therapy holds significant promise for furthering cancer therapy.

3.9 Conclusions

The benefits of combining hyperthermia and radiation have long been appreciated. The precepts of thermal biology as applied in combination with radiation provide a compelling rationale for this therapeutic strategy in oncology. Despite challenges in defining, monitoring, and delivering treatment apparent in early clinical trials of a generation ago, thermal therapy subsequently has been proven a valuable asset in the oncologic armamentarium. Multiple positive clinical trials have made this fact clear. Nanomedicine is a rapidly emerging field with ever-expanding roles in oncology. Within this realm, the application of nanotechnology to thermal medicine holds great promise for furthering the benefits to cancer patients across the therapeutic temperature spectrum.

References

Asea, A., S. K. Kraeft, E. A. Kurt-Jones et al. 2000. HSP70 stimulates cytokine production through a CD14-dependant pathway, demonstrating its dual role as a chaperone and cytokine. *Nature Medicine* 6:435–442.

Bardwell, J. C. A., and E. A. Craig. 1984. Major heat shock gene of *Drosophila* and the *Escherichia coli* heat inducible *dnaK* gene are homologous. *Proceedings of the National Academy of Sciences of the United States of America* 81:848–852.

Beckman, R. P., L. A. Mizzen, and W. J. Welch. 1990. Interaction of HSP70 with newly synthesized proteins: Implications for protein folding and assembly. *Science* 248:850–854.

Berger, E. M., and M. P. Woodward. 1983. Small heat shock proteins in Drosophila may confer thermal tolerance. *Experimental Cell Research* 147:437–442.

Bernardi, R. J., A. R. Lowery, P. A. Thompson, S. M. Blaney, and J. L. West. 2008. Immunonanoshells for targeted photothermal ablation in medulloblastoma and glioma: An in vitro evaluation using human cell lines. *Journal of Neurooncology* 86:165–172.

Biaglow, J. E., Y. Manevich, D. Leeper et al. 1998. MIBG inhibits respiration: Potential for radio- and hyperthermic sensitization. *International Journal of Radiation Oncology, Biology & Physics* 4:871–876.

Biris, A. S., D. Boldor, J. Palmer et al. 2009. Nanophotothermolysis of multiple scattered cancer cells with carbon nanotubes guided by time-resolved infrared thermal imaging. *Journal of Biomedical Optics* 14:021007.

Brizel, D. M., S. P. Scully, J. M. Harrelson et al. 1996. Radiation therapy and hyperthermia improve the oxygenation of human soft tissue sarcomas. *Cancer Research* 56:5347–5350.

Burke, A., X. Ding, R. Singh et al. 2009. Long-term survival following a single treatment of kidney tumors with multiwalled carbon nanotubes and near-infrared radiation. *Proceedings of the National Academy of Sciences of the United States of America* 106:12897–12902.

Busch, W. 1866. Uber den Einfluss welche heftigere Erysipeln zuweilig auf organisierte Neubildungenausuben. *Verhandlungen des Naturhistorischen Vereines der Preussischen Rheinlande und Westphalens* 23:28–30.

Chakravarty, P., R. Marches, N. S. Zimmerman et al. 2008. Thermal ablation of tumor cells with antibody-functionalized single-walled carbon nanotubes. *Proceedings of the National Academy of Sciences of the United States of America* 105:8697–8702.

Chanel, J., and E. Maury E. 1961. Study of thermal shock on the blood proteins in the hog-fish and tench. *Journal of Physiology (Paris)* 53:291–292.

Chen, B., D. Zhong, and A. Monteiro. 2006. Comparative genomics and evolution of the HSP90 family of genes across all kingdoms of organisms. *BMC Genomics* 7:156.

Cheng, F. Y., C. T. Chen, and C. S. Yeh. 2009. Comparative efficiencies of photothermal destruction of malignant cells using antibody-coated silica@Au nanoshells, hollow Au/Ag nanospheres and Au nanorods. *Nanotechnology* 20:425104.

Coley, W. B. 1893. The treatment of malignant tumors by repeated inoculations of erysipelas, with a report of ten original cases. *American Journal of Medical Sciences* 105:488–511.

Dewey, W. C. 1994. Arrhenius relationships from the molecule and cell to the clinic. *International Journal of Hyperthermia* 10:457–483.

Dewey, W. C., L. E. Hopwood, L. A. Sapareto et al. 1977. Cellular response to combinations of hyperthermia and radiation. *Radiology* 123:463–474.

Dewhirst, M. W. 1995. Thermal dosimetry. In: *Thermoradiotherapy and Thermochemotherapy*, ed. M. Seegenschmiedt, P. Fessenden, C. C. Vernon, 123–128. Berlin: Springer Verlag.

Dewhirst, M. W., and D. A. Sim. 1984. Utility of thermal dose as a predictor of tumor and normal tissue responses to combined radiation and hyperthermia. *Cancer Research* 44 (Suppl):4772s–4780s.

Diagaradjane, P., A. Shetty, J. C., Wang et al. 2008. Modulation of in vivo tumor radiation response via gold nanoshell-mediated vascular-focused hyperthermia: Characterizing an integrated antihypoxic and localized vascular disrupting targeting strategy. *NanoLetters* 8:1492–1500.

Dickerson, E. B., E. C. Dreaden, X. Huang et al. 2008. Gold nanorod assisted near-infrared plasmonic photothermal therapy (PPTT) of squamous cell carcinoma in mice. *Cancer Letters* 269:57–66.

Dickson, J. A., and S. A. Shah. 1982. Hyperthermia: the immune response and tumor metastasis. *Journal of the National Cancer Institute Monograph* 61:183–192.

Eddy, H. A. 1980. Alterations in tumor microvasculature during hyperthermia. *Radiology* 137:515–521.

Emami, B., C. Scott, C. A. Perez et al. 1996. Phase III study of interstitial thermoradiotherapy compared with interstitial radiotherapy alone in the treatment of recurrent or persistent human tumors. A prospectively controlled randomized study by the Radiation Therapy Group. *International Journal of Radiation Oncology, Biology & Physics* 34:1097–1104.

Engin, K., D. B. Leeper, A. J. Thistlethwaite et al. 1994. Tumor extracellular pH as a prognostic factor in thermoradiotherapy. *International Journal of Radiation Oncology, Biology & Physics* 29:125–132.

Gabai, V. L., I. V. Zamulaeva, A. F. Mosin et al. 1995. Resistance of Ehrlich tumor cells to apoptosis can be due to accumulation of heat shock proteins. *FEBS Letters* 375:21–26.

Gabai, V. L., A. B. Meriin, J. A. Yaglom et al. 1998. Role of Hsp70 in regulation of stress-kinase JNK: implications in apoptosis and aging. *FEBS Letters* 438:1–4.

Gannon, C. J., P. Cherukuri, B. I. Yakobson et al. 2007. Carbon nanotube-enhanced thermal destruction of cancer cells in a noninvasive radiofrequency field. *Cancer* 110: 2654–2665.

Gazeau, F., M. Levy, and C. Wilhelm. 2008. Optimizing magnetic nanoparticle design for nanothermotherapy. *Nanomedicine* 3:831–844.

Gerweck, L. 1977. Modification of cell lethality at elevated temperatures. The pH effect. *Radiation Research* 70:224–235.

Gerweck, L. E., T. G. Nygaard, and M. Burlett. 1979. Response of cells to hyperthermia under acute and chronic hypoxic conditions. *Cancer Research* 9:966–972.

Gerweck, L. E., B. Richards, and H. B. Michaels. 1982. Influence of low pH on the development and decay of 42 degrees C thermotolerance in CHO cells. *International Journal of Radiation Oncology, Biology & Physics* 8:1935–1941.

Gething, M. J., and J. Sambrook. 1992. Protein folding in the cell. *Nature* 355:33–45.

Gilchrist, R. K., R. Medal, W. D. Shorey et al. 1957. Selective inductive heating of lymph nodes *Annals of Surgery* 146:596–606.

Griffin, R. J., and P. M. Corry. 2009. Commentary on classic paper in hyperthermic oncology 'Tumour oxygenation is increased by hyperthermia at mild temperatures' by CW Song et al., 1996. *International Journal of Hyperthermia* 25:96–98.

Hahn, G. M., and C. Shiu. 1986. Adaptation to low pH modifies thermal and thermochemical responses of mammalian cells. *International Journal of Hyperthermia* 2:379–387.

Hall, E. J. 1994. *Radiobiology for the radiologist*, 4th ed. Philadelphia, PA: Lippincott and Company.

Hand, J. W., D. Machin, C. C. Vernon et al. 1997. Analysis of thermal parameters obtained during phase III trials of hyperthermia as an adjunct to radiotherapy in the treatment of breast carcinoma. *International Journal of Hyperthermia* 13:343–364.

Henle, K. J., and D. B. Leeper. 1976. Combination of hyperthermia (40°, 45°C) with radiation. *Radiology* 121:451–454.

Hirsch, L. R., R. J. Stafford, J. A. Bankson et al. 2003. Nanoshell-mediated near-infrared thermal therapy of tumors under magnetic resonance guidance. *Proceedings of the National Academy of Sciences of the United States of America* 100:13549–13554.

Horkan, C., K. Dalal, J. A. Coderre et al. 2005. Reduced tumor growth with combined radiofrequency ablation and radiation therapy in a rat breast tumor model. *Radiology* 235:81–88.

Hu, M., J. Chen, Z. Y. Li et al. 2006. Gold nanostructures: Engineering their plasmonic properties for biomedical applications. *Chemical Society Reviews* 35:1084–1094.

Huang, X., I. H. El-Sayed, W. Qian et al. 2006. Cancer cell imaging and photothermal therapy in the near-infrared region by using gold nanorods. *Journal of the American Chemical Society* 128:2115–2120.

Huang, X., P. K. Jain, I. H. El-Sayed et al. 2008a. Plasmonic photothermal therapy (PPTT) using gold nanoparticles. *Lasers in Medical Science* 23:217–228.

Huang, Y. F., K. Sefah, S. Bamrungsap, H. T. Chang, W. Tan. 2008b. Selective photothermal therapy for mixed cancer cells using aptamer-conjugated nanorods. *Langmuir* 24:11860–11865.

Huff, T. B., L. Tong, Y. Zhao, M. N. Hansen, J. X. Cheng, and A. Wei. 2007. Hyperthermic effects of gold nanorods on tumor cells. *Nanomedicine (London)* 2:125–132.

Hurwitz, M. D., P. Kaur, G. M. Nagaraja et al. 2010. Radiation therapy induces circulating serum Hsp72 in patients with prostate cancer. *Radiotherapy & Oncology* 95:350–358.

Issels, R. D., and M. Schlemmer. 2002. Current trials and new aspects in soft tissue sarcoma of adults. *Cancer Chemotherapy & Pharmacology* 49(Suppl 1):S4–S8.

Jernberg, A., M. R. Edgren, R. Lewensohn et al. 2001. Cellular effects of high-intensity focused continuous wave ultrasound alone and in combination with X-rays. *International Journal of Radiation Biology* 77:127–135.

Johannsen, M., U. Gneveckow, K. Taymoorian et al. 2007a. Morbidity and quality of life during thermotherapy using magnetic nanoparticles in locally recurrent prostate cancer: results of a prospective phase I trial. *International Journal of Hyperthermia* 23:315–323.

Johannsen, M., U. Gneveckow, B. Thiesen et al. 2007b. Thermotherapy of prostate cancer using magnetic nanoparticles: Feasibility, imaging and three-dimensional temperature distribution. *European Urology* 52:1653–1661.

Johannsen, M., B. Thiesen, P. Wust, and A. Jordan. 2010. Magnetic nanoparticle hyperthermia for prostate cancer. *International Journal of Hyperthermia* 26(8):790–795.

Jones, E. L., M. J. Zhao, M. A. Stevenson et al. 2004. The 70 kilodalton heat shock protein is an inhibitor of apoptosis in prostate cancer. *International Journal of Hyperthermia* 20:835–849.

Jones, E. L., J. R. Oleson, L. R. Prosnitz et al. 2005. Randomized trial of hyperthermia and radiation for superficial tumors. *Journal of Clinical Oncology* 23:3079–3085.

Jordan, A., R. Scholz, K. Maier-Hauff et al. 2006. The effect of thermotherapy using magnetic nanoparticles on rat malignant glioma. *Journal of Neuro-Oncology* 78:7–14.

Kam, N. W., M. O'Connell, J. A. Wisdom et al. 2005. Carbon nanotubes as multifunctional biological transporters and near-infrared agents for selective cancer cell destruction. *Proceedings of the National Academy of Sciences of the United States of America* 102:11600–11605.

Kampinga, H. H. 2006. Cell biological effects of hyperthermia alone or combined with radiation or drugs: A short introduction to newcomers in the field. *International Journal of Hyperthermia.* 22:191–196.

Kampinga, H. H., and E. Dikomey. 2001 Hyperthermic radiosensitization: Mode of action and clinical relevance. *International Journal of Radiation Biology* 77:399–408.

Kano, E. 1985. Hyperthermia and drugs. In: *Hyperthermic Oncology,* ed. E. Overgaard, 277–282. London, UK: Taylor & Francis.

Kapp, D. S., and R. S. Cox. 1995. Thermal treatment parameters are most predictive of outcome in patients with single tumor nodules per treatment field in recurrent adenocarcinoma of the breast. *International Journal of Radiation Oncology, Biology & Physics* 33:887–899.

Kawano, T., Y. Niidome, T. Mori et al. 2009. PNIPAM gel-coated gold nanorods for targeted delivery responding to a near-infrared laser. *Bioconjugate Chemistry* 20:209–212.

Kelkar, S. S., and T. M. Reineke. 2011. Theranostics: Combining imaging and therapy. *Bioconjugate Chemistry.*

Kim, J. H., S. H. Kim, P. Dutta et al. 1992. Preferential killing of glucose-depleted HeLa cells by menadione and hyperthermia. *International Journal of Hyperthermia* 8:139–146.

Kim, S. H., J. H. Kim, and E. W. Hahn. 1976. The enhanced killing of irradiatied HeLa cells in synchronous culture by hyperthermia. *Radiation Research* 66:337–345.

Koutcher, J. A., D. Barnett, A. B. Kornblith et al. 1990. Relationship of changes in pH and energy status to hypoxic cell fraction and hyperthermia sensitivity. *International Journal of Radiation Oncology, Biology & Physics* 18:1429–1435.

Krishnan, S., P. Diagaradjane, and S. H. Cho. 2010. Nanoparticle-mediated thermal therapy: Evolving strategies for prostate cancer therapy. *International Journal of Hyperthermia* 26:775–789.

Kumar, S., J. Aaron, and K. Sokolov. 2008. Directional conjugation of antibodies to nanoparticles for synthesis of multiplexed optical contrast agents with both delivery and targeting moieties. *Nature Protocols* 3:314–320.

Law, M. P., R. G. Ahier, and S. B. Field. 1977. The response of mouse skin to combined hyperthermia and X-rays. *International Journal of Radiation, Biology & and Related Studies in Physics, Chemistry, and Medicine* 32:153–163.

Lee, H. K., A. G. Antell, C. A. Perez et al. 1998. Superficial hyperthermia and irradiation for recurrent breast carcinoma of the chest wall: Prognostic factors in 196 tumors. *International Journal of Radiation Oncology, Biology & Physics* 40:365–375.

Leeper, D. B., K. Engin, A. J. Thistlethwaite et al. 1994. Human tumor extracellular pH as a function of blood glucose concentration. *International Journal of Radiation Oncology, Biology & Physics* 28:935–943.

Lepock, J. R. 2004. Role of nuclear protein denaturation and aggregation in thermal radiosensitization. *International Journal of Hyperthermia* 20:115–130.

Lepock, J. R., H. E. Frey, and K. P. Ritchie. 1993. Protein denaturation in intact hepatocytes and isolated cellular organelles during heat shock. *Journal of Cell Biology* 122:1267–1276.

Lepock, J. R., H. E. Frey, M. L. Heynen et al. 2001. The nuclear matrix is a thermolabile cellular structure. *Cell Stress and Chaperones* 6:136–147.

Li, G. C., N. S. Petersen, and H. K. Mitchell. 1982. Induced thermal tolerance and heat shock protein synthesis in Chinese hamster ovary cells. *International Journal of Radiation Oncology, Biology & Physics* 8:63–67.

Lin, J. C, S. H. Levitt, and C. W. Song. 1992. Relationship between vascular thermal tolerance and intratumor pH. *International Journal of Radiation Oncology, Biology & Physics* 22(1):123–129.

Lindquist, S., and E. A. Craig. 1988. The heat shock proteins. *Annual Review in Genetics* 22:631–677.

Liu, R. Y., X. Li, L. Li et al. 1992. Expression of human hsp70 in rat fibroblasts enhances cell survival and facilitates recovery from translational and transcriptional inhibition following heat shock. *Cancer Research* 1;52(13):3667–3673.

Liu, S. Y., Z. S. Liang, F. Gao et al. 2010. In vitro photothermal study of gold nanoshells functionalized with small targeting peptides to liver cancer cells. *Journal of Materials Science Materials in Medicine* 21:665–674.

Lowe, D. G., W. D. Fulford, and L. A. Moran. 1983. Mouse and Drosophila genes encoding the major heat shock protein (hsp70) are highly conserved. *Molecular and Cellular Biology* 3:1540–1543.

Lüdemann, L., G. Sreenivasa, H. Amthauer, R. Michel, J. Gellermann, and P. Wust. 2009. Use of H(2) (15)O-PET for investigating perfusion changes in pelvic tumors due to regional hyperthermia. *International Journal of Hyperthermia* 25:299–308.

Ma, L. L., M. D. Feldman, J. M. Tam et al. 2009. Small multifunctional nanoclusters (nanoroses) for targeted cellular imaging and therapy. *ACS Nano* 3:2686–2696.

Mackay, J. A., and A. Chilkoti. 2008. Temperature sensitive peptides: Engineering hyperthermia-directed therapeutics. *International Journal of Hyperthermia* 24:483–495.

Mahmood, M., A. Karmakar, A. Fejleh et al. 2009. Synergistic enhancement of cancer therapy using a combination of carbon nanotubes and anti-tumor drug. *Nanomedicine (London)* 4:883–893.

McKenzie, S. L., S. Henikoff, and M. Meselson. 1975. Localization of RNA from heat induced polysomes at puff sites in *Drosophila melanogaster*. *Proceedings of the National Academy of Sciences of the United States of America* 72:1117–1121.

Mittal, B., B. Emami, S. A. Sapareto, F. H. Taylor, and F. G. Abrath. 1984. Effects of sequencing of the total course of combined hyperthermia and radiation on the RIF-1 murine tumor. *Cancer* 54:2889–2897.

Mivechi, N. F., and W. C. Dewey. 1985. DNA polymerase alpha and beta activities during the cell cycle and their role in heat radiosensitization in Chinese hamster ovary cells. *Radiation Research* 103:337–350.

Moon, H. K., S. H. Lee, and H. C. Choi. 2009. In vivo near-infrared mediated tumor destruction by photothermal effect of carbon nanotubes. *ACS Nano* 3:3707–3713.

Muller, C. 1910. Eine neue Behandlungsmethode bosartiger. *Geschwulste* 57:1490–1493.

Mueller-Klieser, W., S. Walenta, D. K. Kelleher et al. 1996. Tumour-growth inhibition by induced hyperglycaemia/hyperlactacidaemia and localized hyperthermia. *International Journal of Hyperthermia* 12:501–511.

Netzer, W. J., and F. U. Hartl. 1998. Protein folding in the cytosol: Chaperonin-dependent and -independent mechanisms. *Trends in Biochemical Sciences* 23:68–73.

O'Connell, M. J., S. M. Bachilo, C. B. Huffman et al. 2002. Band gap fluorescence from individual single-walled carbon nanotubes. *Science* 297:593–596.

O'Neal, D. P., L. R. Hirsch, N. J. Halas et al. 2004. Photo-thermal tumor ablation in mice using near infrared-absorbing nanoparticles. *Cancer Letters* 209:171–176.

Oleson, J. R., T. V. Samulski, K. A. Leopold et al. 1993. Sensitivity of hyperthermia trial outcomes to temperature and time: implications for thermal goals of treatment. *International Journal of Radiation Oncology, Biology & Physics* 25:289–297.

Overgaard, J. 1989. The current and potential role of hyperthermia in radiotherapy. *International Journal of Radiation Oncology, Biology & Physics* 16:535–549.

Peñagarícano, J. A., E. Moros, P. Novák et al. 2008. Feasibility of concurrent treatment with the scanning ultrasound reflector linear array system (SURLAS) and the helical tomotherapy system. *International Journal of Hyperthermia* 24:377–388.

Perez, C. A., T. Pajak, B. Emami et al. 1991. Randomized phase III study comparing irradiation and hyperthermia with irradiation alone in superficial measurable tumors. Final report by the Radiation Therapy Oncology Group. *American Journal of Clinical Oncology* 14:133–141.

Raaphorst, G. P., D. P. Yang, and C. E. Ng. 1994. Effect of protracted mild hyperthermia on polymerase activity in a human melanoma cell line. *International Journal of Hyperthermia* 10:827–834.

Raaphorst, G. P., C. E. Ng, and D. P. Yang. 1999. Thermal radiosensitization and repair inhibition in human melanoma cells: A comparison of survival and DNA double strand breaks. *International Journal of Hyperthermia* 15:17–27.

Sapareto, S. A., and W. C. Dewey. 1984. Thermal dose determination in cancer therapy. *International Journal of Radiation Oncology, Biology & Physics* 10:787–800.

Schwartz, J. A., A. M. Shetty, R. E. Price et al. 2009. Feasibility study of particle-assisted laser ablation of brain tumors in orthotopic canine model. *Cancer Research* 69:1659–1667.

Seegenschmiedt, M. H., and C. C. Vernon. 1995. A historical perspective on hyperthermia in oncology. In: *Thermoradiotherapy and Thermochemotherapy*, ed. M. H. Seegenschmiedt, P. Fessenden, and C. C. Vernon, Vol. 1. Berlin: Springer Verlag.

Seegenschmiedt, M. H., P. Martus, R. Fietkau et al. 1994. Multivariate analysis of prognostic parameters using interstitial thermoradiotherapy (IHT-IRT): Tumor and treatment variables predict outcome. *International Journal of Radiation Oncology, Biology & Physics* 29:1049–1063.

Sherar, M., F. F. Liu, M. Pintilie et al. 1997. Relationship between thermal dose and outcome in thermoradiotherapy treatments for superficial recurrences of breast cancer: data from a phase III trial. *International Journal of Radiation Oncology, Biology & Physics* 39:371–380.

Skrabalak, S. E., L. Au, X. Lu et al. 2007. Gold nanocages for cancer detection and treatment. *Nanomedicine (London)* 2:657–668.

Szmigielski, S., J. Sobczynski, G. Sokolska et al. 1991. Effects of local prostatic hyperthermia on human NK and T cell function. *International Journal of Hyperthermia* 7:869–880.

Sneed, P. K., P. R. Stauffer, M. W. McDermott et al. 1998. Survival benefit of hyperthermia in a prospective randomized trial of brachytherapy boost +/− hyperthermia for glioblastoma multiforme. *International Journal of Radiation Oncology, Biology & Physics* 40:287–295.

Snyder, S. A., J. L. Lanzen, R. D. Braun et al. 2001. Simultaneous administration of glucose and hyperoxic gas achieves greater improvement in tumor oxygenation than hyperoxic gas alone. *International Journal of Radiation Oncology, Biology & Physics* 51:494–506.

Sokolov, K., M. Follen, J. Aaron et al. 2003. Real-time vital optical imaging of precancer using anti-epidermal growth factor receptor antibodies conjugated to gold nanoparticles. *Cancer Research* 63:1999–2004.

Song, C. W. 1984. Effect of local hyperthermia on blood flow and microenvironment: A review. *Cancer Research* 44(suppl 10):S4721–S4730.

Song, C. W., G. E. Kim, J. C. Lyons et al. 1994. Thermosensitization by increasing intracellular acidity with amiloride and its analogs. *International Journal of Radiation Oncology, Biology & Physics*;30:1161–1169.

Sostman, H. D., D. M. Prescott, M. W. Dewhirst et al. 1994. MR imaging and spectroscopy for prognostic evaluation in soft-tissue sarcomas. *Radiology* 190:269–275.

Srivastava, P. K., and R. G. Maki. 1991. Stress-induced proteins in immune response to cancer. *Current Topics in Microbiology and Immunology* 167:109–123.

Stawarz, B., H. Zielinski, S. Szmigielski et al. 1993. Transrectal hyperthermia as palliative treatment for advanced adenocarcinoma of prostate and studies of cell-mediated immunity. *Urology* 41:548–553.

Streffer, C. 1985. Metabolic changes during and after hyperthermia. *International Journal of Hyperthermia* 1:305–319.

Torti, S. V., F. Byrne, O. Whelan et al. 2007. Thermal ablation therapeutics based on CN(*x*) multi-walled nanotubes. *International Journal of Nanomedicine* 2:707–14.

Udono, H., and P. K. Srivastava. 1993. Heat shock protein 70-associated peptides elicit specific cancer immunity. *Journal of Experimental Medicine* 178:1391–1396.

Udono, H., and P. K. Srivastava. 1994. Comparison of tumor-specific immunogenicities of stress induced proteins gp96, hsp90, and hsp70. *Journal of Immunology* 152:5398–5403.

Udono, H., D. L. Levy, and P. K. Srivastava. 1994. Cellular requirements for tumor-specific immunity elicited by heat shock proteins: tumor rejection antigen gp96 primes CD8_T cells in vivo. *Proceedings of the National Academy of Sciences of the United States of America* 91:3077–3081.

Urano, M. 1986. Kinetics of thermotolerance in normal and tumor issues: A review. *Cancer Research* 46:474–482.

Valdagni, R., and M. Amichetti. 1994. Report of long-term follow-up in a randomized trial comparing radiation therapy and radiation therapy plus hyperthermia to metastatic lymph nodes in stage IV head and neck patients. *International Journal of Radiation Oncology, Biology & Physics* 28:163–169.

van der Zee, J., D. González González, G. C. van Rhoon et al. 2000. Comparison of radiotherapy alone with radiotherapy plus hyperthermia in locally advanced pelvic tumours: A prospective, randomised, multicentre trial. Dutch Deep Hyperthermia Group. *Lancet* 355:1119–1125.

van der Zee J. 2002. Heating the patient: A promising approach? *Annals of Oncology* 13:1173–1184.

Vaupel, P. W., and D. K. Kelleher. 2010. Pathophysiological and vascular characteristics of tumours and their importance for hyperthermia: Heterogeneity is the key issue. *International Journal of Hyperthermia* 26:211–223.

Vidair, C. A., S. J. Doxsey, and W. C. Dewey. 1995. Thermotolerant cells possess an enhanced capacity to repair heat-induced alterations to centrosome structure and function. *Journal of Cellular Physiology* 163:194–203.

von Maltzahn, G., J. H. Park, A. Agrawal et al. 2009. Computationally guided photothermal tumor therapy using long-circulating gold nanorod antennas. *Cancer Research* 69:3892–3900.

Wahl, M. L., S. B. Bobyock, D. B. Leeper et al. 1997. Effects of 42 degrees C hyperthermia on intracellular pH in ovarian carcinoma cells during acute or chronic exposure to low extracellular pH. *International Journal of Radiation Oncology, Biology & Physics* 39:205–212.

Wang, C. H., Y. J. Huang, C. W. Chang et al. 2009. In vitro photothermal destruction of neuroblastoma cells using carbon nanotubes conjugated with GD2 monoclonal antibody. *Nanotechnology* 20:315101.

Ward, K. A., D. J. DiPette, T. N. Held et al. 1991. Effects of intravenous versus intraperitoneal glucose injection on system hemodynamics and blood flow rate in normal and tumor tissues in rats. *Cancer Research* 51:3612–3616.

Warren, S. L. 1935. Preliminary study of the effects of artificial fever upon hopeless tumor cases. *American Journal of Roentgenology* 33:75–87.

Westra, A., and W. C. Dewey. 1971. Variation in sensitivity to heat shock during the cell cycle of Chinese hamster cells in vitro. *International Journal of Radiation and Biology* 19:467–477.

Wu, X., T. Ming, X. Wang, P. Wang, J. Wang, and J. Chen. 2010. High-photoluminescence-yield gold nanocubes: For cell imaging and photothermal therapy. *ACS Nano* 4:113–120.

Xiao, L., X. Lu, and D. M. Ruden. 2006. Effectiveness of hsp90 inhibitors as anti-cancer drugs. *Mini-Reviews in Medicinal Chemistry* 6:1137–1143.

Zhang, L., F. X. Gu, J. M. Chan et al. 2008. Nanoparticles in medicine: Therapeutic applications and developments. *Clinical Pharmacology and Therapeutics* 83:761–769.

Zhou, R., N. Bansal, D. B. Leeper et al. 2000. Intracellular acidification of human melanoma xenografts by the respiratory inhibitor *m*-iodobenzylguanidine plus hyperglycemia: A 31P magnetic resonance spectroscopy study. *Cancer Research* 60:3532–3536.

II

Synthesis, Safety, and Imaging of Nanomaterials for Cancer Applications

4

Synthesis and Surface Modification of Nanomaterials for Tumor Targeting

Glenn Goodrich
Nanospectra Biosciences

4.1 Introduction

In the past decade, there has been an explosion in research efforts looking to apply the unique properties of nanoparticles to the treatment of a host of various diseases including cancer. In order to create a therapeutic nanoparticle, one must first fabricate a nanoparticle with the specific properties (optical absorption, magnetic properties, etc.) necessary to provide the therapeutic effect. The particle must then be effectively delivered to the tumor site. Most nanoparticle materials require some type of surface modification in order to facilitate delivery to tumor sites, especially if intravenous delivery is desired. It would be a nearly impossible task to deliver a comprehensive list of all the various permutations of nanoparticle fabrication methods, nanoparticle material, geometries, and properties being developed in the ever-expanding field of nanoparticle-based therapies. In this chapter, the fabrication methods for a number of the most popular nanoparticle systems and the techniques for subsequently modifying these nanoparticles to enable their delivery to the tumor will be described.

4.2 Synthesis and Surface Functionalization of Noble Metal Particles

Nanoparticles composed of the noble metals gold and silver are particularly interesting for biological application because of their low reactivity, low toxicity profiles (Connor et al. 2005; Hauck et al. 2008), and relative ease in functionalization (Davis et al. 2008; Hu et al. 2006; Huff et al. 2007; Zou et al. 2008), and have been used for years in various therapeutic applications. Gold and silver particles are capable of supporting size-dependent surface plasmon resonances that provide the characteristic vivid colors for suspensions made from these particles. Their strong optical resonances have made them interesting for thermal ablation applications and enhanced imaging agents for surface-enhanced Raman scattering and fluorescent assays (Kho et al. 2005; Nah et al. 2009; Thomas and Khamat 2003). Because these particles are also electron dense, they have also been used as contrast agents for electron microscopy (Hashizume et al. 2001; Mayer and Bendayan 1999) and to enhance the efficacy of radiation therapy (Chithrani et al. 2010; Atkinson et al. 2010). In the past few years, there has been an explosion in the number of synthetic methods for gold and silver nanoparticles, yielding a wide variety of size and shapes including spherical, rod-shaped, prisms, and cubes (Gole and Murphy 2004; Khalavka et al. 2009; Kou et al. 2006; Leontidis et al. 2002; Li et al. 2011a; Métraux et al. 2003; Thakor et al. 2011; Yamamoto et al. 2005; Zhang 2010). Silver particles have demonstrated interesting antimicrobial properties and have been applied in wound dressings (Eby et al. 2009; Fan et al. 2001; Jain et al. 2009; Yoon et al. 2008; Yuan et al. 2008). For other *in vivo* applications, the increased stability, and low toxicity of gold nanoparticles have made them the most popular noble metal particle. For therapeutic applications where tumor accumulation is desired, the particle size should be between ~12 and ~200 nm. Smaller-diameter particles <10 nm can be rapidly removed from the bloodstream through the kidneys. Large-diameter particles >250 nm may be too large to pass into the interstitial space around tumors through the leaky vasculature of the tumors. As there is not space enough to cover fabrication methods for all types of metal nanoparticles, this section will focus on the preparation of gold particles most widely used for therapeutic applications.

4.2.1 Spherical Gold

Spherical gold is perhaps the most widely used metal nanoparticle. Spherical gold nanoparticles have been widely used and studied starting with Faraday's pioneering work in 1857 (Faraday 1857). The increased interest in utilizing gold nanoparticles in a variety of biological applications has resulted in a number of companies selling a wide variety of sizes of gold particles. In this section, some of the more popular preparations for gold colloid will be discussed. Although the number of published synthetic methods for spherical particles is large, the basic principles remain constant for most, if not all, of these methods. In these methods, a gold containing salt, usually $HAuCl_4$, is dissolved in water. To initiate particle formation, a reducing agent is added. The size of the resulting nanoparticles can be controlled by varying the amount of salt in solution and by changing the strength of the reducing agent. The majority of these particle preparations are done completely in an aqueous phase. There is a well-established field based on preparing nanoparticles in a two-phase system, where the particles are stabilized by monolayers of alkanethiols dissolved in the organic solvent (Goulet et al. 2010; Li et al. 2011b). These monolayer protected clusters can be useful in biological applications, but the use of organic solvents may provide a barrier to some laboratories. For *in vivo* work, any residual organic solvents would need to be removed to prevent any toxicity. As a result, this chapter will focus on the aqueous-based preparations.

4.2.2 Sodium Citrate

One of the most popular preparations for spherical gold colloid is based on the reduction of the gold salt using sodium citrate. This preparation can be easily modified to make particles 12–50 nm in diameter. The most common preparations are based on the method of Frens (1973) and Turkevich (Kimling et al. 2006; Turkevich et al. 1951). In a typical preparation for 12 nm particles, $HAuCl_4$ solution is brought to a rapid boil with vigorous stirring. To initiate particle growth, sodium citrate is added rapidly. The solution will change color from the initial yellow of the $HAuCl_4$ solution through a black intermediate, finally settling on the red burgundy color. It is important to maintain excellent mixing during the addition of the reducing agent to produce particles with a tight size distribution. This preparation typically produces a size distribution of ~10% in diameter. The overall diameter of the particles can be tailored by adjusting the ratio of gold salt to reducing agents (Bastus et al. 2011).

4.2.3 Near Infrared Absorbing Particles for Generating Hyperthermia

Although some studies have shown that the presence of metal nanoparticles in the tumor can have an enhancing effect on radiation therapies, a large amount of work is currently focused on using the nanoparticles to generate localized hyperthermia to enhance radiation efficacy as well as for direct thermal ablation of the tumors. The two most commonly used near-infrared (NIR) absorbing particles are gold nanoshells and gold nanorods. Both types of particles provide strong, tunable optical resonances in the NIR that is dependent on the geometry of the particle.

4.2.4 Gold Nanoshells

The silica–gold nanoshell, which consists of a silica core covered with a thin shell of gold, was initially developed by Oldenburg et al. (1998, 1999). The tunability of the optical resonance for these particles is a result of the interaction of the surface plasmon on the outer surface of the shell with the surface plasmon on the inner surface of the gold shell (Prodan et al. 2003). This plasmon hybridization allows for the optical resonance of the particle to be tuned by simply changing the thickness of the gold layer around the particle (Prodan et al. 2003).

For optical resonances in the NIR, which are most commonly used in hyperthermia applications, a silica core of 120 nm can be used. This core is prepared using the Stöber method, in which

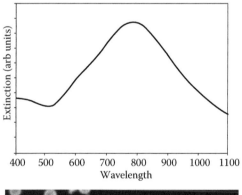

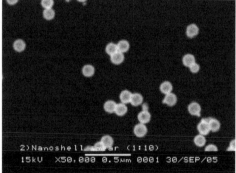

FIGURE 4.1 Optical spectrum of gold nanoshells for NIR absorption (top). TEM of gold nanoshell particles (bottom).

tetraethyl orthosilicate undergoes a base-catalyzed hydrolysis then condensation into SiO_2 (Stöber et al. 1968). In order to facilitate the gold layer deposition on the silica surface, the silica cores are aminated by the addition of amino propyltriethoxysilane to a suspension of the silica particles. Excess silane is removed by centrifugation and resuspension in fresh ethanol. The surface of the silica particles is then decorated with small gold colloid (1–3 nm in diameter) prepared using the method described by Duff and Baiker (1993) by addition of ~3.5 mL of silanized silica particles to 1 L of gold colloid solution. The gold colloid is then adsorbed to the amine groups on the surface of the silica core form nucleating sites, which can then be grown into a complete gold shell by the addition of a gold plating solution (~400 μM $HAuCl_4$, 2 mM K_2CO_3 in water) in the presence of formaldehyde. The final gold shell thickness and the resulting optical resonance can be controlled by varying the concentration of silica seed particles added to the plating solution. For biological applications where NIR resonances are desirable, a 12- to 15–nm-thick gold shell will result in optical absorption peak between 780 and 800 nm (Figure 4.1).

4.2.5 Gold Nanorods

In the past decade, the use of nonspherical metal particles has increased. One of the major advantages of nonspherical particles is the ability to more readily tune the optical properties of the particles by controlling the geometry. One of the most popular nonspherical particles being used today is the gold nanorod. These particles are rod-shaped solid gold particles. The optical resonance of the nanorod can be controlled by varying the diameter and the length of the rod known as the aspect ratio. These particles can be tuned to absorb light throughout the visible and NIR regions of the spectrum using fairly simple changes in the synthetic methods.

The majority of the preparations for gold nanorods are based on a seeded growth method starting with small, spherical gold particles (Gole and Murphy 2004; Smith and Korgel 2008; Jana et al. 2001). In order to grow the particles into rod shapes, it is generally necessary to use the surfactant cetyltrimethylammonium bromide (CTAB). In the absence of this surfactant, addition of more gold and reductant to the solution will produce larger diameter spherical particles. The presence of the surfactant causes the particles to grow along specific crystal facets, resulting in a rod-shaped particle. Although the surfactant is necessary to produce the rod-shaped particles, it also provides a challenge for further use particularly *in vitro* and *in vivo*. CTAB has been shown to have a significant cytotoxicity. In addition, the gold nanorods are coated with a bilayer of the surfactant molecules, which makes further functionalization difficult. In order to use the gold nanorods for biological applications, the vast majority of the CTAB solution must be removed.

As with spherical gold particles, a large number of synthetic variations have been published. The initial work preparing nanorods was developed by Jana et al. (2001). In this preparation, $HAuCl_4 \cdot 3H_2O$, CTAB, $NaBH_4$, $AgNO_3$, and ascorbic acid are

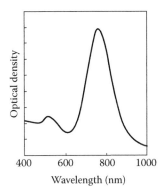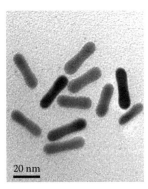

FIGURE 4.2 Optical spectrum of gold nanorods (left panel) and TEM of gold nanorods (right panel).

used as received with no further modification or purification. The gold "seed" particles are first prepared by adding $HAuCl_4 \cdot 3H_2O$ to a 100-mM CTAB solution with brief, gentle mixing. Then, 600 μL of freshly prepared, ice-cold 10 mM $NaBH_4$ solution is added, followed by mixing for 2 min. The nanorod growth solution is prepared by adding 40 mL of 100 mM CTAB, 1.7 mL of 10 mM $HAuCl_4 \cdot 3H_2O$, 250 μL of 10 mM $AgNO_3$ followed by 270 μL of 100 mM ascorbic acid. To initiate nanorod growth, the seed solution was added to a previously prepared solution containing CTAB, $HAuCl_4 \cdot 3H_2O$, $AgNO_3$, and ascorbic acid. The resulting solution is mixed gently, and allowed to stand still for 40 min. At this point, the excess reagents must be removed from the solution to prevent both particle "aging," where the dimensions of the particles continue to change over time, and the cytotoxicity associated with the CTAB (Thakor et al. 2011; Parab et al. 2009; Alkilany et al. 2010; Huff et al. 2007; Leonov et al. 2008; Murphy et al. 2010; Bartneck et al. 2010). Excess reagents can be removed by successive pelleting of the nanorods and resuspension in fresh buffer. Dialysis, tangential flow filtration, and two-phase organic aqueous separations have also been used to reduce the CTAB content in the rod solution. Care must be taken to not remove all of the CTAB from the nanorod surface themselves by excess cleaning. The surfactant layer on the particles acts to stabilize the colloids and prevent aggregation. This preparation results in particles ~10 nm in diameter and ~40 nm long with an optical resonance around 780 nm (Figure 4.2).

4.2.6 Surface Modification of Noble Metal Particles

Surface modification of nanoparticles is done for two primary reasons: (1) to protect the particle from the environment (i.e., the body's immune system) and (2) for targeting the particle to a specific cell or tissue type. These two effects are both important and at times are at odds with each other. Particles designed to resist the immune system generally have no specific targeting moiety, and particles designed for efficient targeting of a specific cell or tissue generally are identified rapidly by the immune system and cleared. The optimal surface coating provides just enough protection from the immune system to allow the particles to

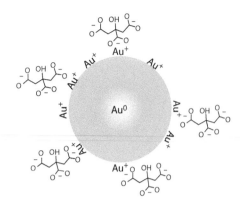

FIGURE 4.3   Illustration of the electrostatic charge around a bare gold particle.

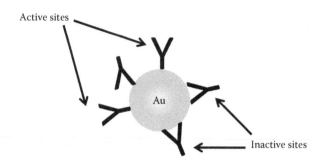

FIGURE 4.4   Direct adsorption of proteins to gold particle surface.

circulate and accumulate in the desired location, while presenting enough targeting moiety to enable targeting.

The two most common methods for surface modification of gold surfaces involve using electrostatic interactions or through the thiol–gold bond. Electrostatic methods take advantage of the native surface charge of gold nanoparticles to adhere molecules to the surface. Incompletely reduced gold ions on the metal surface provide a positive charge on the surface of the gold. Negative ions present in the growth solutions (citrate anion in the case of a citrate reduction) are attracted to this positive charge and adsorb to the surface (Figure 4.3). This net negative charge of the gold particles acts to stabilize the colloidal solution. It is thus possible to modify the surface of the particle using positively charged molecules. This electrostatic method has been used extensively in the attachment of proteins to the gold surface. More recently, there has been significant work in taking advantage of this surface charge to coat the gold particles in polyelectrolyte polymer layers.

4.2.7 Direct Adsorption of Proteins to Gold Nanoparticles

Methods for the direct attachment of proteins to gold nanoparticles have been available for many years. Protein-labeled gold colloids have been used extensively as immunohistochemical labels. Conjugates with common proteins such as avidin/streptavidin, albumin, protein A/G, and various antibodies are available commercially from a number of providers. As the attachment of the protein in this case is due to the charge interaction between the gold particle and the protein, the efficiency of attachment of protein to the gold surface depends on the pH of the solution. Optimal results for protein attachment generally occur when the pH of the solution is near the isoelectric point of the protein of interest. When adjusting the pH of colloidal solutions, care must be taken to not introduce too many ions into the solution, because this may cause aggregation. In order to determine the optimal solution conditions for binding of a specific protein, a flocculation assay can be performed. In this assay, the pH of the colloid solution is varied, and equal amounts of protein are added to each solution. After a set period of time (usually 30–60 min),

excess protein is removed by centrifugation of the particles and resuspension in fresh buffer. Upon the addition of NaCl to the colloid solution, particles with poor protein adhesion to the surface will aggregate and drop from the solution.

The relative ease of the direct adsorption of protein to the particles can make this an attractive method for *in vitro* studies, where clearance by the immune system is not an issue. As there is no "shielding" of the particle from the immune system, particles prepared using this method generally have low circulation half-lives *in vivo*. With direct adsorption of proteins onto the gold particle surface, one must determine that the protein is still biologically active after adsorption (Figure 4.4). Because there is no control over protein orientation on the surface, not all proteins may be active. Also, owing to the constraints on solution ionic strength and pH with regard to colloid stability, denaturation of some more delicate proteins may be an issue when using this method.

4.2.8 Thiol-Based Modification of Gold Surfaces

Perhaps the most commonly used method for attaching molecules to the surface of gold particles is via the use of thiol moieties. The strong interaction between the sulfur atom in the thiol moiety and the gold surface has been well studied on planar gold surfaces and on nanoparticles (Laibinis et al. 1991, 1992; Nuzzo et al. 1990; Bain et al. 1989a, 1989b; Love et al. 2005). Thiols have been shown to self-assemble into well-ordered monolayers on crystalline gold surfaces. The degree of surface coverage and the order of the monolayer are dependent on the properties of the thiol composing the monolayer. Relatively small alkane thiols (mercaptoundecanol, etc.) will form a very ordered monolayer on the gold surface. Larger molecules such as thiol-modified polyethylene glycols form less well organized monolayers because of the size and lack of rigidity in the PEG molecule.

The self-assembly of thiols onto gold surfaces is a powerful tool for surface modification. The use of thiol attachment provides a number of advantages over electrostatic interactions: (1) orientation control of the attached molecule, (2) ability to form well-defined mixed monolayers, and (3) ability to attach a wide variety of molecules independent of charge, hydrophobicity, etc. Because the thiol–gold interaction is one of the stronger interactions a gold surface can take part in, thiols will readily displace other molecules, such as electrostatically adsorbed ions, in the process of self-assembly. Although the thiol–gold bond is quite strong, it is also very labile, which is to say, one thiol containing molecule can displace another on the gold surface. This lability can be used to create mixed monolayers for multifunctional surfaces (e.g., targeting and protection from immune response).

4.2.9 PEGylation of Gold Nanoparticles

For biological applications requiring some sort of protection from the immune system, coating gold surfaces with polyethylene glycol is extremely common. Previous work on liposome-based drug delivery systems has shown that the circulating half-life of nanoparticles in the bloodstream can be greatly increased by the addition of a PEG layer (Lee et al. 2011; Sugiyama et al. 2009; Momekova et al. 2010; Yoshizawa et al. 2011; Milla et al. 2011; Immordino et al. 2006; Kim et al. 2009). The increase in circulating half-life is attributed to the ability of the PEG layer to inhibit opsonization by the body, which acts to hide nanoparticles from the reticuloendothelial system (RES). The addition of a PEG layer also greatly increases colloid stability in high ionic strength environments by providing steric stabilization of the particles to prevent aggregation. Because of the inherent "stickiness" of bare gold surfaces with respect to biomolecules, which was discussed in the section on attaching proteins directly to the surface of gold, opsonizing proteins in the bloodstream will readily attach to the surface of a bare particle, and the circulating half-life of bare particles is measured in minutes.

PEGylation of gold surfaces can be carried out using a thiol-modified PEG molecule of the form $CH_3O(CH_2CH_2O)_n$ CH_2CH_2SH such as the commercially available mPEG-SH MW 5000 (Laysan Bio). This molecule will self-assemble onto the bare gold surface through the thiol–gold interaction described earlier. After assembly, the molecule presents a nonreactive methoxy ($CH_3O–$) group to the solution. The length of the PEG chain required to (1) stabilize the particles and (2) prevent RES activation depends somewhat on the size of the particles being used. For example, gold nanoshells (particle diameter ~150 nm) require a PEG molecule with an average MW of approximately 5000 Da to provide stability and to hide the particles from the RES. The stability and circulating half-life of gold nanoshells prepared with a 2000-MW mPEG-thiol coating are significantly lower than particles prepared using a 5000-MW mPEG-thiol. Much smaller PEG chain lengths can be used to stabilize small colloid preparations, although care must be taken to not use very small PEG molecules (<1500 MW), which have been shown to have some level of toxicity *in vivo.*

PEGylation of the bare gold particles can be carried out by adding an excess of mPEG-SH to the solution containing the bare gold particles. Generally, a large excess of mPEG-SH is used to drive the assembly of molecules on the surface. The amount of PEG excess required can be determined by the total surface area of the gold to be covered. Because of the flexibility in the PEG chain, the footprint—or the effective surface area—taken up by a 5000-MW PEG molecule is rather large. The PEG molecules can adopt either a brush (extended) configuration or a mushroom (compacted) configuration. Previous studies have estimated the footprint of 5000 MW PEG-SH to be approximately 20 nm^2. Using this value, the approximate number of PEG molecules needed to cover a single particle can be determined. Using a 5- to 6-fold excess of mPEG-SH provides good PEG monolayer coverage in 2–3 h. Leaving the solution for longer periods can improve the surface coverage of mPEG-SH as the molecules rearrange on the surface for more efficient packing.

Although PEGylation of the gold surface provides protection from the RES system, thereby increasing the circulating half-life in the bloodstream, it does not provide any specific targeting to cancer cells. Particles coated with PEG will accumulate in tumors through the enhanced permeability and retention (EPR) effect (Maeda 2001; Maeda et al. 2001, 2003). The leaky vasculature of most tumors allows for the gold particles to leave the bloodstream and accumulate in the area around the tumor. This "passive" targeting of tumors has been used to deliver particles for photothermal therapy and for drug delivery applications. Depending on the intended application, it may be desirable to provide the particles with the ability to specifically target cancer cells. Recent work has shown that, although addition of a targeting molecule (antibody, peptide, etc.) has little effect on the number of particles that accumulate in the tumor, they do have a strong effect on the localization and distribution of the particles inside the tumor. This increased uptake by cancer cells is very important for certain applications, including drug delivery and gene therapy. It may also provide enhanced efficacy for radiation dose enhancement and photothermal therapies by more evenly distributing the effects throughout the tumor.

There are several methods for attaching a targeting agent to a PEGylated particle in order to provide the increased specificity for cancer cells while keeping a degree of protection from the RES, and particle stability provided by the PEG layer. Perhaps the two main techniques for this type of surface functionalization include (1) mixed surface preparation and (2) functionalization of the PEG molecules. In mixed surface preparation, the targeting moiety, usually a protein is first directly attached to the surface of the gold particles as described earlier in this chapter. The remaining space on the particle surface is then covered with mPEG-SH to fill in the spaces in between the proteins. This method has some significant drawbacks. All of the orientation issues detailed earlier still apply in this case, and the addition of the PEG layer can serve as a steric "shield" for the protein reducing its functionality *in vivo.* The primary advantage to this method is the relatively low amount of protein required for functionalization, which can be a factor in the cost of preparing the conjugated particles. The second method, where the PEG molecules themselves are modified to contain targeting moieties, is somewhat more robust. In this method, a bifunctional PEG molecule that contains both a thiol group and a second chemically reactive group is used. The thiol group provides the attachment of the molecules to the gold surface, whereas the other reactive group can then be coupled to the targeting moiety.

In conjugation of the targeting molecules to the bifunctional PEG, we can take advantage of the wide variety of chemical methods used in cross-linking and derivitizing proteins. There are a wide variety of conjugation protocols and reagents available commercially, capable of cross-linking thiols to thiols, amines to amines, and amines to carboxyl groups among others. Although there are a wide variety of possible methods for conjugating a chemically reactive PEG to a targeting moiety, perhaps the most common reactive group used in conjugations is the NHS ester (*n*-hydroxysuccinimide). This group is used to form a chemical bond with primary and secondary amine groups. This can be used to conjugate proteins, peptides, and a vast number of small, biologically active molecules such as folate. Depending on the particular targeting moiety, the bifunctional PEG can be modified before or after attachment to the gold surface (Figure 4.5). In most cases, if it is possible, prefunctionalization of the PEG is preferred because of the very short half-life of the NHS ester in aqueous solution (~30 min). Owing to the time required to achieve decent PEG layer coverage on the particles, and the subsequent removal of unattached NHS–PEG-SH from solution, there is usually a significant loss in the reactivity of the NHS–PEG layer. However, there are some possible complications that should be considered when using the pre-surface attachment functionalization method. For larger macromolecules, such as antibodies, it is possible to have multiple attachment sites for the bifunctional PEGs. In this case, the activity of the protein could be severely affected. Another concern is the amount of material that would be used in the actual PEGylation process. Because an excess of PEG is usually added to the particles to effectively drive assembly on the surface, a large amount of the targeting moiety may be lost during the PEGylation process, which can be a major drawback for expensive macromolecules.

In cases where prefunctionalization is not feasible, the relatively short half-life of the NHS ester can be avoided by using a thiol-PEG-carboxyl bifunctional PEG. In this case, the bifunctional PEG can be added in excess to the particle solution to provide excellent surface coverage. After removal of unbound PEG, the carboxyl group can be converted into an NHS ester by using standard EDC–NHS coupling chemistry. In this reaction, the EDC molecule (1-ethyl-3-[3-dimethylaminopropyl] carbodiimide hydrochloride) is added to the PEGylated particles. The EDC reacts with the carboxyl group on the PEG to form an amine-reactive intermediate. If this intermediate does not encounter an amine, it will hydrolyze and regenerate the carboxyl group. In the presence of *N*-hydroxysulfosuccinimide (Sulfo-NHS), EDC can be used to convert carboxyl groups to amine-reactive Sulfo-NHS esters. This Sulfo-NHS ester has a reactive half-life of 30–45 min.

4.2.10 Other Surface Functionalization Methods for Gold Nanoparticles

Although the two methods listed above are the most commonly used methods for surface modification of gold nanoparticles,

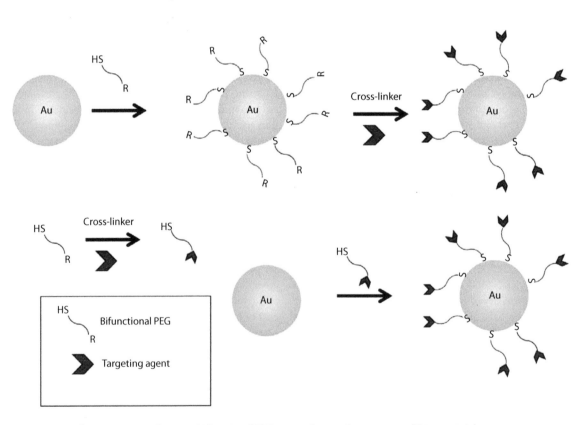

FIGURE 4.5 Functionalization strategy for using bifunctional PEG to attach targeting agent to gold nanoparticle.

there are a number of other methods that have been developed. Methods involving self-assembly of charged polymer layers, such as polystyrene sulfonate and polyallylamine, have been developed and provide a negatively or positively charged surface depending on the polymer used. The polyallylamine layer can also be functionalized using many of the cross-linking methods described for functionalizing PEGs above.

The gold surface can also be coated in a layer of silica. The silica layer stabilizes the gold particles in high ionic strength solutions and has been shown to improve thermal stability in some applications. The surface of the silica layer can subsequently be modified using the well-established silane chemistry. An organosilane is a molecule containing a silicon-to-carbon bond R3–Si–C–R3. The Si in the organosilane can bind to surface hydroxyl in the silica layer forming silicon esters. In this way, the silica surface can be modified to have a wide array of properties including chemical reactivity (for cross-linking proteins or PEGs), hydrophobicity, and hydrophiliccity. The most common applications would involve the addition of a chemically reactive "handle" to the silica surface allowing for further modification with the targeting agent of interest.

4.2.11　Synthesis of Iron Oxide Nanoparticles

Along with noble metal nanoparticles, superparamegnetic iron nanoparticles (SPIOs) are perhaps the most widely applied nanoparticle for biological applications. In the past decade, there has been an intensive development of SPIO nanoparticles for many medical applications including MRI contrast and localized hyperthermia. The SPIO nanoparticles are magnetically active, which is useful for imaging contrast, separations, and for generation of heat under alternating magnetic fields.

As with noble metal particles, there is a wide variety of synthetic routes that have been developed for preparing SPIO nanoparticles. The methods include laser ablation, molecular beam epitaxy, sputtering and arc discharge, classic "wet" chemistry methods, high-temperature thermolysis, sol–gel, polyol, and sonolysis methods. In the scope of this chapter, the focus will be on the wet chemical methods since those methods are available to more researchers without specialized equipment. The classical method for preparing SPIO particles involves the coprecipitation of aqueous ferric our ferrous salts in aqueous medium to form magnetite:

$$Fe^{2+} + 2Fe^{3+} + 8OH \rightarrow Fe_3O_4 + 4H_2O \qquad (4.1)$$

The resulting magnetite can subsequently be converted in to maghemite (γFe_2O_3) by oxidation in the presence of oxygen (Laurent et al. 2008):

$$Fe_3O_4 + 2H^+ \rightarrow \gamma\text{-}Fe_2O_3 + 2Fe^+ + H_2O \qquad (4.2)$$

The synthesis of SPIO without the addition of stabilizing molecules was first performed by Massart (1981). Using this method, it is possible to prepare SPIO with a size ranging from 16.6 to 4.2 nm. The coprecipitation method for synthesizing SPIO forms roughly spherical particles. The size and shape of the particles can be controlled by adjusting several parameters, including pH, ionic strength, or the Fe^{II}/Fe^{III} ratio. The major advantage of the coprecipitation method is the ability to make large quantities of the nanoparticles. However, size distribution of the particles is difficult to control. The polyol method is a versatile method for fabricating nanoparticles with well-defined shapes and sizes. A method for the synthesis of SPIO with controlled size distribution was published by Cai and Wang (2007). In this method, triethylene glycol was reacted with $Fe(acac)_3$ at an elevated temperature. This method produced nonaggregated SPIO with a uniform shape and a narrow size distribution. The polyol method also produces particles coated in hydrophilic polyol ligand and are easily dispersed in aqueous media. The elevated temperature of the synthesis produced particles with high crystallinity and magnetization.

4.2.12　Stabilization and Surface Modification of SPIOs

In order to produce SPIO nanoparticles that are stable in biological media as well as in magnetic fields, it is necessary to add stabilizing agents to the SPIO surface. The stabilizing agents prevent aggregation and can also provide chemical reactivity for conjugation to targeting agents. The stabilizing agents can be monomeric (such as carboxylates or phosphates), polymers such as dextran and polyvinyl alcohol, or inorganic in the case of silica.

SPIO nanoparticles can be stabilized by the adsorption of citric acid to the particle surface. The acid is adsorbed by complexing one or more of the carboxyl groups to the surface, which leaves one carboxyl group free for further reaction, and providing a net negative surface charge for the particle, which provides the increased stability. Additional targeting moieties can be coupled to carboxyl-stabilized SPIO particles using the EDC–NHS coupling chemistry described earlier in this chapter. The free carboxyl group can be coupled to primary or secondary amines in many proteins, peptides, or to functionalized PEGs.

A method for the PEGylation of the SPIO has also been developed, which provides the additional "stealth" capacity for long circulation times in the bloodstream. In this method, the SPIO are treated with a poly(ethyleneglycol)–polyaspartic acid copolymer (Kumagai et al. 2007). The resulting particles showed good stability in biological conditions. Owing to the use of a block copolymer to provide the PEGylated surface, it is difficult to add targeting function to the PEG in this system. As a result, particles functionalized in this manner are better suited to applications relying on passive accumulation in tumor through the EPR effect.

In cases where targeted SPIO nanoparticles are desired, chemical functionalization, using methods described above for conjugation of noble metal particles, can be used to couple antibodies, functionalized PEG molecules, or small molecule targets to the surface of dextran coated particles. Dextran is a polysaccharide composed of α-D-glucopyanosyl units and is perhaps the most commonly used method for stabilization of the SPIO for use in medical applications. The major advantage for using dextran for stabilization is its biocompatibility (Laurent et al. 2004). The dextran layer can then be conjugated to targeting molecules through an oxidative conjugation strategy that produces aldehydes on the dextran. This strategy using periodate oxidation followed by reduction of the Schiff base has been used to couple peptides, proteins, and antibodies to the SPIO.

References

Alkilany, A. M., P. N. Sisco, S. P. Boulos, S. T. Sivapalan, J. A. Yang, D. J. Chernak, and J. Huang. 2010. The many faces of gold nanorods. *Journal of Physical Chemistry Letters* 1(19):2867–2875

Atkinson, R. L., M. Zhang, P. Diagaradjane, S. Peddibhotla, A. Contreras, S. G. Hilsenbeck, W. A. Woodward, S. Krishnan, J. C. Chang, and J. M. Rosen. 2010. Thermal enhancement with optically activated gold nanoshells sensitizes breast cancer stem cells to radiation therapy. *Science Translational Medicine* 55:55–79.

Bain, C. D., H. A. Biebuyck, and G. M. Whitesides. 1989a. Comparison of self-assembled monolayers on gold: Coadsorption of thiols and disulfides. *Langmuir* 5(3):723–727.

Bain, C. D., J. Evall, and G. M. Whitesides. 1989b. Formation of monolayers by the coadsorption of thiols on gold: Variation in the head group, tail group, and solvent. *Journal of the American Chemical Society* 111(18):7155–7164.

Bartneck, M., H. A. Keul, S. Singh, K. Czaja, J. R. Bornemann, M. Bockstaller, M. Moeller, G. Zwadlo-Klarwasser, J. R. Groll. 2010. Rapid uptake of gold nanorods by primary human blood phagocytes and immunomodulatory effects of surface chemistry. *ACS Nano* 4(6): 3073–3086.

Bastus, N. G., J. Comenge, and V. Puntes. 2011. Kinetically controlled seeded growth synthesis of citrate-stabilized gold nanoparticles of up to 200 nm: Size focusing versus Ostwald ripening. *Langmuir* 27(17):11098 –11105.

Cai, W., and J. Wang. 2007. Facile synthesis of superparamagnetic magnetite nanoparticles in liquid polyols. *Journal of Colloid and Interface Science*, 305(2):366-370.

Chithrani, D. B., S. Jelveh, F. Jalali, M. van Prooijen, C. Allen, R. G. Bristow, R. P. Hill, and D. A. Jaffray. 2010. Gold nanoparticles as radiation sensitizers in cancer therapy. *Radiation Research* 173(6):719–728.

Connor, E. E., J. Mwamuka, A. Gole, C. J. Murphy, and M. D. Wyatt. 2005. Gold nanoparticles are taken up by human cells but do not cause acute cytotoxicity. *Small* 1(3): 325–327.

Davis, P. H., C. P. Morrisey, S. M. V. Tuley, and C. I. Bingham, 2008. Synthesis and stabilization of colloidal gold nanoparticle suspensions for SERS. In *Nanoparticles: Synthesis, Stabilization, Passivation, and Functionalization,* vol. 996, pp. 16–30. American Chemical Society, Washington DC.

Duff, D. G., and A. Baiker. 1993. A new hydrosol of gold clusters: 1. Formation and particle size variation. *Langmuir* 9:2301–2309.

Eby, D. M., H. R. Luckarift, and G. R. Johnson. 2009. Hybrid antimicrobial enzyme and silver nanoparticle coatings for medical instruments. *ACS Applied Materials & Interfaces* 1(7):1553–1560.

Fan, F.-R. F., and A. J. Bard. 2001. Chemical, electrochemical, gravimetric, and microscopic studies on antimicrobial silver films. *Journal of Physical Chemistry B* 106(2):279–287.

Faraday, M. 1857. The Bakerian lecture: Experimental relations of gold (and other metals) to light. *Philosphical Transactions of the Royal Society London* 147:145–181.

Frens, G. 1973. Controlled nucleation for the regulation of the particle size in monodisperse gold suspensions. *Nature Physical Science* 241:20–22.

Gole, A., and C. J. Murphy. 2004. Seed-mediated synthesis of gold nanorods: Role of the size and nature of the seed. *Chemistry of Materials* 16(19):3633–3640.

Goulet, P. J. G., and R. B. Lennox. 2010. New insights into Brust–Schiffrin metal nanoparticle synthesis. *Journal of the American Chemical Society* 132(28):9582–9584.

Hashizume, K., Y. Hatanaka, Y. Kamihara, and Y. Tani. 2001. Automated immunohistochemical staining of formalin-fixed and paraffin-embedded tissues using a catalyzed signal amplification method. *Applied Immunohistochemistry & Molecular Morphology* 9(1):54–60.

Hauck, T. S., A. A. Ghazani, and W. C. 2008. Assessing the effect of surface chemistry on gold nanorod uptake, toxicity, and gene expression in mammalian cells. *Small* 4(1):153–159.

Hu, M., J. Chen, Z. Y. Li, L. Au, G. V. Hartland, X. Li, M. Marquez, and Y. Xia. 2006. Gold nanostructures: Engineering their plasmonic properties for biomedical applications. *Chemical Society Reviews* 35(11):1084–1094.

Huff, T. B., M. N. Hansen, Y. Zhao, J. X. Cheng, and A. Wei. 2007. Controlling the cellular uptake of gold nanorods. *Langmuir* 23(4):1596–1599.

Immordino, M. L., F. Dosio, and L. Cattel. 2006. Stealth liposomes: Review of the basic science, rationale, and clinical applications, existing and potential. *International Journal of Nanomedicine* 1(3):297–315.

Jain, J., S. Arora, J. M. Rajwade, P. Omray, S. Khandelwal, and K. M. Paknikar. 2009. Silver nanoparticles in therapeutics: Development of an antimicrobial gel formulation for topical use. *Molecular Pharmaceutics* 6(5):1388–1401.

Jana, N. R., L. Gearheart, and C. J. Murphy. 2001. Wet chemical synthesis of high aspect ratio cylindrical gold nanorods. *Journal of Physical Chemistry B* 105: 4065–4067.

Khalavka, Y., J. Becker, and C. Sönnichsen. 2009. Synthesis of rod-shaped gold nanorattles with improved plasmon sensitivity and catalytic activity. *Journal of the American Chemical Society* 131(5):1871–1875.

Kho, K. W., Z. X. Shen, H. C. Zeng, K. Soo, and M. Olivo. 2005. Deposition method for preparing SERS-active gold nanoparticle substrates. *Analytical Chemistry* 77(22):7462–7471.

Kim, J. Y., J. Kim, J. S., Park, Y. Byun, and C. K. Kim. 2009. The use of PEGylated liposomes to prolong circulation lifetimes of tissue plasminogen activator. *Biomaterials* 30(29):5751–5756.

Kimling, J., M. Maier, B. Okenve, V. Kotaidis, H. Ballot, and A. Plech. 2006. Turkevich method for gold nanoparticle synthesis revisited. *Journal of Physical Chemistry B* 110(32):15700–15707.

Kou, X., S. Zhang, C.-K. Tsung, M.-H. Yeung, Q. Shi, G. D. Stucky, L. Sun, J. Wang, and C. Yan. 2006. Growth of gold nanorods and bipyramids using CTEAB surfactant. *Journal of Physical Chemistry B* 110(33):16377–16383.

Kumagai, M., Y. Imai, T. Nakamura, Y. Yamasaki, M. Sekino, S. Ueno, K. Hanaoka, K. Kikuchi, T. Nagano, E. Kaneko, K. Shimokado, and K. Kataoka. 2007. Iron hydroxide nanoparticles coated with poly(ethylene glycol)–poly(aspartic acid) block copolymer as novel magnetic resonance contrast agents for in vivo cancer imaging. *Colloids and Surfaces B: Biointerfaces* 56(1-2):174–181.

Laibinis, P. E., G. M. Whitesides, D. L. Allara, Y. T. Tao, A. N. Parikh, and R. G. Nuzzo. 1991. Comparison of the structures and wetting properties of self-assembled monolayers of *n*-alkanethiols on the coinage metal surfaces, copper, silver, and gold. *Journal of the American Chemical Society* 113(19):7152–7167.

Laibinis, P. E., R. G. Nuzzo, and G. M. Whitesides. 1992. Structure of monolayers formed by coadsorption of two *n*-alkanethiols of different chain lengths on gold and its relation to wetting. *Journal of Physical Chemistry* 96(12):5097–5105.

Laurent, S., C. Nicotra, Y. Gossuin, A. Roch, A. Ouakssim, L. Vander Elst, M. Cornant, P. Soleil, and R. N. Muller. 2004. Influence of the length of the coating molecules on the nuclear magnetic relaxivity of superparamagnetic colloids. *Physica Status Solidi C* 1(12):3644–3650.

Laurent, S., D. Forge, M. Port, A. Roch, C. Robic, L. Vander Elst, and R. N. Muller. 2008. Magnetic iron oxide nanoparticles: Synthesis, stabilization, vectorization, physicochemical characterizations, and biological applications. *Chemical Reviews* 108(6):2064–2110.

Lee, J. S., M. Ankone, E. Pieters, R. M. Schiffelers, W. E. Hennink, and J. Feijen. 2011. Circulation kinetics and biodistribution of dual-labeled polymersomes with modulated surface charge in tumor-bearing mice: Comparison with stealth liposomes. *Journal of Controlled Release* 155(2):282–288.

Leonov, A. P., J. Zheng, J. D. Clogston, S. T. Stern, A. K. Patri, and A. Wei. 2008. Detoxification of gold nanorods by treatment with polystyrenesulfonate. *ACS Nano* 2(12):2481–2488.

Leontidis, E., K. Kleitou, T. Kyprianidou-Leodidou, V. Bekiari, and P. Lianos. 2002. Gold colloids from cationic surfactant solutions: 1. Mechanisms that control particle morphology. *Langmuir* 18(9):3659–3668.

Li, J., J. Wu, X. Zhang, Y. Liu, D. Zhou, H. Sun, H. Zhang, and B. Yang. 2011a. Controllable synthesis of stable urchin-like gold nanoparticles using hydroquinone to tune the reactivity of gold chloride. *Journal of Physical Chemistry C* 115(9):3630–3637.

Li, Y., O. Zaluzhna, B. Xu, Y. Gao, J. M. Modest, and Y. J. Tong. 2011b. Mechanistic insights into the Brust–Schiffrin two-phase synthesis of organo-chalcogenate–protected metal nanoparticles. *Journal of the American Chemical Society* 133(7):2092–2095.

Love, J. C., L. A. Estroff, J. K. Kriebel, R. G. Nuzzo, and G. M. Whitesides. 2005. Self-assembled monolayers of thiolates on metals as a form of nanotechnology. *Chemical Reviews* 105(4):1103–1170.

Maeda, H. 2001. The enhanced permeability and retention (EPR) effect in tumor vasculature: The key role of tumor-selective macromolecular drug targeting. *Advances in Enzyme Regulation* 41:189–207.

Maeda, H., T. Sawa, and T. Konno. 2001. Mechanism of tumor-targeted delivery of macromolecular drugs, including the EPR effect in solid tumor and clinical overview of the prototype polymeric drug SMANCS. *Journal of Controlled Release* 74(1–3):47–61.

Maeda, H., J. Fang, T. Inutsuka, and Y. Kitamoto. 2003. Vascular permeability enhancement in solid tumor: Various factors, mechanisms involved and its implications. *International Immunopharmacology* 3(3):319–328.

Massart, R. 1981. Preparation of aqueous magnetic liquids in alkaline and acidic media. *IEEE Transactions on Magnetics* 28(3):1247–1248.

Mayer, G., and M. Bendayan. 1999. Immunogold signal amplification: Application of the CARD approach to electron microscopy. *Journal of Histochemistry and Cytochemistry* 47(4):421–430.

Métraux, G. S., Y. C. Cao, R. Jin, and C. A. Mirkin. 2003. Triangular nanoframes made of gold and silver. *Nano Letters* 3(4):519–522.

Milla, P., F. Dosio, and L. Cattel. 2011. PEGylation of proteins and liposomes: A powerful and flexible strategy to improve the drug delivery. *Current Drug Metabolism* 13(1):105–119.

Momekova, D., S. Rangelov, and N. Lambov. 2010 Long-circulating, pH-sensitive liposomes. *Methods in Molecular Biology (Clifton, N.J.)* 605:527–544.

Murphy, C. J., L. B. Thompson, A. M. Alkilany, P. N. Sisco, S. P. Boulos, S. T. Sivapalan, J. A. Yang, D. J. Chernak, and J. Huang. 2010. The many faces of gold nanorods. *Journal of Physical Chemistry Letters* 1(19):2867–2875.

Nah, S., L. Li, and J. T. Fourkas. 2009. Field-enhanced phenomena of gold nanoparticles. *Journal of Physical Chemistry A* 113(16):4416–4422.

Nuzzo, R. G., L. H. Dubois, and D. L. Allara. 1990. Fundamental studies of microscopic wetting on organic surfaces: 1. Formation and structural characterization of a self-consistent series of polyfunctional organic monolayers. *Journal of the American Chemical Society* 112(2):558–569.

Oldenburg, S. J., R. D. Averitt, S. L. Westcott, and N. J. Halas. 1998. Nanoengineering of optical resonances. *Chemical Physics Letters* 288:243–247.

Oldenburg, S. J., J. B. Jackson, S. L. Westcott, and N. J. Halas. 1999. Infrared extinction properties of gold nanoshells. *Applied Physics Letters* 75(19):2897–2899.

Parab, H. J., H. M. Chen, T.-C. Lai, J. H. Huang, P. H. Chen, R.-S. Liu, M. Hsiao, C.-H. Chen, D.-P. Tsai, and Y.-K. Hwu. 2009. Biosensing, cytotoxicity, and cellular uptake studies of surface-modified gold nanorods. *Journal of Physical Chemistry C* 113(18):7574–7578.

Prodan, E., C. Radloff, and N. J. Halas, and P. Nordlander. 2003. A hybridization model for the plasmon response of complex nanostructures. *Science* 302(5644):419–422.

Stöber, W., A. Fink, and E. Bohn. 1968. Controlled growth of monodisperse silica spheres in the micron size range. *Journal of Colloid and Interface Science* 26(1):62–69.

Smith, D. K., and B. A. Korgel. 2008. The importance of the CTAB surfactant on the colloidal seed-mediated synthesis of gold nanorods. *Langmuir* 24(3):644–649.

Sugiyama, I., T. Sonobe, and Y. Sadzuka. 2009. Effect of hybridized liposome by novel modification with some polyethyleneglycol-lipids. *International Journal of Pharmaceutics* 372(1–2):177–183.

Thakor, A. S., J. Jokerst, C. Zavaleta, T. F. Massoud, and S. S. Gambhir. 2011. Gold nanoparticles: A revival in precious metal administration to patients. *Nano Letters* 11(10):4029–4036.

Thomas, K. G., and P. V. Kamat. 2003. Chromophore-functionalized gold nanoparticles. *Accounts of Chemical Research* 36(12): 888–898.

Turkevich, J., P. C. Stevenson, and J. Hillier. 1951. A study of the nucleation and growth processes in the synthesis of colloidal gold. *Discussions of the Faraday Society* 11:55–75.

Yamamoto, M., Y. Kashiwagi, T. Sakata, H. Mori, and M. Nakamoto. 2005. Synthesis and morphology of star-shaped gold nanoplates protected by poly(*N*-vinyl-2-pyrrolidone). *Chemistry of Materials* 17(22):5391–5393.

Yoon, K. Y., J. H. Byeon, C. W. Park, and J. Hwang. 2008. Antimicrobial effect of silver particles on bacterial contamination of activated carbon fibers. *Environmental Science & Technology* 42(4):1251–1255.

Yoshizawa, Y., Y. Kono, K. Ogawara, T. Kimura, and K. Higaki. 2011. PEG liposomalization of paclitaxel improved its in vivo disposition and anti-tumor efficacy. *International Journal of Pharmaceutics* 412(1–2):132–141.

Yuan, W., G. Jiang, J. Che, X. Qi, R. Xu, M. W. Chang, Y. Chen, S. Y. Lim, J. Dai, and M. B. Chan-Park. 2008. Deposition of silver nanoparticles on multiwalled carbon nanotubes grafted with hyperbranched poly(amidoamine) and their antimicrobial effects. *Journal of Physical Chemistry C* 112(48):18754–18759.

Zhang, J. Z. 2010. Biomedical applications of shape-controlled plasmonic nanostructures: A case study of hollow gold nanospheres for photothermal ablation therapy of cancer. *Journal of Physical Chemistry Letters* 1(4):686–695.

Zou, J., Q. Dai, R. Guda, X. Liu, G. Worden James, T. Goodson, and Q. Huo. 2008. Controlled chemical functionalization of gold nanoparticles. In *Nanoparticles: Synthesis, Stabilization, Passivation, and Functionalization*, Vol. 996, pp. 31–40. American Chemical Society, Washington DC.

5

Qingxin Mu
St. Jude Children's Research Hospital

Yi Zhang
St. Jude Children's Research Hospital

Aijuan Liu
Shandong University

Nana Liang
Shandong University

Yanyan Liu
Shandong University

Yan Mu
Shandong University

Gaoxing Su
St. Jude Children's Research Hospital

Bing Yan
St. Jude Children's Research Hospital

Safety Concerns for Nanomaterials in Nanomedicinal Applications

5.1 Introduction

Owing to their unique optical, magnetic, and thermal properties, nanomaterials (NMs) have been widely explored in radiation diagnosis and therapy. For instance, carbon nanotubes (CNTs) are used for Raman and photoacoustic imaging and thermal therapy (Chakravarty et al. 2008; De La Zerda et al. 2008; Liu et al. 2007); magnetic nanoparticles (MNPs) for magnetic resonance imaging (MRI) and thermal therapy (Duguet et al. 2006); and gold nanoparticles (GNPs) for x-ray computed tomography (CT) contrast agents (Popovtzer et al. 2008). However, with recent nanomedicine applications, there is a growing concern about the toxic effects generated by NMs (Nel et al. 2006). Because these theranostic nanoparticles will be inevitably injected into the human body when they are used in clinics, the related toxicity issues must be addressed before any human use. The toxicity of NMs in living cells and mammals has been frequently reported; the results of preliminary studies reveal that multiple biological mechanisms may be involved (Deng et al. 2011; Manna et al. 2005; Mu et al. 2009b). However, our knowledge about the toxicologic mechanism of NMs and the correlation between their physicochemical properties and toxicity is very limited.

NMs have small size (1–100 nm) and extremely large surface areas; thus, large proportions of their atoms or molecules are exposed on the surface (Oberdörster et al. 2005). This renders them with strong surface energy and makes them be prone to react with surrounding molecules, including biomolecules. Therefore, unlike for bulk materials or small-molecule toxicants, the dimensions of NMs are an important variant determining their toxicity.

NMs can enter cells by energy-dependent endocytosis or phagocytosis and by direct cell membrane penetration (Gratton et al. 2008; Kostarelos et al. 2007; Mu et al. 2009a). NMs are also prone to aggregate and accumulate in various organs, resulting in a low elimination rate (Qu et al. 2009). They can also translocate to other regions from the portal of entry and cause a wide range of damages (Ryman-Rasmussen et al. 2009). Because of their dimensions, they may escape the innate immune system and macrophages (Poland et al. 2008). All these traits have raised more concerns about NMs' potential toxicities.

The shape and surface chemistry of NMs are crucial factors that determine their interactions with biological molecules, capability to penetrate cell membranes, and *in vivo* distributions. Therefore, these properties control their potential toxicities. Examples include gold NMs with various shapes (Wang et al. 2008) and CNTs with various surface chemistries (Gao et al. 2011).

5.2 Toxicity of NMs to Living Cells

To understand the cytotoxicity of NMs, their cellular uptake must first be understood. Additionally, one must understand NMs' perturbations on cellular oxidative stress level and cellular signaling events.

5.2.1 Cell Uptake and Cellular Translocation

NMs can enter living cells through endocytosis (Mu et al. 2010; Sahay et al. 2010). During the process, NMs are engulfed into endosomes formed by cell membrane invaginations and then redistributed into various cell organelles. Phagocytosis is one of the endocytic pathways for NMs' cellular uptake. This process occurs in phagocytes, such as macrophages and monocytes, and to a much lower extent in fibroblasts, epithelial cells, and endothelial cells. The uptake of particles by phagocytes does not depend on the size of the particles. Yet, the particle's shape at the attaching point is a crucial factor for macrophage uptake (Mitragotri et al. 2007). NMs usually enter cells through more than one pathway, including clathrin- and caveolae-dependent endocytosis, clathrin- and caveolae-independent endocytosis, and macropinocytosis (Schmid and Conner 2003). Among these, macropinocytosis is the least used for cellular uptake of NMs (Figure 5.1).

Size, shape, charge, and functional groups collectively determine the cellular uptake of NMs. The uptake of GNPs by HeLa cells heavily depends on the particles' size, and the maximum amount of uptake was found for GNPs with a diameter of 50 nm (Chan et al. 2006). Particles between 14 and 100 nm were trapped in vesicles inside the cell and did not enter the nucleus. Furthermore, NMs with higher aspect ratios enter cells faster compared with their more symmetrical, cylindrical particle counterparts (Gratton et al. 2008). Additionally, sphere-shaped nanostructures enter cells compared with rod-shaped ones, showing the effects of curvature of NMs (Chan et al. 2006).

Surface charge also plays an important role in the uptake of NMs. Nanorods coated with a negatively charged layer are not taken up by HeLa cells as quickly as are nanorods with a positively charged layer (Hauck et al. 2008). This outcome may be related to electrostatic interactions with the negatively charged cell surface (Slowing et al. 2006). Because the functional groups on the surface of NMs may change the particles' interactions with cells, chemically modified NMs might have different cellular uptake rates than unmodified ones do. Additionally, having anionic surfactants and polyethylene glycol on the surface may prevent the cellular uptake of nanorods (Huff et al. 2007).

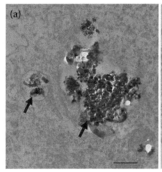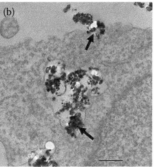

FIGURE 5.1 TEM characterization of Fe@CNPs' uptake into C33A cells. (a) CNPP; (b) CNPA. Arrows indicate Fe@CNPs and scale bars represent 100 nm. (Reprinted with permission from Mu, Q. et al., *Biomaterials*, 31, 5083–5090, 2010. Copyright (2010) Elsevier.)

Serum protein adsorption may also influence cellular uptake. One study shows that serum proteins adsorbed on the surface of NMs may enhance the uptake (Chan et al. 2006). However, another study shows that cellular uptake of carbon NMs is much higher in serum-free culture medium than in culture medium with serum (Zhu et al. 2009).

Most NMs are too big to enter the nucleus. For example, GNPs that enter the human dermal fibroblasts gather in the lysosome (Pernodet et al. 2006), and gold nanorods become trapped in vesicles in cells (Hauck et al. 2008). However, peptide-BSA GNPs can enter the nuclei of HeLa cells (Franzen et al. 2004). Single- and multi-walled CNTs (SWCNTs and MWCNTs) can also enter cell nuclei (Cheng et al. 2008; Mu et al. 2009a).

5.2.2 Oxidative Stress Perturbation by NMs

To counteract the effects of reactive oxygen species (ROS), organisms have several distinct antioxidants, including superoxide dismutase, catalase, ascorbic acid, and glutathione. But when the antioxidant defense is overwhelmed, abnormal oxidative stress will be the consequence (Li et al. 2008). The oxidative stress–induced damages include cellular membrane injury (Sohaebuddin et al. 2010), DNA damage (Singh et al. 2009), protein denaturation (Yoon et al. 2007), mitochondrial perturbation (Palmeira 2008), cell apoptosis, and necrosis (Foldbjerg et al. 2009). Perturbation of the oxidative stress balance has been proposed to be a general mechanism for nanotoxicity (Nel et al. 2006).

Several types of fullerenes (C60) generate superoxide anions in water, possibly causing oxidative damage to cell membranes and subsequent cell death (Sayes et al. 2004). Additionally, pristine SWCNTs (Manna et al. 2005) and MWCNTs (Ye et al. 2009) cause oxidative stress to cells and the inflammatory response. Silver nanoparticles (SNPs) also cause cell damage by increasing ROS production and reducing ATP generation, leading to DNA damage (Valiyaveettil et al. 2009).

5.2.3 Cellular Perturbations of NMs

A 4-day incubation of GNPs and dermal fibroblasts significantly reduces the cells' proliferation, partly because of the GNPs' effects on actin fibrils (Pernodet et al. 2006). Additionally, a maximal dose or chronic low dose of C60(OH)$_{24}$ inhibits the growth of human umbilical vein endothelial cells, causing autophagic cell death (Iwai and Yamawaki 2006). Furthermore, three different types of CNTs (SWCNTs, 50% SWCNTs + 30% MWCNTs + 20% C60, MWCNTs) strongly affect the proliferation of U937 monocytic cells, with little effect on the cells' viability (Ghibelli et al. 2007). COOH-functionalized SWCNTs and MWCNTs also significantly affect mesenchymal stem cells' (MSCs) proliferation, which might be related to the binding of CNTs to nutrients in the culture medium (Mooney et al. 2008) and their effects on the bone morphogenetic protein (BMP) signaling pathway (Liu et al. 2010).

Once GNPs enter human dermal blasts, actin fibers disappear (Pernodet et al. 2006). Carboxylated MWCNTs and SWCNTs inhibit the osteogenic differentiation of MSCs by inhibiting alkaline phosphatase (ALP) activity. During adipocyte differentiation of MSCs, ALP activity is also strongly inhibited on day 14. Quantitative polymerase chain reaction analysis shows that several differentiation genes are down-regulated after treatment with CNTs (Liu et al. 2010). Studies on the effects of several CNTs (i.e., SWCNTs, double-walled CNTs, MWCNTs) on osteoblasts revealed that CNTs inhibit mineralized nodule formation during the final stage of cell differentiation (Yang et al. 2007a).

5.2.4 Effects on Cellular Signaling

NMs have been repeatedly reported to activate or affect cellular signaling pathways, although their direct cellular targets are unknown.

Nuclear factor-kappa B (NF-κB). Several NMs can generate oxidative stress in cells, thus influencing signaling pathways, such as those of MAP kinases and NF-κB. When human keratinocytes are treated with SWCNTs, NF-κB is activated in a dose-dependent manner, which might relate to the activation of stress-related kinases (Manna et al. 2005). MWCNTs have also been reported to induce cell ROS and IL-8, along with the activation of NF-κB, which may lead to the death of A549 cells (Ye et al. 2009).

MAPK. CNTs can activate MAPK/ERK signal transduction, thus promoting neurite outgrowths in DRG neurons and PC12h cells (Matsumoto et al. 2009). Similar results are obtained when PC12 cells are treated with iron oxide NMs (Park et al. 2011). TiO_2 NMs induced phosphorylation of p38 MAPK and Erk-1/2 and inhibited human polymorphonuclear neutrophils apoptosis (Girard et al. 2010). MNPs coated with specific ligands activate the MAPK signaling pathway when a magnetic field is applied (Sniadecki 2010). Furthermore, GNPs modulate osteogenic and adipocytic differentiation of MSCs through the p38 MAPK signaling pathway (Figure 5.2) (Yi et al. 2010).

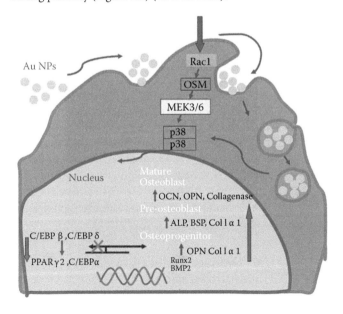

FIGURE 5.2 Molecular mechanism of the modulation of osteogenic and adipocytic differentiation of MSCs by AuNPs through p38 MAPK signaling pathway. (Reprinted with permission from Yi, C. et al., *ACS Nano*, 4, 6439–6448, 2010. Copyright (2010) American Chemical Society.)

BMP. SWCNT-COOH suppress the Smad-dependent BMP signaling pathway and Id protein expression and inhibit cell proliferation by arresting the cell cycle at the G_1/S transition (Mu et al. 2009b). CNTs inhibit cell proliferation and osteoblast differentiation by disturbing BMP signal transduction (Liu et al. 2010).

Mac-1. The uptake of superparamagnetic iron oxide is reportedly mediated by Mac-1 (von zur Muhlen et al. 2007). Negatively charged NMs activate the Mac-1 receptor pathway via unfolding of fibrinogen and cause the release of inflammatory cytokines (Deng et al. 2011).

5.3 Potential Hazards of NMs to Mammals

The theranostic NMs enter the human body through intravenous injection or inhalation. NMs reportedly cross key physical barriers that provide protection for the vulnerable organs, such as the blood–brain barrier (Borm and Kreyling 2004), blood–testis barrier (Kim et al. 2006), and blood–placental barrier (Tsuchiya et al. 1996). These abilities may help overcome the difficulty of delivering therapeutic agents to these protected organs (e.g., for treatment of brain disorders). The other side of this coin is that these abilities may cause toxicities. For example, cationic NMs have an immediate toxic effect on the blood–brain barrier (Lockman et al. 2004). Additionally, MWCNTs entering the testis of male mice cause organ damage and induce ROS although they do not affect fertility (Bai et al. 2010). Thus, NMs reaching the organs affect normal functions of the organs and cause immune dysfunctions.

5.3.1 Biodistribution and Elimination of NMs

NMs have shown great potential for tumor targeting and targeted drug delivery; they can deliver proteins (Kam and Dai 2005), nucleic acids (Liu et al. 2005), and short interfering RNA (Kam et al. 2005). They also have imaging and therapy capabilities after entering the human body and accumulating in tumors. Therefore, a better understanding of NMs' biodistribution in animal models may provide guidance for their eventual applications in humans.

MNPs. Oleic acid–Pluronic-coated iron oxide MNPs in serum and tissue were analyzed after intravenous administration to rats. The biodistribution of the particles in tissues changed with time, more so in the liver and spleen than in the brain, heart, kidney, and lung. The amount of MNPs in the liver and spleen decreased after 3 weeks, suggesting that elimination occurred. The kidneys had markedly lower overall magnetizations, suggesting a lower accumulation of MNPs in kidneys (Jain et al. 2008). In another study, fluorescent MNPs (FMNPs) were exposed to mice through nose-only exposure. FMNPs were distributed in the liver, testis, spleen, lung, and brain. Fewer FMNPs were found in the nasal cavity, heart, kidney, and ovary. Overall, the liver had the strongest fluorescence intensity throughout the whole organ; however, the FMNPs were only observed in specific regions of the spleen and testes (Kwon et al. 2008).

Quantum dots (QDs). Several near-infrared (NIR)-fluorescent QD-based NMs with varied chemical compositions, shapes, sizes, and surface charges were synthesized, and their biodistribution and elimination rates from rat lungs were quantified. NMs having a hydrodiameter less than 6 nm were rapidly translocated from the lungs to the lymph nodes and the bloodstream, and they were subsequently cleared by the kidneys (Choi et al. 2010). CdSe core–QDs and ZnS shell–QDs coated with anionic, zwitterionic, or neutral molecules were synthesized with precise size-series and then intravenously administered to rodents to study the renal filtration threshold. Zwitterionic or neutral organic coatings prevented renal excretion, whereas QDs with a hydrodynamic diameter of 5.5 nm resulted in rapid and efficient urinary excretion and elimination from the body (Choi et al. 2007b).

GNPs. In a biodistribution study of PEG-coated GNPs, non-targeted organs eliminated NPs slowly; however, the percentages of GNPs in the liver and spleen increased slightly after 7 days. Additionally, PEG-coated GNPs were trapped in liver Kupffer cells and spleen macrophages, and the number of PEG-coated GNPs in the cells increased in a time-dependent manner after treatment. This accumulative property does not seem to be related to dosage (Cho et al. 2009).

CNTs. SWCNTs coated with PEG2000 and PEG5400 having radioactive ^{64}Cu were studied. SWCNT-PEG5400 had a longer blood circulation time. Both SWCNT conjugates have prominent uptake rates in liver and spleen and low uptake rates in the tumor, muscle, bone, skin, and other organs. In comparison, SWCNT-PEG5400 has a lower uptake rate in liver than SWCNT-PEG2000 does (Liu et al. 2007). The biodistribution of diethylenetriamine-pentaacetic dianhydride–functionalized MWCNTs (DTPA-MWCNTs) with Indium-111 (^{111}In) radiolabels was tracked *in vivo*. Most of the NMs were in the kidneys and bladder within 30 min. At 6 h, almost all [^{111}In]DTPA-MWCNT was eliminated from the body via the renal excretion route (Lacerda et al. 2008). The biodistribution of pristine SWCNTs differs from that of chemically modified ones. Unlike the chemically modified/functionalized ones, which are cleared from the animal mostly through the renal excretion route, the pristine nanotubes are barely detectable in urine and feces. The pristine SWCNTs were mainly distributed internally in different organs. The ^{13}C-SWCNTs were apparently cleared from the bloodstream quickly and distributed into various organs within 24 h (Yang et al. 2007b).

C60. The biodistribution of C60(OH)x with 99mTc-labeling was assessed in mice and rabbits. In mice, 99mTc-C60(OH)x distributed in all organs or tissues quickly, except for those whose tissue has limited blood flow, such as the brain and muscle. All tissues slowly cleared the particles except for bone, which had a slight increase in particles within 24 h. Relatively high radioactivity was found in liver, kidney, and intestines, indicating that 99mTc-C60(OH)x might be excreted through the urine and gut. In rabbits, 99mTc-C60(OH)x particles were found mainly in liver, kidney, and bone, especially cortical bone, spine, and bone joints; most 99mTc-C60(OH)x was excreted through urine (Qingnuan 2002). Carboxylic acid–derivatized C60s are retained in muscle and fur and can penetrate the blood–brain barrier (Yamago et al. 1995). However, C60(OH)x particles are distributed mainly in bone, liver, and spleen, with little uptake in muscle and fur,

and cannot penetrate the blood–brain barrier. Carboxylic acid derivative was excreted mainly through the intestinal tract, and C60(OH)x was excreted mainly through urine.

Graphene oxide (GO). Intravenously administered ^{188}Re-GO is cleared from the bloodstream of mice rapidly and distributed to most of the organs within 48 h (Zhang et al. 2011). Particles are mainly found in the lungs, liver, and spleen; fewer particles are found in the brain, heart, and bones. The amount of GO in organs decreases with time, except in the liver and spleen. This finding demonstrates the rapid uptake of GO by the mononuclear phagocytes in the reticuloendothelial system (RES). A relatively high amount of GO was found in urine within 12 h. The half-life of GO in mice is longer than that of SWCNT (Liu et al. 2007; Singh et al. 2006) and C60 (Ji et al. 2006; Qingnuan 2002). This difference may be attributable to the different surface chemistry of GO or to its different physicochemical properties, such as size, water dispersion, and structure (Zhang et al. 2011).

5.3.2 Immune Responses of Mammals to Foreign NMs

The immunotoxicity of NMs should be well evaluated before biomedical applications in humans. Inhaled MWCNTs can suppress systemic immune function via activation of cyclooxygenase enzymes in the spleen, which results from the activation of TGF-β in the lungs (Mitchell et al. 2009). Bound proteins on NMs result in different immune responses. Proteins around the particle determine the NM uptake by various cells of the immune system and influence how they interact with blood components (Gref et al. 1994; Leu et al. 1984). If NMs' surfaces are not modified to prevent the adsorption of opsonins, then a rapid removal by macrophage cells will occur. Other immune cells and removal mechanisms would be simultaneously stimulated to remove NMs from the bloodstream.

Physicochemical properties of NMs influence their immunotoxicity. Smaller NMs (~25 nm) travel through the lymphatic system more readily than do larger particles (~100 nm), and they accumulate in the lymph nodes' resident dendritic cells (Reddy et al. 2007). The surface charge of liposomes also makes a difference. Cationic liposomes generate a greater immune response than do anionic or neutral liposomes (Nakanishi et al. 1999). Solid lipid nanoparticle (SLN)–encapsulated antisense oligodeoxyribonucleotide G3139 had greater immunostimulatory and antitumor activity than did free (i.e., nonencapsulated) G3139. Because of SLNs' small size, tumor resident macrophages and dendritic cells take them up efficiently (Pan et al. 2008).

5.3.3 Organ Damage Induced by NM Administration

NMs can penetrate various barriers and enter different organs. The oxidative stress signaling activated by NMs will directly affect the functions of organs in which they are located and may cause systematic toxicity through blood circulation of these stress signaling molecules.

MNPs. The toxicity of MNPs can be decreased by modification. For example, the toxicity of MNPs was effectively lowered by encapsulating them in poly(D,L-lactide) (Gajdosíková et al. 2006). Many MNPs have shown potential as tumor targeting and imaging molecules without much toxicity (Muldoon et al. 2005; Weissleder et al. 1989); however, some of them cause acute toxicity (Jain et al. 2008).

GNPs. GNPs ranging from 8 to 37 nm induce severe sickness in mice (Chen et al. 2009). Pathologic examination of the major organs of the mice revealed an increase of Kupffer cells in the liver, a loss of structural integrity in the lungs, and diffusion of white pulp in the spleen. The pathologic abnormality was associated with the presence of GNPs at the diseased sites. After modifying the surface of the GNPs by incorporating immunogenic peptides, the toxicity of the GNPs was reduced. The toxicity of 13.5 nm GNPs in

mice was evaluated by using different administration models and various doses. At low concentrations, GNPs caused no toxicity *in vivo*. At high concentrations, they caused decreases in body weight, red blood cell numbers, and hematocrit values. The mice that received oral and intraperitoneal doses experienced higher toxicity than those dosed through tail vein injections (Zhang et al. 2010).

CNTs. CNTs can reach the pleural cavity or the peritoneum, resulting in chronic granulomatous inflammation, which might be the forerunner of mesothelioma (Poland et al. 2008; Takagi et al. 2008). When mice inhaled CNTs, dose-dependent epithelioid granulomas were found (Lam et al. 2004). Furthermore, the inhaled CNTs induce secondary platelet activation in the systemic circulation and promote atherosclerosis (Erdely 2008; Li et al. 2007). CNTs could also induce reversible reproductive toxicity in adult males (Figure 5.3) (Bai et al. 2010). As is the case

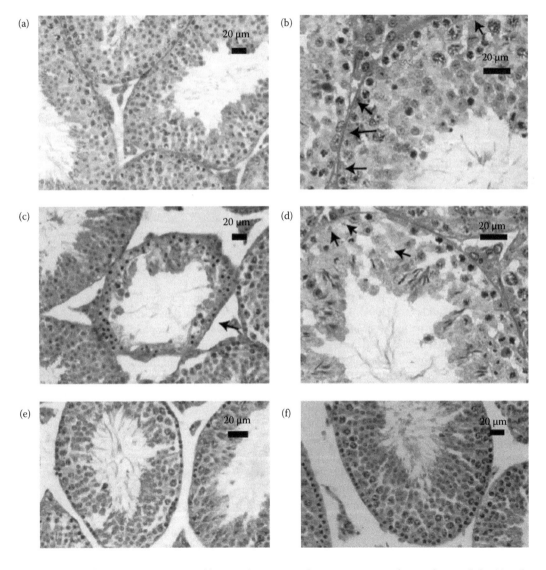

FIGURE 5.3 **(See color insert.)** MWCNTs cause reversible testis damage. Histology cross sections of seminiferous tubules (a) and an enlarged view (b) from testes of control mice show normal Sertoli cells. The decreased germinative layer thickness (c) and vacuolization of Sertoli cells (arrows in d) were observed on day 15 after five doses of MWCNT-COOH treatment. On days 60 (e) and 90 (f), most alterations disappeared, indicating a general recovery from early damage. (Reprinted with permission from Bai, Y. et al., *Nat. Nanotechnol.*, 5, 683–689, 2010. Copyright (2010) Nature Publishing Group.)

with other NMs, the toxicity of CNTs is related to their surface modification. When properly functionalized, CNTs can be used in the body as imaging agents with no severe toxicity for more than a month (Schipper et al. 2008).

C60. Aqueous C60 suspensions without organic solvent cause no acute or subacute toxicity in rats; thus they may have potential as powerful liver-protective agents (Gharbi et al. 2005). When 2000 mg/kg of C60 is administered to rats, no deaths, abnormalities, or body weight differences are observed (Mori et al. 2006). Study shows that C60 derivatives are more toxic than pristine C60 (Rancan et al. 2002; Schuster et al. 1996).

GO. GO does not cause any obvious toxicity in mice at a dose between 0.1 and 0.25 mg. However, a 0.4-mg dose causes lung granuloma andchronic toxicity with death (Wang et al. 2011).

5.4 Methods for Assessing Toxicity of NMs

The methods used to evaluate the toxicity of NMs are different from those used for small molecule toxicants because of NMs' unique physicochemical properties. Methods have been developed to study the *in vitro* and *in vivo* toxicity of NMs. For instance, the imaging approaches in medical use are also applied in toxicity studies, and the inorganic nature of some NMs enables various elemental analysis methods to be used. However, more efficient methods still need to be developed for fast and thorough evaluation of nanotoxicity in biological matrices.

5.4.1 Evaluating the *In Vitro* Toxicity of NMs

5.4.1.1 Cell Uptake

Transmission electron microscopy (TEM) is a method used for qualitative analysis. Detailed ultrastructural information, such as the size and morphology of NMs and the location of NMs inside cells, can be obtained (Shukla et al. 2005). TEM also provides information on cell uptake pathways (Nan et al. 2008). However, its application is limited to electron-dense NMs.

Inductively coupled plasma atomic emission spectroscopy (ICP-AES) is a method used to quantify internalized NMs according to each element's emission spectrum. It can be used with high sensitivity to identify the elements in NMs (Hauck et al. 2008; Matuszewski et al. 2005). However, this technique cannot provide any spatial information.

Inductively coupled plasma mass spectrometry (ICP-MS) is highly sensitive and capable of detecting the presence of a range of metals and several nonmetals at ppt concentrations. Similar to ICP-AES, sample preparation before NM component quantification is required (Stayton 2009). Compared with several other quantitative analytical methods, ICP-MS has a lower detection limit, wider dynamic range, and higher precision. This method has the advantage of enabling the simultaneous determination of multiple elements.

Fluorescence detection can be used to quantify NM uptake and observe NMs' location. Quantitative assessment can be achieved in a manner similar to that used for ICP-AES by using bulk fluorescence or, on a single cell, confocal fluorescence. It is also possible to correlate NMs counting directly with cell numbers or sort cells on the basis of NM uptake by using fluorescence-activated cell sorting (FACS). Detection is limited by the fluorescent properties of the NM species and the collection efficiency of the instrument. The spatial resolution is diffraction-limited, 200 nm at best (Fernandez-Suarez and Ting 2008). The inherent fluorescent properties of certain NMs allow simple analysis such as that of the uptake of QDs into human MSCs (Seleverstov et al. 2006).

5.4.1.2 Cell Proliferation and Apoptosis

Cellular reduction of tetrazolium salts to produce formazan dyes is widely used as an *in vitro* nanotoxicity assessment. The production of formazan-based dyes is monitored by optical absorbance as a measurement of cellular metabolism, which is used to assess the percentage of metabolically active cells. 3-(4,5-Dimethylthiazol-2-yl)-2,5-diphenyltetrazolium bromide (MTT) is one of the most commonly used dyes for this purpose (Tim 1983). The MTT method, however, suffers from one disadvantage—that formazan is insoluble but should be dissolved before the absorbance measurement—which seems to limit its application. Similar to MTT, XTT (Scudiero et al. 1988), MTS (Berg et al. 1994), and WST-1 (Ishiyama et al. 1995) are substrates of mitochondrial dehydrogenase and produce a highly water-soluble formazan from metabolically active cells, allowing a direct and user-friendly colorimetric measurement of cell viability and proliferation. The alamar blue assay has been used in colorimetric metabolic assay used to ascertain cell proliferation via the bioreduction of the nonfluorescent alamar blue dye to a pink fluorescent dye. The active cell number is counted by either optical absorbance measurements or fluorescence detection (Shvedova et al. 2003).

Cell apoptosis assays include the annexin-V assay, DNA laddering, Comet assay, and TUNEL assay. Annexin-V (Koopman et al. 1994) is a phosphatidylserine (PS)-specific binding substrate that translocates to the exterior of cells in the early stage of apoptosis because of the restructuring of the plasma membrane. When labeled with a fluorescent tag, such as FITC, annexin-V can be used as a probe to detect PS exposed in combination with propidium iodide. The apoptotic cell amount can then be measured by flow cytometry. Inspection of morphologic changes during apoptosis is the least instrumentally intensive method for characterizing apoptosis, requiring only a light microscope and visual inspection. Despite the low cost, this method is not used widely in nanotoxicology because of its time-intensive nature.

5.4.1.3 High-Content Screening Assay

High-content screening assay is useful for screening toxicity in potential drug applications and has been recently applied to screening of NMs. This high-throughput technique uses automated, commercially available instruments to perform computer-assisted microscopic image analysis (Jan et al. 2008).

5.4.2 Evaluating *In Vivo* Toxicity of NMs

5.4.2.1 Biodistribution and Circulation

Suitable *in vivo* detection methods of NMs include radioactive tracing, ICP-MS, atomic absorption spectroscopy, and MRI.

The radioactive tracing technique, having the advantages of high sensitivity, credibility, and freedom from interference, has been widely used to obtain information about the behavior of NMs *in vivo*. By using GO radiolabeled with ^{188}Re, the radioactivity of each tissue has been measured with a gamma-ray counter to demonstrate the pharmacokinetics of GO in mice (Zhang et al. 2011). This strategy has been applied to many NMs. Using a similar method, researchers have uncovered the distribution and other characteristics of a variety of carbon NMs, including C60 (Xu et al. 2007), SWCNT (Yang et al. 2007b), MWCNT (Gao et al. 2011).

ICP-MS is also used to determine the metal NM content and to analyze its biodistribution and blood circulation (Jain et al. 2008). Using atomic absorption spectroscopy, researchers have investigated the biodistribution of GNPs and Au/SiO$_2$ and the circulation kinetics of GNPs in the bloodstream of rabbits (Terentyuk et al. 2009). Additionally, MRI results have revealed the distribution of iron oxide agents (Muldoon et al. 2005). Furthermore, real-time intraoperative NIR fluorescence imaging provides highly sensitive and real-time images, enabling monitoring of the translocation and distribution of NMs (Choi et al. 2010).

5.4.2.2 Immune Responses and Organ Functions

The immunotoxicity of traditional drugs has been evaluated by using the local lymph node assay (LLNA) and plaque-forming cell (PFC) assay (Jack 1997). However, the different portals of entry inevitably enable NMs to target different populations of cells (Dobrovolskaia and McNeil 2007) and, thus, the standard LLNA test is not recommended for NMs. The lymph node proliferation assay has been used to predict drug immunotoxicity in humans (Weaver et al. 2005) and is recommended for NMs. The effects of NMs on the immune system may be well predicted by the PFC assay. Reticuloendothelial uptake and tests of macrophage function are also useful methods for evaluating NMs (Dobrovolskaia et al. 2009). The systemic immunological responses of mice injected with CNTs containing impurities were revealed by monitoring changes in peripheral T-cell subset and peripheral cytokine levels (Koyama 2009). In this study, sampled blood was FACS-sorted to separate CD4 and CD8 T lymphocytes, and then nine different cytokines related to inflammatory reactions were measured by ELISA.

In assessing organ toxicity, histologic examination is a well-accepted method. The level of oxidative stress is also a toxicity indicator in organs. Both assays can be considered conventional and all-purpose strategies for evaluating organ and tissue toxicity. Corresponding assays should be used to assess effects within specific organs. For example, liver function tests including ALP, alanine transaminase, and aspartate transaminase were used to assess the liver's function (Sahu 2009). The lungs of vehicle- and particle-exposed rats have been assessed by using bronchoalveolar lavage fluid biomarkers, oxidant and glutathione endpoints, and airway and lung parenchymal cell proliferation methods and histopathological evaluation (Sayes et al. 2007). The results of complete blood count tests may reflect the NMs' influence on blood circulation (Schipper et al. 2008).

5.5 Strategies to Reduce Adverse Effects of NMs

Understanding nanotoxicity will not control the adverse effects of NMs; strategies to reduce the toxicity are needed. Because the toxicity of NMs is related to their chemical compositions, sizes, shapes, and surface chemistry, regulating these parameters should help reduce their toxicity (Hussain et al. 2009).

5.5.1 Chemical Compositions, Sizes, and Shapes

The composition of NMs plays an important role in producing toxicity. Four typical NMs have been used to explore the interrelationship between particle size, shape, chemical composition and toxicity: carbon black (CB), SWCNTs, silicon dioxide (SiO$_2$) and zinc oxide (ZnO) NMs (Yang et al. 2009). ZnO causes high oxidative stress and much greater cytotoxicity compared to nonmetal NMs. Compared with ZnO NMs, SWCNTs are moderately cytotoxic but induce more DNA damage. CB and SiO$_2$ are less toxic.

Size-dependent cellular interactions with SNPs (15, 30, 55 nm) have also been evaluated. Cell viability significantly decreases after 24 h of exposure of 15 and 30 nm SNPs at doses ranging from 10 to 75 µg/mL, and ROS levels increased 10-fold in cells exposed to 15 nm SNPs at a dose of 50 µg/mL. These results indicate that size-dependent toxicity induced by SNPs is mediated through oxidative stress (Carlson 2008).

The cytotoxicity of SWCNTs appears to be greater than that of MWCNTs, which is greater than that of quartz, which is greater than that of C60. SWCNTs significantly impair phagocytosis at the low dose of 0.38 µg/cm^2, whereas MWCNTs and C60 induce injury only at the high dose of 3.06 µg/cm^2. Exposure to SWCNTs or MWCNTs induces necrosis and apoptosis. These results show that carbon NMs with different shapes exhibit quite different cytotoxicity profiles *in vitro*, and that spherical NMs have lower cytotoxicity (Jia et al. 2005). However, gold nanorods are more cytotoxic than gold nanospheres when used at the same concentration in a human cancer cell line (Alkilany et al. 2009).

5.5.2 Surface Chemistry

Altering the surface chemistry of NMs is an effective method of affecting toxicity. Conjugation with proteins, polymers, or small molecules alters the biological effects of NMs by altering the surface-dominant material. Furthermore, surface functionalization is often easier than changing chemical compositions, sizes, and shapes.

In addition to a ZnS shell, capping molecules have been used to reduce toxicity of QDs in many cases. *N*-Acetylcysteine–coated CdTe-QDs reduce CdTe-induced Fas up-regulation and apoptosis and decrease cytotoxicity in neuroblastoma cells (Choi et al. 2007a). CdTe-QDs coated with mercaptopropionic acid and cysteamine are less toxic to PC12 cells than are uncoated QDs (Lovric et al. 2005). Coating CdSe/ZnS with dihydroxylipoic acid also reduces the toxicity of QDs in several cell lines (Voura et al. 2004). These effective capping molecules all appear to be good antioxidants, supporting a role for oxidative stress in QD toxicity (Rzigalinski and Strobl 2009).

PEGylation is a common strategy used to improve water solubility and biocompatibility of NMs (Yang et al. 2008). After SWCNTs, gold nanospheres, and gold nanorods are conjugated with PEG-grafted branched polymers, they are highly stable in aqueous solutions at different pH values, at elevated temperatures, and in serum. PEG-coated SWCNTs have a half-life of 22.1 h in the blood after intravenous injection into mice, exceeding the previous record of 5.4 h (Prencipe et al. 2009). The long blood circulation time suggests that PEGylation delays the clearance of NMs by the RES of mice.

Combinatorial and high-throughput approaches have shown great promise in drug discovery. Recently, they were applied to the field of nanotechnology to minimize the unwanted toxicity of NMs (Zhou et al. 2008). Combinatorial modifications of the NMs' surface enable us to map unknown chemical space more effectively, rapidly discover NMs with reduced toxicity (Figure 5.4), and reveal quantitative nanostructure–activity relationships (QNARs) at the same time.

5.6 Perspectives

The unique optical, magnetic, and thermal properties and the multifunctionalization capability of NMs make them promising candidates for use in the diagnosis and therapy of catastrophic diseases. However, the reasonable concerns about the toxicity of NMs demand that these tools be developed rationally and cautiously to prevent nanotoxicity. Unlike most traditional

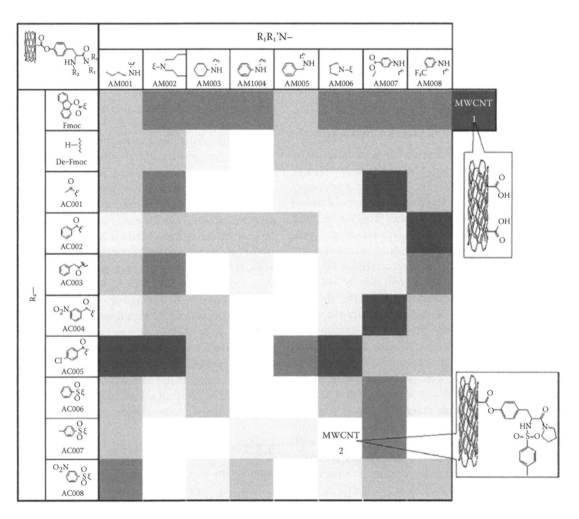

FIGURE 5.4 Heat map for nitrogen monoxide (NO) generation of MWCNT 1 (MWCNT-COOH) and an 80-member combinatorial MWCNT library. MWCNT-2 has much less NO generation than MWCNT-1. (Reprinted with permission from Gao, N. et al., *ACS Nano*, 5, 4581–4591, 2011. Copyright (2011) American Chemical Society.)

small-molecule contrast agents, NMs are three-dimensional, bind multiple biomacromolecules, and penetrate various biological barriers. Therefore, understanding and eliminating nanotoxicity will require the collaboration of scientists in material science, chemistry, and biology. Work in the near future should be focused on four major areas (discussed below).

The first priority is to fully characterize the integrity of NMs outside and inside biologic systems. Unlike small molecules, NMs are nonuniform in nature; they can have a variety of sizes, shapes, and surfaces. The same component tends to aggregate or degrade in aqueous solutions depending on environment. Therefore, it is imperative to confirm the integrity and alterations of NMs. Furthermore, NMs encounter an abundant amount of proteins and other biomolecules, making the determination of NM-biomolecule complexes as a novel biological identity also necessary for understanding their behavior *in vivo*.

The second priority is to speed up the evaluation of nanotoxicity. Traditional approaches for evaluating *in vitro* and *in vivo* effects of NMs are slow and may generate artifacts because of interference of NMs with the assay systems. Therefore, it is urgent to develop novel approaches such as RT-CES technology and other high-throughput and high-content screening methods. Using high-throughput screening systems, the *in vitro* and *in vivo* (such as in zebrafish) effects of various types of NMs can be determined quickly. Based on the biological effects of NMs with various components, size, and surface, etc., the QNAR can be established to obtain the relationship of biological activities with NMs' structural parameters. Therefore, prediction of nanotoxicity will be possible by using the QNAR approach.

The third priority of future NM studies is to correlate nanotoxicity at the cellular level to that at the system level. It is not surprising that the cellular results thus far are not consistent with the activities in live animals. Correlations of genotypic changes with phenotypic alterations, molecular interactions (e.g., plasma protein binding) with cellular effects, and all such properties with *in vivo* biodistribution and organ damages are required for an accelerated evaluation of nanotoxicity. Such information will provide much-needed understanding about the mechanisms of NMs' bioactivities.

The fourth priority of NM studies in the near future is to determine how to control the toxicity of NMs. If the research tasks mentioned above are completed, then we should know enough about the toxicity of NMs to modulate their physicochemical parameters in such a way as to minimize toxicity. For example, the combinatorial approach, which is well developed in drug discovery and is now used with NMs, could be used to modify NMs' surface to reduce their cytotoxicity and immunotoxicity. From such efforts, a QNAR relationship can be established. Ultimately, determining the safety of NMs and controlling nanotoxicity will minimize hazards to humans while enabling the healthy growth of medicinal applications of nanotechnology.

Acknowledgments

This work was supported by the American Lebanese Syrian Associated Charities (ALSAC), the National Basic Research Program of China (2010CB933504), and National Natural Science Foundation of China (90913006, 21077068, and 21137002).

References

Alkilany, A. M., P. K. Nagaria, C. R. Hexel et al. 2009. Cellular uptake and cytotoxicity of gold nanorods: Molecular origin of cytotoxicity and surface effects. *Small* 5:701–708.

Bai, Y., Y. Zhang, J. Zhang et al. 2010. Repeated administrations of carbon nanotubes in male mice cause reversible testis damage without affecting fertility. *Nature Nanotechnology* 5:683–689.

Berg, K., L. Zhai, M. Chen, A. Kharazmi, and T. C. Owen. 1994. The use of a water-soluble formazan complex to quantitate the cell number and mitochondrial function of *Leishmania major* promastigotes. *Parasitology Research* 80: 235–239.

Borm, P. J. A., and W. Kreyling. 2004. Toxicological hazards of inhaled nanoparticles—potential implications for drug delivery. *Journal of Nanoscience and Nanotechnology* 4:521–531.

Carlson, C., S. M. Hussain, A. M. Schrand et al. 2008. Unique cellular interaction of silver nanoparticles: Size-dependent generation of reactive oxygen species. *Journal of Physical Chemistry B* 112:13608–13619.

Chakravarty, P., R. Marches, N. S. Zimmerman et al. 2008. Thermal ablation of tumor cells with antibody-functionalized single-walled carbon nanotubes. *Proceedings of the National Academy of Sciences* 105:8697–8702.

Chan, W. C. W., B. D. Chithrani, and A. A. Ghazani. 2006. Determining the size and shape dependence of gold nanoparticle uptake into mammalian cells. *Nano Letters* 6:662–668.

Chen, Y. S., Y. C. Hung, I. Liau, and G. S. Huang. 2009. Assessment of the in vivo toxicity of gold nanoparticles. *Nanoscale Research Letters* 4:858–864.

Cheng, J., K. A. S. Fernando, L. M. Veca et al. 2008. Reversible accumulation of PEGylated single-walled carbon nanotubes in the mammalian nucleus. *ACS Nano* 2:2085–2094.

Cho, W. S., M. Cho, J. Jeong et al. 2009. Acute toxicity and pharmacokinetics of 13 nm-sized PEG-coated gold nanoparticles. *Toxicology and Applied Pharmacology* 236:16–24.

Choi, A., S. J. Cho, J. Desbarats, J. Lovric, and D. Maysinger. 2007a. Quantum dot-induced cell death involves Fas upregulation and lipid peroxidation in human neuroblastoma cells. *Journal of Nanobiotechnology* 5:1.

Choi, H. S., W. Liu, P. Misra et al. 2007b. Renal clearance of quantum dots. *Nature Biotechnology* 25:1165–1170.

Choi, H. S., Y. Ashitate, J. H. Lee et al. 2010. Rapid translocation of nanoparticles from the lung airspaces to the body. *Nature Biotechnology* 28:1300–1303.

De La Zerda, A., C. Zavaleta, S. Keren et al. 2008. Carbon nanotubes as photoacoustic molecular imaging agents in living mice. *Nature Nanotechnology* 3:557–562.

Deng, Z. J., M. T. Liang, M. Monteiro, I. Toth, and R. F. Minchin. 2011. Nanoparticle-induced unfolding of fibrinogen promotes Mac-1 receptor activation and inflammation. *Nature Nanotechnology* 6:39–44.

Dobrovolskaia, M. A., and S. E. McNeil. 2007. Immunological properties of engineered nanomaterials. *Nature Nanotechnology* 2:469–478.

Dobrovolskaia, M. A., D. R. Germolec, and J. L. Weaver. 2009. Evaluation of nanoparticle immunotoxicity. *Nature Nanotechnology* 4:411–414.

Duguet, E., S. Vasseur, S. Mornet, and J.-M. Devoisselle. 2006. Magnetic nanoparticles and their applications in medicine. *Nanomedicine* 1:157–168.

Erdely, A., T. Hulderman, R. Salmen et al. 2008. Cross-talk between lung and systemic circulation during carbon nanotube respiratory exposure. Potential biomarkers. *Nano Letters* 9:36–43.

Fernandez-Suarez, M., and A. Y. Ting. 2008. Fluorescent probes for super-resolution imaging in living cells. *Nature Reviews Molcular Cell Biology* 9:929–943.

Foldbjerg, R., P. Olesen, M. Hougaard et al. 2009. PVP-coated silver nanoparticles and silver ions induce reactive oxygen species, apoptosis and necrosis in THP-1 monocytes. *Toxicology Letters* 190:156–162.

Franzen, S., A. G. Tkachenko, H. Xie et al. 2004. Cellular trajectories of peptide-modified gold particle complexes: Comparison of nuclear localization signals and peptide transduction domains. *Bioconjugate Chemistry* 15:482–490.

Gajdosíková, A., A. Gajdosík, M. Koneracká et al. 2006. Acute toxicity of magnetic nanoparticles in mice. *Neuroendocrinology Letters Supplement* 2:96–99.

Gao, N., Q. Zhang, Q. Mu et al. 2011. Steering carbon nanotubes to scavenger receptor recognition by nanotube surface chemistry modification partially alleviates NFκB activation and reduces its immunotoxicity. *ACS Nano* 5:4581–4591.

Gharbi, N., M. Pressac, M. Hadchouel et al. 2005. [60]Fullerene is a powerful antioxidant in vivo with no acute or subacute toxicity. *Nano Letters* 5:2578–2585.

Ghibelli, L., M. De Nicola, D. M. Gattia et al. 2007. Effect of different carbon nanotubes on cell viability and proliferation. *Journal of Physics: Condensed Matter* 19:395013.

Girard, D., D. M. Goncalves, and S. Chiasson. 2010. Activation of human neutrophils by titanium dioxide (TiO(2)) nanoparticles. *Toxicology In Vitro* 24:1002–1008.

Gratton, S. E. A., P. A. Ropp, P. D. Pohlhaus et al. 2008. The effect of particle design on cellular internalization pathways. *Proceedings of the National Academy of Sciences* 105:11613–11618.

Gref, R., Y. Minamitake, M. T. Peracchia et al. 1994. Biodegradable long-circulating polymeric nanospheres. *Science* 263:1600.

Hauck, T. S., A. A. Ghazani, and W. C. W. Chan. 2008. Assessing the effect of surface chemistry on gold nanorod uptake, toxicity, and gene expression in mammalian cells. *Small* 4:153–159.

Huff, T. B., M. N. Hansen, Y. Zhao, J. X. Cheng, and A. Wei. 2007. Controlling the cellular uptake of gold nanorods. *Langmuir* 23:1596–1599.

Hussain, S. M., L. K. Braydich-Stolle, A. M. Schrand et al. 2009. Toxicity evaluation for safe use of nanomaterials: Recent achievements and technical challenges. *Advanced Materials* 21:1549–1559.

Ishiyama, M., H. Tominaga, M. Shiga et al. 1995. Novel cell proliferation and cytotoxicity assays using a tetrazolium salt that produces a water-soluble formazan dye. *Toxicology In Vitro* 8:187–190.

Iwai, N., and H. Yamawaki. 2006. Cytotoxicity of water-soluble fullerene in vascular endothelial cells. *American Journal of Physioliogy – Cell Physiology* 290:C1495–C1502.

Jack, H. D. 1997. Issues with introducing new immunotoxicology methods into the safety assessment of pharmaceuticals. *Toxicology* 119: 95–101.

Jain, T. K., M. K. Reddy, M. A. Morales, D. L. Leslie-Pelecky, and V. Labhasetwar. 2008. Biodistribution, clearance, and biocompatibility of iron oxide magnetic nanoparticles in rats. *Molecular Pharmaceutics* 5:316–327.

Jan, E., S. J. Byrne, M. Cuddihy et al. 2008. High-content screening as a universal tool for fingerprinting of cytotoxicity of nanoparticles. *ACS Nano* 2:928–938.

Ji, Z. Q., H. Sun, H. Wang et al. 2006. Biodistribution and tumor uptake of C60(OH)x in mice. *Journal of Nanoparticle Research* 8:53–63.

Jia, G., H. Wang, L. Yan et al. 2005. Cytotoxicity of carbon nanomaterials: Single-wall nanotube, multi-wall nanotube, and fullerene. *Environmental Science & Technology* 39:1378–1383.

Kam, N. W. S., and H. Dai. 2005. Carbon nanotubes as intracellular protein transporters: Generality and biological functionality. *Journal of the American Chemical Society* 127:6021–6026.

Kam, N. W. S., Z. Liu, and H. Dai. 2005. Functionalization of carbon nanotubes via cleavable disulfide bonds for efficient intracellular delivery of siRNA and potent gene silencing. *Journal of the American Chemical Society* 127: 12492–12493.

Kim, J. S., T. J. Yoon, K. N. Yu et al. 2006. Toxicity and tissue distribution of magnetic nanoparticles in mice. *Toxicological Sciences* 89:338.

Koopman, G., C. P. M. Reutelingsperger, G. A. M. Kuijten et al. 1994. Annexin-V for flow cytometric detection of phosphatidylserine expression on B-cells undergoing apoptosis. *Blood* 84:1415–1420.

Kostarelos, K., L. Lacerda, G. Pastorin et al. 2007. Cellular uptake of functionalized carbon nanotubes is independent of functional group and cell type. *Nature Nanotechnology* 2: 108–113.

Koyama, S., Y. A. Kim, T. Hayashi et al. 2009. In vivo immunological toxicity in mice of carbon nanotubes with impurities. *Carbon* 47:1365–1372.

Kwon, J. T., S. K. Hwang, H. Jin et al. 2008. Body distribution of inhaled fluorescent magnetic nanoparticles in the mice. *Journal of Occupational Health* 50:1–6.

Lacerda, L., A. Soundararajan, R. Singh et al. 2008. Dynamic imaging of functionalized multi-walled carbon nanotube systemic circulation and urinary excretion. *Advanced Materials* 20:225–230.

Lam, C. W., J. T. James, R. McCluskey, and R. L. Hunter. 2004. Pulmonary toxicity of single-wall carbon nanotubes in mice 7 and 90 days after intratracheal instillation. *Toxicological Sciences* 77:126.

Leu, D., B. Manthey, J. Kreuter, P. Speiser, and P. P. Deluca. 1984. Distribution and elimination of coated polymethyl [2-14C] methacrylate nanoparticles after intravenous injection in rats. *Journal of Pharmaceutical Sciences* 73:1433–1437.

Li, N., T. Xia, and A. E. Nel. 2008. The role of oxidative stress in ambient particulate matter-induced lung diseases and its implications in the toxicity of engineered nanoparticles. *Free Radical Biology and Medicine* 44:1689–1699.

Li, Z., T. Hulderman, R. Salmen et al. 2007. Cardiovascular effects of pulmonary exposure to single-wall carbon nanotubes. *Environmental Health Perspectives* 115:377.

Liu, D., C. Yi, D. Zhang, J. Zhang, and M. Yang. 2010. Inhibition of proliferation and differentiation of mesenchymal stem cells by carboxylated carbon nanotubes. *ACS Nano* 4:2185–2195.

Liu, Y., D. C. Wu, W. D. Zhang et al. 2005. Polyethylenimine grafted multiwalled carbon nanotubes for secure noncovalent immobilization and efficient delivery of DNA. *Angewandte Chemie* 117:4860–4863.

Liu, Z., W. Cai, L. He et al. 2007. in vivo biodistribution and highly efficient tumour targeting of carbon nanotubes in mice. *Nature Nanotechnology* 2:47–52.

Lockman, P. R., J. M. Koziara, R. J. Mumper, and D. D. Allen. 2004. Nanoparticle surface charges alter blood–brain barrier integrity and permeability. *Journal of Drug Targeting* 12:635–641.

Lovric, J., S. J. Cho, F. M. Winnik, and D. Maysinger. 2005. Unmodified cadmium telluride quantum dots induce reactive oxygen species formation leading to multiple organelle damage and cell death. *Chemistry & Biology* 12:1227–1234.

Manna, S. K., S. Sarkar, J. Barr et al. 2005. Single-walled carbon nanotube induces oxidative stress and activates nuclear transcription factor-kappaB in human keratinocytes. *Nano Letters* 5:1676–1684.

Matsumoto, K., C. Sato, R. L. D. Whitby, and N. Shimizu. 2009. Carbon nanotubes stimulate neurite outgrowths of neurons by activation of MAPK/ERK signal transduction. *New Biotechnology* 25:S27–S28.

Matuszewski, L., T. Persigehl, A. Wall et al. 2005. Cell tagging with clinically approved iron oxides: Feasibility and effect of lipofection, particle size, and surface coating on labeling efficiency 1. *Radiology* 235:155.

Mitchell, L., F. Lauer, S. Burchiel, and J. McDonald. 2009. Mechanisms for how inhaled multiwalled carbon nanotubes suppress systemic immune function in mice. *Nature Nanotechnology* 4:451–456.

Mitragotri, S., J. A. Champion, and Y. K. Katare. 2007. Making polymeric micro- and nanoparticles of complex shapes. *Proceedings of the National Academy of Sciences of the United States of America* 104:11901–11904.

Mooney, E., P. Dockery, U. Greiser, M. Murphy, and V. Barron. 2008. Carbon nanotubes and mesenchymal stem cells: Biocompatibility, proliferation and differentiation. *Nano Letters* 8:2137–2143.

Mori, T., H. Takada, S. Ito et al. 2006. Preclinical studies on safety of fullerene upon acute oral administration and evaluation for no mutagenesis. *Toxicology* 225:48–54.

Mu, Q., D. L. Broughton, and B. Yan. 2009a. Endosomal leakage and nuclear translocation of multiwalled carbon nanotubes: Developing a model for cell uptake. *Nano Letters* 9:4370–4375.

Mu, Q., G. Du, T. Chen, B. Zhang, and B. Yan. 2009b. Suppression of human bone morphogenetic protein signaling by carboxylated single-walled carbon nanotubes. *ACS Nano* 3:1139–1144.

Mu, Q., L. Yang, J. C. Davis et al. 2010. Biocompatibility of polymer grafted core/shell iron/carbon nanoparticles. *Biomaterials* 31:5083–5090.

Muldoon, L. L., M. Sàndor, K. E. Pinkston, and E. A. Neuwelt. 2005. Imaging, distribution, and toxicity of superparamagnetic iron oxide magnetic resonance nanoparticles in the rat brain and intracerebral tumor. *Neurosurgery* 57:785.

Nakanishi, T., J. Kunisawa, A. Hayashi et al. 1999. Positively charged liposome functions as an efficient immunoadjuvant in inducing cell-mediated immune response to soluble proteins. *Journal of Controlled Release* 61:233–240.

Nan, A., X. Bai, S. J. Son, S. B. Lee, and H. Ghandehari. 2008. Cellular uptake and cytotoxicity of silica nanotubes. *Nano Letters* 8:2150–2154.

Nel, A., T. Xia, L. Mädler, and N. Li. 2006. Toxic potential of materials at the nanolevel. *Science* 311:622–627.

Oberdörster, G., E. Oberdörster, and J. Oberdörster. 2005. Nanotoxicology: An emerging discipline evolving from studies of ultrafine particles. *Environmental Health Perspectives* 113:823–839.

Palmeira, C. M. 2008. In vitro assessment of silver nanoparticles toxicity in hepatic mitochondrial function. EOARD, Unit 4515 BOX 14, APO AE 09421.

Pan, X., L. Chen, S. Liu et al. 2008. Antitumor activity of G3139 lipid nanoparticles (LNPs). *Molecular Pharmaceutics* 6:211–220.

Park, T. H., J. A. Kim, N. H. Lee et al. 2011. Enhancement of neurite outgrowth in PC12 cells by iron oxide nanoparticles. *Biomaterials* 32:2871–2877.

Pernodet, N., X. H. Fang, Y. Sun et al. 2006. Adverse effects of citrate/gold nanoparticles on human dermal fibroblasts. *Small* 2:766–773.

Poland, C. A., R. Duffin, I. Kinloch et al. 2008. Carbon nanotubes introduced into the abdominal cavity of mice show asbestos-like pathogenicity in a pilot study. *Nature Nanotechnology* 3:423–428.

Popovtzer, R., A. Agrawal, N. A. Kotov et al. 2008. Targeted gold nanoparticles enable molecular CT imaging of cancer. *Nano Letters* 8:4593–4596.

Prencipe, G., S. M. Tabakman, K. Welsher et al. 2009. PEG branched polymer for functionalization of nanomaterials with ultralong blood circulation. *Journal of the American Chemical Society* 131:4783–4787.

Qingnuan, L. 2002. Preparation of ^{99m}Tc-C$_{60}$ (OH)$_x$ and its biodistribution studies. *Nuclear Medicine and Biology* 29:707–710.

Qu, G., Y. Bai, Y. Zhang et al. 2009. The effect of multiwalled carbon nanotube agglomeration on their accumulation in and damage to organs in mice. *Carbon* 47:2060–2069.

Rancan, F., S. Rosan, F. Boehm et al. 2002. Cytotoxicity and photocytotoxicity of a dendritic C60 mono-adduct and a malonic acid C60 tris-adduct on Jurkat cells. *Journal of Photochemistry and Photobiology B: Biology* 67:157–162.

Reddy, S. T., A. J. Van Der Vlies, E. Simeoni et al. 2007. Exploiting lymphatic transport and complement activation in nanoparticle vaccines. *Nature Biotechnology* 25:1159–1164.

Ryman-Rasmussen, J. P., M. F. Cesta, A. R. Brody et al. 2009. Inhaled carbon nanotubes reach the subpleural tissue in mice. *Nature Nanotechnology* 4:747–751.

Rzigalinski, B. A., and J. S. Strobl. 2009. Cadmium-containing nanoparticles: Perspectives on pharmacology and toxicology of quantum dots. *Toxicology and Applied Pharmacology* 238:280–288.

Sahay, G., D. Y. Alakhova, and A. V. Kabanov. 2010. Endocytosis of nanomedicines. *Journal of Controlled Release* 145:182–195.

Sahu, S. C. 2009. Hepatotoxic potential of nanomaterials. In *Nanotoxicity*, Sahu, S. C. and Casciano, D. A., eds., 183–189. Chichester, UK: John Wiley & Sons.

Sayes, C. M., J. D. Fortner, W. Guo et al. 2004. The differential cytotoxicity of water-soluble fullerenes. *Nano Letters* 4:1881–1887.

Sayes, C. M., A. A. Marchione, K. L. Reed, and D. B. Warheit. 2007. Comparative pulmonary toxicity assessments of C60 water suspensions in rats: Few differences in fullerene toxicity in vivo in contrast to in vitro profiles. *Nano Letters* 7:2399–2406.

Schipper, M. L., N. Nakayama-Ratchford, C. R. Davis et al. 2008. A pilot toxicology study of single-walled carbon nanotubes in a small sample of mice. *Nature Nanotechnology* 3:216–221.

Schmid, S. L., and S. D. Conner. 2003. Regulated portals of entry into the cell. *Nature* 422:37–44.

Schuster, D. I., S. R. Wilson, and R. F. Schinazi. 1996. Anti-human immunodeficiency virus activity and cytotoxicity of derivatized buckminsterfullerenes. *Bioorganic & Medicinal Chemistry Letters* 6:1253–1256.

Scudiero, D. A., R. H. Shoemaker, K. S. Paull et al. 1988. Evaluation of a soluble tetrazolium/formazan assay for cell growth and drug sensitivity in culture using human and other tumor cell lines. *Cancer Research* 48:4827–4833.

Seleverstov, O., O. Zabirnyk, M. Zscharnack et al. 2006. Quantum dots for human mesenchymal stem cells labeling. A size-dependent autophagy activation. *Nano Letters* 6:2826–2832.

Shukla, R., V. Bansal, M. Chaudhary et al. 2005. Biocompatibility of gold nanoparticles and their endocytotic fate inside the cellular compartment: A microscopic overview. *Langmuir* 21:10644–10654.

Shvedova, A., V. Castranova, E. Kisin et al. 2003. Exposure to carbon nanotube material: assessment of nanotube cytotoxicity using human keratinocyte cells. *Journal of Toxicology and Environmental Health Part A* 66:1909–1926.

Singh, N., B. Manshian, G. J. S. Jenkins et al. 2009. Nano-Genotoxicology: The DNA damaging potential of engineered nanomaterials. *Biomaterials* 30:3891–3914.

Singh, R., D. Pantarotto, L. Lacerda et al. 2006. Tissue biodistribution and blood clearance rates of intravenously administered carbon nanotube radiotracers. *Proceedings of the National Academy of Sciences of the United States of America* 103:3357.

Slowing, I., B. G. Trewyn, and V. S. Lin. 2006. Effect of surface functionalization of MCM-41-type mesoporous silica nanoparticles on the endocytosis by human cancer cells. *Journal of the American Chemical Society* 128:14792–14793.

Sniadecki, N. J. 2010. A tiny touch: Activation of cell signaling pathways with magnetic nanoparticles. *Endocrinology* 151:451–457.

Sohaebuddin, S., P. Thevenot, D. Baker, J. Eaton, and L. Tang. 2010. Nanomaterial cytotoxicity is composition, size, and cell type dependent. *Particle and Fibre Toxicology* 7:22.

Stayton, I., J. Winiarz, K. Shannon, and Y. Ma. 2009. Study of uptake and loss of silica nanoparticles in living human lung epithelial cells at single cell level. *Analytical and Bioanalytical Chemistry* 394:1595–1608.

Takagi, A., A. Hirose, T. Nishimura et al. 2008. Induction of mesothelioma in p53+/-mouse by intraperitoneal application of multi-wall carbon nanotube. *Journal of Toxicological Sciences* 33:105.

Terentyuk, G. S., G. N. Maslyakova, L. V. Suleymanova et al. 2009. Circulation and distribution of gold nanoparticles and induced alterations of tissue morphology at intravenous particle delivery. *Journal of Biophotonics* 2:292–302.

Tim, M. 1983. Rapid colorimetric assay for cellular growth and survival: Application to proliferation and cytotoxicity assays. *Journal of Immunological Methods* 65:55–63.

Tsuchiya, T., I. Oguri, Y. N. Yamakoshi, and N. Miyata. 1996. Novel harmful effects of [60]fullerene on mouse embryos in vitro and in vivo. *FEBS Letters* 393:139–145.

Valiyaveettil, S., P. V. AshaRani, G. L. K. Mun, and M. P. Hande. 2009. Cytotoxicity and genotoxicity of silver nanoparticles in human cells. *ACS Nano* 3:279–290.

von zur Muhlen, C., D. von Elverfeldt, N. Bassler et al. 2007. Superparamagnetic iron oxide binding and uptake as imaged by magnetic resonance is mediated by the integrin receptor Mac-1 (CD11b/CD18): Implications on imaging of atherosclerotic plaques. *Atherosclerosis* 193:102–111.

Voura, E. B., J. K. Jaiswal, H. Mattoussi, and S. M. Simon. 2004. Tracking metastatic tumor cell extravasation with quantum dot nanocrystals and fluorescence emission-scanning microscopy. *Nature Medicine* 10:993–998.

Wang, K., J. Ruan, H. Song et al. 2011. Biocompatibility of graphene oxide. *Nanoscale Research Letters* 6:8.

Wang, S., W. Lu, O. Tovmachenko et al. 2008. Challenge in understanding size and shape dependent toxicity of gold nanomaterials in human skin keratinocytes. *Chemical Physics Letters* 463:145–149.

Weaver, J. L., J. M. Chapdelaine, J. Descotes et al. 2005. Evaluation of a lymph node proliferation assay for its ability to detect pharmaceuticals with potential to cause immune-mediated drug reactions. *Journal of Immunotoxicology* 2:11–20.

Weissleder, R., D. Stark, B. Engelstad et al. 1989. Superparamagnetic iron oxide: Pharmacokinetics and toxicity. *American Journal of Roentgenology* 152:167.

Xu, J.-Y., Q.-N. Li, J.-G. Li et al. 2007. Biodistribution of 99mTc-C60(OH)x in Sprague–Dawley rats after intratracheal instillation. *Carbon* 45:1865–1870.

Yamago, S., H. Tokuyama, E. Nakamura et al. 1995. In vivo biological behavior of a water-miscible fullerene: 14C labeling, absorption, distribution, excretion and acute toxicity. *Chemistry & Biology* 2:385–389.

Yang, H., C. Liu, D. Yang, H. Zhang, and Z. Xi. 2009. Comparative study of cytotoxicity, oxidative stress and genotoxicity induced by four typical nanomaterials: The role of particle size, shape and composition. *Journal of Applied Toxicology* 29:69–78.

Yang, M. S., S. J. Qi, C. Q. Yi et al. 2007a. Effects of silicon nanowires on HepG2 cell adhesion and spreading. *ChemBioChem* 8:1115–1118.

Yang, S., W. Guo, Y. Lin et al. 2007b. Biodistribution of pristine single-walled carbon nanotubes in vivo. *Journal of Physical Chemistry* C111:17761–17764.

Yang, S.-T., K. A. S. Fernando, J.-H. Liu et al. 2008. Covalently PEGylated carbon nanotubes with stealth character in vivo. *Small* 4:940–944.

Ye, S. F., Y. H. Wu, Z. Q. Hou, and Q. Q. Zhang. 2009. ROS and NF-kappa B are involved in upregulation of IL-8 in A549 cells exposed to multi-walled carbon nanotubes. *Biochemical and Biophysical Research Communications* 379: 643–648.

Yi, C., D. Liu, C.-C. Fong, J. Zhang, and M. Yang. 2010. Gold nanoparticles promote osteogenic differentiation of mesenchymal stem cells through p38 MAPK pathway. *ACS Nano* 4:6439–6448.

Yoon, K.-Y., J. Hoon Byeon, J.-H. Park, and J. Hwang. 2007. Susceptibility constants of *Escherichia coli* and *Bacillus subtilis* to silver and copper nanoparticles. *Science of the Total Environment* 373: 572–575.

Zhang, X. D., H. Y. Wu, D. Wu, Y. Y. Wang et al. 2010. Toxicologic effects of gold nanoparticles in vivo by different administration routes. *International Journal of Nanomedicine* 5:771–781.

Zhang, X., J. Yin, C. Peng et al. 2011. Distribution and biocompatibility studies of graphene oxide in mice after intravenous administration. *Carbon* 49: 986–995.

Zhou, H., Q. Mu, N. Gao et al. 2008. A nano-combinatorial library strategy for the discovery of nanotubes with reduced protein-binding, cytotoxicity, and immune response. *Nano Letters* 8:859–865.

Zhu, Y., W. Li, Q. Li et al. 2009. Effects of serum proteins on intracellular uptake and cytotoxicity of carbon nanoparticles. *Carbon* 47:1351–1358.

6

Imaging with Nanoparticles

Amit Joshi
Baylor College of Medicine

6.1 Introduction

Noninvasive and minimally invasive biomedical imaging techniques are valuable tools for clinical diagnostics. The field of medical imaging has been dominated by structural or anatomical imaging provided by x-ray, magnetic resonance, and ultrasound (US) imaging. In recent years, molecular or functional imaging techniques represented by nuclear [single-photon emission computed tomography (SPECT)/positron emission tomography (PET)] imaging, optical molecular imaging, and contrast enhanced variants of x-ray, magnetic resonance imaging (MRI), US, and multiple hybrid imaging methods [e.g., photoacoustic imaging (PAI)] have rapidly advanced, and are in various stages of progress toward widespread clinical translation. Molecular imaging methods are expected to dominate clinical diagnostics in future. Current molecular imaging research is focused toward developing sensitive and highly specific means of visualizing cellular biochemical events for applications in early-stage cancer detection/staging, image-guided chemotherapy, guided stem cell therapies, image-guided gene therapies, and image-guided surgery/thermoablative therapies (Tallury et al. 2008; Hahn et al. 2010). The primary limitations of current medical imaging techniques include poor spatial resolution, low sensitivity, insufficient signal penetration, and inability to multiplex either the imaging targets or contrast agents (Jokerst and Gambhir 2011). At the same time, the field of nanoparticles (NPs) for medical applications has grown by leaps and bound, and it is increasingly obvious that most limitations of biomedical imaging can be alleviated by NP-mediated methods. As a result, NPs have been applied to all imaging areas, and in this chapter, we will concisely review their impact on medical imaging. NPs for biomedical applications range in sizes from 5 to 1000 nm, are bigger than proteins but smaller than typical cells, and thus exhibit

different *in vivo* pharmacokinetics and pharmacodynamics than conventional imaging and therapeutic agents (Jain et al. 2008; Jokerst and Gambhir 2011). NPs are similar in size and share functionalities with subcellular organelles such as ribosomes, proteasomes, and transport vesicles, and this has been exploited to create unique imaging and therapeutic applications (Debbage and Jaschke 2008). Compared to conventional contrast agents, NPs provide improved *in vivo* detection and enhanced molecular targeting efficiencies via long and engineered circulation times, designed size-dependent clearance and trapping pathways [e.g., enhanced permeability and retention (EPR)-based tumor accumulation], and have multimeric binding capacities to target multiple bioevents of interest and integrate multiple signaling agents of varying types in a single vehicle. Diagnosis with NPs in molecular imaging requires the correlation of the imaging signal with a disease phenotype (Jokerst and Gambhir 2011). The location or intensity of NP signals emerging from the site of interest can then indicate the size and state of the disease. The accumulations of contrast agents can be efficiently increased by confining the contrast in a nanoscale structure and exploit its favorable biodistribution and clearance profiles. The binding of NPs at sites of interest can be further increased by actively targeting to cell surface receptors or molecular phenotypes of the disease under investigation, and here the large surface provided by NPs allows for the engineering of multiple bioadhesive sites for target recognition and binding. For weak contrasts such as fluorophores or short-lived contrasts such as PET tracers, NPs allow signal amplification by storing thousands of signaling entities within their structure and make them available at the site of interest. Furthermore, NPs allow efficient combinations of differing and complementary contrasts such as MR–PET, MR–optical, or photoacoustic–optical, thus combining the wide variations in relative resolution-sensitivity properties of these

TABLE 6.1 Impact of Nanoparticles on Current Imaging Modalities

Modality	Resolution	Depth	Sensitivity (mol)	Limitations	NP solutions	NP Types
CT	50 μm	Unlimited	10^{-6}	Low soft tissue contrast, Low sensitivity	Increase contrast, Molecular targeting	Iodinated carrier NPs, Gold/silver/metal alloy NPs
Optical	1–3 mm	~1 cm	10^{-12} to 10^{-15}	Poor depth penetration, poor multiplexing, poor contrast stability	Multimodal contrast, signal amplification, stability and photobleaching resistance	Fluorophore loaded NPs, Quantum dots, carbon NPs, gold/silver NPs
MRI	50 μm	Unlimited	10^{-9} to 10^{-6}	Low sensitivity, poor multiplexing	Enhance contrast, Multimodal imaging	Gd^{3+}/Fe_3O_4 loaded NPs, Gd/Mn oxide NPs, Fe/Co/Fe_3O_4 NPs
PET/SPECT	1–2 mm	Unlimited	10^{-14} to 10^{-15}	Low sensitivity (SPECT), Poor multiplexing, Poor spatial resolution	Multimodal contrast, favorable pharmacokinetics and biodistribution	Radionuclide loaded NPs
US	50 μm	~10 cm	10^{-8}	Low sensitivity, Poor image contrast	Enhance contrast	Nanobubbles, silica NPs, Polystyrene NPs
PAI/PAT	50 μm	~5 cm	10^{-12}	Suboptimal image contrast	Enhance contrast	Gold/silver NPs, carbon NPs, dye loaded silica NPs

Source: Data in the table have been adapted from Baker, M., *Nature*, 463, 977–980, 2010; Hahn, M. et al., *Anal. Bioanal. Chem.*, 300, 3–27, 2010; Jokerst, J.V., Gambhir, S.S. et al., *Acc. Chem. Res.*, 44, 1050–1060, 2011; Sosnovik, D., and Weissleder, R. *Prog. Drug Res.*, 62, 83–115, 2005.

Note: CT, computed tomography; MRI, magnetic resonance imaging; PAI, photoacoustic imaging; PAT, photoacoustic tomography; SPECT, single-photon emission computed tomography; US, ultrasound.

imaging modalities. The impact of NPs on various imaging modalities is summarized in Table 6.1.

6.1.1 NP Classification

NPs for biomedical imaging applications can be broadly classified into three classes according to the physical origin of contrast: (1) nanocarriers, which encapsulate contrast agents; examples include liposomes, micelles, polymeric NPs, dendritic NPs, and silica NPs (Liong et al. 2008; Torchilin 2007); (2) NPs with native contrast, viz. carbon NPs, quantum dots (QDs), iron oxide, gold/silver based NPs, and photoluminescent polymers (Arbab et al. 2003; de la Zerda et al. 2008; Soo Choi et al. 2007; Sun et al. 2009); (3) NPs that enhance external contrasts such as near-infrared (NIR) fluorescence enhancing nanoshells (Bardhan et al. 2009a), Raman signal–enhancing gold NPs (Jokerst et al. 2011), MR contrast–enhancing gadolinium-loaded carbon nanotubes,

and dye-loaded calcium phosphate nanoshells (Altinoglu et al. 2008). These three types of NPs are summarized in Table 6.2. Type I NPs are typically designed to maximize payload space to act as drug carriers, and the imaging modality is added to enable visualization and optimization of NP biodistribution and for performing image-guided therapies. Type II NPs may be specifically developed for imaging applications, and the mechanism providing native image contrast also frequently allows for therapeutic action such as photothermal heating of carbon nanotubes or silver/gold NPs (Cole and Halas 2009) and magneto-thermal heating of iron oxide NPs (Sonvico et al. 2005). Type III NPs are recently emerging variants of type II and rarely type I NPs where nanoscale engineering of surface features is exploited to create a synergistic enhancement of conventional imaging agents, for example, the plasmonic enhancement of organic fluorescent dyes on gold nanoshell and nanorod surfaces (Bardhan et al. 2009b), enhancement of Raman signal in the vicinity of surface roughness of silver

TABLE 6.2 Classification of Nanoparticles for Imaging

Type	Description	Application	Examples
I	Carrier NPs, NP acts as a passive carrier vehicle of the imaging contrast agent	Improve pK or biodistribution of the contrast agent; allow multiplexing/multimodal imaging; shield the contrast agent to improve stability; image guided drug delivery	Contrast loaded Liposomes/micelles, polymeric NPs, dye loaded silica NPs
II	NPs with native contrast	Enhanced/stable image contrast; higher sensitivity; externally modulated therapy	Carbon NPs, Gold/silver NPs, Iron/Gd/Mn oxide NPs, Quantum dots
III	NPs enhancing signal of external contrast agents	Improve stability/intensity of conventional contrast agents	Dye conjugated gold NPs, Gd^{3+}/dye loaded Carbon NPs, Surface enhanced Raman probes

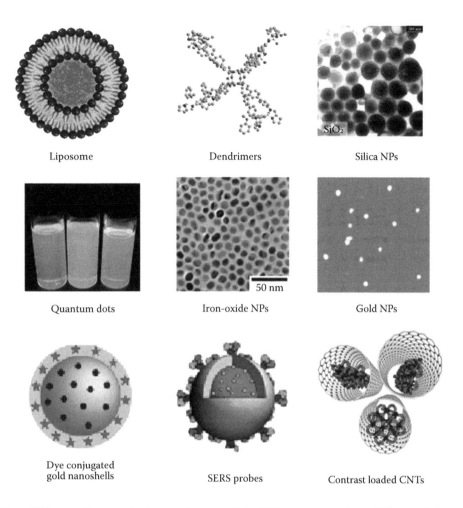

FIGURE 6.1 Illustration of NP types. Top row depicts type I nanoparticles (NPs) or nanocarriers, middle row depicts type II NPs or native contrast providers, and bottom row indicates type III NPs that enhance the contrast of conventional contrast agents including gold nanoshell conjugated fluorescent dyes, Raman substrates on gold/silver NPs, and optical/MR contrast agents loaded in carbon nanotubes.

and gold NPs (Von Maltzahn et al. 2009a), and T1 relaxation enhancement of gadolinium when trapped in carbon nanotubes (Tang et al. 2010) or in mesoporous silica NPs (Ananta et al. 2010). Figure 6.1 depicts the representative NPs from the aforementioned three classes.

6.2 Imaging Modalities

In the following sections, we summarize the recent developments in NP-mediated methods developed for major biomedical imaging modalities. We discuss the impact on imaging modalities first in isolation, followed by a description of emerging multimodal imaging applications.

6.2.1 Optical Imaging

Biomedical optical imaging includes a broad gamut of techniques utilizing the visible and NIR spectrum of electromagnetic radiation for investigating the functional state and molecular phenotypes of tissue, frequently in combination with exogenous optical molecular probes. Although they provide high resolution only for superficial imaging, optical techniques are nonetheless important as they provide a way to interrogate the disease at a molecular level and with high sensitivity without using ionizing radiation. NPs are impacting optical imaging via the development of enhanced contrast probes, the incorporation of multiplexed signals, and probes enabling multimodal combinations with structural imaging techniques to overcome the penetration limitations of optical imaging. In this chapter, we will restrict the discussion to NP-mediated NIR fluorescence imaging and Raman imaging, as these are the field most impacted by development of high quantum yield NP-based probes.

6.2.1.1 NIR Fluorescence Imaging

NIR light can travel multiple centimeters in tissue by repeated scattering, as the absorption by tissue chromophores and water is minimized in the 700–900 nm wavelength window. NIR-excitable and NIR-emitting fluorophores can thus interrogate significant tissue volumes by acting as beacons for targeted bio-molecular events. The primary limitations of conventional fluorophores, which are mostly composed of organic dye molecules conjugated to a targeting ligand, are their low quantum yield, low photostability, and limited circulation time of the imaging agents in the body. NP-based fluorescence imaging methods can address all the aforementioned shortcomings and have been the subject of extensive research in the past two decades. NP applications for optical imaging are thoroughly reviewed by Atinoğlu and Adair (2010), Debbage and Jaschke (2008), He et al. (2010), and Ntziachristos (2010).

The first NPs to strongly impact the field of optical molecular imaging were QDs. QDs have broad excitation spectra, large and tunable absorption cross sections, high quantum yields, and high resistance to photobleaching. Furthermore, QDs enable easy multiplexing because of their sharp size-dependent emission spectra, allowing the use of multiple QDs in a single assay. The primary limitation of early QDs was their restriction to the visible spectrum, which limited their role to microscopy and superficial imaging. However, in recent years, multiple NIR emitting QDs have been synthesized from combinations of group II–VI, IV–VI, and III–V elements (e.g., CdSe, CdTe, HgTe, PbS, PbSe, PbTe, InAs, InP, and GaS), alloys, and core–shell structures (Hahn et al. 2010). QDs have been used for image-guided sentinel lymph node (SLN) resection procedures in clinically relevant large animal models (Kim et al. 2004). All reported QDs incorporate heavy metal ions, and the associated toxicity concerns have limited their applications to microscopic and preclinical *in vivo* studies.

A second example of fluorescent NP that is biocompatible and potentially suitable for clinical applications is provided by NIR dye doped silica NPs. The advantages of silica NPs include optical transparency, excellent aqueous dispersibility, biological inertness, and low toxicity (Hahn et al. 2010). Furthermore, the methods for decorating silica NPs' surface with targeting ligands such as proteins, peptides, and oligonucleotides are well developed and use robust silane chemistry techniques. The porous structure of amorphous silica NPs is an effective carrier for NIR fluorophores, which can increase the quantum yield of fluorophores and reduce their degradation by shielding them from the aqueous environment. Loading silica NPs with dyes also enables signal amplification by concentrating a large number of dye molecules in a confined location, and favorably impacts biodistribution to the tumor tissue by EPR-based accumulation. The dyes reported to be encapsulated *in silica* NPs include the only Food and Drug Administration (FDA)–cleared NIR dye Indocyanine Green (ICG), cyanine dyes Cy5, Cy7, Alexa Fluor-750, and IRdye780 among others. The mesoporous silica structure can carry therapeutic compounds in addition to imaging probes, and in this application, the optical imaging signal can act as a surrogate marker for drug delivery to desired locations (Tallury et al. 2008).

Another type II class of fluorescent NPs is composed of up-converting NPs (Bachmann et al. 2008; Yu et al. 2008). These are rare earth compounds doped with rare earth metal ions, that absorb light in the far NIR region beyond 900 nm, and as opposed to conventional fluorophores emit light at a high frequency/longer wavelength in the green to far red/NIR region. Up-converting NPs have long fluorescent lifetimes, which can stretch into milliseconds. The most reported systems consist of yttrium oxide NPs doped with erbium and yttrium, and these have shown excellent photostability and biocompatibility (Hilderbrand et al. 2009). Zhang and coworkers have reported polyethylenimine-coated NaYF4:Yb, Er, and NaYF4:Yb,Tm NPs that excite in the NIR spectrum and emit in the visible region, and they demonstrated imaging through skin and muscle tissue for up to 1 cm depth (Chatterjee et al. 2008). To further improve the tissue penetration, efforts are underway to develop NIR exciting and NIR emitting up-converting NPs. These systems include systems based on $YF_3:Yb^{3+}/Er^{3+}$ NPs and NdF_3/SiO_2 core–shell NPs for which efficient deep tissue imaging in small animal models has been demonstrated (Wang et al. 2010). Up-converting NPs are aqueous dispersible and easy to conjugate to targeting ligands. These are novel NPs, and exhaustive safety and toxicity evaluation remains to be done; hence, the clinical translation is yet unclear.

Carbon nanomaterials also comprise the type II class NPs for optical imaging applications. Single-wall carbon nanotubes fluoresce in the second IR window (1000–1350 nm) of enhanced tissue penetration (Welsher et al. 2009). Carbon dots passivated with polymer coatings and colloidal nanodiamonds have also been reported to have visible to far-red emission (Barnard 2009; Yang et al. 2009). Graphene or one-atom-thick 2-D graphite layers have weak NIR fluorescence emission (Yang et al. 2011). The advantage of carbon-based nanomaterials is their highly reactive surface, which allows conjugation of targeting ligands and makes them effective carriers for both conventional drugs and gene therapy agents. The current limitations in carbon-based NPs are attributed to lack of scalable and reproducible synthesis techniques and the absence of convincing safety and toxicity information for future clinical translation.

Type III NP constructs for optical molecular imaging are also emerging. Currently, only organic fluorescence dyes have been approved for human use. Organic dyes such as ICG have low quantum yields and poor photostability. Recently reported constructs such as ICG containing gold nanoshells (Bardhan et al. 2009b) and calcium phosphate shells (Muddana et al. 2009) increase the brightness and stability of ICG by directly increasing the radiative relaxation rate of dye molecules after excitation and/or enhancing the available excitation light by concentrating the incident energy on the NP surface. With these enhancements, organic fluorophore performance can be raised to QD levels, but without the related toxicity concerns. Figure 6.2 illustrates representative examples of optical imaging with NP agents.

6.2.1.2 Raman Imaging

Although conceptually understood and appreciated for decades, Raman imaging is finally taking rapid strides toward clinical

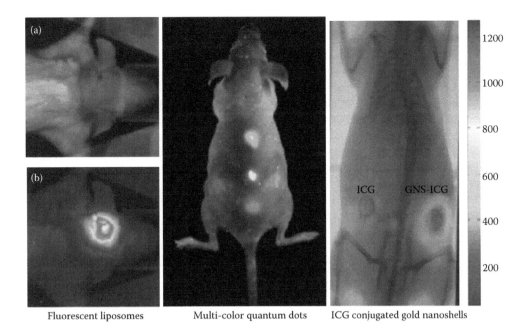

			1200
ICG	GNS-ICG		1000
			800
			600
			400
			200

Fluorescent liposomes   Multi-color quantum dots   ICG conjugated gold nanoshells

FIGURE 6.2 **(See color insert.)** Representative NPs for optical imaging. Left column illustrates white light and fluorescence images for dye loaded liposomes targeted to edema in mouse ear (from Deissler, V. et al., *Small*, 4, 1240–1246, 2008, with permission; Copyright © John Wiley and Sons, 2008); middle column depicts multicolor quantum dots (from Gao, X. et al., *Nat. Biotechnol.*, 22, 969–976, 2004, with permission; Copyright © Nature Publishing Group, 2004); right column indicates the contrast enhancement obtained by conjugating the dye ICG to gold nanoshells.

translation and widespread biomedical imaging application with the development of surface enhanced Raman probes, which typically comprise fluorescent dyes adsorbed on metallic gold, silver, or platinum NPs (Doering et al. 2007). In Raman spectroscopy, similar to fluorescence spectroscopy, the biomolecules are excited by optical photons, but the incident energy need not correspond to the electronic transition states of the molecule; rather, Raman excitation creates a virtual state, leading to annihilation of the incoming photon and instantaneous emission of a second photon at a fixed frequency shift to the incident photon. The emitted photon spectrum is specific to the scattering molecule; hence, Raman imaging can allow direct interrogation of molecular structure of tissue, with the caveat that Raman scattering process is extremely weak compared to fluorescence emission. As a classic example of type III NPs, in the presence of metal surfaces with roughness features on 10–100 nm scales, Raman signal has been reported to be enhanced by million-fold, and under special conditions while detecting single molecules trillion-fold enhancements have been reported (Nie and Emory 1997) Hence, a variety of Raman probes, ranging from bare metal NPs, silica core gold nanoshells, and hollow gold nanoshells, are being proposed both for biosensing and imaging. Raman signal is immune to photobleaching; hence, unlike fluorescence reporters, Raman probes can be used for prolonged imaging sessions and high incident photon fluxes. Whereas most applications have focused on *ex vivo* biosensing, a few reports on *in vivo* imaging have been emerging in recent years (Jokerst et al. 2011), and with further progress in tuning the plasmonic

resonances of metal nanostructures, widespread *in vivo* Raman-based biosensing applications are possible in the future.

6.2.2 US Imaging

US imaging relies primarily on mechanical contrast between different tissue types and layers and uses the pulse–echo principle, where sound waves with frequencies beyond the audible range of 20 kHz are launched into the tissue and the echoes are captured by transducers at the tissue boundary, which convert the detected sound pressure signal to an electric signal followed by computational processing to produce an echo image of the tissue interior. Clinical applications use sound waves in the range of 2–3 MHz for pediatric imaging and 5–12 MHz for adult imaging to provide resolutions ranging from 0.2 to 1 mm, but at frequencies higher than 30 MHz, are required for detecting NPs smaller than 1000 nm, and these frequencies do not penetrate more than few millimeters in tissue (Debbage and Jaschke 2008). Hence, efforts for molecular imaging or contrast-enhanced US imaging have primarily relied on microscale technologies for development of gaseous-phase contrast agents to increase the echogenicity of vasculature or other targeted regions. Microbubbles are typically in the 3–5 μm range and composed of surfactant or protein/polymer layers containing gaseous cores of air, perfluorocarbons, or nitrogen (Dayton and Rychak 2007). Perfluorocarbons and nitrogen cores are preferred because of their limited serum solubility, hence increasing the circulation time and improved longitudinal imaging performance. In recent years, similar techniques have been applied to produce nanobubbles and nanoemulsions, which typically range in the 150- to 1000-nm size (Gessner and Dayton 2010). The performance of nanoscale agents with clinical US frequencies is inferior to microbubbles because of their lower scattering cross sections and the less-than-optimum mechanical properties of the shell. However, since nanoscale materials can exploit EPR-based tumor accumulation and have other biodistribution advantages, there are continued efforts to improve the echogenicity and stability of these structures. Targeted perfluorocarbon NPs were the first reported molecular imaging agents for US imaging, and Lanza et al. (2000) demonstrated more than 2 orders of magnitude contrast enhancement of fibrin thrombi with them. Reflective liposomes have also been reported by this group for targeting endothelial integrins.

6.2.2.1 Photoacoustic Imaging

PAI is a relatively new hybrid imaging modality, which enables the imaging of optical absorption contrast with US resolution (Wang 2009). Tissue is typically excited by short pulses of NIR radiation, and tissue chromophores and exogenous contrast agents absorb this light and undergo a transient increase in temperature on the order of ~10 mK and subsequently relax by thermoelastic expansion of the absorbers, while generating an ultrasonic acoustic signal that can be detected by wide-band transducers positioned on the tissue periphery. Whereas thin tissue sections can be imaged with photoacoustic microscopy (where coupled focused ultrasonic detector and confocal optical illumination are used to scan through tissue slices), deeper tissue are imaged with photoacoustic tomography (PAT) techniques—where the laser-illuminated tissue volume is interrogated by circumferential detectors, and boundary measurements are converted to interior absorber distribution by inverse imaging algorithms. For incorporating molecular imaging ability, exogenous targeted absorbers such as fluorescent dyes are used. NPs are attractive exogenous agents for PAI, as compared to free organic dyes, where very high optical absorption cross sections can be engineered. Type I NPs for PAI include multiple formulations for dye-doped constructs. Loading multiple dye molecules into the protective NP matrix enables signal amplification and provides additional NIR fluorescence contrast for intraoperative use or to facilitate *ex vivo* analysis, apart from other benefits of reduced degradation and improved targeting as discussed in the "Optical Imaging" section. ICG is the most commonly used dye for PAI in NP formulations. PAT has been used to demonstrate imaging of ICG in tissue phantoms up to the depth of 5 cm (Wang 2009). Encapsulation of ICG in carrier NPs constructed from organically modified silica, polylactic-*co*-glycolic acid (PLGA), and calcium phosphate has been demonstrated to improve circulation time and stability for *in vivo* imaging applications (Wang et al. 2004).

Type II NPs or NPs with intrinsic contrast for PAI are primarily composed of gold-based nanomaterials. PAT has been demonstrated with spherical gold NPs, gold nanorods, nanocages, agglomerates, hollow nanoshells, and composite nanomaterials such as silica core nanoshells, cobalt shells with gold cores, gold speckled silica, and polymer–gold hybrids (Hahn et al. 2010) The reason for the popularity of gold-based agents is their size- and shape-dependent plasmon resonance properties, allowing for precise engineering of absorption cross sections most favorable to PAI. Although tunable plasmon resonance can also be obtained with silver NPs, the benign toxicity profile of gold-based NPs and ongoing clinical trials increase the probability gold-based NPs for widespread clinical applications. Conjugation chemistry methods for attaching targeting ligands to gold surface are well established, and gold-based NPs can be easily modified to include multimodal capabilities in addition to PAI. Nanorods and nanocages have been reported to have higher absorption cross sections compared to nanoshells and are thus increasingly being used for PAI applications in place of nanoshells (Hu et al. 2006).

Single-walled nanotubes (SWNTs) have been extensively used for PAI imaging because of their high NIR absorption cross section and ease of surface functionalization for molecular targeting (de la Zerda et al. 2008). Figure 6.3 illustrates the application of SWNTs for PAI. Antibody conjugated SWNTs have been demonstrated for imaging multiple tumors and detection of SLNs by PAI (Pramanik et al. 2009). Although widely reported for PAI, the NIR absorption of SWNTs are relatively low compared to gold-based NPs, and hence hybrids such as gold plated SWNTs have been proposed. Golden carbon nanotubes (GNTs) have demonstrated 100-fold increase in PAI contrast (Kim et al. 2009). Higher PA sensitivity of GNTs has been exploited for imaging magnetically captured circulating tumor cells and *in vivo* imaging of lymphatic vessels. In addition to gold plating, and as an example of type III NPs (enhanced exogenous contrast), the absorbance of SWNTs has also been enhanced up to 20 times by covalently attaching ICG dye molecules on the SWNT surface (de la Zerda et al. 2010).

6.2.3 Nuclear Imaging

Nuclear imaging is the oldest and clinically most advanced molecular imaging modality. Nuclear imaging relies on either the detection of gamma ray photons produced by unstable isotopes such as Technicium-99 (SPECT imaging), or the coupled gamma photons produced by the decay of positrons produced by isotopes such as F-18, or Cu-64 (PET imaging). The contrast in nuclear imaging depends on the concentration of the radioactive isotopes achieved in the desired regions of the body. Nuclear imaging is the least impacted imaging modality by NPs, because—unlike optical, MR, or computed tomography (CT) imaging—nuclear contrast cannot be directly nanoengineered, and the only way NPs can enhance nuclear imaging is by encapsulation of multiple radionuclides in a confined space and by exploiting the favorable biodistribution of NPs. Even this process has disadvantages because of the short half-lives of many radionuclides where addition of processing steps before *in vivo* use is counterproductive to signal intensity, and additional challenges to formulation processes are introduced by ionizing radiation hazards. There have been very few reports of

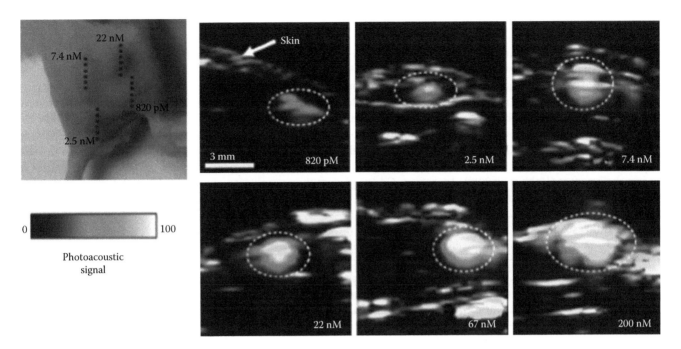

FIGURE 6.3 **(See color insert.)** Photoacoustic detection of single-walled nanotube-Indocyanine Green (SWNT-ICG) in living mice. (Reproduced with permission from de la Zerda, A. et al., *Nano Lett.*, 10, 2168–2172, 2010.) (a) Mice were injected subcutaneously with SWNT-ICG at concentrations of 0.82–200 nM. The images represent ultrasound (gray) and photoacoustic (green) vertical slices through the subcutaneous injections (dotted black line). The skin is visualized in the ultrasound images, whereas the photoacoustic images show the SWNT-ICG distribution. The white dotted lines on the images illustrate the approximate edges of each inclusion (Copyright © American Chemical Society, 2010).

PET imaging of targeted NPs. Radionuclides used for SPECT imaging have relatively longer half-lives, and NPs have been reported for multiplexed imaging (Cai and Chen 2007). SPECT imaging is primarily used for noninvasive assessment of NP biodistribution. The pharmacokinetics, tumor uptake, and therapeutic efficacy of In-111 labeled monoclonal antibody attached to iron oxide NPs was reported for nude mice bearing human breast cancer xenografts for combined SPECT–MRI imaging (DeNardo et al. 2005). Liposomes encapsulating Technicium-99 and SWNTs loaded with In-125 have also been reported for combined SPECT–CT imaging (Cai and Chen 2007).

6.2.4 MRI Imaging

MRI provides physiological and pathological information about the living tissue by primarily measuring the water proton relaxation rates. The entity to be imaged is placed in a strong magnetic field, and protons are excited by pulsed radiofrequency (RF) radiation. The relaxation rates of protons after RF excitation are measured, and they constitute the imaging signal, which varies with the local physiological environment. The strength of magnetic field governs the resolution, and although images can be achieved by sub 1-T fields, typical clinical scanners use 1.5- or 3-T fields, and preclinical scanners can range up to 9.4 T. In addition to magnetic field strength, the image resolution can depend on the imaging sequence, sampling rate, and signal sampling strategies (Debbage and Jaschke

2008). MRI is a noninvasive and nonionizing method and can provide high soft tissue contrast with high resolution of up to 50 μm. The chief drawback of MRI is its low inherent sensitivity, as millimolar concentrations of protons are needed for detection of their relaxation rates. Hence, MRI often requires the use of exogenous agents for sensitive imaging, and it can achieve significant signal amplification by the use of NP-based contrast agents. MR contrast agents are classified into two types: longitudinal relaxation rate enhancers, which increase the T1 signal in T1-weighted images, and transverse relaxivity enhancers, which increase T2 contrast in T2-weighted images. T1 agents produce bright contrast, with higher image intensity in the targeted region, whereas T2 agents produce dark contrast, with lower intensities in the targeted regions. T1 contrast is easy to visualize but at the cost of lower sensitivity, while T2 agents such as iron oxide NPs can provide exquisite sensitivities enough to track single cells, but have a lower and difficult-to-interpret image quality.

NP-based T1 contrast agents are chiefly based on gadolinium. All three NP types—type I or carrier NPs comprising liposomes, polymers, or silica for carrying chelated Gd; type II or native contrast NPs composed of gadolinium oxide, gadolinium fluoride, or gadolinium phosphate; and type III NPs with dramatically enhanced T1 contrast obtained by geometric confinement of Gd in conjugation with silica, gold, or carbon nanostructures—have been proposed (Debbage and Jaschke 2008; Hahn et al. 2010). There are three requirements for designing highly sensitive paramagnetic NPs with

T1 contrast: (1) large number water protons in coordination with metal (Gd), (2) optimum residence lifetime at the metal site, and (3) slow tumbling motion of NP. These factors are exploited for type III contrast enhancing NPs, where Gd is incorporated in structures such as silica or perfluorocarbon NPs, carbon nanotubes, and nanodiamonds, which all yield high MR contrast because of high Gd payload and slow tumbling motion of NPs (Manus et al. 2010; Na et al. 2009). Type II NPs composed of Gd_2O_3, GdF_3, or $GdPO_4$ also yield high magnetic moments because of the abundance of paramagnetic ions on their surface (Hahn et al. 2010). Apart from Gd-based NPs, other transition metal oxides such as MnO-based NPs have also been proposed for T1-based imaging of brain lesions, liver, and kidneys (Shin et al. 2009).

NPs providing T2 contrast are predominantly based on iron oxide. For MR imaging applications, iron oxide NPs are typically coated with dextran, PEG, or other polymers. These NPs are clinically approved under many trade names such as Resoist, Feridex, Ferrumoxtran-10, or Gastromark (Hahn et al. 2010). The polymer coating allows long circulation times, and combined with low toxicity and high sensitivity, iron oxide NPs are agents of choice for T2 contrast-based clinical applications and *in vivo* cell tracking. Iron oxide NPs are further subclassified according to their size, with micron-size particles referred to as magnetic iron oxide nanoparticles, ~100 nm particles referred as superparamagnetic iron oxide (SPIO) NPs, and sub-50 nm size NPs referred to as ultrasmall superparamagnetic

iron oxide (USPIO) NPs (Hahn et al. 2010). SPIO NPs are clinically used for imaging liver disease and have also been used for stem cell tracking applications, apart from lymph node imaging, angiography, and blood pool imaging (Harisinghani et al. 2003; Weissleder et al. 1988). Neural precursor cell tracking application is illustrated in Figure 6.4. To further improve the sensitivity and T2 contrast, metal alloy–based NPs have also been proposed. These include $CoFe_2O_4$-, $MnFe_2O_4$-, and $NiFe_2O_4$-based NPs (Hahn et al. 2010); however, as the long-term fate and toxicities associated with these particles are unknown, only iron oxide–based NPs demonstrate near-term clinical promise.

6.2.5 X-Ray/CT Imaging

X-ray imaging is the oldest medical imaging modality. The contrast in x-ray imaging and x-ray CT is created by the differential attenuation of x-ray photons between tissue types. The contrast also depends on the energy of x-rays used for imaging; hence, contrast agents developed for fluoroscopy may not perform well for CT scanning and vice versa. X-ray imaging contrast agents have high radio opacity. The first known NP-based x-ray agents were composed of 3- to 10-nm thorium dioxide NPs and used under the trade name Thorotrast (Becker et al. 2008). Thorotrast was discontinued because of radiation hazards and carcinogenic properties of Thorium-232. The most prevalent x-ray contrast agents today are composed of hydrophilic iodinated molecules. To exploit the favorable biodistribution of NPs, iodinated agents are being reformulated as nano-sized agents by trapping iodinated compounds in liposomes, emulsions, or other polymeric NPs both to increase the circulation time and to increase the local concentrations of accumulated iodine (Hahn et al. 2010). Iodine-based contrast agents have been compared with gold NPs and were found to have similar contrasts for typical fluoroscopy applications. However, gold NPs exceed the performance of iodinated contrasts for mammography and CT applications (Jackson et al. 2010). Figure 6.5 indicates the application of gold nanorods for CT guided photothermal therapy (Von Maltzahn et al. 2009b). The fate, transport, clearance, and toxicity profiles of systemically injected gold NPs are well understood, and they are the most promising nanotechnology-based agents for x-ray CT imaging. Apart from iodine, newer NPs with bismuth sulfide and lanthanide materials have been proposed (Ajeesh et al. 2010; Rabin et al. 2006). Because of unanswered toxicity concerns about these materials and lack of radical contrast enhancement over gold NPs, they may not translate clinically.

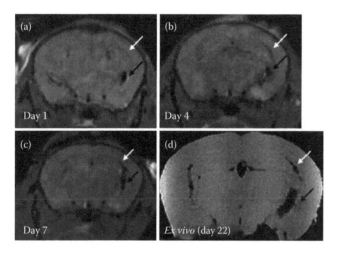

FIGURE 6.4 Serial *in vivo* MRI tracking of intracerebroventricularly (ICV) transplanted neuroprecursor cells (NPCs) in encephalomyelitis (EAE). (Reproduced with permission from Cohen, M.E. et al., *J. Neurosci. Res.*, 88, 936–944, 2010.) Ferumoxides-labeled NPCs were transplanted to the right ventricle of EAE mice (black arrow). At day 1 after ICV transplantation (a), cells indicated by hypointense (black) MRI signal are found exclusively within the cerebral ventricles and are absent within the corpus callosum (white arrow). At 4 (b) and 7 days (c) after ICV transplantation, some cells had migrated into the corpus callosum (white arrow). *Ex vivo* MRI at day 22 post-transplantation confirmed this pattern of migration (d). (Copyright © John Wiley and Sons, 2010.)

6.2.6 Multimodal Imaging

Since all imaging modalities have specific strengths and weaknesses, which can often be complementary, multimodal imaging is emerging as an attractive option for sensitive and high-resolution detection of pathologies. Conventional multimodal imaging approaches (e.g., PET–CT) are driven primarily by device integration, and contrasts for multiple modalities are

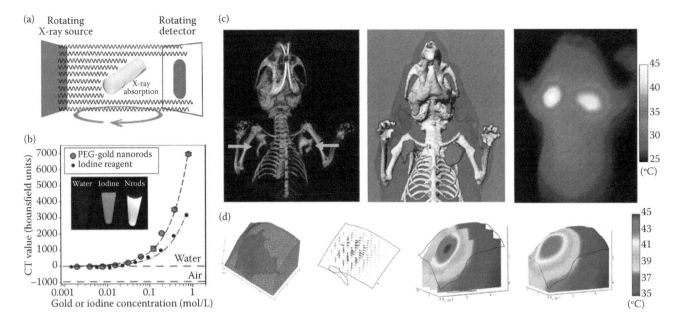

FIGURE 6.5 **(See color insert.)** X-ray computed tomography (CT), quantitative photothermal modeling, and near-IR photothermal heating of gold NRs *in vivo*. (a) Schematic of x-ray absorption by gold NRs in x-ray CT. (b) X-ray CT number of PEG-NRs compared with an iodine standard (Isovue-370). (c) PEG-NRs were intratumorally given to mice bearing bilateral MDA-MB-435 tumors and imaged using x-ray CT to visualize three-dimensional PEG-NR distribution in tumors (left). A three-dimensional solid model of the complete geometry was rapidly reconstructed by image processing for use with computational photothermal modeling (middle). Red, PEG-NRs. Experimental thermographic surveillance of NIR irradiation after x-ray CT (*f*0.75 W/cm², 1 min; right). (d) Meshed geometry of the left tumor chosen as the computational domain (left). Plot of theoretical heat flux propagation inside the tumor upon irradiation (middle left). Predicted internal temperature distribution at three different planes inside the tumor (middle right) along with surface temperature map (right) matching the left tumor in (c). (Adapted and reprinted with permission from the American Association of Cancer Research, Von Maltzahn, G. et al., *Cancer Res.*, 69, 3892–3900, 2009. Copyright © American Association of Cancer Research, 2009.)

injected separately. NPs provide the unique advantage of being able to integrate multiple contrasts in a single structure, thus simplifying the potential multimodal image acquisition procedures. Most attractive multimodal combinations combine a structural imaging modality with high resolution such as CT or MRI, with a highly sensitive but lower resolution technique such as optical or PET imaging. Combined MR–optical NP agents have seen the most progress, as both of them are nonionizing modalities, and MR provides deep tissue penetration with high resolution, whereas optical imaging can allow detection of small metastasis or tumor margins in an intraoperative setting, and longitudinal imaging. MR–optical agents have been reported for monitoring enzyme activity, tumor imaging, apoptosis detection and monitoring, and atherosclerosis (Jennings and Long 2009). MR–optical contrast agents include iron oxide NPs conjugated with fluorescent dyes or QDs, hybrid gold nanoshells with iron oxide and ICG (Figure 6.6), liposome and other carrier NPs containing iron oxide of Gd contrasts along with fluorescence dyes or QDs (Hahn et al. 2010). One concern in the design of MR–optical NPs is the relative concentrations of MR and optical contrasts. MRI is orders-of-magnitude less sensitive than optical imaging; hence to obtain an equivalent performance, careful calibration of relative ratios of MR and optical agents might be required. Other multimodal imaging combinations include

combined PET–NIR fluorescence imaging; the objective here is to combine the depth penetration of PET imaging for clinical use with the ease of use for preclinical studies and longitudinal imaging provided by NIR optical means. Similar motivations are behind the development of NPs with SPECT-fluorescence contrasts. The fluorescence contrasts are provided by ICG (FDA-cleared dye) or QDs. As a structure–function combination,

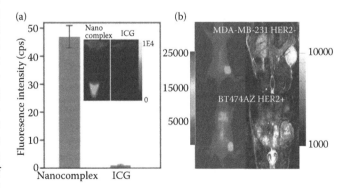

FIGURE 6.6 **(See color insert.)** Illustration of NIR and MRI enhancing gold nanoshell-based nanocomplex. (a) ×50 Enhancement of ICG fluorescence after NS conjugation. (b) NIR and MR contrast between HER2+/– tumors.

PET–MRI agents are becoming increasingly popular and have a high translational potential. Other modalities combined with MRI are PAI by using Gd-loaded SWNTs and US imaging. Apart from dual modalities, triple modality, and even quadruple modality, NPs have also been proposed by combining MRI, NIR, bioluminescence resonance transfer, and PET imaging (Hwang et al. 2009; Jennings and Long 2009).

6.3 Biomedical Applications for Imaging NPs

NP-based imaging is finding diverse applications in the biomedical research arena, from accelerating the understanding of biological mechanisms to clinical imaging of disease. The application can be broadly divided into cancer-related imaging and noncancer vascular imaging.

6.3.1 Cancer Imaging and Theranostics

In vivo cancer imaging is the area where nanoscale research can make the most impact in the coming decade. The cost of cancer on the society is staggering. According to National Cancer Institute (NCI) estimates, cancer-related expenditures exceed $100 billion a year, and NCI spends more than $3 billion a year on funding cancer research (NCI 2010). As the U.S. population ages, these numbers will show an increasing trend. NP-based methods can make a dent in the mortality and morbidity associated with cancer by deploying early detection, image-guided interventions, and theranostic imaging, and by monitoring and tailoring the therapy response with image guidance. The first clinical area to witness NP-based methods is SLN imaging. SLN mapping typically involves injecting contrast agents in peritumoral space to identify the first lymph node draining from the region. The sentinel node is dissected and analyzed for the presence of malignancy to support decisions on complete regional lymph node removal. SLN mapping is especially useful in breast cancer for its potential to avoid complications, such as lymphedema, which result from complete axillary lymph node removal (Chen et al. 2007). SLN mapping is also becoming the standard of care for cutaneous melanoma (Morton et al. 2005). Conventional SLN imaging contrasts are composed of Technicium-99–based radio colloid for noninvasive mapping of lymph node location or isosulfan blue, which is a visible dye for guiding surgical resection of the SLN. Imaging SLNs with radiocolloid is plagued with poor spatial resolution and low sensitivity, whereas the isosulfan dye is restricted to intraoperative settings. An agent that can directly image the status of SLN with high sensitivity can improve the staging of disease to guide therapy and avoid unnecessary surgery. NP-based agents can be tailored for efficient lymphatic transport and high contrasts for clearly delineating the ambiguous lymphatic drainage patterns (Khullar et al. 2009). NP-based agents have been proposed for SLN mapping with multiple imaging modalities to provide both high resolution and sensitivity. Ravizzini et al. (2009) have reviewed the use of CT, US, MRI, and optical imaging for SLN mapping.

Apart from imaging, the second application of NP-based imaging is to provide simultaneous therapy, monitor the payload drug delivery, and possibly track the response to therapy. NPs with simultaneous imaging and therapy capabilities have started entering clinical trials (Jokerst and Gambhir 2011). Nanospectra Biosciences Inc. Houston is translating silica core gold nanoshells for photothermal therapy of cancer. PLGA-based NPs carrying siRNA for melanoma therapy and a fluorophore for *ex vivo* detection have been deployed in initial trials by Davis et al. (2008).

6.3.2 Vascular Imaging

Imaging of vascular pathologies is the second major area of NP applications after cancer. About 700,000 cardiac-related deaths are reported in the United States, with the majority occurring without advance warning of the disease (Wickline et al. 2007). Artherosclerosis and/or vulnerable unstable plaque deposits are often detected only after acute emergencies or fatal events and are implicated in 70% of heart attacks (Zheng et al. 2001). Hence, imaging artherosclerosis or plaque deposits, and identification of their pathological state early and with high sensitivity can have a major impact on patient mortality. Like cancer, artherosclerosis growth also involves angiogenic events and permeable vasculature; hence, NP-based agents are an attractive option for sensitive imaging (Wickline et al. 2007). Since vascular pathologies are in deep tissue, surface weighted imaging modalities such as optical or PAI can only be applied on the preclinical level, and clinical focus has been on US, x-ray, CT, MRI, or nuclear imaging. Low resolution limits the utility of nuclear medicine and US for arterial imaging and x-ray CT and MRI are the modalities of choice. NP-based agents can tap into artherosclerosis deposits either by EPR-based retention, by using low-density or high-density cholesterol to penetrate the plaque (Cormode et al. 2008), or by targeting specific molecular signatures of inflammation such as MMP overexpression often present at these sites (Mulder et al. 2008). NPs rapidly taken up by macrophages are also favored, since a high incidence of apoptosis and necrosis and a high macrophage burden are associated with vulnerable plaques (Wickline et al. 2007).

Wickline and Lanza (2002, 2003) were the first to report the imaging of plaque disruption by imaging small fibrin deposits with US and paramagnetic contrast containing perfluorocarbon-based NP agents. Macrophage imaging with USPIOs was demonstrated by Schmitz et al. (1999) and Ruehm et al. (2001) in rabbits. Clinically, Trivedi et al. (2004) and Kooi et al. (2003) have reported USPIO accumulations in macrophages in plaque in patients undergoing carotid endarterectomy. For preclinical studies, combined optical and MR contrast agents have been proposed by Mulder et al. (2006), who encapsulated Gd^{3+} and QDs in a phosoholipid layer.

Since MRI acquisition is time-intensive and respiratory motion artifacts can degrade resolution enough to limit the visualization of vessel walls, NP-based x-ray CT contrast agents are also under active development (Cormode et al. 2010). Both iodinated NPs and lipid-coated gold NPs have been proposed for spectrally sensitive CT imaging and staging of artherosclerosis. CT scanners can acquire high spatial resolution images of the entire heart within 5 s, and developments in multicolor CT with NP-based agents targeted to specific plaque state can significantly affect the state of care for vascular pathologies.

6.4 Challenges and Future Outlook

NPs have been approved by the FDA for more than 15 years as delivery vehicles, but only liposomes have seen widespread clinical applications (Jokerst and Gambhir 2011). Other NPs are undergoing development or initial clinical trials. There are two predominant factors limiting the translation of NP-based imaging: (1) safety and toxicity concerns and (2) challenges in scaling up the production of imaging NPs for clinical use. Translation of therapeutic NPs is more advanced as compared to imaging NPs, because the safety standards and toxicity concerns are higher for routine diagnostic imaging applications. The first safety concerns arise directly from the NP geometry since NPs have very high surface/volume ratios (Debbage and Jaschke 2008; Hahn et al. 2010). The large reactive surfaces NPs present in the biological environment can potentiate unforeseen risks. The second concern arises from the often-unknown degradation pathways of NPs in biological environment, especially for imaging NPs, which may carry multiple external components for providing contrast and deliver therapy. The NPs can release their imaging or therapeutic payloads prematurely or the degradation products and released contrast agents may have a pharmacokinetic profile of their own, requiring additional testing. The clearance profiles of multiple fragments may interfere with each other to further complicate the matters, for example NPs carrying chelated Gd may release the Gd^{3+} ions in kidneys, which may harm the patients with sensitive renal system (Perazella 2008). There is no general consensus on a set of standard toxicological assays that is applicable to all imaging NPs. The state of the art relies on a patchwork of *in vitro* assays, reactive oxygen species production assays, *in vivo* biodistribution studies, membrane permeation, and multiple apoptosis assays. NCI's Nanotechnology Characterization Laboratory (NCL; http://ncl.cancer.gov) is working with the FDA to develop standard tests for NP safety. The task is challenging because of the enormous variety in the available NPs. Currently, the NCL groups NPs according to size, surface charge, and solubility. The complexity in assessing NP safety translates into higher costs and prolonged time for getting NP-based imaging agents into clinical trials. To accelerate the translation process, strategies such as coupling the NP to an approved marker or combining with approved therapeutics are being used (Bawa 2008). The second challenge for translation of imaging NPs is production scale-up. Most published reports on imaging NPs relies on milliliter-scale production for *in vitro* and *in vivo* proof of principle studies in small animals. Since an average man is roughly 3500 more massive than an average nude mouse, a 200-μl dose of an imaging NP will translate to a 700-ml dose in a human

being at the same concentration. Hence, running even small-scale clinical trials may require hundred-liter scale manufacture. Scaled-up production techniques are easy to achieve for molecular pharmaceuticals but is challenging for NPs, particularly for imaging NPs with exogenous contrast agents or multimodal imaging and therapy attributes. Apart from physical attributes such as size, shape, surface functionalization, coating thickness, porosity, elemental composition, crystalline structure, and allowable impurity levels, imaging NPs also require precise measurement and calibration of contrast, which may require spectroscopic or fluoroscopic calibration, optimization of absorption/scattering cross sections and fluorescence yield/lifetimes for optical contrasts, magnetic susceptibility and relaxivity for magnetic contrast, thermoelasticity for photoacoustic contrast, and analogous metrics for other modalities (Hahn et al. 2010). Only imaging NPs to be clinically approved universally are based on iron oxide NPs. As bionanotechnology advances and currently ongoing initial clinical trials on imaging NPs show success, larger companies such as Siemens, General Electric, Phillips, and other major pharmaceutical firms are expected to fill in the gaps on scale-up of imaging NPs. Established scale-up procedures for NP synthesis will also help in clearing the regulatory hurdles.

In summary, the field of medical imaging with NPs is at an interesting juncture, where the application of basic nanoscience has made substantial enhancements to most imaging modalities, but large-scale production techniques and regulatory strategies are still in infancy. As the field transitions from academia to industry, the next two decades should witness rapid strides toward clinical translation.

References

Ajeesh, M., B. F. Francis, J. Annie et al. 2010. Nano iron oxide–hydroxyapatite composite ceramics with enhanced radiopacity. *Journal of Materials Science: Materials in Medicine* 21:1427–1434.

Altinoğlu, E. I., and J. H. Adair. 2010. Near infrared imaging with nanoparticles. *Wiley Interdisciplinary Reviews: Nanomedicine and Nanobiotechnology* 2:461–477.

Altinoglu, E. I., T. J. Russin, J. M. Kaiser et al. 2008. Near-infrared emitting fluorophore-doped calcium phosphate nanoparticles for in vivo imaging of human breast cancer. *ACS Nano* 2:2075–2084.

Ananta, J. S., B. Godin, R. Sethi et al. 2010. Geometrical confinement of gadolinium-based contrast agents in nanoporous particles enhances T1 contrast. *Nature Nanotechnology* 5:815–821.

Arbab, A. S., L. A. Bashaw, B. R. Miller et al. 2003. Characterization of biophysical and metabolic properties of cells labeled with superparamagnetic iron oxide nanoparticles and transfection agent for cellular MR imaging. *Radiology* 229:838–846.

Bachmann, P. K., H. Hummel, T. Jüstel et al. 2008. Near-infrared luminescent nanomaterials for in-vivo optical imaging. *Journal of Nanophotonics* 2:021920.

Baker, M. 2010. Whole-animal imaging: The whole picture. *Nature* 463:977–980.

Bardhan, R., W. Chen, C. Perez-Torres et al. 2009a. Nanoshells with targeted simultaneous enhancement of magnetic and optical imaging and photothermal therapeutic response. *Advanced Functional Materials* 19:3901–3909.

Bardhan, R., N. K. Grady, J. R. Cole et al. 2009b. Fluorescence enhancement by Au nanostructures: Nanoshells and nanorods. *ACS Nano* 24:744–752.

Barnard, A. S. 2009. Diamond standard in diagnostics: Nanodiamond biolabels make their mark. *Analyst* 134:1751–1764.

Bawa, R. 2008. Nanoparticle-based therapeutics in humans: A survey. *Nanotechnology Law and Business* 5:135–155.

Becker, N., D. Liebermann, H. Wesch et al. 2008. Mortality among Thorotrast-exposed patients and an unexposed comparison group in the German Thorotrast study. *European Journal of Cancer* 44:1259–1268.

Cai, W., and X. Chen. 2007. Nanoplatforms for targeted molecular imaging in living subjects. *Small* 3:1840–1854.

Chatterjee, D. K., A. J. Rufaihah, and Y. Zhang. 2008. Upconversion fluorescence imaging of cells and small animals using lanthanide doped nanocrystals. *Biomaterials* 29:937–943.

Chen, S. L., F. M. Hoehne, and A. E. Giuliano. 2007. The prognostic significance of micrometastases in breast cancer: A SEER population-based analysis. *Annals of Surgical Oncology* 14:3378–3384.

Cole, J., and N. J. Halas. 2009. Photothermal efficiencies of nanorods and nanoshells. *Journal of Physical Chemistry C* 113:12090–12094.

Cohen, M. E., N. Muja, N. Fainstein et al. 2010. Conserved fate and function of ferumoxides-labelled neural precursor cells in vitro and in vivo. *Journal of Neuroscience Research* 88:936–944.

Cormode, D. P., T. Skajaa, M. M. van Schooneveld et al. 2008. Nanocrystal core high-density lipoproteins: A multimodality contrast agent platform. *Nano Letters* 8:3715–3723.

Cormode, D. P., E. Roessl, A. Thran et al. 2010. Atherosclerotic plaque composition: Analysis with multicolor CT and targeted gold nanoparticles. *Radiology* 256:774–782.

Davis, M. E., Z. Chen, and D. M. Shin. 2008. Nanoparticle therapeutics: An emerging treatment modality for cancer. *Nature Reviews Drug Discovery* 7:771–782.

Dayton, P. A., and J. J. Rychak. 2007. Molecular ultrasound imaging using microbubble contrast agents. *Frontiers in Bioscience* 12:5124–5142.

de la Zerda, A., C. Zavaleta, S. Keren et al. 2008. Carbon nanotubes as photoacoustic molecular imaging agents in living mice. *Nature Nanotechnology* 3:557–562.

de la Zerda, A., Z. Liu, S. Bodapati et al. 2010. Ultrahigh sensitivity carbon nanotube agents for photoacoustic molecular imaging in living mice. *Nano Letters* 10:2168–2172.

Debbage, P., and W. Jaschke. 2008. Molecular imaging with nanoparticles: Giant roles for dwarf actors. *Histochemistry and Cell Biology* 130:845–875.

Deissler, V., R. Ruger, W. Frank et al. 2008. Fluorescent liposomes as contrast agents for in vivo optical imaging of edemas in mice. *Small* 4:1240–1246.

DeNardo, S. J., G. L. DeNardo, L. A. Miers et al. 2005. Development of tumor targeting bioprobes (111In-chimeric L6 monoclonal antibody nanoparticles) for alternating magnetic field cancer therapy. *Clinical Cancer Research* 11:7087s–7092s.

Doering, W. E., M. E. Piotti, M. J. Natan et al. 2007. SERS as a foundation for nanoscale, optically detected biological labels. *Advanced Materials* 19:3100–3108.

Gao, X., Y. Cui, R. M. Levenson et al. 2004. in vivo cancer targeting and imaging with semiconductor quantum dots. *Nature Biotechnology* 22:969–976.

Gessner, R., and P. A. Dayton. 2010. Advances in molecular imaging with ultrasound. *Molecular Imaging* 9:117–127.

Hahn, M., A. Singh, and P. Sharma. 2010. Nanoparticles as contrast agents for in-vivo bioimaging: Current status and future perspectives. *Analytical and Bioanalytical Chemistry* 300:3–27.

Harisinghani, M. G., J. Barentsz, P. F. Hahn et al. 2003. Noninvasive detection of clinically occult lymph-node metastases in prostate cancer. *New England Journal of Medicine* 348:2491–2499.

He, X., K. Wang, and Z. Cheng. 2010. In vivo near-infrared fluorescence imaging of cancer with nanoparticle-based probes. *Wiley Interdisciplinary Reviews: Nanomedicine and Nanobiotechnology* 2:349–366.

Hilderbrand, S. A., F. Shao, C. Salthouse et al. 2009. Upconverting luminescent nanomaterials: Application to in vivo bioimaging. *Chemical Communications* :4188–4190.

Hu, M., J. Chen, Z. Y. Li et al. 2006. Gold nanostructures: Engineering their plasmonic properties for biomedical applications. *Chemical Society Reviews* 35:1084–1094.

Hwang, D. W., H. Y. Ko, S. K. Kim et al. 2009. Development of a quadruple imaging modality by using nanoparticles. *Chemistry — A European Journal* 15:9387–9393.

Jackson, P. A., W. N. Rahman, C. J. Wong et al. 2010. Potential dependent superiority of gold nanoparticles in comparison to iodinated contrast agents. *European Journal of Radiology* 75:104–109.

Jain, P. K., X. Huang, I. H. El-Sayed et al. 2008. Noble metals on the nanoscale: Optical and photothermal properties and some applications in imaging, sensing, biology, and medicine. *Accounts of Chemical Research* 41:1578–1586.

Jennings, L. E., and N. J. Long. 2009. 'Two is better than one' — Probes for dual-modality molecular imaging. *Chemical Communications* 24:3511–3524.

Jokerst, J. V., and S. S. Gambhir. 2011. Molecular imaging with theranostic nanoparticles. *Accounts of Chemical Research* 44:1050–1060.

Jokerst, J. V., Z. Miao, C. Zavaleta et al. 2011. Affibody-functionalized gold–silica nanoparticles for Raman molecular imaging of the epidermal growth factor receptor. *Small* 7:625–633.

Khullar, O., J. V. Frangioni, M. Grinstaff et al. 2009. Image-guided sentinel lymph node mapping and nanotechnology-based nodal treatment in lung cancer using invisible near-infrared fluorescent Light. *Seminars in Thoracic and Cardiovascular Surgery* 21:309–315.

Kim, J. W., E. I. Galanzha, E. V. Shashkov et al. 2009. Golden carbon nanotubes as multimodal photoacoustic and photothermal high-contrast molecular agents. *Nature Nanotechnology* 4:688–694.

Kim, S., Y. T. Lim, E. G. Soltesz et al. 2004. Near-infrared fluorescent type II quantum dots for sentinel lymph node mapping. *Nature Biotechnology* 22:93–97.

Kooi, M. E., V. C. Cappendijk, and K. B. Cleutjens et al. 2003. Accumulation of ultrasmall superparamagnetic particles of iron oxide in human atherosclerotic plaques can be detected by in vivo magnetic resonance imaging. *Circulation* 107: 2453–2458.

Lanza, G. M., D. R. Abendschein, C. S. Hall et al. 2000. In vivo molecular imaging of stretch-induced tissue factor in carotid arteries with ligand-targeted nanoparticles. *Journal of the American Society of Echocardiography* 13:608–614.

Liong, M., J. Lu, M. Kovochich et al. 2008. Multifunctional inorganic nanoparticles for imaging, targeting, and drug delivery. *ACS Nano* 2:889–896.

Manus, L. M., D. J. Mastarone, E. A. Waters et al. 2010. Gd(III)-nanodiamond conjugates for MRI contrast enhancement. *Nano Letters* 10:484–489.

Morton, D. L., A. J. Cochran, J. F. Thompson et al. 2005. Sentinel node biopsy for early-stage melanoma: Accuracy and morbidity in MSLT-I, an international multicenter trial. *Annals of Surgery* 242:302–313.

Muddana, H. S., T. T. Morgan, J. H. Adair et al. 2009. Photophysics of Cy3-encapsulated calcium phosphate nanoparticles. *Nano Letters* 9:1559–1566.

Mulder, W. J., D. P. Cormode, S. Hak et al. 2008. Multimodality nanotracers for cardiovascular applications. Nature clinical practice. *Cardiovascular Medicine* 5(Suppl 2):S103–S111.

Mulder, W. J. M., R. Koole, R. J. Brandwijk et al. 2006. Quantum dots with a paramagnetic coating as a bimodal molecular imaging probe. *Nano Letters* 6:1–6.

Na, H. B., I. C. Song, and T. Hyeon. 2009. Inorganic nanoparticles for MRI contrast agents. *Advanced Materials* 21:2133–2148.

National Cancer Institute (NCI) 2010. Cancer Trends Progress Report 2009/2010 Update. National Cancer Institute, NIH DHHS, Bethesda, MD. http://progressreport.cancer.gov.

Nie, S., and S. R. Emory. 1997. Probing single molecules and single nanoparticles by surface-enhanced raman scattering. *Science* 275:1102–1106.

Ntziachristos, V. 2010. Going deeper than microscopy: The optical imaging frontier in biology. *Nature Methods* 7: 603–614.

Perazella, M. A. 2008. Gadolinium-contrast toxicity in patients with kidney disease: Nephrotoxicity and nephrogenic systemic fibrosis. *Current Drug Safety* 3:67–75.

Pramanik, M., K. H. Song, M. Swierczewska et al. 2009. In vivo carbon nanotube-enhanced non-invasive photoacoustic mapping of the sentinel lymph node. *Physics in Medicine and Biology* 54:3291–3301.

Rabin, O., J. M. Perez, J. Grimm et al. 2006. An x-ray computed tomography imaging agent based on long-circulating bis muth sulphide nanoparticles. *Nature Materials* 5: 118–122.

Ravizzini, G., B. Turkbey, T. Barrett et al. 2009. Nanoparticles in sentinel lymph node mapping. *Wiley Interdisciplinary Reviews. Nanomedicine and Nanobiotechnology* 1: 610–623.

Ruehm, S. G., C. Corot, P. Vogt et al. 2001. Magnetic resonance imaging of atherosclerotic plaque with ultrasmall superparamagnetic particles of iron oxide in hyperlipidemic rabbits. *Circulation* 103:415–422.

Schmitz, S. A., T. Albrecht, and K. J. Wolf. 1999. MR angiography with superparamagnetic iron oxide: Feasibility study. *Radiology* 213:603–607.

Shin, J., R. M. Anisur, M. K. Ko et al. 2009. Hollow manganese oxide nanoparticles as multifunctional agents for magnetic resonance imaging and drug delivery. *Angewandte Chemie - International Edition* 48:321–324.

Sonvico, F., S. Mornet, S. Vasseur et al. 2005. Folate-conjugated iron oxide nanoparticles for solid tumor targeting as potential specific magnetic hyperthermia mediators: Synthesis, physicochemical characterization, and in vitro experiments. *Bioconjugate Chemistry* 16:1181–1188.

Soo Choi, H., W. Liu, P. Misra et al. 2007. Renal clearance of quantum dots. *Nature Biotechnology* 25:1165–1170.

Sosnovik, D., and R. Weissleder. 2005. Magnetic resonance and fluorescence based molecular imaging technologies. *Progress in Drug Research* 62:83–115.

Sun, I. C., D. K. Eun, J. H. Na et al. 2009. Heparin-Coated gold nanopartieles for liver-specific CT imaging. *Chemistry – A European Journal* 15:13341–13347+13276.

Tallury, P., K. Payton, and S. Santra. 2008. Silica-based multimodal/multifunctional nanoparticles for bioimaging and biosensing applications. *Nanomedicine* 3:579–592.

Tang, A. M., J. S. Ananta, H. Zhao et al. 2010. Cellular uptake and imaging studies of gadolinium-loaded single-walled carbon nanotubes as MRI contrast agents. *Contrast Media & Molecular Imaging* 6:93–99.

Torchilin, V. 2007. Targeted pharmaceutical nanocarriers for cancer therapy and imaging. *The AAPS Journal* 9:E128–E147.

Trivedi, R. A., J. M. U-King-Im, M. Graves et al. 2004. In vivo detection of macrophages in human carotid atheroma: Temporal dependence of ultrasmall superparamagnetic particles of iron oxide-enhanced MRI. *Stroke, a Journal of Cerebral Circulation* 35:1631–1635.

Von Maltzahn, G., A. Centrone, J. H. Park et al. 2009a. SERS-coded cold nanorods as a multifunctional platform for densely multiplexed near-infrared imaging and photothermal heating. *Advanced Materials* 21:3175–3180.

Von Maltzahn, G., J. H. Park, A. Agrawal et al. 2009b. Computationally guided photothermal tumor therapy using long-circulating gold nanorod antennas. *Cancer Research* 69:3892–3900.

Wang, L., Y. Zhang, and Y. Zhu. 2010. One-pot synthesis and strong near-infrared upconversion luminescence of poly(acrylic acid)-functionalized YF_3:Yb^{3+}/Er^{3+} nanocrystals. *Nano Research* 3:317–325.

Wang, L. V. 2009. Multiscale photoacoustic microscopy and computed tomography. *Nature Photonics* 3:503–509.

Wang, X., G. Ku, M. A. Wegiel et al. 2004. Noninvasive photoacoustic angiography of animal brains in vivo with near-infrared light and an optical contrast agent. *Optics Letters* 29:730–732.

Weissleder, R., P. F. Hahn, D. D. Stark et al. 1988. Superparamagnetic iron oxide: Enhanced detection of focal splenic tumors with MR imaging. *Radiology* 169:399–403.

Welsher, K., Z. Liu, S. P. Sherlock et al. 2009. A route to brightly fluorescent carbon nanotubes for near-infrared imaging in mice. *Nature Nanotechnology* 4:773–780.

Wickline, S., and G. Lanza. 2002. Molecular imaging, targeted therapeutics, and nanoscience. *Journal of Cellular Biochemistry* 87:90–97.

Wickline, S. A., and G. M. Lanza. 2003. Nanotechnology for molecular imaging and targeted therapy. *Circulation* 107:1092–1095.

Wickline, S. A., A. M. Neubauer, P. M. Winter et al. 2007. Molecular imaging and therapy of atherosclerosis with targeted nanoparticles. *Journal of Magnetic Resonance Imaging JMRI* 25:667–680.

Yang, K., J. Wan, S. Zhang et al. 2011. In vivo pharmacokinetics, long-term biodistribution, and toxicology of PEGylated graphene in mice. *ACS Nano* 5:516–522.

Yang, S. T., L. Cao, P. G. Luo et al. 2009. Carbon dots for optical imaging in vivo. *Journal of the American Chemical Society* 131:11308–11309.

Yu, X. F., L. D. Chen, M. Li et al. 2008. Highly efficient fluorescence of NdF_3/SiO_2 core/shell nanoparticles and the applications for in vivo NIR detection. *Advanced Materials* 20:4118–4123.

Zheng, Z. J., J. B. Croft, W. H. Giles et al. 2001. Sudden cardiac death in the United States, 1989 to 1998. *Circulation* 104:2158–2163.

Optical Microscopy of Plasmonic Nanoparticles

Varun P. Pattani
The University of Texas at Austin

James W. Tunnell
The University of Texas at Austin

7.1 Introduction

Nanoparticles are widely researched in the field of biotechnology for their use as a contrast and therapeutic agent for several different diseases. They are composed of an assortment of materials fabricated on the nanoscale. In this size range, typical bulk methods to measure the electromagnetic properties do not apply to certain materials, including noble metal particles, known as plasmonic nanoparticles. Plasmonic nanoparticles exhibit a property identified as surface plasmon resonance (SPR) (Jain et al. 2008). In this chapter, we will focus on this class of nanoparticles—specifically gold and silver, as they are the most commonly used nanoparticles in biological applications because of their SPR phenomena found in the visible to the near infrared region of the electromagnetic spectrum (Kreibig and Vollmer 1995).

Plasmonic nanoparticles have been used extensively for biological imaging and detection. The preferred method of visualizing these plasmonic nanoparticles is through optical microscopy, as a result of their absorption and scattering properties. In this chapter, we divide the common optical microscopy techniques into two sections: scattering-based microscopy and luminescence-based microscopy. Some scattering-based microscopy techniques, such as differential interference contrast (DIC) and dark-field microscopy (DFM), are used more commonly because of their ease of use and availability. Optical coherence tomography (OCT), another scattering-based technique, has a larger imaging depth than other typical microscopy techniques (1–2 mm) and can be used for *in vivo* imaging of plasmonic nanoparticles in three dimensions. However, two-photon microscopy (TPM), a luminescence-based technique, implements thin optical sectioning, which allows for higher resolution than the other microscopy techniques in three-dimensional (3-D) images. Thus, the ideal choice of microscopy technique for imaging plasmonic nanoparticles depends on the application and obtainable resources. This chapter reviews the available microscopy techniques for imaging plasmonic nanoparticles in cells and whole tissue. We begin with a description of the physical characteristics of plasmonic nanoparticles that affect the imaging techniques. Then, we discuss the scattering and luminescence-based microscopy techniques along with specific examples from recent literature.

7.2 Plasmonic Nanoparticles

7.2.1 Surface Plasmon Resonance

SPR is a unique property utilized for the detection and imaging of plasmonic nanoparticles, first realized by Gustav Mie in 1908 (Mie 1908). Plasmonic nanoparticles absorb and scatter light regulated by the SPR characteristic frequency. The incident light electric field induces the collective oscillation of the conduction electrons on the nanoparticle surface. Consequently, the nanoparticle electric field is enhanced and displaced by the collective oscillation of the surface conduction electrons allowing for the nanoparticle absorption during the nonradiative decay (Figure 7.1) (Bohren and Huffman 1983; El-Sayed 2001; Kreibig and Vollmer 1995). Adversely, nanoparticle scattering, in the Rayleigh regime, corresponds specifically to the electromagnetic energy dissipated from the nanoparticle as a result of interaction with the oscillating electric field (Kreibig and Vollmer 1995).

The SPR characteristic frequency depends strongly on the physical and electromagnetic properties of the plasmonic nanoparticle. Studies have investigated the effect of metal type, geometry (Jain et al. 2006b; Kelly et al. 2003; Link and El-Sayed 2000), internanoparticle coupling (Jain et al. 2006a; Rechberger et al.

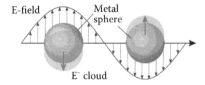

FIGURE 7.1 Schematic of plasmon oscillation for a sphere, showing the displacement of the conduction electron charge cloud relative to the nuclei. (Reproduced with permission from Kelly, K.L. et al., *J. Phys. Chem. B*, 107, 668–677, 2003.)

2003; Su et al. 2003), nanoparticle surface functionalization, and the nanoparticle and surrounding medium dielectric properties on SPR frequency (Ghosh et al. 2004; Underwood and Mulvaney 1994). Utilizing these characteristics, the SPR frequency can be tuned from the ultraviolet (UV) to the infrared (IR). Typically, hemoglobin (absorbing in the visible, ~533 nm), melanin (absorbs in the UV) and water (absorbs from the mid-IR to IR) are the most prominent molecular absorbers in the body. Thus, the near-infrared wavelength (NIR, 600–1000 nm) range, known as the therapeutic window, is the most desired for biological applications because there is less interference, from scattering and absorption, compared to other wavelengths between the UV and IR (Patterson et al. 1989).

The metal nanoparticle electron relaxation is achieved through the nonradiative decay pathway of the plasmon oscillations, leading to two different effects: heat and luminescence. The rapid localized heating uniformly heats the surroundings of the nanoparticle. It is a photothermal effect proportional to the power of the incident light electromagnetic field (Link and El-Sayed 2000). The luminescence, on the other hand, can be used as a contrast mechanism to image plasmonic nanoparticles. The nanoparticle Rayleigh scattering is also effective as a contrast mechanism for imaging and determining the structure of these plasmonic nanoparticles. In this chapter, we have separated the two primary optical microscopy methods of nanoparticles into luminescence-based and scattering-based techniques.

The optical properties and peak SPR frequency can be modeled for gold and silver nanoparticles using solutions to Maxwell's equations. The SPR-dependent values, including nanoparticle geometry and the dielectric properties of the nanoparticle and surrounding media, are necessary to model the nanoparticle effectively. Mie theory is widely used as an exact solution to Maxwell's equations for homogenous concentric spheres (Mie 1908). It has also been expanded to solve for spheres with layered distributions of different materials. According to the Mie theory, small spherical nanoparticles exhibit only one peak plasmon resonance frequency, whereas anisotropic particles exhibit two or more peak plasmon bands based on their shape.

Typically, the discrete dipole approximation (DDA), based on the lattice dispersion relation, is used to determine the optical properties and SPR frequency of nonspherical, anisotropic nanoparticles. The approximation models the geometry of the nanoparticle as a finite array of polarizable points, which act as dipoles because of the incident electric fields (Draine and Flatau

1994; Purcell and Pennypacker 1973). Draine and Flatau (2010) developed the DDSCAT program to use the DDA to solve for the absorption and scattering of electromagnetic waves by targets of arbitrary geometry. This approximation has been used to study colloids of both gold and silver (Felidj et al. 1999) as well as gold nanorods (Brioude et al. 2005; Lee and El-Sayed 2005; Prescott and Mulvaney 2006; Ungureanu et al. 2009).

The resulting values from both Mie theory and DDA are absorption and scattering efficiencies for each wavelength modeled. The wavelength that corresponds to the maximum absorption efficiency is the peak SPR frequency. When added together for each wavelength, the absorption and scattering efficiencies amount to the extinction efficiency, the total attenuation by the nanoparticle. The ratio of the absorption and scattering efficiency values, respectively, with the total extinction values indicate the percentage of light absorbed or scattered at each wavelength. With the scattering and absorption efficiency values, determining the optical scattering and absorption cross section is straightforward:

$$\sigma_s = Q_s * \sigma_g$$

$$\sigma_a = Q_a * \sigma_g$$

where σ_s and σ_g are the optical scattering and absorption cross sections, respectively, Q_s and Q_a are the scattering and absorption efficiencies, respectively, and σ_g is the geometric cross section, which can be measured (Wang and Wu 2007). The wavelengths at which nanoparticles strongly scatter or absorb are directly based on these results, which are exploited for scattering-based and luminescence-based imaging techniques.

7.2.2 Silver Nanoparticles

Silver, when fabricated on the nanoscale, exhibits localized SPR. Because of the SPR peak dependence on the nanoparticle morphology, silver nanoparticles have been fabricated with different geometries (Jain et al. 2006b, 2008). The most common is the silver nanosphere, spherical in shape and usually ranging in size between 1 and 50 nm. Its peak SPR frequency is entrenched in the mid-visible wavelength range and can be tuned slightly by modifying the radius of the nanoparticle. The increase in the peak SPR frequency is minor because it is not directly proportional to large increases in radius; therefore, in comparison with other nanoparticles, silver nanospheres are considered relatively untunable.

Several groups have fabricated silver nanoparticles (Figure 7.2) with SPR frequencies in different parts of the spectrum, including silver nanoprisms (Jin et al. 2001; Sherry et al. 2006), nanopentagons (Mock et al. 2002), and nanocubes (Sherry et al. 2005). For example, silver nanoprisms have, in contrast to the nanospheres, three SPR bands: a strong in-plane dipole plasmon resonance peak in the NIR; an in-plane quadrapole plasmon resonance peak in the mid-visible wavelength; and a weak out-of-plane quadrapole resonance peak in the low-visible wavelength

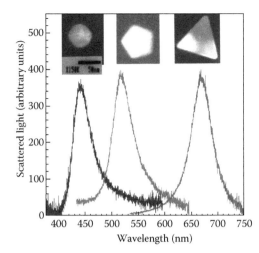

FIGURE 7.2 Typical optical spectroscopy of individual silver nanoparticles. The figure shows the spectrum of individual red (~680 nm), green (~520 nm), and blue (~450 nm) particles. And the high-resolution transmission electron microscopy (TEM) images of the corresponding particle are shown above their respective spectrum. Triangular-shaped particles appear mostly red, pentagon shaped particles appear green, and spherical particles appear blue. (Reproduced with permission from Mock, J. et al., *J. Chem. Phys.*, 116, 6755, 2002.)

(Jin et al. 2001). However, in most optical imaging studies, silver nanospheres have been the nanoparticle geometry of choice.

For biological applications, silver nanoparticles have been found to be harmful and relatively unstable *in vitro* and *in vivo*. In fact, silver ions are known to be strongly toxic to a wide range of microorganisms, having been used as antimicrobial agents since the early Roman era (Chen and Schluesener 2008). The toxicity of silver ions toward microbes, such as bacteria and fungi, has been well characterized (Feng et al. 2000; Gupta 1998; Gupta et al. 1998; Matsumura et al. 2003). There is a higher threshold for silver toxicity in mammalian cells than for microbes; thus, silver nanoparticles have been applied for many wound healing applications for humans. Silver ions can readily bind to sulfur on the cell membrane causing granules to form leading to significant structural changes to the bacterial cell membrane (Gupta et al. 1998). Furthermore, silver ions have been shown to target the mitochondria within cells disrupting the membrane causing pores to form allowing for an increase in the mitochondrial permeability, leading to cell death (Almofti et al. 2003).

As a result, there has been an amplified interest in applying silver nanoparticles as an antibacterial, such as in wound dressings and textiles (Chen and Schluesener 2008). Yet, the mechanism behind the bactericidal effect of the silver nanoparticles, rather than silver ions alone, was not fully understood. To understand this effect, several studies used optical imaging to localize the silver nanoparticles in determining the cause of cytotoxicity (AshaRani et al. 2008, 2009; Carlson et al. 2008; Nallathamby and Xu 2010; Stensberg et al. 2011). Some determined that cytotoxic mechanisms are based on the release of silver ions after internalization of the silver nanoparticle leading to the causes discussed previously. Others found that the cellular uptake of

the silver nanoparticles leads to increased production of radical oxygen species due to mitochondrial membrane disruption causing oxidative stress and cell death (AshaRani et al. 2008; Carlson et al. 2008; Hussain et al. 2005). Nevertheless, numerous studies have still used silver nanoparticles as a contrast agent for optical microscopy *in vitro*.

7.2.3 Gold Nanoparticles

Gold is more widely used in metal nanoparticle fabrication for biological applications because it is inert and biocompatible. For many years, gold was ingested as an anti-inflammatory agent and for arthritis treatment (Messori and Marcon 2004). The gold surface chemistry is also well understood, making the conjugation of biological molecules to the surface straightforward (Burda et al. 2005; Katz and Willner 2004). Specific biomolecules, such as polymers, antibodies and peptides, can be conjugated onto the surface to enhance biocompatibility and residence times in the body as well as actively target diseased tissues.

Many groups have capitalized on the localized heat emission of gold nanoparticles for use in photothermal therapy of cancer (Chen et al. 2007; Dickerson et al. 2008; El-Sayed et al. 2006; Gobin et al. 2007; Hirsch 2003; Huang et al. 2006; Loo et al. 2005; Melancon et al. 2008; O'Neal et al. 2004; Pitsillides et al. 2003; Tong et al. 2007). In photothermal therapy, incident light electromagnetic waves induce localized nanoparticle heat emission to ablate diseased tissue (Anderson and Parrish 1983). Gold nanoparticles are effective photothermal agents that can, either passively or actively (via specific targeting to biomolecules

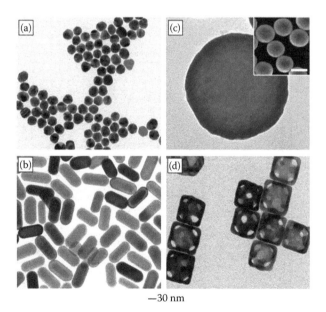

—30 nm

FIGURE 7.3 TEM images showing the structure of most common gold nanoparticles in biological applications: (a) gold nanospheres, (b) gold nanoshells, (c) gold nanorods, (d) gold nanocages. (Reproduced with permission from Cobley, C.M. et al., *Chem. Soc. Rev.*, 40, 44–56, 2011; adapted with permission from Cho, E.C. et al., *Angew. Chem.*, 2020–2024, 2010; Wang, H. et al., *Acc. Chem. Res.*, 40, 53–62, 2007.)

unique to specific tissues), reach diseased tissue. Using optical microscopy, studies have been performed to determine the targeting and binding abilities of the nanoparticles with cancerous tumor cells (Chen et al. 2007; Dickerson et al. 2008; El-Sayed et al. 2006; Gobin et al. 2007; Hirsch 2003; Huang et al. 2006; Loo et al. 2005; Melancon et al. 2008; O'Neal et al. 2004; Pitsillides et al. 2003; Tong et al. 2007).

Gold nanoparticles have been fabricated in many different shapes and sizes. The most basic form, gold nanospheres, is a solid, spherical colloid, which can range in size anywhere in the nanoscale (Figure 7.3a). Nanospheres tend to absorb light in the visible region of the spectra, and their peak SPR frequency is relatively untunable (Figure 7.4b). Similar to the silver nanospheres, the peak SPR frequency depends on the particle radii but large changes in the radii only slightly influence the frequency (Link and El-Sayed 2000). According to the Mie theory, gold nanospheres have a higher absorption component than scattering because of their small size, but are still visualizable by both

scattering-based and luminescence-based imaging techniques. As the nanosphere size increases, the scattering component increases and the absorption component decreases. This trend is seen in all geometries of gold nanoparticles; therefore, it has been stated that, generally, smaller gold nanoparticles (<50 nm) are absorption dominant whereas larger gold nanoparticles (>50 nm) are scattering dominant (Jain et al. 2006b; Kelly et al. 2003; Kreibig and Vollmer 1995).

Several groups have fabricated other nanoparticles with a tunable SPR frequency, including the gold nanoshell. The pioneering work on the gold nanoshell was performed by research groups from Rice University (Halas, Drezek, and West). The nanoshell SPR frequency is tunable because it is composed of a silica core coated with a gold shell (Figure 7.3b) (Averitt et al. 1999; Oldenburg et al. 1998, 1999). Adjusting the shell thickness to total diameter ratio modifies the peak SPR frequency, making it possible to tune the frequency into the NIR (Figure 7.4b). These particles are relatively large, with sizes on the order of

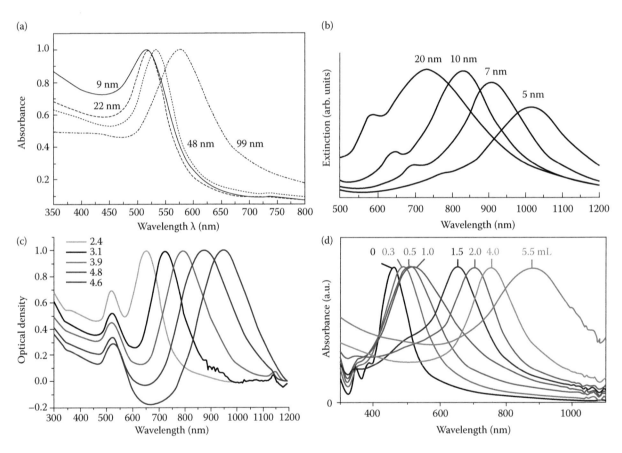

FIGURE 7.4 Tunability of the peak surface plasmon resonance frequency for the most common gold nanoparticles. (a) Optical spectra of gold nanospheres at sizes ranging from 9 to 99 nm, showing that the peak SPR frequency can be modified only slightly and not into the NIR. (c) Extinction spectra of gold nanoshells with different gold shell thicknesses (5–20 nm) at a single core diameter (120 nm), showing that the peak SPR frequency can be tuned from ~720 to ~1050 nm in the NIR. (c) Gold nanorod optical spectra measured at different aspect ratios (2.4–5.6) showing that there are two peak SPR frequencies from the longitudinal and transverse axis and that the stronger longitudinal peak can be tuned from ~640 to ~1000 nm. (d) Optical spectra of gold nanocages prepared with different volumes of HAuCl$_4$, showing that the peak SPR frequency can be tuned from ~500 to ~900 nm. (Adapted with permission from Huang, X.H. et al., *J. Am. Chem. Soc.*, 128, 2115–2120, 2006; Link, S., El-Sayed, M.A., *J. Phys. Chem. B*, 103, 4212–4217, 1999; Oldenburg, S. et al., *Chem. Phys. Lett.*, 288, 243–247, 1998; Skrabalak, S.E. et al., *Acc. Chem. Res.*, 41, 1587–1595, 2008.)

100 nm. Because of their larger size, these nanoparticles have a stronger scattering component than absorbing. Even though the scattering component is dominant, the absorption cross section of nanoshells is still much larger than typical fluorescent molecules (Averitt et al. 1999; Jain et al. 2006b). As a modification on traditional nanoshells, one group has also fabricated hollow gold nanoshells with a hollow core and a gold shell, which are still tunable, but smaller than typical gold nanoshells, on the order of 40–50 nm (Lu et al. 2009; Melancon et al. 2008).

Gold nanorods are another frequently used tunable particle, composed of solid gold in a pill shape, ranging in size typically from 20 to 50 nm (Figure 7.3c). These nanoparticles are smaller than typical nanoshells and, as a result, are absorption dominant, which is essential for luminescence-based imaging techniques (Yu et al. 1997). Even though the scattering is dominated by the absorption, the scattering component is strong enough to be imaged by many scattering-based techniques. Because of their pill shape, gold nanorods have a longitudinal (along the length of the gold nanorod) axis and a transverse (along the width of the gold nanorod) axis (Figure 7.4c). Accordingly, each axis has an SPR peak frequency, but because of their relative sizes, the longitudinal axis SPR peak is much stronger than the transverse axis SPR peak (Kooij and Poelsema 2006). By increasing the aspect ratio (ratio of the length and width dimension), the SPR peak wavelength progresses to longer wavelengths (Figure 7.4c) and can be tuned to the NIR (Brioude et al. 2005; Jain et al. 2006b; Link et al. 1999).

Gold nanocages are tunable, hollow nanostructures surrounded by porous walls, on the order of 10–100 nm (Figure 7.3d; Chen et al. 2005b). The gold nanocages are fabricated by interacting chloroauric acid ($HAuCl_4$) with silver nanocubes. The SPR frequency can be tuned to the NIR by varying the number of silver nanocubes mixed with differing amounts of chloroauric solution, modifying the wall thickness (Figure 7.4d). Because of the variability in size of these nanoparticles, gold nanocages can be fabricated as scattering- or absorption-dominated particles depending on the application (Chen et al. 2005a, 2007).

7.3 Optical Microscopy

In this section, we discuss a variety of optical microscopy techniques divided by the contrast mechanism used for differentiating plasmonic nanoparticles from their surroundings. As discussed earlier, plasmonic nanoparticles have strong scattering and absorption cross sections that can be exploited as contrast mechanisms for imaging. Scattering-based microscopy includes such techniques as bright-field, DIC, dark-field and reflectance confocal microscopy as well as OCT. By using the absorption properties of plasmonic nanoparticle, it is also possible to visualize the resulting luminescence emitted from the plasmonic nanoparticle. Luminescence-based modalities that are discussed include epifluorescence, fluorescence confocal, and TPM.

A common disadvantage found in all optical microscopy techniques is diffraction-limited resolution. Diffraction is essentially the bending and spreading of waves, and occurs when an

electromagnetic wave, such as light, is incident on a circular lens or mirror. As a result, the light does not focus to a singular point but rather an Airy disk. The Airy disk is observed at the best focused point as a circular spot of light. The diameter of the Airy disk and, consequently, the microscope resolution depends directly on the wavelength of light (λ) used to illuminate the sample:

$$d = \frac{\lambda}{2(n \sin \theta)}$$

where d is the resolution (and Airy disk diameter), n is the refractive index of medium from the objective to the sample, and θ is the angle at which the light is converging.

Normally, with visible light the ideal resolution is limited to approximately 0.2 μm (micron to submicron range); therefore, these optical techniques are known to be microscopic. Plasmonic nanoparticles are much smaller than this resolution limit; thus, with traditional microscopy, it is difficult to resolve them individually, as shown in this section. The following optical microscopy techniques include traditional modalities (e.g., brightfield and epifluorescence microscopy) of visualizing plasmonic nanoparticles in aggregated form as well as modalities (laser scanning confocal microscopy and TPM) that push the diffraction-limited boundaries of resolution for single nanoparticle tracking.

7.3.1 Scattering-Based Microscopy

Several light microscopy techniques for imaging biological material use scattered light as the primary source of contrast. Biological materials, including cells and tissue, are optically transparent and relatively invisible under transmitted light microscopy. Whereas, based on the structure of these materials, they are highly scattering; therefore, scattering-based imaging modalities exploit this fact, allowing for the independent tracking of nanoparticles, based on geometry, in biological material.

In this section, we discuss some common scattering-based imaging techniques that have been used with plasmonic nanoparticles as contrast agents. DIC microscopy is an interferometer-based technique that images scattering in optically thin samples with similar resolutions as brightfield microscopy. DFM is the most common technique for imaging plasmonic nanoparticles because of its accessibility and high signal strength. Reflectance laser scanning confocal microscopy (LSCM) is a more complex modality for imaging plasmonic nanoparticles, limiting accessibility, but it has a high spatial (0.5–1.5 μm) and lateral resolution (350 nm), as well as the capability for thin optical sectioning allowing for 3-D imaging. OCT is another complex modality with a resolution in the micron range and an imaging depth of 1–2 mm, allowing for *in vivo* imaging of plasmonic nanoparticles in cross-sectional images.

7.3.1.1 Brightfield Microscopy

Brightfield light microscopy is the most ubiquitous technique used for general microscopy. Brightfield microscopy is based on the transmission of light through an object based on the attenuation of the sample, which is primarily scattering in biological material. Therefore, in this chapter, we consider brightfield microscopy a scattering-based microscopy technique. Light microscopes are composed of several basic components: a light source, condenser lens (for focusing), stage, objective (for collection and magnification), and ocular lens. Brightfield microscopy inherently has low contrast because of its dependence on transmitted light; thus, it is not possible to image optically transparent, thin specimens, such as cellular structures, because their absorbance is too low (Davidson and Abramowitz 2002).

Since nanoparticles are much smaller than the diffraction limit, it is nearly impossible to visualize individual nanoparticles in brightfield microscopy. Nonetheless, it is possible to image plasmonic nanoparticles when aggregated or enhanced in size by a visible stain. One such stain used to image gold nanoparticles is the silver enhanced stain, which is composed of reactive silver nitrate, when catalyzed by the gold nanoparticles, deposits onto the surface of the gold nanoparticle. Visualization of the gold nanoparticles at the microscope magnification is accomplished because of the enhanced size of the gold–silver conjugates. A study used this technique with brightfield microscopy to examine the gold nanoparticle distribution *in vivo*, based on particle size and surface chemistry, in histological sections of breast cancer tumors (Figure 7.5) (Perrault et al. 2009). This technique has also been used to analyze the targeting abilities

of anti-HER2 (human epidermal growth factor receptor 2) antibody conjugated gold nanorods in histological sections of breast cancer tumors (Eghtedari et al. 2009) and Tf (transferrin)-conjugated gold nanospheres in histological sections of neuroblastoma tumors (Choi et al. 2010).

Using brightfield microscopy for imaging plasmonic nanoparticles is advantageous because a standard microscope can be used. However, there are significant disadvantages to this technique including the need for significant processing of the sample and allowing for the possibility of incomplete labeling. Furthermore, the resolution is very low and brightfield microscopy images of plasmonic nanoparticles are not at the single particle level.

7.3.1.2 DIC Microscopy

DIC microscopy is a technique that enhances the contrast of unstained samples with interferometry, which uses interference to determine the sample structure (Davidson and Abramowitz 2002). In this setup, light enters the microscope and is separated by a prism into two orthogonally polarized, mutually coherent beams that travel through the sample, collected by an objective and recombined using another prism, creating an interference pattern. The interference provides contrast based on differences in the refractive index (scattering) and thickness (Murphy 2001). This allows for the imaging of plasmonic nanoparticles, because of their high scattering cross section, and biological material,

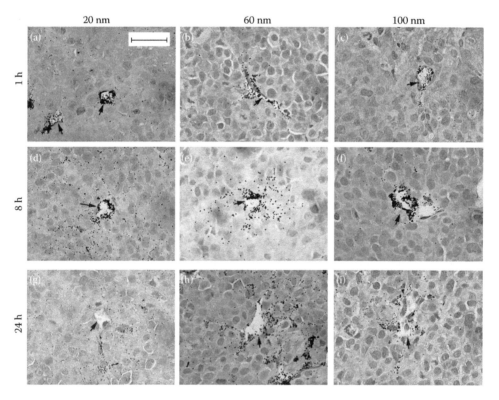

FIGURE 7.5 Particle size-dependent permeation of the tumor interstitial space. (a–i) Histological samples were obtained for 20, 60, and 100 nm particle sizes at 1, 8, and 24 h postinjection. Silver enhancement causes growth of the GNP, allowing their distribution relative to blood vessels (marked by arrows) to be visualized under brightfield microscopy (scale bar in a corresponds to 40 μm in all images). (Reproduced with permission from Perrault, S.D. et al., *Nano Letters*, 9, 1909–1915, 2009.)

such as cells, in one image without the need to correlate two images. A disadvantage of this system is that the resolution is not enhanced compared to traditional brightfield microscopy. Second, DIC microscopy requires a transparent sample with a similar refractive index to its surroundings; therefore, this technique is not preferable for thick tissue samples.

One study used DIC to determine the uptake and cellular targeting efficiency of gold nanospheres conjugated with polyethylene glycol (PEG) and cell penetrating peptides (CPPs) (Oh et al. 2011). PEG is a common "stealth" molecule functionalized onto the gold surface that increases the biocompatibility of the particles within the body. CPPs, typically based on viruses, assist nanoparticle cellular uptake. In this study, the authors used the silver staining technique discussed previously to enhance the size of the nanoparticles for DIC imaging. DIC microscopy allowed the authors to visualize both the nanospheres and the cells simultaneously to correlate and compare the locations of the PEG-functionalized and the CPP-conjugated nanospheres.

Feldheim and coworkers have used a combination of video enhanced color (VEC) and DIC microscopy to image gold nanoparticles. The VEC component allows for higher observed resolutions with nanoparticles as small as 20 nm with spectral information (Inoue 1981). Therefore, when combined with the scattering contrast with DIC, it is possible to distinguish gold nanospheres. By functionalizing gold nanospheres with nuclear localization signal (NLS) peptides, they can be targeted to the nuclei. In comparisons of several possible NLS peptide sequences and combinations, the authors used VECDIC to visualize the nuclei and the pathways that the nanoparticles travel using peptide-conjugated gold nanoparticles (Figure 7.6) (Tkachenko et al. 2003, 2004).

Another study was performed on gold and silver nanospheres with dual-wavelength DIC. Instead of using white light, this group modified their DIC microscope to illuminate the sample at two different wavelengths. In this case, the wavelengths corresponded to the SPR peaks of the gold nanospheres and silver nanospheres, respectively. This study first showed it could image gold nanospheres targeted to cells with a CPP. In addition, the authors determined that it was possible to image both silver and gold nanosphere on the same glass slide, by switching between the wavelengths (Sun et al. 2009).

7.3.1.3 Dark Field Microscopy

DFM is another technique that utilizes the same optical path as light microscopy. Typically, a white light source is used in DFM, but the light encounters a patch stop before the condenser, which blocks the inner circle leaving only an outer ring of light. The directly imaged light continues at an angle, through the sample, and is blocked again, such that only the scattered light is allowed to pass through to the objective. The resolution of DFM is comparable to brightfield microscopy (Sönnichsen et al. 2000).

DFM is the most widely used method of imaging plasmonic nanoparticles because of its simplicity, high signal strength, and temporal resolution. Similar to the DIC, DFM allows for the imaging of live and unstained biological samples with

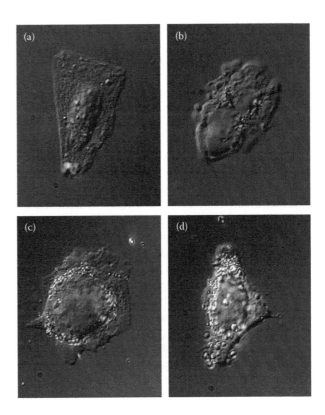

FIGURE 7.6 Nanoparticle–peptide complexes incubated with HepG2 cells for 2 h: adenoviral NLS (a), adenoviral RME (b), adenoviral fiber protein (large peptide containing adenoviral NLS and RME) (c), and combination of adenoviral NLS and RME (d). Cells were examined using a Leica DMLB DIC equipped microscope with 100×/1.3 NA objective. Images were taken and processed with a Nikon DMX-1200 digital color CCD camera. (Reproduced with permission from Tkachenko, A.G. et al., *J. Am. Chem. Soc.*, 125, 4700–4701, 2003.)

plasmonic nanoparticles without the need for correlating two images. Plasmonic nanoparticles can provide access to targets on and within the cell because of their size and surface chemistry, through functionalization of molecular targeting moieties. Therefore, this imaging technique allows for the independent tracking of plasmonic nanoparticles, depending on the tunable SPR frequency, during cellular binding, uptake, and localization within the cell. However, a shortcoming of DFM is that the sample needs to be illuminated strongly because of the low levels of light in the final image, which can cause damage.

One of the pioneering works using plasmonic nanoparticles as a contrast agent in DFM studied the use of ligand coated nanoparticles as target-specific labels for common procedures in biochemistry, cell biology, and medical diagnosis. In this study, the authors fabricated gold nanospheres coated with silver to modify the SPR peak. The authors presented a sandwich immunoassay for goat anti-biotin antibody based on silver nanoparticles, imaged using DFM (Schultz et al. 2000).

Most studies have utilized plasmonic nanoparticles as a targeting mechanism to either track or localize certain biological materials with DFM. For example, Sokolov and coworkers have shown that gold nanospheres can be used as an intracellular label to monitor molecular pathways, such as actin rearrangement in fibroblasts. In one study, the authors show that 20-nm gold nanospheres can be delivered into the cytoplasm using the TAT–HA2 combination peptide, where the TAT peptide allows for endosomal uptake into the cell and the HA2 peptide allows for escape from the endosome. Using DFM, the authors were able to demonstrate that this peptide combination worked in delivering the gold nanospheres into the cytoplasm in living cells. Furthermore, the authors attached an anti-actin antibody onto the nanosphere surface to bind to the actin filaments in fibroblasts, which led to a red-shift in the scattering signal and showed the distribution of these filaments as well as their movements in living cells (Kumar et al. 2007). The authors performed a similar study to monitor epidermal growth factor receptor (EGFR) trafficking in A431 (squamous cell carcinoma) cells over time. Gold nanospheres, 25 nm in diameter, were functionalized with anti-EGFR antibodies to target the overexpressed EGFR on the cell surface. Using DFM combined with hyperspectral imaging, it was possible to localize the gold nanospheres as well as determine their scattering peaks over time. Hyperspectral imaging, when combined with DFM, essentially collects the backscattered light from the sample into a spectrophotometer, to determine the electromagnetic spectra of the image, in addition to the CCD for DFM imaging (Aaron et al. 2009).

El-Sayed and coworkers have also used gold nanospheres and nanorods as intracellular molecular targeting agents to provide contrast in DFM. In one of their first studies, the authors functionalized anti-EGFR antibodies onto gold nanospheres (35 nm in diameter) and incubated them with three different cell lines: (1) nonmalignant keratinocytes (HaCaT), (2) and (3) two malignant human squamous cell carcinomas (HOC 313 and HSC 3). Imaging with DFM, unconjugated gold nanospheres were found to internalize within all three cell types because of nonspecific binding, but the anti-EGFR antibody conjugated nanospheres bound to the cellular membrane of both malignant cell lines at a 600% greater affinity than the nonmalignant cell line. Additionally when bound to the cell membrane, the spectra of the gold nanospheres were red-shifted, which can be exploited in biosensing applications (El-Sayed et al. 2005, 2006).

The authors performed similar selective detection and photothermal studies on cancer cells with anti-EGFR antibody conjugated gold nanorods (aspect ratio of 3.9). Gold nanorods were preferred because their peak SPR frequency is tuned to the NIR. The conjugated gold nanorods were found to have an extremely bright signal in the orange to red color, when visualized with DFM, corresponding to the longitudinal oscillation in the NIR. Shown with DFM, as with the gold nanospheres, the conjugated gold nanorods bind to the surface of the malignant cells at a much higher affinity than the nonmalignant cell line (Figure 7.7). The comparison in Figure 7.7 also shows that with DFM it is possible to distinguish between the two gold nanoparticle types based on the spectral signature corresponding to the maximum SPR frequency, with nanospheres and nanorods in the yellow and red wavelength regions, respectively (Huang et al. 2006).

The next few studies performed by El-Sayed and coworkers used DFM as a method for demonstrating the localization of gold nanospheres and nanorods. Gold nanospheres were targeted to the cytoplasm by conjugation with an arginine–glycine–aspartic acid peptide (RGD) and the nuclei by conjugation with a combination of the RGD and an NLS peptide (Huang et al. 2010). The conjugated nanospheres were found to selectively transport to the desired targets using DFM, and the RGD/NLS conjugated gold nanospheres were found to cause cell death through DNA damage (Kang et al. 2010). Similar studies were performed with nuclear targeted gold nanorods as an intracellular nanotracer, and localization was confirmed with DFM (Oyelere et al. 2007). Furthermore, this group modified an existing inverted light microscope setup by adding an environmental chamber and an angled dark-field illumination setup. Using the peptide-conjugated gold nanorods and this microscope system, the authors were able to monitor and determine the kinetics of the nuclear uptake of the nuclear targeted gold nanorods as well as the tracking of the cancer cell cycle from birth to division (Qian et al. 2010).

The groups from Rice University (discussed above) have also demonstrated that gold nanoshells are highly scattering nanoparticles and, consequently, effective DFM contrast agents. In one study, the authors targeted the gold nanoshells to cancer cells to reduce the photothermal power threshold necessary to kill cells. By using DFM to image, the authors were able to determine the exact location of the nanoshells. The authors showed that anti-HER2 antibody-conjugated nanoshells target and bind to the cell membrane of malignant breast cancer cells (SKBr3) that overexpress HER2 using DFM (Loo et al. 2004, 2005).

Hollow gold nanoshells have also been used as an effective contrast agent for DFM. One *in vitro* study used DFM to image anti-EGFR antibody conjugated hollow gold nanoshells. The DFM images showed preferential binding to A431 cells in comparison to hollow gold nanoshells conjugated with a nonspecific antibody (immunoglobulin G, IgG). In addition, the authors performed an *in vivo* study with the hollow gold nanoshells intravenously injected into tumor bearing mice, demonstrating that it was possible to image tumor slices *ex vivo* using DFM (Melancon et al. 2008).

DFM can also be used *in vivo* as demonstrated by silver nanosphere uptake studies performed on zebrafish embryos. In this study, the authors fabricated optically uniform, purified, and monodisperse silver nanospheres. These nanoparticles were incubated at subnanomolar concentrations with zebrafish embryos for hours, during which they tracked single particles using DFM. Using the silver nanoparticles as contrast agents for the continuous environmental imaging, the authors were able to measure changes in the diffusion coefficients of the silver nanoparticles, which correlated to local gradients in viscosities and flow patterns (Nallathamby et al. 2008).

Silver nanoparticles, as discussed before, are known to have strong antimicrobial properties; consequently, the uptake of silver nanoparticles in different microbial organisms has been studied using DFM. One study measured the uptake of silver

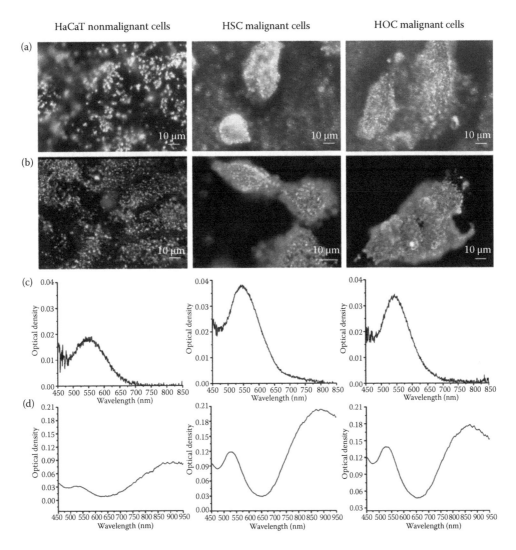

FIGURE 7.7 **(See color insert.)** (a) Light scattering images of anti-EGFR/Au nanospheres after incubation with cells for 30 min at room temperature. (b) Light scattering images of anti-EGFR/Au nanorods after incubation with cells for 30 min at room temperature. (c) Average extinction spectra of anti-EGFR/Au nanospheres from 20 different single cells for each kind. (d) Average extinction spectra of anti-EGFR/Au nanorods from 20 different single cells for each kind. (Reproduced with permission from Huang, X.H. et al., *J. Am. Chem. Soc.*, 128, 2115–2120, 2006.)

nanoparticles in *Pseudomonas aeruginosa*, a bacterial pathogen, imaged using DFM. The authors demonstrated that the silver nanoparticles (ranging up to 80 nm in diameter) are internalized through the inner and outer membranes of the microbial cells. Surprisingly, a low concentration accumulated within the cells at the same order of magnitude as a concentration of single silver nanoparticles, which does not adversely affect the bacteria (Xu et al. 2004).

DFM imaging of silver nanoparticles has also been used for studies analyzing the interactions between the nanoparticles and mammalian cells. One study measured the cytotoxicity of silver nanoparticles with fibrosarcoma cells (L929). Cell growth was found to reduce significantly when incubated with unconjugated silver nanoparticles. The cells showed considerable abnormal geometry after silver nanoparticle incubation for 72 h. Using DFM, the authors were able to determine that the silver

nanoparticles were found in the cytoplasm and nuclei at concentrations increasing over time (Nallathamby and Xu 2010).

Another study used DFM imaging of silver nanoparticles to characterize receptor molecules in fibroblast cells. Silver nanoparticles (11.6 nm) were functionalized with mercaptoundecanoic acid and IgG to be used as biosensors for the receptor molecules. The authors were able to detect individual receptor molecules, map the distribution of receptor molecules, and measure the kinetics of binding from the DFM imaging of the biosensors and cells over time (Huang et al. 2007). In this study, the authors functionalized silver nanoparticles (40 nm) with aptamer molecules, peptides that bind to a specific target protein within the cell, targeting specific proteins in the cell after caveolae-based endocytosis. Using DFM, the localization and distribution of the aptamer-adapted silver nanoparticles within the SK-N-SH (neuroblast) cells was determined. Furthermore, the authors obtained single nanoparticle spectra; thus, there is potential in using DFM and different-sized targeted silver nanoparticles for microenvironment analysis (Chen et al. 2010).

Silver nanoparticles have also been used in conjunction with gold nanorods for the multiplexed DFM imaging of pancreatic cancer cells (Panc-1 and MiaPaCa). Both nanoparticles were targeted to the pancreatic cancer cells by first being coated with a polyelectrolyte, to stabilize the specific nanoparticle, and then conjugated with Tf or specific antibodies (Figure 7.8). Since silver nanospheres scatter in the visible wavelength range and the gold nanorods scatter in the NIR, different locations of both nanoparticles can be distinguished within the same DFM image based on the spectral differences (Hu et al. 2009).

7.3.1.4 Reflectance LSCM

LSCM is a far-field optical imaging modality that enables 3-D sectioning of samples within the diffraction limited resolution. This technique uses point illumination scanned over the field of view instead of illuminating the sample field of view with evenly distributed light, the method used by all previously discussed microscopy modalities. Each point is illuminated, individually creating a component with discrete signal intensity. These components are limited in size based on the spot size of the laser, which depends on the diffraction limit (Corle et al. 1996; Pawley 2006).

The performance of the LSCM has been optimized substantially. An aperture is placed in front of the light source to reduce the higher-order diffraction patterns of the laser. This decreases the diameter of the Airy disk and increases the resolving power. However, the aperture reduces the amount of laser light and, accordingly, the sample signal intensity, allowing for background noise to possibly overwhelm the image. To compensate, it is necessary to increase the initial intensity of the laser and use highly sensitive photodetectors. A pinhole is placed in the detection path of the optical system to reject all light from regions outside the focal volume by clipping them at the aperture stop. As a result, the focal volume permits point resolved detection; thus, the detected signal is produced in a diffraction-limited region at the focal plane. Therefore, the resolution and depths of field range from 0.5 to 1.5 μm axially and down to 350 nm laterally, allowing for depth-resolved optical sectioning in three dimensions at imaging depths up to 500 μm (Corle et al. 1996; Pawley 2006).

There are two ways the confocal setup can be used to image plasmonic nanoparticles: (1) reflectance and (2) fluorescence. The latter is discussed in a forthcoming section. Reflectance LSCM collects backscattered (reflected) light to visualize samples instead of all scattered light, which can affect signal strength. The reflectance setup does not require any confocal microscope modification: the light goes through the optical path and is illuminated point by point, but only the reflected light is captured through the pinhole.

One of the first studies used reflectance LSCM to image gold nanospheres conjugated with anti-EGFR antibody to actively target SiHa (cervical tumor) cells. The images show the gold nanospheres bound to the cell membrane. In this study, the authors compared anti-EGFR–conjugated gold nanospheres targeting on *ex vivo* tumor and normal cervical tissue. The tumor tissue had significantly more binding of nanospheres, resulting in more contrast than the normal tissue under reflectance LSCM. In further experiments, the authors incubated the anti-EGFR–conjugated gold nanospheres and PVP (permeability enhancer for topical applications) with cervical tumor constructs,

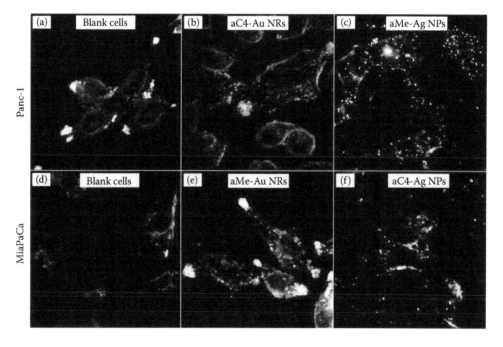

FIGURE 7.8 **(See color insert.)** DFM imaging of Panc-1 and MiaPaCa cells, control without nanoparticles (a) and (d), labeled with antibody conjugated gold nanorods, (b) and (e), and antibody conjugated silver nanospheres, (c) and (f). (Reproduced with permission from Hu, R. et al., *J. Phys. Chem. C*, 113, 2676–2684, 2009.)

composed of multiple layers of cervical tumor tissue, mimicking a tumor. The authors showed that combined incubation of gold nanospheres and PVP increases the number of nanoparticles that can penetrate deeper into the tumor. After sectioning the tumors, the authors were able to visualize the targeted gold nanospheres with reflectance LSCM (Sokolov et al. 2003).

Another group performed a similar study using anti-EGFR antibodies functionalized onto gold nanospheres, approximately 20 nm in size. In this study, the authors incubated the anti-EGFR conjugated gold nanospheres with nasopharyngeal carcinoma (CNE2) cells that overexpress EGFR compared with BSA (bovine serum albumin)-functionalized gold nanospheres as the control. The anti-EGFR antibody conjugated gold nanospheres targeted the carcinoma cells, whereas the BSA nanospheres did not, shown using reflectance LSCM imaging (Figure 7.9). With the information gathered from the reflectance LSCM imaging, the authors were able to generate a map of the relevant expression of biomarkers, such as EGFR, on a cell, which could be beneficial for cancer detection (Kah et al. 2008).

Various cell lines have been analyzed to determine if there is a cell-specific response to incubation with unconjugated gold nanospheres. The cell lines used in this study were derived from baby hamster kidneys (BHK21), human lung carcinomas (A549), and hepatocellular liver carcinomas (HepG2). The gold nanospheres did not affect the BHK21 and HepG2 cell lines, but were shown to induce cell death in A549 cells through an apoptotic pathway. Accumulation of the gold nanospheres in the A549 cells was found adjacent to the cell nuclei using reflectance LSCM, but the authors were unable to determine if this localization lead to the cell death (Patra et al. 2007).

One group fabricated new spherical nanoparticles with an iron oxide core and a gold shell. As discussed in other chapters, iron oxide nanoparticles have a property known as superparamagnetism, which enhances their ability to be imaged using magnetic techniques such as magnetic resonance imaging (MRI). The authors used both gold and iron oxide to capitalize on the gold photothermal properties as well as their high scattering component and the iron oxide magnetic properties for *in vivo* imaging. The gold–iron oxide nanoparticles were conjugated with anti-EGFR antibodies to target breast carcinoma cells (MDA-MB-468). Using reflectance LSCM, the authors were able to show that their nanoparticles preferentially bound to the surface of these cells, colocalizing it with MRI (Larson et al. 2007).

7.3.1.5 Optical Coherence Tomography

OCT uses optical techniques to procure cross-sectional images instead of the typical *en face* images discussed in other microscopy techniques. OCT, which uses low coherence interferometry, utilizes broadband light sources (between 800 and 1300 nm), to improve the resolution because the light interference occurs over microns instead of meters (Huang et al. 1991). This technique has a much higher penetration depth (1–2 mm) than traditional microscopy, allowing for *in vivo* imaging at a much higher resolution (1–10 μm) than common *in vivo* imaging modalities, such as MRI and PET (positron emission tomography). However, a broadband source produces light at several wavelengths simultaneously that can affect the attenuation and signal in the image since it is not limited to a single wavelength. Additionally, as with all reflectance-based imaging techniques, OCT is dependent on only backscattered light instead of all scattered light, which can affect signal strength.

An interferometer, similar to DIC, is used to analyze the backward scattered (reflected) light at submicron to micron resolution. There are two paths the light travels in the interferometer for OCT: the sample arm and the reference arm. The interference pattern depends directly on the reflected light from the sample detailing the structure. The most common image analyzed is a subsurface cross-sectional image created by combining multiple depth scans in the lateral direction and is known as a B-scan. Using these B-scans, it is possible to make 3-D images with OCT. Another imaging method is an A-scan, which contains information in the depth direction. M-scans are multiple A-scans over time, therefore, encompassing information in the depth and time domains (Huang et al. 1991).

Primarily, OCT has been used to image the retina of the eye and its vasculature without any exogenous staining (Huang et al. 1991). However, the use of contrast agents with OCT creates a powerful diagnostic device in other parts of the body. Incoherent

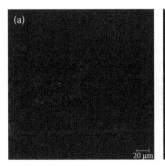  

FIGURE 7.9 Confocal reflectance images of CNE2 cells: (a) before labeling; (b) after labeling with our control BSA-conjugated gold nanospheres; and (c) after labeling with anti-EGFR-conjugated gold nanospheres. Images are cross-sectional slices of cells taken at the mid-focal plane at 20× magnification. Obtained at excitation of 633 nm; scale bar is 20 μm. (Reproduced with permission from Kah, J.C.Y. et al., *Mol. Cell. Probes*, 22, 14–23, 2008.)

scattering signals, such as fluorescence and Raman scattering, are essentially transparent to OCT; however, this technique is highly sensitive to coherent scatterers, such as dyes, proteins, microbubbles, and plasmonic nanoparticles. Because of their high scattering components, gold nanospheres (Skala et al. 2008), nanoshells (Adler et al. 2008; Agrawal et al. 2006; Barton et al. 2004; Gobin et al. 2007; Loo et al. 2004), nanorods (Oldenburg et al. 2006), and nanocages (Cang et al. 2005; Chen et al. 2005a) have been investigated as potential contrast agents for OCT.

As discussed earlier, Halas, West, Drezek and coworkers have contributed many studies toward the use of gold nanoshells in optical imaging, including OCT molecular imaging. In their first study, the authors established gold nanoshells (100 nm silica core diameter and 20 nm gold shell thickness) as a contrast agent for OCT, because of their NIR scattering, by comparing them with 2-μm polystyrene microspheres. The authors acquired separate M-scan images of solutions of gold nanoshells (10^9 particles/mL) and microspheres (at 0.2% solid) dispersed in water, comparing the intensities relative to each other. The gold nanoshells were found to be brighter, causing less attenuation than the microspheres. Furthermore, the authors injected these gold nanoshells into the tail vein of a mouse as well as the skin of a hamster and compared the pre-injection OCT B-scan images with post-injection B-scan images, illustrating plasmonic nanoparticles as contrast agents for OCT for *in vivo* imaging (Barton et al. 2004; Loo et al. 2004).

As a follow-up, this group, in conjunction with the Food and Drug Administration, evaluated and optimized the use of gold nanoshells as a contrast agent for OCT. The nanoshell core diameter, shell thickness, and the overall concentration were varied to find the most backscattering (reflective) particles, compared on the basis of the enhancement in signal in OCT images. All OCT measurements were performed on cuvettes with nanoshells dispersed in water and tissue-simulating phantoms. The B-scan images were composed of 120 A-scans at 40 Hz with an acquisition time of 3 s. The authors used two different core sizes (126 and 291 nm) and modified the shell thickness from 8 to 25 nm. As the concentration increased the signal intensity increased as expected, and to have a noticeable increase in intensity (2 dB) a concentration of 10^9 was necessary. As the core and shell sizes increased at a single concentration of 5×10^9 nanoshells/mL, there were monotonic increases up to 8 dB in signal gain and 6 dB in signal attenuation. The maximum backscattering signal was produced with a nanoshell with 291/25 nm dimensions; however, this particle is too large to be useful for *in vivo* applications (Agrawal et al. 2006).

The most recent paper from this group demonstrated that gold nanoshells can aid in the detection of cancerous tumors using *in vivo* OCT imaging. The authors intravenously injected tumor bearing mice with PEGylated gold nanoshells (119/12 nm) and phosphate-buffered saline (PBS) as a control. OCT images (Figure 7.10) of the tumor tissue in mice injected with nanoshells

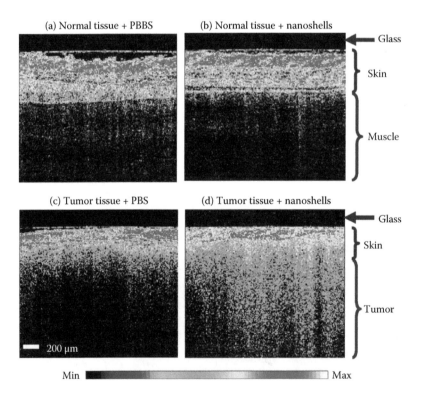

FIGURE 7.10 (**See color insert.**) Representative OCT images from normal skin and muscle tissue areas of mice systemically injected with nanoshells (a) or with PBS (b). Representative OCT images from tumors of mice systemically injected with nanoshells (c) or with PBS (d). Analysis of all images shows a significant increase in contrast intensity after nanoshell injection in the tumors of mice treated with nanoshells, whereas no increase in intensity is observed in the normal tissue. The glass of the probe is 200 μm thick and shows as a dark nonscattering layer. (Reproduced with permission from Gobin, A.M. et al., *Nano Lett.*, 7, 1929–1934, 2007.)

had a much higher contrast than the three controls: (1) normal mice tissue injected with PBS, (2) normal mice tissue with nanoshells, and (3) mice tumor tissue with PBS. The tumors with nanoshells were found to have a 56% higher signal (Figure 7.10) than the normal tissue in the same mice injected with nanoshells (Gobin et al. 2007).

The Li group from the University of Washington has used gold nanocages as a contrast agent for OCT. In their first study, the authors fabricated ~40-nm gold nanocages with an SPR wavelength tuned to ~800 nm. These nanocages were embedded in a gelatin phantom with TiO_2 to mimic the background scattering of soft tissues. OCT images of the phantom demonstrated that the nanocages can be detected as a contrast agent. The authors extracted the scattering and absorption cross sections from the sample and determined that the nanocages are moderately scattering ($8.1 \times 10^{-16} \, m^2$) and highly absorbing ($7.26 \times 10^{-15} \, m^2$). The scattering implies that the nanocages are effective as a contrast agent for OCT, whereas the absorption signifies a possibility for a therapeutic application such as photothermal therapy (Chen et al. 2005a).

Following up on that study, the authors investigated gold nanocages as a contrast agent for spectroscopic OCT. Spectroscopic OCT is a modification on OCT that allows for the measurement of the spectra of the backscattered light over all optical frequencies; thus, spatially resolved spectra can be obtained. These differences in imaging gold nanocages were shown by imaging with the spectroscopic OCT and traditional OCT. Because of the tunability of gold nanocages, it is possible to distinguish different-sized nanocages using the spectroscopic data in the spatial domain. In their study, Cang et al. (2005) demonstrated that the gold nanocages can be used as an effective contrast agent in not only OCT but also spectroscopic OCT.

Gold nanorods have also been investigated as a powerful contrast agent for OCT. One study used backscattered albedo as a contrast agent for OCT. Backscattered albedo is the ratio of the backscatter to the total extinction, which is the combination of both the absorption and scattering components. The authors imaged the gold nanorods in tissue phantom solutions to demonstrate that the low backscattering albedo of the gold nanorods provided sufficient contrast to be visualized. Human tissue is primarily forward-scattering; therefore, it is necessary to use a backscattering contrast agent to visualize a desired target. The authors found the concentration threshold of gold nanorods (30 ppm) necessary for detection in a tissue-simulating phantom (Oldenburg et al. 2006).

7.3.2 Luminescence-Based Microscopy

There are three primary imaging modalities that exploit the luminescence properties of plasmonic nanoparticles: epifluorescence microscopy, fluorescence confocal microscopy, and TPM. Plasmonic nanoparticles are usually not inherently fluorescent; therefore, they have been tagged with other fluorescence molecules to image in some of these modalities. Epifluorescence microscopy is the most ubiquitous fluorescence technique for

imaging biological materials, but there is no benefit regarding spatial or lateral resolution from brightfield microscopy. Fluorescence confocal microscopy limits the amount of out-of-focus light collected, increasing the resolution laterally (350 nm) and spatially (0.5–1.5 μm) allowing for thin optical sectioning and 3-D imaging. TPM capitalizes on plasmonic nanoparticle's innate ability to luminesce when irradiated with two photons of light eliminating the need for extrinsic tags. Furthermore, TPM also rejects more out-of-focus light than confocal microscopy, such that there is no superfluous absorption outside of the focal plane, limiting sample photodamage and photobleaching. In addition, this allows for enhanced excitation light penetration for obtaining high-resolution 3-D images in thick samples with light penetration 3–4 times more than confocal microscopy.

Fluorescence, a form of luminescence, is exploited as imaging contrast for these microscopy techniques. Molecules that exhibit this fluorescent property absorb light strongly at a specific characteristic wavelength of light. When illuminated at the characteristic wavelength, the electrons in the molecule become excited. As the electrons in the molecule relax back to its ground state, the molecule will emit a photon of light as a method of radiative decay. Typically, the emitted photon is at a longer wavelength, and lower energy, than the excitation wavelength, because of nonradiative losses such as heat (Lakowicz 2006).

7.3.2.1 Epifluorescence Microscopy

Epifluorescence microscopy is a technique that uses endogenous or exogenous molecules to add contrast to an image. Broadband light travels through a filter illuminating the sample at a single excitation wavelength. The light emitted from the sample is collected from the same side as the excitation path, but through a different filter for the emission wavelength to reach the eyepiece, eliminating all superfluous wavelengths of light. Even though the resolution depends on the wavelength, the overall resolution of an epifluorescent microscope is still limited by diffraction with the whole field of view illuminated uniformly (Lichtman and Conchello 2005).

Commonly, plasmonic nanoparticles are labeled using an extrinsic fluorophore to be imaged with this modality. As an example, one study focused on comparing targeted and untargeted hollow gold nanoshells for cancer detection and therapy. The authors tagged hollow gold nanoshells with FITC to visualize the nanoparticle *in vitro* for making an accurate comparison. They found that the targeted nanoshells bound to cell surface at a much higher rate than untargeted nanoshells (Lu et al. 2009).

Dickson and coworkers, however, have been able to fabricate small (<2 nm) silver nanoclusters that exhibit intrinsic fluorescence. In their first study, these fluorescent silver nanoclusters were first fabricated in the nucleoli of cells using ambient temperature photoactivation. Nucleolin, a naturally occurring protein in the nucleoli, binds silver together creating the nanoclusters. The authors were able to confirm the emission fluorescence of the silver nanoclusters through imaging with a fluorescence microscope as well as localizing the nanoclusters to the nucleoli. To fabricate the silver nanoclusters without cells, the authors designed a peptide to mimic the properties of the

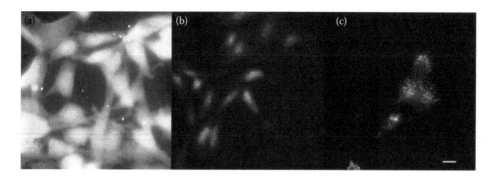

FIGURE 7.11 Fluorescence images of NIH 3T3 cells stained with Avidin-C24-Ag nanoparticles. (a) Fixed cells, biotinylated; (b) fixed cells, non-biotinylated; (c) live cells, biotinylated. Scale bar = 30 μm. (Reproduced with permission from Yu, J. et al., *Photochem. Photobiol.*, 84, 1435–1439, 2008.)

nucleolin and were able to create clusters that exhibited similar photophysics to those in nucleolus with emissions ranging from 610 to 630 nm. These nanoclusters were found to load into fibroblast cells (NIH-3T3) and were imaged using fluorescence microscopy (Yu et al. 2007).

Further studies performed by Dickson and coworkers used silver nanoclusters encapsulated by avidin-conjugated DNA and bound by the peptides determined in the earlier study. Avidin and biotin are commonly used as a strong binding pair. The photostability and emission intensity was shown to be much greater than conventional fluorophores for *in vitro* fluorescence imaging, with an emission maximum centered at 640 nm and excitation maximum centered at 580 nm. These clusters were shown to be effective fluorescent labels on live, biotinylated NIH-3T3 cells, imaged using fluorescent microscopy (Figure 7.11) (Yu et al. 2008). Moreover, the authors varied the fabrication chemistry to reduce the potential for adversely affecting protein function. This enabled the authors to modify the targeting antibodies attached to the silver nanoclusters, allowing the nanoclusters to be used as fluorescent labels for specific cellular components such as actin, microtubule filaments, and specific surface proteins (Yu et al. 2009).

7.3.2.2 Fluorescence LSCM

LSCM is more commonly used in biological applications in fluorescent mode rather than the reflectance-based setup discussed previously. However, plasmonic nanoparticles, as discussed previously, are not typically inherently fluorescent; therefore, there are less studies of plasmonic nanoparticles imaged with fluorescent LSCM than reflectance LSCM. In the fluorescence setup of LSCM, lasers excite the sample through point illumination at a single wavelength and emission filters to collect and separate the emitted light through the pinhole. The optical path is the same as reflectance LSCM, taking advantage of the point illumination and the pinhole to increase axial and spatial resolution. Fluorescent LSCM is very commonly used to obtain high-resolution fluorescent images as well as reconstructed 3-D images because of the thin optical sectioning (Corle et al. 1996; Lakowicz 2006; Pawley 2006).

Fluorescent LSCM has been used in biological applications with plasmonic nanoparticles to visualize their location *in vitro*.

With most other techniques mentioned previously, the axial resolution was limited, and it was not possible to determine if the nanoparticles were internalized within the cell. Using the thin optical sectioning of LSCM, it is possible to localize the nanoparticles as above, below, or within the cell. One study used fluorescent LSCM to image the uptake of transferrin-conjugated gold nanospheres tagged with Texas red (fluorophore) within HeLa (ovarian cancer) cells. By capitalizing on the thin optical sectioning of fluorescence LSCM, the authors imaged at multiple depths, concluding that the fluorescent spots were observed through the HeLa cell (Figure 7.12). Thus, they inferred that the

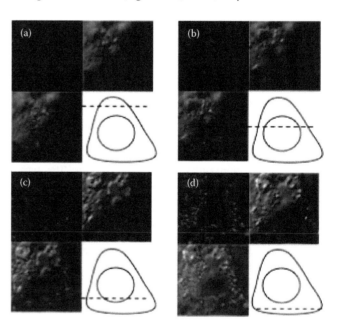

FIGURE 7.12 Confocal images at different *Z*-axis of transferrin-coated spherical gold nanoparticles conjugated with the organic fluorophore Texas red. (a–d) Fluorescence images of cells with transferrin-coated gold nanoparticles and their corresponding DIC images. A schematic depicting the *Z*-axis of the image is shown with each of the four panels. As the *Z*-axis moves from top to bottom of the Hela cell, observe the fluorescence spots throughout the Hela cell, indicating cell uptake. (Reproduced with permission from Chithrani, B.D., Chan, W.C.W., *Nano Lett.*, 7, 1542–1550, 2007.)

nanospheres were internalized within the cell rather than only bound to the surface, which was confirmed using transmission electron microscopy (TEM) (Chithrani and Chan 2007).

Another group performed a similar study, using fluorescence LSCM to image transferrin-conjugated gold nanosphere, tagged with a fluorophore, uptake within SUNE1 cells. The primary goal of this study was to show that the nanospheres were internalized in the cell using atomic force microscopy, but confirmation was performed using fluorescence LSCM. The authors imaged the cells with the nanospheres at multiple depths and were able to conclude that the nanoparticles were internalized within the cells (Yang et al. 2005).

To capitalize on the SPR effects of plasmonic nanoparticles in conjunction with the fluorescence properties of molecules, several studies have focused on metal enhanced fluorescence. Fluorescence can be enhanced by the near-field interaction of the fluorophore bound on a plasmonic nanoparticle. The bound fluorophore can be considered an oscillating dipole, radiating energy when excited. The nanoparticle SPR causes a change to the emission properties, known as radiative decay engineering, through a coupling mechanism when the radiating energy of the bound fluorophore is localized near the particle. Radiative decay engineering can enhance the intensity of the emission by $10–10^3$ times, depending on the coupling between the specific fluorophore and plasmonic nanoparticle, as well as making the whole process of emission more photostable. One group, in particular, has conducted multiple studies into this metal enhanced fluorescence based on the SPR of the nanoparticle (Lakowicz 2005).

One study the authors performed imaged plasmon-coupled probes composed of silver nanoparticles (20 nm) bound with Alexa Fluor 647 (fluorescent molecule)–labeled concanavalin A (con A). The peak SPR wavelength of the silver nanoparticles was at 405 nm, and the con A was used because it can bind to sugars on the cell membrane. As a comparison, the authors imaged the unmodified silver nanoparticles under reflectance LSCM, in which the nanoparticles displayed a "blinking" effect and did not have a full, round shape. The unmodified con A under fluorescence LSCM was also imaged as a comparison, showing full, round spots, but the intensity gradually decayed over time. When the authors imaged the plasmon-coupled probe using fluorescence LSCM, they found the fluorescence to be seven times brighter than the con A and with no visible photobleaching effect. The plasmon-coupled probes were also imaged *in vitro* attached to human embryonic kidney (HEK293A) cell lines, and the comparison against the controls gave similar results. When bound onto the surface, the coupling between the silver nanoparticles and the fluorophores was even more pronounced, causing a 20-fold intensity increase from unmodified con A (Zhang et al. 2007).

As a follow-up study, the authors studied the dependence of the plasmon coupled probe on the size of the silver nanoparticle. A single-stranded oligonucleotide was used as the targeting moiety and Cy5 as the fluorophore, which was bound to silver nanoparticles with diameters of 5, 20, 50, 70 and 100 nm. Using fluorescence LSCM, the 50-nm silver nanoparticles displayed

the highest enhanced intensity. The enhancement was 17 times greater than free fluorophores with a single fluorophore labeled on the surface of the nanoparticle and up to 400 times greater with multiple fluorophores labeled on the surface (Zhang et al. 2008).

7.3.2.3 Two-Photon Microscopy

Two-photon microscopy (TPM) is a laser scanning fluorescence imaging modality that provides improvements over conventional widefield and single-photon fluorescence microscopy. In TPM, the basic idea is that at sufficiently high photon densities, two photons with half the energy difference of an electronic transition—exciting fluorescent molecules—can be absorbed simultaneously, in a nonlinear process. Since the probability of the two photons to be absorbed simultaneously is quadratically dependent on the incident intensity, a femtosecond pulsed laser is used such that enough excitation photons are provided to induce two photon absorption. Additionally, the two photons need to be tightly focused such that the excitation and resulting fluorescence generation is limited to an extremely small focal volume, usually accomplished using high numerical aperture objectives. This spatial confinement of both two photon absorption and subsequent emission results in the property of 3-D resolvability, owing to the reduction in out-of-focus signal generation and increase in the spatial resolution (Denk et al. 1990).

Owing to the use of longer wavelength light, typically in the NIR (700–1000 nm), and the absorption volume being spatially confined to the focal region, TPM is well suited for higher depth imaging in optically thick biological samples. TPM generates less out-of-focus signal than either wide-field or confocal fluorescence, resulting in improved sectioning for 3-D imaging and limited photobleaching. Typically, the resolution spatially and laterally is the same as a perfectly aligned confocal microscope without needing a pinhole, allowing for flexibility in detection geometry. However, this imaging modality is expensive and complex because of the need for high NA objectives and ultrafast pulsed lasers with sufficient peak powers and pulse widths; as a result, it is not widely accessible (Denk et al. 1990; Helmchen and Denk 2002).

Plasmonic nanoparticles were found to strongly absorb the two photons simultaneously in a nonlinear process, because the coupling of the localized SPRs, exciting the nanoparticle electrons to a higher energy state. In the nonradiative relaxation process, these nanoparticles emit an extremely strong photoluminescence effect; thus, plasmonic nanoparticles have a strong two-photon absorption cross section. One of the first studies used TPM to image gold nanospheres. The authors imaged different-sized nanospheres (2.5, 15, 60 and 125 nm) under TPM and determined that the 60-nm particles exhibited the strongest emission, whereas the 2.5-nm particles were the weakest by a factor of 3–4. The particles were found to be photostable and devoid of "blinking" effects. To illustrate that aggregation was artificially enhancing the two-photon emission, the authors used DFM to image 15-nm gold nanospheres coated with silica to reduce nanoparticle coupling (Farrer et al. 2005).

One group in particular, Wei, Cheng and coworkers, has pioneered the use of gold nanorods as a contrast agent for TPM. In one of their first studies, the authors characterized single nanorods using a sample of nanorods with different excitation wavelengths imaged using TPM. They determined that the nanorod two-photon excitation correlated well with the longitudinal SPR band, and the emission was in the 400–650 nm region, nearly 60 times brighter than a fluorescent rhodamine molecule. Since nanorod absorption is polarization-dependent because of its shape, the authors also established that the nanorods have a cos^4 dependence on the excitation polarization. *In vivo* TPM imaging of intravenous tail vein injected gold nanorods was performed for the first time on blood vessels in the mouse ear lobes (Wang et al. 2005).

The next studies presented by this group involved the *in vitro* imaging of gold nanorods using TPM as a method to determine the nanorod localization. In one of their studies, the authors compared the cellular uptake of nanorods coated with cetyltrimethylammounium bromide (CTAB)—a typical surfactant used in the fabrication of gold nanorods—and nanorods coated with hydrophilic surfactants such as bis(p-sulfonatophenyl) phenylphosphine (BSP) and PEG. TPM was used to visualize the nanorod position and colocalized with phase contrast microscopy to visualize the oral epithelium cancer (KB) cells. Using single nanorod tracking, the CTAB nanorod uptake kinetics were determined and found to cause no adverse effects on the cell within a 5-day period. The CTAB nanorods were internalized at a much higher rate than the BSP and PEG nanorods (Huff et al. 2007). The authors furthered their studies by targeting the gold nanorods to the KB cells using folate conjugated onto the gold surface. TPM images of the folate-conjugated nanorods show their preferential binding to the cell surface of KB cells, which overexpress the folate receptor, compared to the negative control of NIH-3T3 cells, which do not express the folate receptor (Figure 7.13; Tong et al. 2007).

This group has also accomplished intravital TPM imaging of surface blood vessels to compare the pharmacokinetics of linear PEG and branched PEG functionalized gold nanorods *in vivo*. Because of the gold nanorod SPR frequency tuned to the NIR, it is possible to image deeper *in vivo* without interference from endogenous sources. In addition, the authors used TPM to determine the biodistribution of the nanorods in explanted organs, using autofluorescence to visualize the cells, from mice (Tong et al. 2009).

Several other groups have used TPM as a method of visualizing plasmonic nanoparticles to determine their distribution and localization *in vitro*. One group imaged hamster ovary (CHO) cells, visualized using two-photon autofluorescence, targeted by con A conjugated 10-nm gold nanospheres (Yelin et al. 2003). Another group imaged, using TPM, anti-EGFR antibody conjugated nanorods targeted to A431 cells, imaged using autofluorescence, *in vitro* and embedded in a collagen matrix at increasing depths. The authors also characterized the difference in power necessary to image at the same signal level: autofluorescence in cells (~9 mW) and the gold nanorods (140 µW), showing that the gold nanorods are ~4000 times brighter (Durr et al. 2007).

Tunnell and coworkers used TPM to image the gold nanoshell distribution *ex vivo*, in excised tumors. First, the authors characterized nanoshells (135 nm in diameter) as an effective contrast agent for TPM. In the study, the authors intravenously injected gold nanoshells into the tail veins of tumored mice, with subcutaneous colorectal tumors (HCT 116), and excised the tumors to image the distribution of the gold nanoshells within the tumor. Intratumoral localization was performed by staining the blood vessels [immunohistochemistry (IHC)] and nuclei (YOYO). 3-D imaging was also performed, exploiting the thin optical sectioning capabilities of TPM, to illustrate that the nanoshells were found surrounding blood vessels in the tumor (Figure 7.14; Park et al. 2008).

This same group followed up on their study by using TPM to analyze and determine the cellular level biodistribution of gold nanoshell and nanorods (41 nm in length and 10 nm in width) in tumored mice. The authors excised the tumors with either gold nanoshells or nanorods and imaged multiple sections stained with hematoxylin and eosin (H&E, a common histopathological stain), YOYO for nuclei, or IHC for blood vessels. In addition, the authors imaged H&E-stained liver and spleen slices with either gold nanoshells or nanorods. Through all of the TPM images, the authors determined that the nanoshells and nanorods had a heterogeneous distribution in the tumor with most

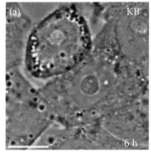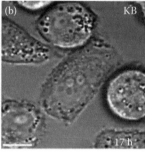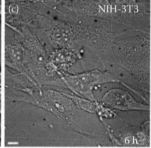

FIGURE 7.13 (**See color insert.**) Targeted adsorption and uptake of folate-conjugated GNRs (F-NRs, red) by KB cells overexpressing folate receptors (imaged in transmission mode, gray). (a) A high density of F-NRs was observed on the surface of KB cells after 6 h incubation at 37°C. (b) F-NRs were internalized into KB cells and delivered to the perinuclear region after 17 h incubation. (c) No binding was observed of F-NRs to NIH-3T3 cells, which express folate receptors at a low level. Bar = 10 µm. (Reproduced with permission from Tong, L. et al., *Adv. Mater.*, 19, 3136–3141, 2007.)

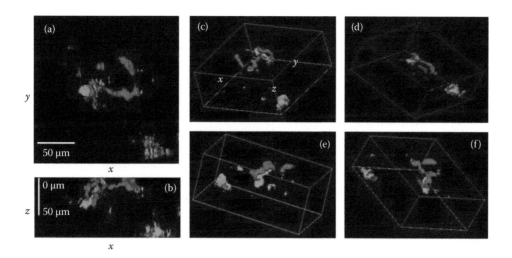

FIGURE 7.14 **(See color insert.)** 3-D visualization of nanoshells (green) and blood vessels (red) in tumor. (a) *z*-projection of *x*–*y* images from tumor (red: fluorescein in blood vessels, green: gold nanoshells); (b) y-projection of x–z images with field of view 198 × 80 μm; (c)–(f) 3-D images of nanoshells in tumor at different rotational angles. Scale bar = 50 μm (a,b). (Reproduced with permission from Park, J. et al., *Optics Express*, 16, 1590–1599, 2008.)

accumulation occurring near the tumor edge and unique patterns close to vasculature. The nanorods had a higher accumulation in the tumor core than the nanoshells, possibly because of their smaller size. However, in the liver and spleen, based on TPM images, there was a significant accumulation of both nanoshells and nanorods with no discernible difference between the two particle types (Park et al. 2010).

Another study performed by Dickson and coworkers, characterized their silver nanoclusters encapsulated in oligonucleotides as a contrast agent in TPM. The authors fabricated three different silver nanoclusters (emitting at 660, 680, or 710 nm, respectively) for comparison. They found that the emission spectrum of all three nanoclusters was the same under two-photon excitation as it was with one-photon excitation. Additionally, the two-photon absorption cross section of the three nanoclusters was determined to be 2 orders of magnitude higher than typical fluorophores (Patel et al. 2008).

One group used bifunctional heterodimer nanoparticles composed of iron oxide nanoparticle coupled with a silver nanoparticle. Iron oxide nanoparticles are superparamagnetic, and were used because they can be manipulated magnetically. The silver nanoparticle was used for its ability to be imaged using TPM. These heterodimer nanoparticles were coated with a TMAH and GSH on the iron oxide and silver surfaces, respectively, to allow for water solubility. The authors were successful in imaging the heterodimer nanoparticles internalized in RAW 247.6 (mouse macrophage) cells with TPM, showing excellent contrast (Jiang et al. 2008).

7.4 Conclusion

In conclusion, it is readily apparent that the optical properties of plasmonic nanoparticles make them strong contrast agents for optical microscopy in biological imaging. These plasmonic nanoparticles differ in size and shape and even material,

ranging from silver nanospheres to gold nanorods and nanocages. Plasmonic nanoparticles are a more effective contrast agent for cellular imaging in comparison with traditional sources, because they are detectable at low concentrations owing to their intense scattering and luminescence properties without any photobleaching.

Scattering-based imaging techniques, including DIC, darkfield, reflectance LSCM, and OCT, capitalize on the large scattering cross section of plasmonic nanoparticles. In contrast, luminescence-based imaging techniques, including epifluorescence, fluorescence LSCM, and TPM, take advantage of the plasmon resonance absorption and resulting luminescence of plasmonic nanoparticles. DFM is the most widely used method of visualizing the location of plasmonic nanoparticles *in vitro*; however, OCT has the ability to penetrate deeper into tissue (1–2 mm) at micron resolution, which is advantageous for *in vivo* imaging. LSCM has a high resolution laterally and spatially allowing for 3-D imaging of the plasmonic nanoparticles. However, TPM can image plasmonic nanoparticles in optically thick samples with an identical high 3-D resolution with a tighter focal plane. Depending on application and accessibility, different modalities of optical microscopy can be more advantageous for imaging plasmonic nanoparticles.

References

Aaron, J., K. Travis, N. Harrison, and K. Sokolov. 2009. Dynamic imaging of molecular assemblies in live cells based on nanoparticle plasmon resonance coupling. *Nano Letters* 9:3612–3618.

Adler, D. C., S.-W. Huang, R. Huber, and J. G. Fujimoto. 2008. Photothermal detection of gold nanoparticles using phase-sensitive optical coherence tomography. *Optics Express* 16: 4376–4393.

Agrawal, A., S. Huang, A. W. H. Lin et al. 2006. Quantitative evaluation of optical coherence tomography signal enhancement with gold nanoshells. *Journal of Biomedical Optics* 11:041121.

Almofti, M. R., T. Ichikawa, K. Yamashita, H. Terada, and Y. Shinohara. 2003. Silver ion induces a cyclosporine a-insensitive permeability transition in rat liver mitochondria and release of apoptogenic cytochrome C. *Journal of Biochemistry* 134:43.

Anderson, R. R., and J. A. Parrish. 1983. Selective photothermolysis: Precise microsurgery by selective absorption of pulsed radiation. *Science* 220:524.

AshaRani, P., G. Low Kah Mun, M. P. Hande, and S. Valiyaveettil. 2008. Cytotoxicity and genotoxicity of silver nanoparticles in human cells. *ACS Nano* 3:279–290.

AshaRani, P. V., M. P. Hande, and S. Valiyaveettil. 2009. Antiproliferative activity of silver nanoparticles. *BMC Cell Biology* 10:14.

Averitt, R. D., S. L. Westcott, and N. J. Halas. 1999. Linear optical properties of gold nanoshells. *Journal of the Optical Society of America B* 16:1824–1832.

Barton, J. K., N. J. Halas, J. L. West, and R. A. Drezek. 2004. Nanoshells as an optical coherence tomography contrast agent. In *Conference on Coherence Domain Optical Methods and Optical Coherence Tomography in Biomedicine VIII*, San Jose, CA, pp. 99–106.

Bohren, C. F., and D. R. Huffman. 1983. *Absorption and Scattering of Light by Small Particles*. Research supported by the University of Arizona and Institute of Occupational and Environmental Health. New York: Wiley-Interscience, 541 pp., p. 1.

Brioude, A., X. C. Jiang, and M. P. Pileni. 2005. Optical properties of gold nanorods: DDA simulations supported by experiments. *Journal of Physical Chemistry B* 109:13138–13142.

Burda, C., X. Chen, R. Narayanan, and M. A. El-Sayed. 2005. Chemistry and properties of nanocrystals of different shapes. *Chemical Reviews* 105:1025–1102.

Cang, H., T. Sun, Z. Y. Li et al. 2005. Gold nanocages as contrast agents for spectroscopic optical coherence tomography. *Optics Letters* 30:3048–3050.

Carlson, C., S. Hussain, A. Schrand et al. 2008. Unique cellular interaction of silver nanoparticles: Size-dependent generation of reactive oxygen species. *Journal of Physical Chemistry B* 112:13608–13619.

Chen, J., F. Saeki, B. J. Wiley et al. 2005a. Gold nanocages: Bioconjugation and their potential use as optical imaging contrast agents. *Nano Letters* 5:473–477.

Chen, J., B. Wiley, Z. Y. Li et al. 2005b. Gold nanocages: Engineering their structure for biomedical applications. *Advanced Materials* 17:2255–2261.

Chen, J. Y., D. L. Wang, J. F. Xi et al. 2007. Immuno gold nanocages with tailored optical properties for targeted photothermal destruction of cancer cells. *Nano Letters* 7:1318–1322.

Chen, L. Q., S. J. Xiao, L. Peng et al. 2010. Aptamer-based silver nanoparticles used for intracellular protein imaging and single nanoparticle spectral analysis. *Journal of Physical Chemistry B* 114:3655–3659.

Chen, X., and H. J. Schluesener. 2008. Nanosilver: A nanoproduct in medical application. *Toxicology Letters* 176:1–12.

Chithrani, B. D., and W. C. W. Chan. 2007. Elucidating the mechanism of cellular uptake and removal of protein-coated gold nanoparticles of different sizes and shapes. *Nano Letters* 7:1542–1550.

Cho, E. C., Y. Liu, and Y. Xia. 2010. A simple spectroscopic method for differentiating cellular uptakes of gold nanospheres and nanorods from their mixtures. *Angewandte Chemie* 122:2020–2024.

Choi, C. H. J., C. A. Alabi, P. Webster, and M. E. Davis. 2010. Mechanism of active targeting in solid tumors with transferrin-containing gold nanoparticles. *Proceedings of the National Academy of Sciences of the United States of America* 107:1235–1240.

Cobley, C. M., J. Chen, E. C. Cho, L. V. Wang, and Y. Xia. 2011. Gold nanostructures: A class of multifunctional materials for biomedical applications. *Chemical Society Reviews* 40:44–56.

Corle, T. R., G. S. Kino, and ScienceDirect. 1996. *Confocal Scanning Optical Microscopy and Related Imaging Systems*. Boston: Academic Press.

Davidson, M., and M. Abramowitz. 2002. Optical microscopy. In *Encylcopedia of Imaging Science and Technology*, ed. J. Hornak, 1106–1141. New York: Wiley-Interscience.

Denk, W., J. H. Strickler, and W. W. Webb. 1990. Two-photon laser scanning fluorescence microscopy. *Science* 248:73.

Dickerson, E. B., E. C. Dreaden, X. H. Huang et al. 2008. Gold nanorod assisted near-infrared plasmonic photothermal therapy (PPTT) of squamous cell carcinoma in mice. *Cancer Letters* 269:57–66.

Draine, B. T., and P. J. Flatau. 1994. Discrete-dipole approximation for scattering calculations. *Journal of the Optical Society of America A Optics Image Science and Vision* 11:1491–1499.

Draine, B. T., and P. J. Flatau. 2010. User Guide for the Discrete Dipole Approximation Code DDSCAT 7.1. arXiv:1002.1505. Available at: http://arXiv.org/labs/1002.1505v1.

Durr, N. J., T. Larson, D. K. Smith, B. A. Korgel, K. Sokolov, and A. Ben-Yakar. 2007. Two-photon luminescence imaging of cancer cells using molecularly targeted gold nanorods. *Nano Letters* 7:941–945.

Eghtedari, M., A. V. Liopo, J. A. Copland, A. A. Oraevslty, and M. Motamedi. 2009. Engineering of hetero-functional gold nanorods for the in vivo molecular targeting of breast cancer cells. *Nano Letters* 9:287–291.

El-Sayed, I. H., X. H. Huang, and M. A. El-Sayed. 2005. Surface plasmon resonance scattering and absorption of anti-EGFR antibody conjugated gold nanoparticles in cancer diagnostics: Applications in oral cancer. *Nano Letters* 5:829–834.

El-Sayed, I. H., X. Huang, and M. A. El-Sayed. 2006. Selective laser photo-thermal therapy of epithelial carcinoma using anti-EGFR antibody conjugated gold nanoparticles. *Cancer Letters* 239:129–135.

El-Sayed, M. A. 2001. Some interesting properties of metals confined in time and nanometer space of different shapes. *Accounts of Chemical Research* 34:257–264.

Farrer, R. A., F. L. Butterfield, V. W. Chen, and J. T. Fourkas. 2005. Highly efficient multiphoton-absorption–induced luminescence from gold nanoparticles. *Nano Letters* 5:1139–1142.

Felidj, N., J. Aubard, and G. Levi. 1999. Discrete dipole approximation for ultraviolet–visible extinction spectra simulation of silver and gold colloids. *Journal of Chemical Physics* 111:1195–1208.

Feng, Q., J. Wu, G. Chen, F. Cui, T. Kim, and J. Kim. 2000. A mechanistic study of the antibacterial effect of silver ions on *Escherichia coli* and *Staphylococcus aureus*. *Journal of Biomedical Materials Research* 52:662–668.

Ghosh, S. K., S. Nath, S. Kundu, K. Esumi, and T. Pal. 2004. Solvent and ligand effects on the localized surface plasmon resonance (LSPR) of gold colloids. *Journal of Physical Chemistry B* 108:13963–13971.

Gobin, A. M., M. H. Lee, N. J. Halas, W. D. James, R. A. Drezek, and J. L. West. 2007. Near-infrared resonant nanoshells for combined optical imaging and photothermal cancer therapy. *Nano Letters* 7:1929–1934.

Gupta, A. 1998. Silver as a biocide: Will resistance become a problem? *Nature Biotechnology* 16:888–888.

Gupta, A., M. Maynes, and S. Silver. 1998. Effects of halides on plasmid-mediated silver resistance in *Escherichia coli*. *Applied and Environmental Microbiology* 64:5042.

Helmchen, F., and W. Denk. 2002. Deep tissue two-photon microscopy. *Nature Methods* 2(12):932–940.

Hirsch, L. R., R. J. Stafford, J. A. Bankson, S. R. Sershen, B. Rivera, R. E. Price, J. D. Hazle, N. J. Halas, and J. L. West. 2003. Nanoshell-mediated near-infrared thermal therapy of tumors under magnetic resonance guidance. *Proceedings of the National Academy of Sciences of the United States of America* 100:13549–13554.

Hu, R., K.-T. Yong, I. Roy, H. Ding, S. He, and P. N. Prasad. 2009. Metallic nanostructures as localized plasmon resonance enhanced scattering probes for multiplex dark-field targeted imaging of cancer cells. *Journal of Physical Chemistry C* 113:2676–2684.

Huang, D., E. A. Swanson, C. P. Lin et al. 1991. Optical coherence tomography. *Science* 254:1178.

Huang, T., P. D. Nallathamby, D. Gillet, and X.-H. N. Xu. 2007. Design and synthesis of single-nanoparticle optical biosensors for imaging and characterization of single receptor molecules on single living cells. *Analytical Chemistry* 79:7708–7718.

Huang, X., B. Kang, W. Qian et al. 2010. Comparative study of photothermolysis of cancer cells with nuclear-targeted or cytoplasm-targeted gold nanospheres: Continuous wave or pulsed lasers. *Journal of Biomedical Optics* 15:058002.

Huang, X. H., I. H. El-Sayed, W. Qian, and M. A. El-Sayed. 2006. Cancer cell imaging and photothermal therapy in the near-infrared region by using gold nanorods. *Journal of the American Chemical Society* 128:2115–2120.

Huff, T. B., M. N. Hansen, Y. Zhao, J. X. Cheng, and A. Wei. 2007. Controlling the cellular uptake of gold nanorods. *Langmuir* 23:1596–1599.

Hussain, S., K. Hess, J. Gearhart, K. Geiss, and J. Schlager. 2005. In vitro toxicity of nanoparticles in BRL 3A rat liver cells. *Toxicology In Vitro* 19:975–983.

Inoue, S. 1981. Video image-processing greatly enhances contrast, quality, and speed in polarization-based microscopy. *Journal of Cell Biology* 89:346–356.

Jain, P. K., S. Eustis, and M. A. El-Sayed. 2006a. Plasmon coupling in nanorod assemblies: Optical absorption, discrete dipole approximation simulation, and exciton-coupling model. *Journal of Physical Chemistry B* 110:18243–18253.

Jain, P. K., K. S. Lee, I. H. El-Sayed, and M. A. El-Sayed. 2006b. Calculated absorption and scattering properties of gold nanoparticles of different size, shape, and composition: Applications in biological imaging and biomedicine. *Journal of Physical Chemistry B* 110:7238–7248.

Jain, P. K., X. Huang, I. H. El-Sayed, and M. A. El-Sayed. 2008. Noble metals on the nanoscale: Optical and photothermal properties and some applications in imaging, sensing, biology, and medicine. *Accounts of Chemical Research* 41:1578–1586.

Jiang, J., H. Gu, H. Shao, E. Devlin, G. C. Papaefthymiou, and J. Y. Ying. 2008. Manipulation bifunctional Fe(3)O(4)-Ag heterodimer nanoparticles for two-photon fluorescence imaging and magnetic manipulation. *Advanced Materials* 20:4403–4407.

Jin, R., Y. W. Cao, C. A. Mirkin, K. Kelly, G. C. Schatz, and J. Zheng. 2001. Photoinduced conversion of silver nanospheres to nanoprisms. *Science* 294:1901.

Kah, J. C. Y., M. C. Olivo, C. G. L. Lee, and C. J. R. Sheppard. 2008. Molecular contrast of EGFR expression using gold nanoparticles as a reflectance-based imaging probe. *Molecular and Cellular Probes* 22:14–23.

Kang, B., M. A. Mackey, and M. A. El-Sayed. 2010. Nuclear targeting of gold nanoparticles in cancer cells induces DNA damage, causing cytokinesis arrest and apoptosis. *Journal of the American Chemical Society* 132:1517–1519.

Katz, E., and I. Willner. 2004. Integrated nanoparticle–biomolecule hybrid systems: Synthesis, properties, and applications. *Angewandte Chemie International Edition* 43:6042–6108.

Kelly, K. L., E. Coronado, L. L. Zhao, and G. C. Schatz. 2003. The optical properties of metal nanoparticles: The influence of size, shape, and dielectric environment. *Journal of Physical Chemistry B* 107:668–677.

Kooij, E. S., and B. Poelsema. 2006. Shape and size effects in the optical properties of metallic nanorods. *Physical Chemistry Chemical Physics* 8:3349–3357.

Kreibig, U., and M. Vollmer. 1995. Optical properties of metal clusters. Springer, Basel, Switzerland.

Kumar, S., N. Harrison, R. Richards-Kortum, and K. Sokolov. 2007. Plasmonic nanosensors for imaging intracellular biomarkers in live cells. *Nano Letters* 7:1338–1343.

Lakowicz, J. R. 2005. Radiative decay engineering: 5. Metal-enhanced fluorescence and plasmon emission. *Analytical Biochemistry* 337:171–194.

Lakowicz, J. R. 2006. *Principles of Fluorescence Spectroscopy*. New York: Springer.

Larson, T. A., J. Bankson, J. Aaron, and K. Sokolov. 2007. Hybrid plasmonic magnetic nanoparticles as molecular specific agents for MRI/optical imaging and photothermal therapy of cancer cells. *Nanotechnology* 18:325101–325108.

Lee, K. S., and M. A. El-Sayed. 2005. Dependence of the enhanced optical scattering efficiency relative to that of absorption for gold metal nanorods on aspect ratio, size, end-cap shape, and medium refractive index. *Journal of Physical Chemistry B* 109:20331–20338.

Lichtman, J. W., and J. A. Conchello. 2005. Fluorescence microscopy. *Nature Methods* 2:910.

Link, S., and M. A. El-Sayed. 1999. Size and temperature dependence of the plasmon absorption of colloidal gold nanoparticles. *Journal of Physical Chemistry B* 103:4212–4217.

Link, S., and M. A. El-Sayed. 2000. Shape and size dependence of radiative, non-radiative and photothermal properties of gold nanocrystals. *International Reviews in Physical Chemistry* 19:409–454.

Link, S., M. Mohamed, and M. El-Sayed. 1999. Simulation of the optical absorption spectra of gold nanorods as a function of their aspect ratio and the effect of the medium dielectric constant. *Journal of Physical Chemistry B* 103:3073–3077.

Loo, C., A. Lin, L. Hirsch et al. 2004. Nanoshell-enabled photonics-based imaging and therapy of cancer. *Technology in Cancer Research & Treatment* 3:33–40.

Loo, C., A. Lowery, N. Halas, J. West, and R. Drezek. 2005. Immunotargeted nanoshells for integrated cancer imaging and therapy. *Nano Letters* 5:709–711.

Lu, W., C. Y. Xiong, G. D. Zhang et al. 2009. Targeted photothermal ablation of murine melanomas with melanocyte-stimulating hormone analog-conjugated hollow gold nanospheres. *Clinical Cancer Research* 15:876–886.

Matsumura, Y., K. Yoshikata, S. Kunisaki, and T. Tsuchido. 2003. Mode of bactericidal action of silver zeolite and its comparison with that of silver nitrate. *Applied and Environmental Microbiology* 69:4278.

Melancon, M. P., W. Lu, Z. Yang et al. 2008. In vitro and in vivo targeting of hollow gold nanoshells directed at epidermal growth factor receptor for photothermal ablation therapy. *Molecular Cancer Therapeutics* 7:1730–1739.

Messori, L., and G. Marcon. 2004. Gold complexes in the treatment of rheumatoid arthritis. In *Metal Ions and Their Complexes in Medication*. Marcel Dekker, p. 279, New York.

Mie, G. 1908. Considerations on the optics of turbid media, especially colloidal metal sols. *Annalen der Physik* 25:377–442.

Mock, J., M. Barbic, D. Smith, D. Schultz, and S. Schultz. 2002. Shape effects in plasmon resonance of individual colloidal silver nanoparticles. *Journal of Chemical Physics* 116:6755.

Murphy, D. 2001. Differential interference contrast (DIC) microscopy and modulation contrast microscopy. *Fundamentals of Light Microscopy and Digital Imaging*. New York: Wiley-Liss.

Nallathamby, P. D., K. J. Lee, and X.-H. N. Xu. 2008. Design of stable and uniform single nanoparticle photonics for in vivo dynamics Imaging of nanoenvironments of zebrafish embryonic fluids. *ACS Nano* 2:1371–1380.

Nallathamby, P. D., and X.-H. N. Xu. 2010. Study of cytotoxic and therapeutic effects of stable and purified silver nanoparticles on tumor cells. *Nanoscale* 2:942–952.

O'Neal, D. P., L. R. Hirsch, N. J. Halas, J. D. Payne, and J. L. West. 2004. Photo-thermal tumor ablation in mice using near infrared-absorbing nanoparticles. *Cancer Letters* 209:171–176.

Oh, E., J. B. Delehanty, K. E. Sapsford et al. 2011. Cellular uptake and fate of PEGylated gold nanoparticles is dependent on both cell-penetration peptides and particle size. *ACS Nano* 5:6434–6448.

Oldenburg, A. L., M. N. Hansen, D. A. Zweifel, A. Wei, and S. A. Boppart. 2006. Plasmon-resonant gold nanorods as low backscattering albedo contrast agents for optical coherence tomography. *Optics Express* 14:6724–6738.

Oldenburg, S., R. Averitt, S. Westcott, and N. Halas. 1998. Nanoengineering of optical resonances. *Chemical Physics Letters* 288:243–247.

Oldenburg, S. J., J. B. Jackson, S. L. Westcott, and N. Halas. 1999. Infrared extinction properties of gold nanoshells. *Applied Physics Letters* 75:2897.

Oyelere, A. K., P. C. Chen, X. Huang, I. H. El-Sayed, and M. A. El-Sayed. 2007. Peptide-conjugated gold nanorods for nuclear targeting. *Bioconjugate Chemistry* 18:1490–1497.

Park, J., A. Estrada, K. Sharp et al. 2008. Two-photon-induced photoluminescence imaging of tumors using near-infrared excited gold nanoshells. *Optics Express* 16:1590–1599.

Park, J., A. Estrada, J. A. Schwartz et al. 2010. Intra-organ biodistribution of gold nanoparticles using intrinsic two-photon–induced photoluminescence. *Lasers in Surgery and Medicine* 42:630–639.

Patel, S. A., C. I. Richards, J.-C. Hsiang, and R. M. Dickson. 2008. Water-soluble Ag nanoclusters exhibit strong two-photon-induced fluorescence. *Journal of the American Chemical Society* 130:11602–11603.

Patra, H. K., S. Banerjee, U. Chaudhuri, P. Lahiri, and A. K. Dasgupta. 2007. Cell selective response to gold nanoparticles. *Nanomedicine: Nanotechnology Biology Medicine* 3:111–119.

Patterson, M. S., B. Chance, and B. C. Wilson. 1989. Time resolved reflectance and transmittance for the non-invasive measurement of tissue optical properties. *Applied Optics* 28:2331–2336.

Pawley, J. B. 2006. *Handbook of Biological Confocal Microscopy*. New York: Springer Verlag.

Perrault, S. D., C. Walkey, T. Jennings, H. C. Fischer, and W. C. W. Chan. 2009. Mediating tumor targeting efficiency of nanoparticles through design. *Nano Letters* 9:1909–1915.

Pitsillides, C. M., E. K. Joe, X. B. Wei, R. R. Anderson, and C. P. Lin. 2003. Selective cell targeting with light-absorbing microparticles and nanoparticles. *Biophysical Journal* 84: 4023–4032.

Prescott, S. W., and P. Mulvaney. 2006. Gold nanorod extinction spectra. *Journal of Applied Physics* 99:123504–123507.

Purcell, E. M., and C. R. Pennypacker. 1973. Scattering and absorption of light by nonspherical dielectric grains. *Astrophysical Journal* 186:705–714.

Qian, W., X. Huang, B. Kang, and M. A. El-Sayed. 2010. Dark-field light scattering imaging of living cancer cell component from birth through division using bioconjugated gold nanoprobes. *Journal of Biomedical Optics* 15:046025.

Rechberger, W., A. Hohenau, A. Leitner, J. Krenn, B. Lamprecht, and F. Aussenegg. 2003. Optical properties of two interacting gold nanoparticles. *Optics Communications* 220:137–141.

Schultz, S., D. R. Smith, J. J. Mock, and D. A. Schultz. 2000. Single-target molecule detection with nonbleaching multi-color optical immunolabels. *Proceedings of the National Academy of Sciences of the United States of America* 97: 996–1001.

Sherry, L. J., S. H. Chang, G. C. Schatz, R. P. Van Duyne, J. Benjamin, and Y. Xia. 2005. Localized surface plasmon resonance spectroscopy of single silver nanocubes. *Nano Letters* 5:2034–2038.

Sherry, L. J., R. Jin, C. A. Mirkin, G. C. Schatz, and R. P. Van Duyne. 2006. Localized surface plasmon resonance spectroscopy of single silver triangular nanoprisms. *Nano Letters* 6:2060–2065.

Skala, M. C., M. J. Crow, A. Wax, and J. A. Izatt. 2008. Photothermal optical coherence tomography of epidermal growth factor receptor in live cells using immunotargeted gold nanospheres. *Nano Letters* 8:3461–3467.

Skrabalak, S. E., J. Chen, Y. Sun et al. 2008. Gold nanocages: Synthesis, properties, and applications. *Accounts of Chemical Research* 41:1587–1595.

Sokolov, K., M. Follen, J. Aaron et al. 2003. Real-time vital optical imaging of precancer using anti-epidermal growth factor receptor antibodies conjugated to gold nanoparticles. *Cancer Research* 63:1999–2004.

Sönnichsen, C., S. Geier, N. Hecker et al. 2000. Spectroscopy of single metallic nanoparticles using total internal reflection microscopy. *Applied Physics Letters* 77:2949.

Stensberg, M. C., Q. Wei, E. S. McLamore, D. M. Porterfield, A. Wei, and M. S. Sepulveda. 2011. Toxicological studies on silver nanoparticles: challenges and opportunities in assessment, monitoring and imaging. *Nanomedicine* 6:879–898.

Su, K. H., Q. H. Wei, X. Zhang, J. Mock, D. Smith, and S. Schultz. 2003. Interparticle coupling effects on plasmon resonances of nanogold particles. *Nano Letters* 3:1087–1090.

Sun, W., G. F. Wang, N. Fang, and E. S. Yeung. 2009. Wavelength-dependent differential interference contrast microscopy: Selectively imaging nanoparticle probes in live cells. *Analytical Chemistry* 81:9203–9208.

Tkachenko, A. G., H. Xie, D. Coleman et al. 2003. Multifunctional gold nanoparticle–peptide complexes for nuclear targeting. *Journal of the American Chemical Society* 125:4700–4701.

Tkachenko, A. G., H. Xie, Y. L. Liu et al. 2004. Cellular trajectories of peptide-modified gold particle complexes: Comparison of nuclear localization signals and peptide transduction domains. *Bioconjugate Chemistry* 15:482–490.

Tong, L., Y. Zhao, T. B. Huff, M. N. Hansen, A. Wei, and J. X. Cheng. 2007. Gold nanorods mediate tumor cell death by compromising membrane integrity. *Advanced Materials* 19:3136–3141.

Tong, L., W. He, Y. Zhang, W. Zheng, and J. X. Cheng. 2009. Visualizing systemic clearance and cellular level biodistribution of gold nanorods by intrinsic two-photon luminescence. *Langmuir* 25:12454–12459.

Underwood, S., and P. Mulvaney. 1994. Effect of the solution refractive index on the color of gold colloids. *Langmuir* 10:3427–3430.

Ungureanu, C., R. G. Rayavarapu, S. Manohar, and T. G. van Leeuwen. 2009. Discrete dipole approximation simulations of gold nanorod optical properties: Choice of input parameters and comparison with experiment. *Journal of Applied Physics* 105:102032.

Wang, H., T. B. Huff, D. A. Zweifel et al. 2005. In vitro and in vivo two-photon luminescence imaging of single gold nanorods. *Proceedings of the National Academy of Sciences of the United States of America* 102:15752.

Wang, L. V., and H. Wu. 2007. *Biomedical Optics: Principles and Imaging.* New York: Wiley-Blackwell.

Wang, H., D. W. Brandl, P. Nordlander, and N. J. Halas. 2007. Plasmonic nanostructures: artificial molecules. *Accounts of Chemical Research* 40:53–62.

Xu, X. H. N., W. J. Brownlow, S. V. Kyriacou, Q. Wan, and J. J. Viola. 2004. Real-time probing of membrane transport in living microbial cells using single nanoparticle optics and living cell imaging. *Biochemistry* 43:10400–10413.

Yang, P. H., X. S. Sun, J. F. Chiu, H. Z. Sun, and Q. Y. He. 2005. Transferrin-mediated gold nanoparticle cellular uptake. *Bioconjugate Chemistry* 16:494–496.

Yelin, D., D. Oron, S. Thiberge, E. Moses, and Y. Silberberg. 2003. Multiphoton plasmon-resonance microscopy. *Optics Express* 11:1385–1391.

Yu, J., S. A. Patel, and R. M. Dickson. 2007. In vitro and intracellular production of peptide-encapsulated fluorescent silver nanoclusters. *Angewandte Chemie-International Edition* 46:2028–2030.

Yu, J., S. Choi, C. I. Richards, Y. Antoku, and R. M. Dickson. 2008. Live cell surface labeling with fluorescent Ag nanocluster conjugates. *Photochemistry and Photobiology* 84:1435–1439.

Yu, J., S. Choi, and R. M. Dickson. 2009. Shuttle-based fluorogenic silver-cluster biolabels. *Angewandte Chemie-International Edition* 48:318–320.

Yu, Y. Y., S. S. Chang, C. L. Lee, and C. R. C. Wang. 1997. Gold nanorods: Electrochemical synthesis and optical properties. *Journal of Physical Chemistry* B 101:6661–6664.

Zhang, J., Y. Fu, and J. R. Lakowicz. 2007. Single cell fluorescence imaging using metal plasmon-coupled probe. *Bioconjugate Chemistry* 18:800–805.

Zhang, J., Y. Fu, M. H. Chowdhury, and J. R. Lakowicz. 2008. Single-molecule studies on fluorescently labeled silver particles: Effects of particle size. *Journal of Physical Chemistry C* 112:18–26.

8

X-Ray Fluorescence Computed Tomography Imaging of Nanoparticles

Bernard L. Jones
University of Colorado
School of Medicine

Sang Hyun Cho
Georgia Institute of Technology

8.1 Introduction

8.1.1 Tomographic Imaging

Since the discovery of the practical applications of the Radon transform by Hounsfield in the mid-twentieth century, tomographic imaging has been a driving force in the advancement of medical science and research. X-ray computed tomography (CT) scans have given physicians and engineers a noninvasive and nondestructive tool to acquire high-resolution images of the internal structure of a patient or material sample. Emission tomography scans such as single-photon emission CT (SPECT) and positron emission tomography (PET) have also given us the ability to track chemical tracers within a living organism. Now that nanoparticles are finding a plethora of biomedical applications, an imaging modality able to acquire a tomographic image of nanoparticle distributions *in vivo* could assist in the development of new therapeutic or diagnostic nanoagents.

Tomographic imaging techniques generally fall into two categories: transmission scans and emission scans. Transmission scanning, such as x-ray CT, is generally used to create a map of attenuation within different parts of an object, allowing one to distinguish between differing regions (e.g., bone, muscle, and fat) and generate an image of the anatomy or structure of an object. Emission scanning involves the detection and localization of radiation emitted by tracers injected or inserted into an object. When used in living subjects, these tracers are designed to seek a certain region of the body because of their biological or chemical nature, and are thus used to generate functional images of different biological processes.

Emission tomography has a long history of successful application in the field of radiology. In general, this technique involves generating a tomographic image of activity inside an object by detecting radiation emitted by the substance in question. Typically, the substance is a radioisotope, which emits either a single photon (SPECT) or a positron, which annihilates and emits two photons (PET). This imaging modality hinges on the successful detection of radiation that is specific to the radioisotope, such as the 140-keV photon in ^{99m}Tc SPECT or the dual 511-keV annihilation photons in PET. An image can also be reconstructed by capturing photons stimulated by irradiation with an external beam of radiation, instead of injecting radioactivity directly. Such imaging modalities can be called *stimulated* emission tomography. For instance, it is possible in principle to verify dose delivery in proton radiotherapy by performing PET imaging to detect ^{11}C and [15]O produced during interactions of protons with tissue (Oelfke et al. 1996). This principle of stimulated emission tomography forms the basis of x-ray fluorescence CT (XFCT), which involves the detection of characteristic fluorescence x-rays emitted from an object under irradiation by an external x-ray beam.

8.1.2 Imaging Challenges in Nanoparticle Applications

The wide array of possible properties for various nanoparticles makes them a diverse platform for functional imaging. These properties can be inherent of the nanoparticles themselves, or derived from some attached functional agent. Through a phenomenon known as enhanced permeability and retention (Maeda et al. 2003), nanoparticles of sufficiently small size are able to penetrate the tumor vasculature and interstitium, resulting in an elevated concentration of nanoparticles in the tumor as compared to that of normal tissue. In addition, nanoparticles can be conjugated to antibodies of tumor-specific biomarkers under an "active" targeting scenario. These potential targets include tumor markers such as epidermal growth factor receptor (El-Sayed et al. 2005) and human epidermal growth factor receptor-2 (Copland et al. 2004), and mediators of angiogenesis such as vascular endothelial growth factor (Mukherjee et al. 2007). Additionally, metabolic agents such as deoxyglucose can also be used for the purpose of active targeting and internalization of nanoparticles via a mechanism similar to ^{18}F deoxyglucose positron emission tomography (FDG-PET) imaging (Li et al. 2010).

In the past decade, some notable applications utilizing nanoparticles have emerged for cancer imaging, radiation therapy, and thermal therapy, many of which are detailed in this book. Imaging, both *in vitro* and *in vivo*, has played an important role for the development of these applications. Various imaging challenges associated with *in vitro* studies appear to have been adequately handled so far. On the other hand, previous research efforts were often hindered or delayed because of the lack of an effective *in vivo* assay or imaging tool to determine the biodistribution of nanoparticles injected into animals. Without such a tool, the biodistribution needs to be determined via *ex vivo* analysis after sacrificing animals. In principle, it would be possible to determine the biodistribution of nanoparticles *in vivo* if the spatial distribution and amount of nanoparticles within a tumor and other critical organs can be quantified by an imaging modality.

8.2 X-Ray Fluorescence Imaging

One unique physical property controlled by the base material of the metallic nanoparticle itself is its fluorescence x-rays (or characteristic x-rays). Thus, fluorescence x-rays provide a powerful tool to assess the material composition of a sample. In the electron shell model, the electrons of each atom reside in different energy levels. Electrons can transition between these energy states via absorption or emission of a photon whose energy

exactly matches the difference between shells. Variations in the charge of the nucleus lead to different configurations of these energy levels for each value of the atomic number Z. As a result, the fluorescence x-ray energies of each element are unique, and an atom can be identified by the energy of the fluorescence x-rays it emits.

Analytic techniques using fluorescence x-rays are commonly used when one wishes to identify the material composition of a sample. In practice, such methods include x-ray fluorescence, where the sample is bombarded with photons, or energy-dispersive x-ray spectroscopy, which is performed with electrons in a method similar to electron microscopy. In XFCT, these techniques are coupled with methods of tomographic imaging in order to determine the location and concentration of some fluorescing substance within an object. The overview of this method is as follows:

- A nanoparticle-loaded object is irradiated by a beam of photons, leading to the emission of fluorescence x-rays.
- The fluorescence x-rays are measured by an energy-sensitive photon detector system, which can determine the energy of fluorescence x-rays, thereby enabling the identification of the base material of nanoparticles.
- Based on the intensity of the fluorescence signal at different positions and angles relative to the object, the distribution of nanoparticles is calculated using tomographic reconstruction techniques.

These steps are illustrated in Figure 8.1. An optional step is to reconstruct a tomographic transmission image using the direct

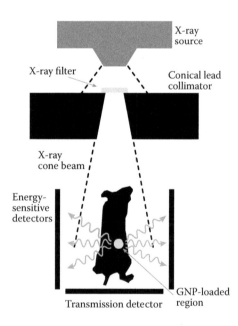

FIGURE 8.1 X-ray fluorescence computed tomography of an object loaded with gold nanoparticles (GNPs). A beam of x-rays irradiates an object, stimulating production of fluorescence photons from GNPs. Detection of these photons allows for tomographic reconstruction of location and concentration of GNPs.

x-rays passing through the object. This image can identify the internal structure of the object or be used in a later step for the attenuation correction of XFCT image.

8.3 Geometry of XFCT

8.3.1 Direct Measurement of Fluorescence X-Rays by Collimation of Source and Detector

In the simplest formulation of XFCT, one can directly measure the location of areas where fluorescence x-rays originate. In other words, using a thin, pencil beam of x-rays, it is possible to excite fluorescence x-rays in a very narrow region of an object. Then, a tightly collimated detector views the fluorescence x-rays emitted along some line perpendicular to the original x-ray beam. In this way, small regions of the object are successively interrogated to determine the nanoparticle concentration in each. This method, as illustrated in Figure 8.2, is simple to implement, and image reconstruction is trivial (requiring no knowledge of tomographic methods). However, it becomes more difficult to accurately correct for attenuation, and this method is hindered by the inability to collect more than one sinogram element per irradiation of the object (increasing the time required for a scan).

8.3.2 Measurement of Fluorescence X-Rays by Collimation of Source or Detector

More sophisticated methods can allow for decreases in the amount of scan time required while preserving image quality. In order to reconstruct a tomographic image of the nanoparticle distribution, the measured fluorescence signal must somehow be spatially encoded. For image reconstruction techniques based on inverting the Radon transform, the measured signal consists of line integrals through the object. In transmission imaging, this is the attenuation of x-rays between the source and one pixel of the detector. For XFCT, this means that the measured fluorescence signal must be constrained to lie along some ray that passes through the object. This can be accomplished at either the x-ray source or detector. In the first case (source collimation), the incident x-rays are collimated into a thin ray, or pencil beam, and the measured fluorescence signal is constrained to lie along the line this ray makes as it passes through the object. In the second case (detector collimation), the detector is placed behind a parallel-pinhole collimator. Thus, the measured signal originates from the line traced by the view of the detector through the object.

One advantage of source collimation is that an x-ray beam can be made small, allowing for very fine spatial resolution (Figure 8.3). However, there is a tradeoff between beam diameter and scan time, since each ray through the object must be irradiated separately. Source collimation techniques are common when XFCT is performed using a synchrotron beam to excite fluorescence x-rays within the sample. Synchrotron beams have many advantages for XFCT, as they can be made monochromatic and very bright (i.e., high photon fluence). Monochromatic beams

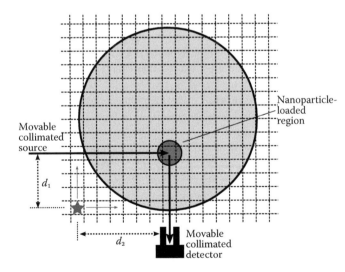

FIGURE 8.2 Direct measurement geometry (source and detector collimator). A collimated pencil beam of x-rays (at d_1) is scanned through an object. A moveable collimated detector (at d_2) sees scattered and fluorescence photons emitted from pixel at intersection of pencil beam and detector view. Projections are formed by assembling the total fluorescence signal $P(d_1,d_2)$ recorded. Iterative attenuation correction begins in bottom left pixel (star), which experiences no attenuation. The measured value is used to correct for attenuation in two neighboring pixels, and this correction continues iteratively through the phantom.

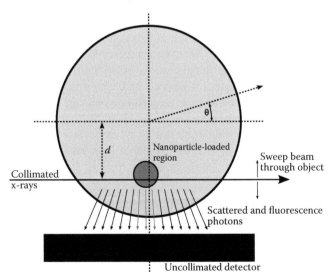

FIGURE 8.3 Source collimation geometry. x-rays are collimated into a thin pencil beam, which is scanned over the phantom in successive iterations (d). Line projections are formed by assigning total fluorescence signal $P(d,\theta)$ recorded by uncollimated detector to the path of beam through the phantom at each rotation position (θ).

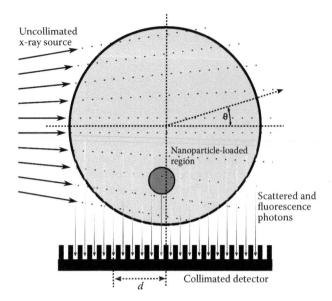

FIGURE 8.4 Detector collimation geometry. A fan or cone beam of x-rays is incident onto an object. Each detector element (at position d) sees scattered and fluorescence photons emitted along its line of view. Projections are formed by assigning total fluorescence signal $P(d,\theta)$ recorded by collimated detector element to line of view through the phantom at each rotation position (θ).

have a high dose/signal ratio (discussed below), and there is a direct relationship between photon fluence and scan time. Thus, synchrotron beams are able to perform very high resolution XFCT with reasonable scan time.

Detector collimation results in a similar imaging geometry as SPECT, where a line or area detector records many rays through the object simultaneously (Figure 8.4). This has the advantage of allowing for parallelization of data collection, since all data for one projection can be taken concurrently. This allows for the scan time and imaging dose (two of the limiting factors for *in vivo* XFCT imaging) to be decreased. However, the limiting factor for spatial resolution of the image becomes the size of each pixel element of the detector. For small-animal polychromatic XFCT, this can restrict image quality because of size limitations on energy-sensitive detectors. Research is ongoing to reduce these problems.

8.4 Measurement of XFCT Signal

There are two main effects within the object that affect the fluorescence detection process. The first is obviously fluorescence x-ray emission by the nanoparticles, which occurs when a source photon interacts with an atom in a nanoparticle, creating a vacancy in the electron shell and resulting in fluorescence x-ray emission. The second effect is Compton scatter of the source photons within the object. The first effect constitutes the signal of XFCT, and the scattered photons make up the noise. It is necessary to use an energy-sensitive detector in order to

separate the fluorescence photons from those photons scattered in the object.

The physics of XFCT makes scatter correction much more difficult than other emission tomography modalities. In SPECT, a radioactive tracer is injected into the bloodstream, and detectors placed around the patient measure the intensity of radiation emitted in different directions. Since scattering and other effects at these energies can only decrease the energy of the photons emitted by the tracer, the highest energy photons in the patient are those that have not interacted. Scatter can be separated from the primary by using a simple energy window to exclude lower energies.

In XFCT, on the other hand, the highest energy photons are found in the x-ray source. This is because any fluorescence photon will have less energy than the photon that has undergone photoelectric absorption in the originating atom. For instance, the gold K and L edges occur at 80.7 and 11.9–14.4 keV, respectively. Thus, for the prominent K_α (L-to-K) fluorescence x-ray emission, a source photon of energy 81 keV will result in a fluorescence photon of 66 or 68 keV. That same 81 keV photon could also undergo Compton scattering at 90°, which would also result in a photon of 66 keV! Therefore, it is impossible to differentiate fluorescence photons from scattered photons by energy alone. It is necessary to estimate the magnitude of the scatter background, and subtract it from the measured signal at the fluorescence photon energies.

In order to maximize the efficiency of XFCT data acquisition, the energy spectrum of the x-ray source must be carefully chosen. In particular, the spectrum must be optimized to maximally excite the chosen energy shell of the nanoparticles used. For instance, consider the K-shell of gold. The two innermost electrons of the gold atom have roughly 81 keV of binding energy, and any photon that is intended to cause K-shell fluorescence x-ray emission must exceed that energy. Since the interaction cross section of gold is peaked at 81 keV (because of K-shell photoelectric absorption), a monoenergetic beam of 81 keV photons would cause the greatest amount of fluorescence photons to be emitted.

In the absence of a monochromatic beam, XFCT can be performed with a polychromatic beam. However, for standard bremsstrahlung-heavy photon beams, the x-rays must be heavily filtered to harden the beam spectrum. Any photon that lies below the targeted absorption edge of the nanoparticles cannot induce fluorescence x-ray emission, and adds only unnecessary dose and noise. For imaging of gold nanoparticle (GNP)-loaded regions within small animal–sized objects, where the intended target is the K_α fluorescence line, one must weight the x-ray beam as heavily as possible above 81 keV.

The drawback from heavy filtration is a decrease in the photon fluence rate of the source, which increases the amount of scanning time required to acquire a useable image. Thus, in designing a suitable x-ray filter for polychromatic XFCT, one must balance the tradeoffs among hardness in the source spectrum, x-ray dose, and photon fluence of the source. Lead, which has been used for some GNP-based XFCT applications as a filter, has

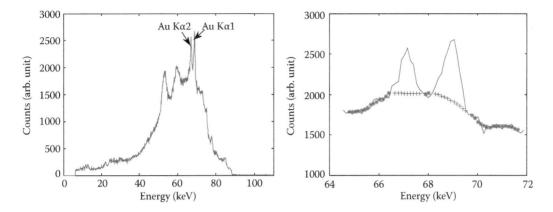

FIGURE 8.5 Measured fluorescence spectrum with scatter background. Full spectrum measured from a GNP-loaded object under irradiation by a 110 kVp source is shown on left. At right, fluorescence signal emitted by GNPs is shown. Signal is extracted by fitting a polynomial to measured Compton background (*) and subtracting scattered photons from fluorescence channels (+). (Reprinted with permission from Cheong, S.K. et al., *Phys. Med. Biol.*, 55, 647–662, 2010.)

a transmission band between the *L* and *K* shells (16–88 keV) that results in a 110 kVp x-ray beam that is peaked between 50 and 90 keV. This is usable for GNP-based XFCT, but the decreased transmission above 90 keV and high fluence below 80 keV results in a low signal/dose ratio. A lead-filtered beam would be most suited for XFCT applications that use nanoparticles with an atomic number in the range of 50 (tin) to 70 (ytterbium).

One x-ray filter that had applications in the days of orthovoltage radiation therapy is the Thoraeus filter, which is composed of stacked layers of tin, copper, and aluminum. For 100- to 150-kVp x-rays, the spectrum from this filter is peaked strongly in the 90- to 100-keV range. However, the overall fluence is greatly reduced.

One other possible approach involves quasi-monochromatization of a polychromatic beam. For beams above 50 keV, where this becomes difficult to accomplish with filtration, a single energy can be extracted from a beam using a material such as highly oriented pyrolitic graphite (HOPG). HOPG has a very closely packed crystal structure and homogeneous interatomic spacing, which allows one to extract a monochromatic beam using Bragg diffraction. In theory, a 95 keV beam can be extracted from a polychromatic beam at a scattering angle of roughly 1°. However, there are practical considerations such as the low intensity of a resulting pencil beam or the geometrical complexity of using Bragg diffraction to form a fan/cone beam, which might pose some difficulty in applying this technique to GNP-based XFCT.

To determine the intensity of the fluorescence signal, the fluorescence peak must be separated from the Compton scatter background. This is possible because fluorescence x-ray emission occurs at discrete energies, whereas Compton scattered photons manifest as a continuum spectrum. However, an energy-sensitive detector with fairly good energy resolution is required, as the scatter-to-primary ratio for a polychromatic beam is fairly high. If a good signal can also be acquired for the neighboring energy channels around the fluorescence peak, the fluorescence intensity can be determined by fitting a polynomial to the scatter

background (Figure 8.5). For gold K_α fluorescence photons generated using a lead- or tin-filtered polychromatic beam, a third-degree polynomial performs this task well. For other energies or spectrums, a lower-order fit may be appropriate. It may also be possible to acquire a better signal by using more advanced spectrum filtering or curve fitting methods.

8.5 Reconstruction of XFCT Image

As in most other tomographic imaging methods, the goal of image reconstruction is to invert the Radon transform of the imaged object. In transmission imaging, where the Radon transform is densely sampled (large number of projected rays) and the required resolution of the reconstructed image is very high (large number of voxels in the reconstructed image), Fourier back-projection methods are most commonly used. However, emission tomography differs in that there is typically a lower density in both the signal and the reconstructed image, as well as more noise in the measured sinogram. For these reasons, it is more common to use iterative reconstruction methods, as these are more robust against undersampled and noisy projection data.

The reconstruction problem in XFCT is identical to that of SPECT and is well suited to using iterative methods to solve. For imaging of a sparse nanoparticle distribution, such as a small animal injected with nanoparticles targeted to a particular tumor or organ, maximum likelihood/expectation maximization methods perform well, as they place a greater weight on voxels containing higher signal. The low signal projection elements have a much higher scatter-to-primary ratio (SPR), and reconstruction performs better when these spurious signals are ignored.

It is also likely for compressed sensing (CS) reconstruction algorithms to see greater use in XFCT image reconstruction. These algorithms are already beginning to see use in applications such as 4-D cone-beam CT, where the high SPR and significant undersampling leads to very poor image quality when using traditional back-projection reconstruction. Because of the difficulty in acquiring a measurable fluorescence signal while meeting constraints on scanning time and imaging dose, only a limited number of projections would be allowed for *in vivo* animal imaging. CS algorithms may allow for an acceptable XFCT image to be reconstructed from a noisy, undersampled dataset.

8.5.1 XFCT Reconstruction Algorithm: Example

Tomographic image reconstruction is a very vibrant field, and countless researchers are working on improving the methods used to build a 3-D image from projected 2-D data. In principle, XFCT images could be reconstructed using any number of methods. Here, we present one example reconstruction method that has been applied to XFCT. This is far from the only way to reconstruct an XFCT image; however, in reviewing the current simple example, one may gain further understanding of the problems and challenges specific to XFCT reconstruction.

Using the Maximum Likelihood/Expectation Maximization (ML-EM) (Cherry et al. 2003; Lange and Carson 1984; Shepp and Vardi 2007) formulation of iterative image reconstruction, one can construct a relationship between the measured sinogram, p, the reconstructed image, a, and the image response matrix, $\mathbf{M}$. Consider a set of measured sinograms (or projection dataset) p. Each element p_j describes the gold fluorescence signal seen by one detector at a specific projection angle, and the number of elements in p is equal to the product of the number of detectors and the number of projection angles. One wishes to use p to reconstruct the image dataset a, whose element a_i represents the intensity of each pixel in the reconstructed image. To do this, one must construct the system response matrix $\mathbf{M}$, where each element $M_{i,j}$ is the probability that a fluorescence photon will be created at the pixel a_i and detected in the projection element p_j. Note that p denotes any measured sinogram and could be acquired under a source- or detector- collimation geometry.

The operation of multiplying the matrix $\mathbf{M}$ with the image a denotes the forward projection operation, or the Radon transform, of the object along the rays defined in the imaging geometry, and the result is the projection dataset p (Equation 8.1). By calculating $\mathbf{M}$ and measuring p, the goal of image reconstruction is to invert this relationship to determine the reconstructed image a. It has been shown that one can iteratively converge upon a solution by performing a forward projection of the calculated image and comparing this to the measured projection. The full operating equation of the ML-EM algorithm is given in (Equation 8.2).

$$p_j = \sum_i M_{i,j} a_i \tag{8.1}$$

$$a_i^{k+1} = \frac{a_i^k}{\sum_j M_{i,j}} \times \sum_j M_{i,j} \frac{p_j}{\sum_j M_{i,j} a_i^k} \tag{8.2}$$

8.5.2 Attenuation Correction for Fan/Cone Beam Geometry: Example

In order to reconstruct an accurate image, the system response matrix $\mathbf{M}$ must be carefully constructed, as it includes all factors that influence the signal from the x-ray source to the detector. This includes all the various attenuation effects. The ability of XFCT to accurately reconstruct the concentration of nanoparticles within an object relies on proper attenuation correction. In fan/cone beam XFCT, there are four main attenuation effects one must consider:

- Attenuation of the primary beam as it travels through the object
- Attenuation of emitted fluorescence photons by the object as they travel to the detector
- Attenuation of emitted fluorescence photons by other nanoparticles, as they travel to the detector
- Inverse square fall-off of isotropically emitted fluorescence photon fluence from a source (nanoparticle) with distance

In addition to the list above, the system response matrix also includes the effects of source geometry, collimation, and image voxelization.

The image response matrix can be constructed through several means. It could be based on measurements taken specifically for this purpose, as is usually the case in SPECT. However, this may require specialized sources created explicitly for this task. It could also be based on Monte Carlo simulations of certain parameters. Finally, one could calculate the magnitude of these different effects from first principles.

8.6 Developments and Applications of XFCT

8.6.1 Synchrotron XFCT

Almost all applications of XFCT have centered on the use of a synchrotron as a source of x-rays. As mentioned in the previous section, synchrotron x-ray beams are very bright and monochromatic, which allows one to acquire XFCT data quickly with a low SPR. Researchers had long been using x-ray fluorescence techniques to analyze the material composition of a sample, and extension of this technique to tomography was a natural progression.

Much of the early work on synchrotron XFCT for biological applications focused on detection of x-ray contrast agents already in use, such as iodine and gadolinium. Isotopes of iodine can also be used for SPECT imaging, such as ^{123}I-based imaging of cerebral blood flow. The ability to detect a fluorescence signal depends on the energy of the emitted fluorescence photon, and so these high-Z contrast agents were a logical choice for XFCT investigations. Whereas transmission imaging is used to detect very high concentrations of contrast agents in blood, the advantage of XFCT over transmission CT is the ability to detect much smaller concentrations of iodine using x-ray fluorescence. However, the short penetration depth of iodine K_α photons makes this technique possible only in small objects.

One of the first demonstrations of fluorescence tomography was published in 1989 by Cesareo and Mascarenhas. Their method was not based on inversion of the Radon transform; instead, they performed a direct measurement of each voxel in their sample by collimating both the x-ray source and detector as discussed earlier in this chapter (see Figure 8.2). As a proof-of-principle, they imaged a 2×2 cm² plexiglass object containing regions loaded with iodine at a concentration of 5 mg/mL, and demonstrated that fluorescence tomography is a nondestructive method to localize high-Z substances within an object. Takeda et al. (1995) also published a similar investigation in 1995 and were able to detect concentrations of iodine as low as 50 ng/mL. Their experiment also used the "direct measurement" technique (source and detector collimation as illustrated in Figure 8.2), and additionally they showed that the linear relationship between iodine concentration and fluorescence intensity could be used to measure the concentration of high-Z substances as well as location.

One problem noted from the studies mentioned above was that it was difficult to account for attenuation in the "direct measurement" geometry. The attenuation of the signal in each pixel depended on the concentration of tracer in each nearby pixel. The images were corrected by summing the attenuation along the lines from pixel-to-source and pixel-to-detector; thus, any error in a pixel was propagated to all pixels behind it. This was addressed by Hogan et al. in 1991, showing through computer simulations that rotational tomography could be used to reconstruct an XFCT image using the source collimation geometry (see Figure 8.3).

In 1996, Takeda et al. expanded their method to include Radon inversion by removing the pencil-beam collimation from the detector. They used Fourier backprojection to reconstruct images of a 5-mm-diameter phantom containing iodine at 0.5–2 mg/mL. Two years later, Rust and Weigelt (1998) again demonstrated the use of Radon inversion for XFCT, using an experimental setup that relied on an iterative reconstruction algorithm to produce an image. They applied their methodology to perform imaging of a biological specimen and were able to acquire images of iodine within a 1.5-cm *ex vivo* thyroid gland at concentrations of 0.6 mg/mL. Takeda et al. performed similar thyroid measurements in 1999 (Takeda et al. 1999).

After its initial development and refinement, synchrotron XFCT was used for a wide range of applications. Some of these applications include mapping trace metals in breast tissue (Pereira et al. 2007; Rocha et al. 2007), determining compartmentalization of metals in plant roots (McNear et al. 2005), and measuring the cerebral perfusion of mice (Takeda et al. 2009). Another inorganic application of XFCT was a study that mapped the distribution of cadmium in fly ash particles (Camerani et al. 2004). Further information on modern synchrotron XFCT and its applications can be found in a recent review paper by de Jonge and Vogt (2010).

8.6.2 Benchtop XFCT

There is no question that the fluence rate and energy spectrum of synchrotron beams make it ideal for XFCT. However, there are issues with synchrotron beams, especially for *in vivo* animal imaging. Imaging with a synchrotron source can be quite difficult simply because it is hard to access such a beam. Currently, there are around 40 synchrotron facilities operating worldwide, hosting research on materials science, physics, chemistry, biology, and other related disciplines. Because of the extreme cost of building a synchrotron facility, it is uncommon to see one operated by an entity below the state or national level. Also, because the finite number of beam lines, access to synchrotron x-rays are limited. Moreover, typical dose rates of high-flux synchrotron beams could be too high for routine *in vivo* animal imaging (Liu et al. 2009).

For the above reasons, the ability to perform XFCT in a benchtop laboratory setting would be beneficial. Benchtop polychromatic XFCT is a new field, and to date there have not been many applications of this technology. However, several proof-of-principle studies have been performed that demonstrate the power and versatility of this XFCT technique. The main difficulty in performing XFCT with a polychromatic source is the extremely high SPR. For instance, with a lead-filtered 110 kVp beam, the number of Compton scattered photons can be more than 100 times higher than the number of fluorescence photons emitted by a low-concentration GNP-loaded sample. Any attempt to acquire a fluorescence signal with a polychromatic beam must carefully consider the best way to avoid the production or detection of scattered photons.

Consider a hypothetical spectrum shown in Figure 8.6. The solid line shows an idealized fluorescence peak visible above a flat background of scattered photons. If a detector were allowed to measure this spectrum for an arbitrarily long amount of time, it would be trivial to subtract the height of the fluorescence peak form the scatter background. However, any realistic imaging modality must acquire a useable signal while exposing the

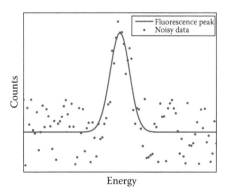

FIGURE 8.6 Hypothetical measured spectrum showing stochastic effects of Compton scatter noise. Solid line shows true spectrum of fluorescence photons above scatter background. Stochastic noise in measured scatter obscures peak, and so for fluorescence photons to be detected, peak height must overcome these fluctuations.

subject to as little dose as possible. For a finite counting time, the measured spectrum will resemble the dotted curve in the figure. Based on the Poisson model of counting statistics, for a scatter background of magnitude N, there will be fluctuations from channel to channel with a magnitude of roughly $\sqrt{N}$. If the fluorescence signal is to be measured accurately, the peak height must exceed this value.

For this reason, one of the main challenges of benchtop XFCT is the generation of a fluorescence signal for low concentrations of GNPs that is able to overcome fluctuations in the scatter background. This can be accomplished through careful choice of the acquisition geometry and x-ray filter. However, it remains a major limitation, and attempts to avoid it by acquiring signal for a longer period can lead to increases in acquisition time and x-ray dose.

In 2010, Cheong et al. were able to successfully acquire an image of low (<2% by weight) concentrations of GNPs with a 110-kVp benchtop x-ray source (Figure 8.7). The experiment was performed in a 5-cm-diameter polymethyl methacrylate (PMMA) phantom using a pencil beam filtered by lead, and the gold concentration was determined by measuring the amount of K_α fluorescence photons emitted by GNPs. Although the fluorescence acquisition geometry resembled the source collimation used by several synchrotron-based studies (see Figure 8.3), it also used a weak detector collimator as well as elements seen in the direct measurement geometry. A cadmium telluride (CdTe) detector was placed at a 90° relative to the pencil-beam central axis, and was shielded by a relatively wide collimator that allowed the detector to observe a ~1-cm-diameter field of view within the phantom. For each pencil beam through the object, the detector was translated in a series of five 1-cm steps. At each step, the measured fluorescence spectrum was subtracted from the Compton background, and each line projection was assembled by summing the response at each of the five steps. The image was reconstructed using an unfiltered back-projection algorithm that included a manual scatter correction based on *a priori* knowledge of the geometry.

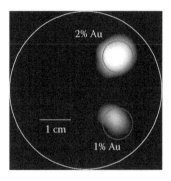

FIGURE 8.7 Reconstructed image of 5-cm PMMA phantom containing GNP-loaded regions at 2% and 1% Au by weight. Image was acquired using a polychromatic benchtop x-ray source, and used source collimation geometry. (Reprinted from Cheong, S.K. et al., *Phys. Med. Biol.*, 55, 647–662, 2010.)

Without this weak detector collimation, the SPR of the measured signal is too high, and the gold fluorescence is indistinguishable from the scatter background. However, the separation of data acquisition into five steps is obviously nonideal, since this increases the scan time and x-ray dose by a factor of 5. In fact, the dose delivered to the phantom during this study was estimated at 2 Gy, which greatly exceeds common imaging doses from other modalities such as micro CT (~20 cGy).

In order to explore reductions in scan time and dose, a computational model was published by the current authors in 2011 detailing a cone-beam (detector collimation) formulation of benchtop polychromatic XFCT (Jones and Cho 2011). This method has several key advantages over the pencil-beam method. First, strong detector collimation significantly decreases the amount of Compton scatter seen by the detector. This allows for a decrease in the amount of fluorescence signal needed to overcome fluctuations in the scatter background (see Figure 8.4). Second, the cone-beam source allows for parallelization of data collection, which decreases scan time by a factor equal to the number of pencil beams.

In this model, simulations were performed with several computational phantoms containing GNP-loaded columns at concentrations ranging from 0.1% to 2.0% by weight. These columns were placed within a 5-cm-diameter phantom in various configurations, leading to the sinograms and reconstructed images shown in Figure 8.8.

The feasibility of a polychromatic cone-beam XFCT device described in this computational study may not be fully established without a successful demonstration by an experimental study. Nevertheless, these results may provide valuable insight into possible design and technical specifications associated with such a device. For example, it can be immediately noticeable that there would be an almost 10-fold reduction in scanning time (total of about 1 h) and x-ray dose (~30 cGy for a 60-projection scan) as a result of a cone-beam implementation, suggesting that routine *in vivo* animal imaging could be feasible with a polychromatic cone-beam XFCT device. The detection limit (0.1% gold by weight) and image resolution (on the order of 1 mm) of this XFCT setup could be further improved by a few additional modifications such as further tailoring/quasi-monochromatization of incident x-ray spectrum, use of novel fluorescence peak selection algorithm, and further optimization of detector collimation. Moreover, x-ray dose could be reduced further given the aforementioned advancements in beam spectrum and fluorescence detection.

An additional benefit of fan/cone beam XFCT is the compatibility of this geometry with existing methods of transmission imaging. Since the fluorescence signal in XFCT is acquired perpendicularly from the beam's central axis, transmission of the source photons through the object may still be used in principle to reconstruct a tomographic image of the radiographic density. The generated data due to transmitted photons are "free" in the sense that there is no additional dose or scan time required. Furthermore, these two imaging modes are complementary to

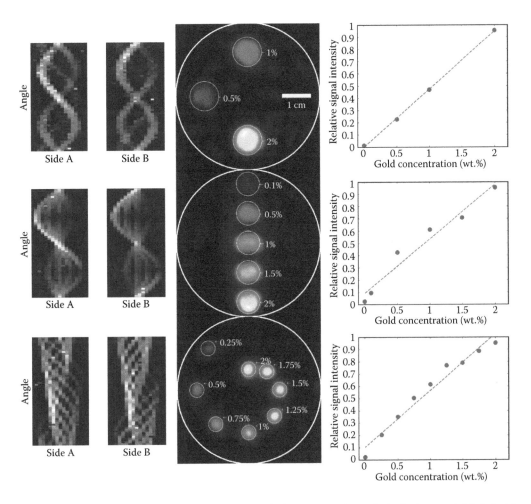

FIGURE 8.8　(**See color insert.**) Results of computational cone beam polychromatic XFCT study. Each phantom is labeled with gold concentration (by weight) of each GNP-loaded column included. Right-side panels show linear relationship between gold concentration and average signal intensity in each column. (Reprinted from Jones, B.L., Cho, S.H., *Phys. Med. Biol.*, 56, 3719–3730, 2011.)

each other, as one provides anatomical imaging, whereas the other provides functional information.

8.7　Conclusions

As evidenced by the exciting developments described in this book, nanoparticles are promising platforms for future diagnostic and treatment techniques in oncology (and other fields of biology and medicine). However, additional tools are needed to help characterize the actions of nanoparticles *in vivo* during preclinical studies. Because of the nature of the atomic structure, fluorescence x-rays provide a powerful technique to assess the material composition of a sample. XFCT combines x-ray fluorescence analysis with the methods of tomographic image reconstruction in order to measure the concentration and location of metallic nanoparticles such as GNPs *in vivo*. With further refinements, it may be possible to accomplish these goals for GNPs present at very low concentrations in small animals with a benchtop apparatus. A device of this kind would open up

a wealth of new analytical techniques for biomedical researchers working with nanoparticles.

Acknowledgment

This work was supported in part by NIH/NCI grant 1R01CA155446.

References

Camerani, M. C., B. Golosio, A. Somogyi et al. 2004. X-ray fluorescence tomography of individual municipal solid waste and biomass fly ash particles. *Analytical Chemistry* 76:1586–1595.

Cesareo, R., and S. Mascarenhas. 1989. A new tomographic device based on the detection of fluorescent X-rays. *Nuclear Instruments and Methods in Physics Research Section A* 277:669–672.

Cheong, S. K., B. L. Jones, A. K. Siddiqi et al. 2010. X-ray fluorescence computed tomography (XFCT) imaging of gold nanoparticle-loaded objects using 110 kVp x-rays. *Physics in Medicine and Biology* 55:647–662.

Cherry, S., J. Sorenson, and M. Phelps. 2003. *Physics in Nuclear Medicine*, 3rd edn. Philadelphia, PA: Saunders.

Copland, J. A., M. Eghtedari, V. L. Popov et al. 2004. Bioconjugated gold nanoparticles as a molecular based contrast agent: Implications for imaging of deep tumors using optoacoustic tomography. *Molecular Imaging and Biology* 6:341–349.

de Jonge, M. D., and S. Vogt. 2010. Hard x-ray fluorescence tomography— an emerging tool for structural visualization. *Current Opinion in Structural Biology* 20:606–614.

El-Sayed, I., X. Huang, and M. El-Sayed. 2005. Surface plasmon resonance scattering and absorption of anti-EGFR antibody conjugated gold nanoparticles in cancer diagnostics: Applications in oral cancer. *Nano Letters* 5:829–834.

Hogan, J. P., R. A. Gonsalves, and A. S. Krieger. 1991. Fluorescent computer tomography: A model for correction of x-ray absorption. *IEEE Transactions on Nuclear Science* 38: 1721–1727.

Jones, B. L., and S. H. Cho. 2011. The feasibility of polychromatic cone-beam x-ray fluorescence computed tomography (XFCT) imaging of gold nanoparticle-loaded objects: A Monte Carlo study. *Physics in Medicine and Biology* 56:3719–3730.

Lange, K., and R. Carson. 1984. EM reconstruction algorithms for emission and transmission tomography. *Journal of Computer Assisted Tomography* 8:306–316.

Li, J., A. Chaudhary, S. J. Chmura et al. 2010. A novel functional CT contrast agent for molecular imaging of cancer. *Physics in Medicine and Biology* 55:4389–4397.

Liu, C.-J., C.-H. Wang, C.-L. Wang et al. 2009. Simple dose rate measurements for a very high synchrotron x-ray flux. *Journal of Synchrotron Radiation* 16:395–397.

Maeda, H., J. Fang, T. Inutsuka, and Y. Kitamoto. 2003. Vascular permeability enhancement in solid tumor: various factors, mechanisms involved and its implications. *International Immunopharmacology* 3:319–328.

McNear, D. H., E. Peltier, J. Everhart et al. 2005. Application of quantitative fluorescence and absorption-edge computed microtomography to image metal compartmentalization in *Alyssum murale*. *Environmental Science & Technology* 39:2210–2218.

Mukherjee, P., R. Bhattacharya, N. Bone et al. 2007. Potential therapeutic application of gold nanoparticles in B-chronic lymphocytic leukemia (BCLL): Enhancing apoptosis. *Journal of Nanobiotechnology* 5:4.

Oelfke, U., G. K. Y. Lam, and M. S. Atkins. 1996. Proton dose monitoring with PET: Quantitative studies in Lucite. *Physics in Medicine and Biology* 41:177.

Pereira, G. R., R. T. Lopes, M. J. Anjos, H. S. Rocha, and C. A. Pérez. 2007. X-ray fluorescence microtomography analyzing reference samples. *Nuclear Instruments and Methods in Physics Research Section A* 579:322–325.

Rocha, H. S., G. R. Pereira, M. J. Anjos et al. 2007. Diffraction enhanced imaging and x-ray fluorescence microtomography for analyzing biological samples. *X-ray Spectrometry* 36:247–253.

Rust, G. F., and J. Weigelt. 1998. X-ray fluorescent computer tomography with synchrotron radiation. *IEEE Transactions on Nuclear Science* 45:75–88.

Shepp, L., and Y. Vardi. 2007. Maximum likelihood reconstruction for emission tomography. *IEEE Transactions on Medical Imaging* 1:113–122.

Takeda, T., M. Akiba, T. Yuasa et al. 1996. Fluorescent x-ray computed tomography with synchrotron radiation using fan collimator. *Medical Imaging 1996: Physics of Medical Imaging* 2708:685–695.

Takeda, T., T. Maeda, T. Yuasa et al. 1995. Fluorescent scanning x-ray tomography with synchrotron radiation. *Review of Scientific Instruments* 66:1471–1473.

Takeda, T., J. Wu, Q. Huo et al. 2009. X-ray fluorescent CT imaging of cerebral uptake of stable-iodine perfusion agent iodo-amphetamine analog IMP in mice. *Journal of Synchrotron Radiation* 16:57–62.

Takeda, T., Q. Yu, T. Yashiro et al. 1999. Human thyroid specimen imaging by fluorescent x-ray computed tomography with synchrotron radiation. *Medical Imaging 1996: Physics of Medical Imaging* 3772:258–267.

III

Nanomaterials for Radiation Therapy

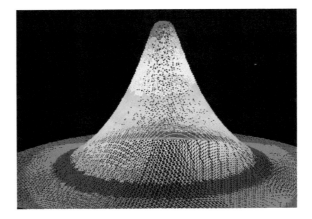

FIGURE 3.2

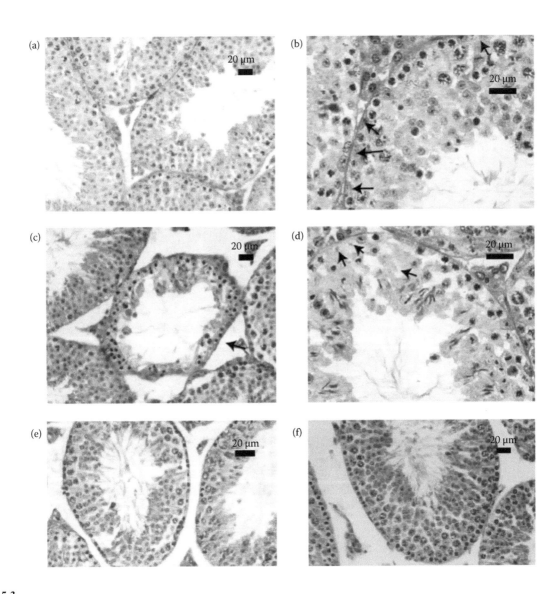

FIGURE 5.3

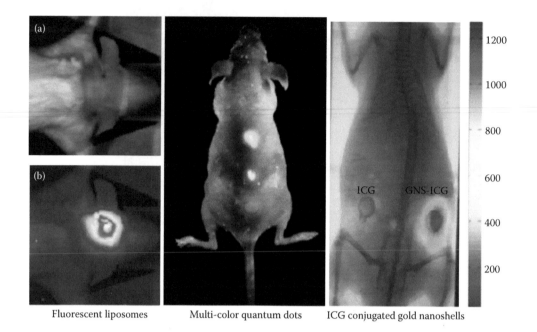

Fluorescent liposomes   Multi-color quantum dots   ICG conjugated gold nanoshells

FIGURE 6.2

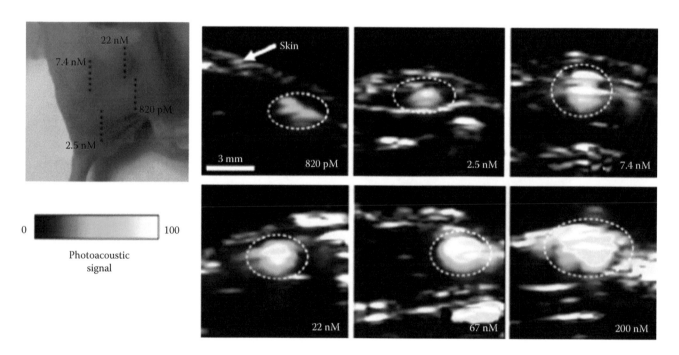

FIGURE 6.3

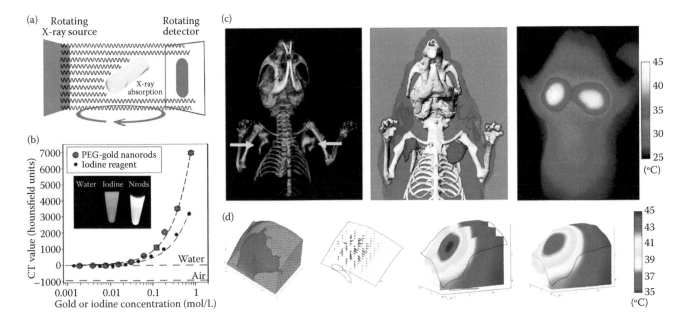

FIGURE 6.5

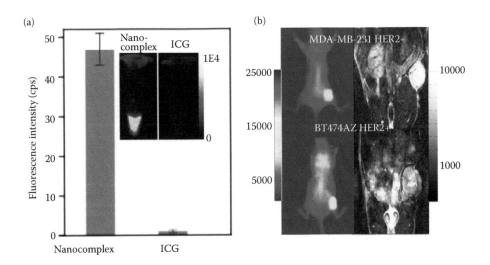

FIGURE 6.6

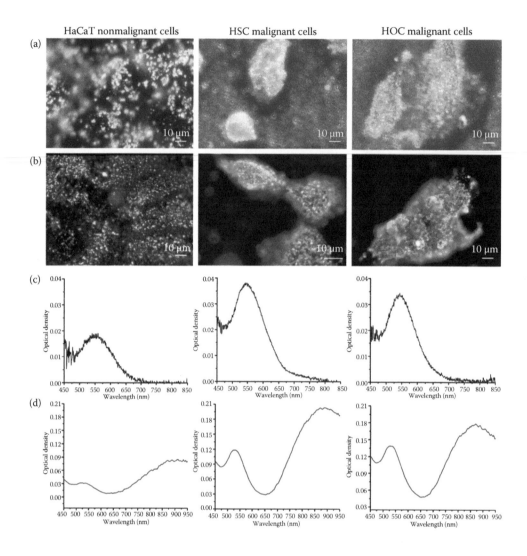

FIGURE 7.7

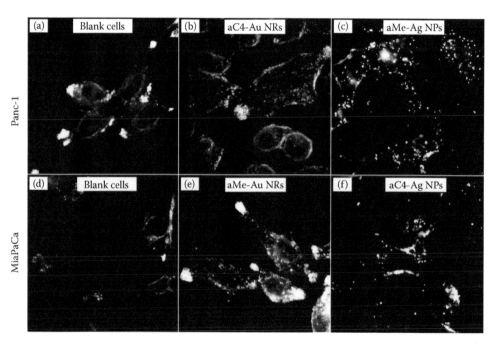

FIGURE 7.8

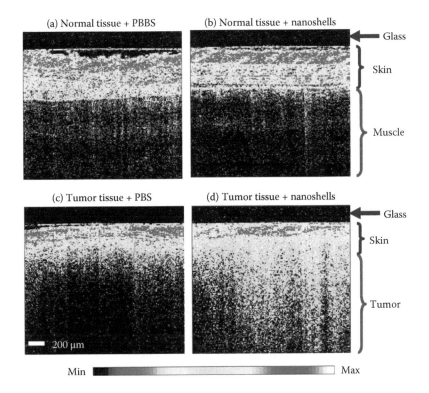

(a) Normal tissue + PBBS   (b) Normal tissue + nanoshells → Glass

Skin

Muscle

(c) Tumor tissue + PBS   (d) Tumor tissue + nanoshells → Glass

Skin

Tumor

200 μm

Min ▭ Max

FIGURE 7.10

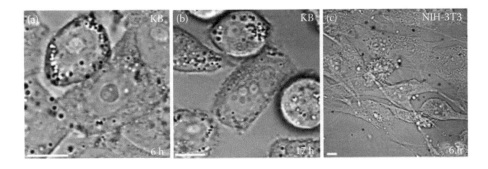

FIGURE 7.13

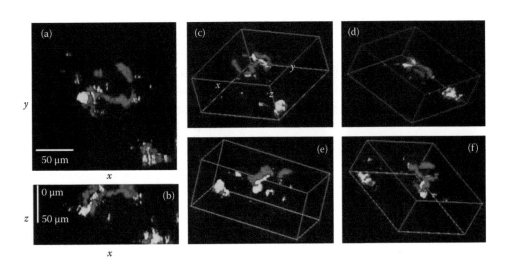

FIGURE 7.14

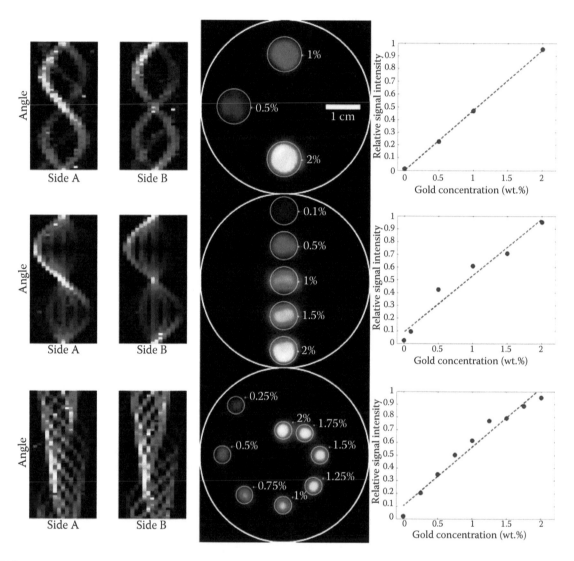

FIGURE 8.8

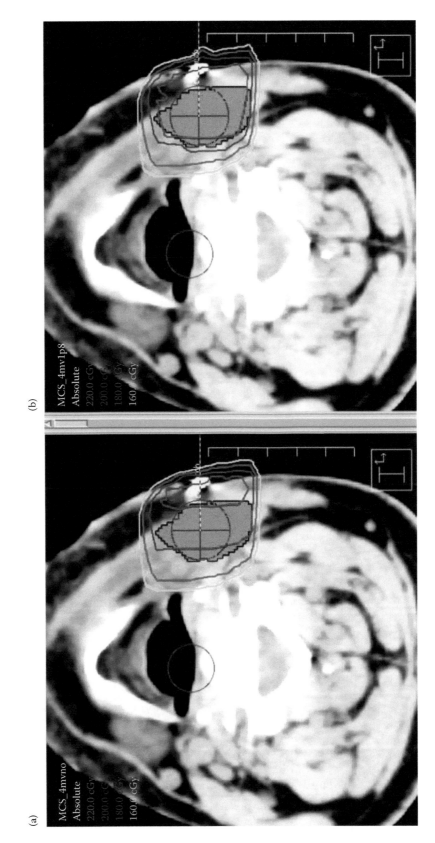

FIGURE 10.3

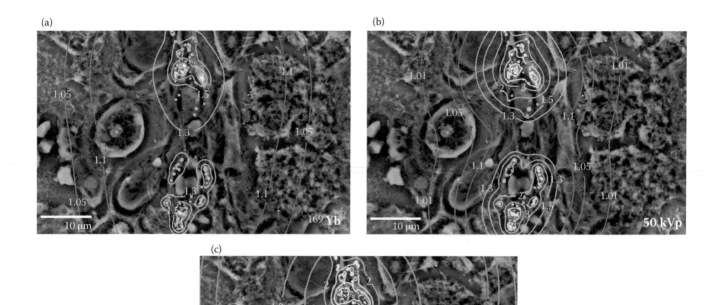

FIGURE 10.12

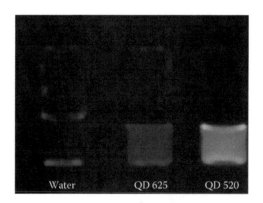

FIGURE 11.2

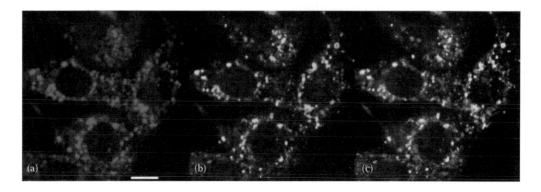

FIGURE 11.11

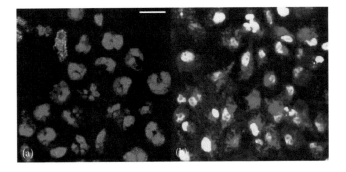

FIGURE 11.12

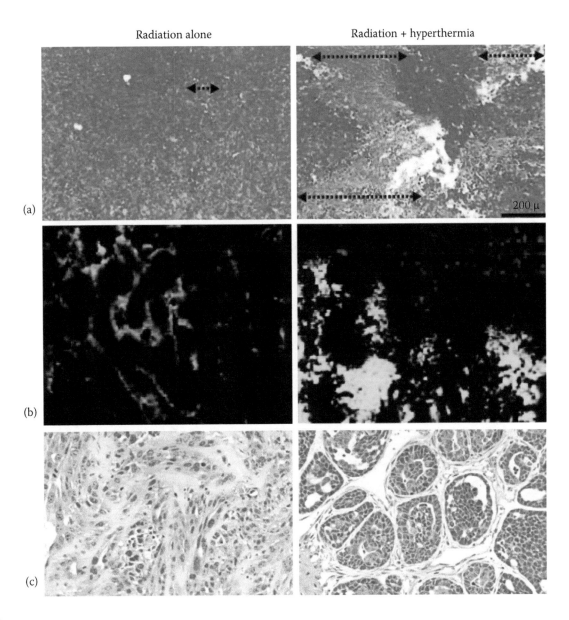

FIGURE 14.4

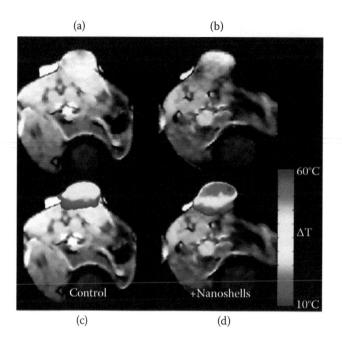

(a)

FIGURE 15.4

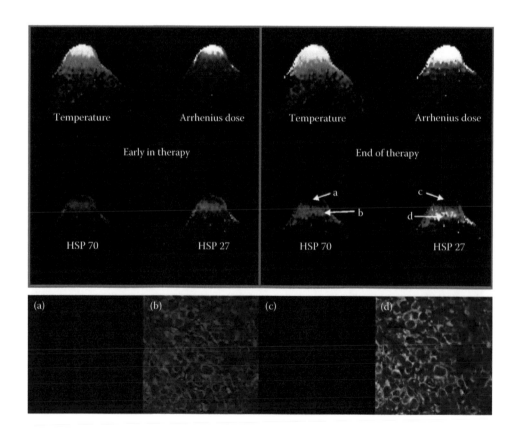

FIGURE 15.5

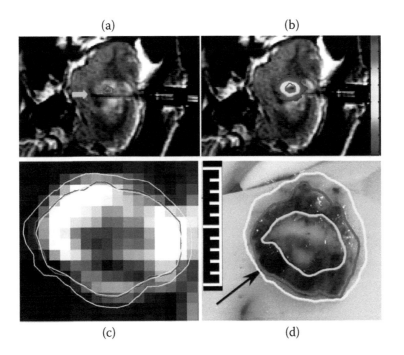

FIGURE 15.6

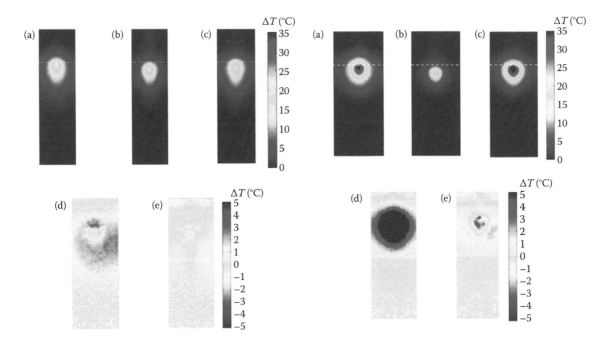

FIGURE 16.2

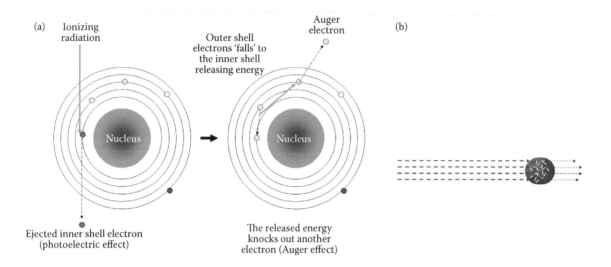

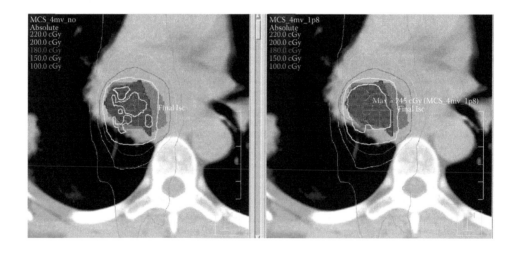

(a)

Ionizing
radiation

Auger
electron

Outer shell
electrons 'falls' to
the inner shell
releasing energy

(b)

Nucleus

Nucleus

Ejected inner shell electron
(photoelectric effect)

The released energy
knocks out another
electron (Auger effect)

FIGURE 19.2

FIGURE 19.3

Gold Nanoparticle–Mediated Radiosensitization

Devika B. Chithrani
Ryerson University

9.1 Introduction

Cancer is the leading cause of death in economically developed countries and the second leading cause of death in developing countries. Millions of people worldwide are diagnosed with cancer in each year, and approximately half of the people who develop cancer each year receive radiation therapy as a component of their treatment. Cancer is also the second leading cause of death in the United States and accounts for approximately one in every four deaths. Many researchers around the world are investigating ways to improve the outcome of current radiation-based therapeutics techniques. As a step forward in this direction, radiation enhancers are being explored for enhancing the therapeutic effects of radiation. Radiation enhancers could cause more tissue damage by increasing the absorption or scattering of the radiation, and cause more local energy deposition. High atomic number (Z) materials are being used as radiation enhancers. These can be introduced into the target material, such as tumor, to increase the probability of ionization events leading to enhanced deposition of energy to destroy tumor tissues. However, delivering a curative dose of radiation to tumor tissues while sparing normal tissues is still a great challenge in radiation therapy. Major milestones achieved toward improved cancer care through enhancing radiation dose will be discussed in the next section (see Figure 9.1).

The concept of using high-Z materials to increase the dose given to a tumor during radiation therapy was advanced more than 20 years ago when iodine was shown by Matsudaira et al. to sensitize cultured cells (Matsudaira et al. 1980; Mello et al. 1983). Mello et al. (1983) found that direct intratumoral injection of iodine with radiation suppressed the growth of 80% of tumors in mice. Nath et al. (1990) demonstrated enhancement of radiosensitivity by a factor of 3 by incorporating iodine into cellular DNA with iododeoxyuridine *in vitro*. Norman et al. (1997) modified a computed tomography scanner to deliver tomographic orthovoltage (140 kVp) x-rays to spontaneous canine brain tumors after intravenous injection with iodine contrast medium, which resulted in 53% longer survival. Radiation oncologists have also noted tissue necrosis around metal implants after therapeutic irradiation with x-rays (Castillo et al. 1988). Das and Chopra (1995) made careful measurements of the dose enhancement factor (DEF) at low-Z/high-Z interfaces irradiated by x-rays.

Experimental x-ray dose enhancement adjacent to bulk metallic gold was reported by Regulla and coworkers (1998). They disclosed a method for treating a site in a human body to inhibit abnormal proliferation of tissue at the site by introducing a metal surface at the site and then directing ionizing irradiation to the metal surface to obtain locally enhanced radiation therapy (US Patent 6,001,054). The metal surface can be solid, e.g., a metallic stent, which is placed in the blood vessels adjacent to the tissue to ablate. Unfortunately, it would be impractical to place bulk metal surfaces throughout all tumor vessels and tissues. In addition, the form of radiation used was restricted to less than 400 keV, which could not treat tumors at depth. Although skin cancers might be treated using this photon energy range, such tumors are more readily removed surgically. The observed dose enhancement at the interface between materials of high and low Z has been attributed to the production of secondary electrons scattering from the surface of the high-Z material in the immediate area of the surrounding tissue

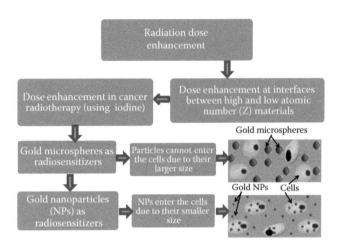

FIGURE 9.1 Schematic diagram explaining the evolvement of gold-mediated sensitization.

(Das 1997; Das and Chopra 1995). This has been confirmed experimentally by several reports using both direct dose measurement (Allal et al. 1998; Melian et al. 1999) and Monte Carlo modeling (Cho 2005; Li et al. 1999).

Herold et al. (2000) reported the use of 1.5- to 3.0-μm-diameter gold particles (1% by weight) in a stirred suspension with living cells during irradiation with 100–240 kVp x-rays and found a DEF from a clonogenic assay to be 1.54. They also injected these particles (1.5–3 μm in diameter, 1% gold suspension) directly into a growing tumor at three sites followed by irradiation (8 Gy, 200 kVp). No tumor remission or shrinkage in the animals was reported, but extracted cells from the tumor were found to have a 0.15 plating efficiency rather than the control value of 0.25. Their histological data showed gold particles predominantly in the interstitial fluid, and "no gold particles were found in zones of tightly packed tumor cells, suggesting that it would be difficult to achieve uniform delivery of particles." This is not surprising since the size of the particles is comparable to the size of the cells and there is less chance of getting them inside the cell. In order to overcome these difficulties, more attention has been given to use gold nanoparticles (GNPs) with size scale from 1 to 100 nm as radiation enhancers (radiosensitizers) in radiation therapy (Butterworth et al. 2010; Chen and Zhang 2006; Herold et al. 2000; Kong et al. 2008; Liu et al. 2010; Rahman et al. 2009). For such applications, GNPs have two interesting properties: (1) they increase the absorption of radiation energy and (2) they can be preferentially targeted to the tumor tissue to spare normal tissue (Anshup et al. 2005; Niidome et al. 2004).

9.2 GNP-Mediated Sensitization— *In Vitro* Studies

The use of GNP as a radiosensitizer seems more promising because of its higher Z number, targeting capability, and biocompatibility (Connor et al. 2005; Lewinski et al. 2008; Matsudaira et al. 1980; Mello et al. 1983; Norman et al. 1997; Shukla et al.

2005). Investigations of the *in vitro* toxicity of GNPs have shown cytotoxicity for smaller GNPs (1–2 nm) but not for the larger GNPs (Lewinski et al. 2008; Pan et al. 2007). Nanoparticles with dimensions up to 100 nm can traverse the cell membrane and may accumulate preferentially in cancer cells (Anshup et al. 2005; Niidome et al. 2004). Such nanoparticles (1–100 nm) are smaller than the typical cutoff size of the pores in tumor vasculature (e.g., up to 400 nm) so they may access cells in tumors (Unezaki et al. 1996). Recent reports suggest that GNP-based sensitization is dependent on size of the NPs, concentration of the NPs, energy of irradiation, and cell type, as discussed in the next section.

9.2.1 Size-Dependent Radiation Response

Recent studies have identified that the size of GNPs is an important factor in their cellular uptake process (Arnida and Ghandehari 2009; Chithrani et al. 2006; Xu et al. 2004). Figure 9.3 illustrates the size-dependent radiation response of the GNPs. It is believed that size-dependent uptake of these NPs could lead to variation in radiation response. TEM images of fixed cells with internalized GNPs are shown in Figure 9.2a; the GNPs are localized in small vesicles of size range, 300–500 nm. GNPs with diameter ~50 nm exhibited significantly higher uptake compared to smaller or larger NPs (see Figure 9.2b) (Arnida and Ghandehari 2009; Chithrani et al. 2006). Aoyama and coworkers have also shown that the NP uptake is strongly size-dependent, and the optimum NP diameter for uptake is ~50 nm (Aoyama et al. 2003; Nakai et al. 2003; Osaki et al. 2004). Theoretical models have been put forward to explain this size-dependent uptake of NPs. According to the Gaos model, the optimal size for the cellular uptake process is a result of competition between the thermodynamic driving force for cell uptake and receptor diffusion kinetics (Gao et al. 2005). For NPs smaller than the optimal size, the increased elastic energy associated with bending of the membrane results in decreased driving force for membrane wrapping. When the particle size is smaller, membrane wrapping causes an increase in free energy and cannot proceed. For particles larger than the optimum size, diffusion of receptors over a longer distance, and thus a longer wrapping time, is required. Several theoretical models have been established to provide insights into the dynamics of size-dependent uptake of NPs (Bao and Bao 2005; Gao et al. 2005; Shi 2008; Zhang et al. 2009).

Chithrani et al. (2006, 2010) have shown that the size of the NPs plays a big role in their uptake at the cellular level leading to different sensitization properties. Figure 9.2c–d summarizes the differences in the sensitization properties of the different-sized GNPs; cells that internalized 50-nm GNPs showed the greatest sensitization. As illustrated in the left panel of Figure 9.2b, this effect appears to be related to the higher number of NPs present in the cells. This was verified by evaluating the variation in radiation response as a function of the number of internalized GNPs by changing the concentration of GNPs in the medium as discussed in the next section.

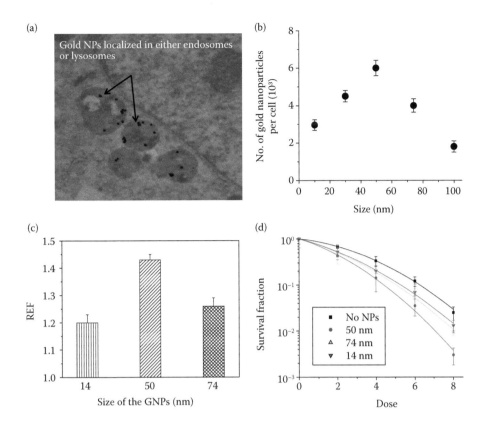

FIGURE 9.2 Gold nanoparticle–mediated sensitization—*in vitro* study. (a) TEM micrograph of a section of a cell showing gold nanoparticles localized in vesicles of size ~500 nm. Particles are observed as black dots. (b) Cellular uptake of nanoparticles is dependent on size of nanoparticles. Nanoparticles of 50 nm diameter show highest uptake. (c) Radiation dose enhancement due to nanoparticles is dependent on size of nanoparticles. (d) Cell survival curves showing size-dependent radiation response due to gold nanoparticles. HeLa cells were incubated with different size nanoparticles and irradiated with 200 kVp x-rays. (Reproduced with permission from Chithrani, B.D. et al., *Nano Lett.*, 6, 662–668, 2006; *Radiat. Res.*, 173, 719–728, 2010.)

9.2.2 Concentration-Dependent Radiation Response

As mentioned before, dose enhancement is also dependent on the NP concentration (Butterworth et al. 2010; Chithrani et al. 2010; Rahman et al. 2009). There were rapid increases in dose enhancement values with GNP concentrations for 80 kVp x-ray compared with 6-MeV electron beams. For example, dose enhancement values at 80 kVp x-rays were increased from 4 to 20 with increasing concentration; this means that dose enhancement value increases five times by doubling the concentration. However, at 6-MeVelectron beam, just a slight increase in DEF occurred with GNP concentration in comparison with that observed with the low-energy x-rays. The dependence of radiosensitization properties of GNPs as function of energy of radiation will be discussed in the next section. These findings indicate that the presence of more absorbing GNPs in the cells can increase the probability of radiation interactions inside the cells. The presence of metallic gold atoms inside the cell generates a larger number of secondary electrons from the radiation interactions in comparison with an absence of GNPs. This increase in the number of secondary electrons and resulting "free radicals"

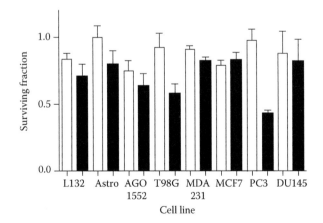

FIGURE 9.3 Effect of gold nanoparticles on clonogenic survival of different cell lines. Cytotoxicity of gold nanoparticles following 1 h exposure to concentrations of 10 μg ml^{-1} (white column) and 100 μg ml^{-1} (solid column). Surviving fractions were normalized to untreated control cells in each of the experiments. (Reproduced with permission from Butterworth, K.T. et al., *Nanotechnology*, 21, 295101, 2010.)

could lead to increase in cell death, because these free radicals can damage the DNA molecules inside the cells. These effects are in agreement with the earlier documented *in vitro* study using iodine compounds similar to those used as radiological contrast media (Corde et al. 2004). However, it is important to mention that rate and extent of NP uptake can be varied among different cell lines as well (Cartiera et al. 2009). Figure 9.3 shows the cell-dependent radiation response of NPs. DU-145 cells showed the highest levels of cytotoxicity compared to MDA-231-MB cells, which showed relatively low levels of cytotoxicity. Different NP-uptake properties of cells could lead to this cell-dependent radiation response.

9.2.3 Beam Energy–Dependent Radiation Response

According to a recently published work by Sanche and coworkers, low-energy electrons are created in large numbers by any kind of ionizing radiation; hence, the radiosensitizing properties of GNPs should be universal and should exist for any type of high-energy radiation, including the 1–18 MeV photon beams commonly used in radiotherapy (Zheng et al. 2008). However, extent of sensitization could vary depending on the energy of

irradiation. Chithrani et al. used 50-nm GNPs, whereas Rahman et al. used 1.9-nm GNPs. Greater radiation sensitization was seen for cells irradiated with the lower energy radiation beams (Figure 9.4a). These findings are further verified by a recent Monte Carlo study. Figure 9.4b compares the number of secondary electrons created because of the presence of the GNP for different photon beams. In addition, it also shows the number of secondary electrons created based on NP sizes. The GEANT4 MC code was used to simulate electron emission from different size NPs irradiated with photon beams of different energies. As a consequence of the strong energy dependence of the photoelectric cross section, irradiating the GNP with a low-energy photon beam greatly increased the number of secondary electrons. Although secondary electrons generated by irradiating the GNP with a high-energy photon beam traveled greater distances, irradiation with a low-energy photon beam generated far more secondary electrons, at 2 to 3 orders of magnitude greater. Also, the majority of the energy deposition in GNP-enhanced radiotherapy is outside of the GNP. In addition, the ratio of the mean electron energy to the mean effective range for high-energy photon beams was less than that in low-energy beams. This would decrease the efficiency of the cell killing. They conclude that for a cell of typical size (10 μm diameter), low-energy photon beam irradiation considerably increases the generation of secondary

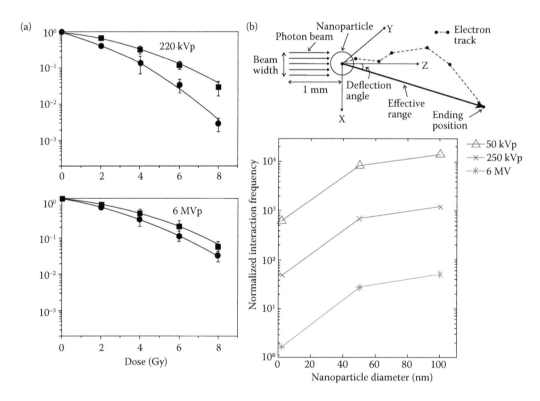

FIGURE 9.4 Radiation dose enhancement effects is dependent on energy of radiation. (a) Cell survival curves for energies of 200 kVp and 6 MVp. Control cell and cells with internalized nanoparticles are represented by filled squares and circles, respectively. Points are means ±SD for three experiments. Solid lines are fit to the Linear Quadratic model. (b) Top: schematic diagrams of Monte Carlo simulation geometry. Effective range and deflection angle of electrons emitted from NP are defined. Bottom: comparison of number of interaction events for secondary electron creation for GNPs with diameters of 2, 50, and 100 nm irradiated at different energies. (Reproduced with permission from Chithrani, B.D. et al., *Radiat. Res.*, 173, 719–728, 2010; Leung, M.K.K. et al., *Med. Phys.*, 38, 624–631, 2011.)

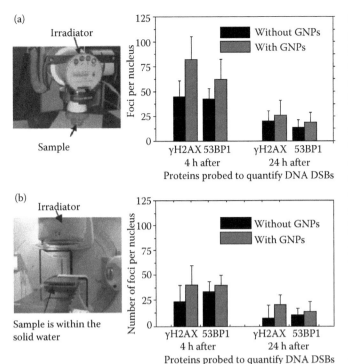

FIGURE 9.5 (a, b) Quantitative analysis of DNA damage using lower-energy X-ray photons (220 kVp) and 6-MV photons, respectively. Left panel: setup used for irradiations. Right panel: quantification of γ-H2AX and 53BP1 radiation-induced foci after 4 Gy of 220 kVp x-rays [cells pretreated with GNPs (gray) and with no GNP pretreatment (black)]. (Reproduced with permission from Chithrani, B.D. et al., *Radiat. Res.*, 173, 719–728, 2010.)

electrons. These electrons have sufficient range to cause damage within the entire cell volume where the NPs are present. Irradiation with low-energy photons will be more effective for cell killing than irradiation with high-energy photons. For example, the interaction enhancement ratio is approximately 10 for the MV beams and up to 2000 for the 50 kVp beam.

Beam energy–dependent radiation response was further monitored by quantification of DNA double strand breaks (DSBs) by Chithrani et al. (2010). Figure 9.5 shows DNA DSBs in cells irradiated with 220 kVp and 6-MVp X-rays after a dose of 4 Gy is given. Two proteins (γ-H2AX and 53BP1) present at the sites of DNA DSBs were probed for quantifying the damage. The experimental setups for the irradiations are also shown in the left panel of Figure 9.5. Right panels of Figure 9.5 shows that there is more DNA damage for lower energy of irradiation than for higher energy. The increase in DSBs in cells with internalized GNPs is consistent with the clonogenic radiation cell survival data for both lower and higher energies presented in Figure 9.4a.

9.3 GNP-Mediated Sensitization— *In Vivo* Studies

An important milestone in the field of radiation therapy was reached when Hainfeld et al. (2004, 2010) conducted the detailed experiments *in vivo* to explore the enhancement effect of GNPs, and the data showed the potential utility of GNPs for cancer x-ray therapy. They demonstrated that EMT-6 mammary tumors implanted in mice that received an intravenous injection of 1.35 g GNPs/kg could be completely eradicated in 30 days after irradiation with 250 kVp x-rays as illustrated in Figure 9.6a

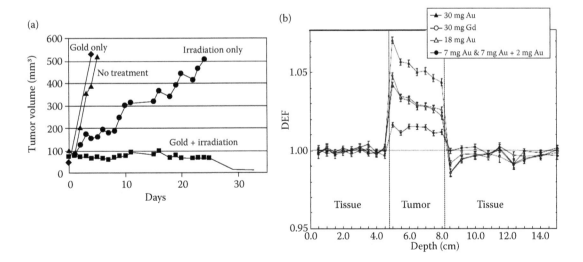

FIGURE 9.6 Gold nanoparticle–mediated sensitization—*in vivo* study. (a) Average tumor volume after: no treatment (triangles); gold only (diamonds); irradiation only (30 Gy, 250 kVp, circles); intravenous gold injection (1.35 g Au/kg) followed by irradiation (squares). (b) Calculated dose enhancement factor (DEF) for unflattened 6MV photon beam. Factors outside tumor region are not truly DEFs but are dose ratios between cases with and without gold (or gadolinium). DEF for entire tumor region loaded with gadolinium at 30 mg/g tumor was 1.032. Error bar represents statistical uncertainty (1σ) of each data point combined in quadrature. (Reproduced with permission from Cho, S.H., *Phys. Med. Biol.*, 50, N163–N173, 2005; Hainfeld, J.F. et al., *Phys. Med. Biol.*, 49, N309–N315, 2004.)

(Hainfeld et al. 2004). However, detailed mechanisms leading to such impressive result are yet not known. It is believed that a larger portion of the energy of the primary ionizing photons is transferred to the tumor because of the increased absorption of x-rays by GNPs (Brun et al. 2009; Carter et al. 2007; Hainfeld et al. 2010; Montenegro et al. 2009). In this experiment, mice were irradiated only 2 min after intravenous injection of GNPs, which is far too short to obtain a significant accumulation in the tumor cells. Recent *in vitro* studies suggest that NP uptake is maximized hours after incubation, and the enhancement of radiosensitization is dependent on the number of internalized NPs (Chithrani et al. 2010). The study by Hainfeld et al. was primarily focused on using low energy irradiations, and the concentration of NPs used was high. The concentration of gold at the tumor site was 0.7%. The dose administered to achieve this level in tumors is too high for translation to humans. A Monte Carlo study was done to estimate the tumor dose enhancement effects due to GNPs as discussed in the Hainfeld et al. study (Cho 2005). As expected, the maximum dose enhancement occurred for lower energy x-rays. For 140 kVp x-rays, a tremendous dose enhancement was seen, ranging from a factor of about 2 at the lowest but practically achievable gold concentration to a factor of almost 6 at the highest gold concentration. For the 4- and 6-MV photon beams, dose enhancement ranging from 1% to 7% was seen, depending on the gold concentrations and beam qualities. Figure 9.5b presents a comparison between gold and gadolinium at the same concentration level and nominal photon beam energy (i.e., 30 mg Au or Gd/g tumor and unflattened 6MV photon beam). As shown in Figure 9.2, the current study found that the tumor dose enhancement from gold was larger than that from gadolinium by about 2% for the case being considered. The magnitude of the increase in DEF from gadolinium to gold was comparable to that from iodine to gadolinium.

Recently, Chang et al. (2008) demonstrated the feasibility of obtaining dose enhancement effects with a lower concentration of GNPs in tumor-bearing mice in combination with clinical electron beams. GNPs were injected intravenously into the mice, and the authors used a GNP concentration of 1 g/kg compared to 2.7 g/kg by Hainfeld et al. (Chang et al. 2008; Hainfeld et al. 2004). Moreover, in the study by Chang et al. the irradiation time-point, 24 h post-GNP injection and they have shown higher accumulation within the tumor compared to tumor periphery. This may be due to the enhanced permeability and retention effect, which takes advantage of the poorly formed tumor vasculature (Regehly et al. 2007).

However, several *in vivo* studies have shown that the NPs were distributed not only in the tumor but also in the tumor periphery, muscle, liver, kidneys, and blood (see Table 9.1) (Chang et al. 2008; Hainfeld et al. 2004). Hence, for future human trials, it is necessary to conjugate GNPs with antibodies for directing them against tumor-specific receptors (e.g., epidermal growth factor receptor) and angiogenesis markers. This would also enable biological intensity modulation of the beam in order to deliver a prescribed dose selectively to the tumor cells only within the target volume. In future studies, GNPs can be surface modified for

TABLE 9.1 Bio-Distribution of Gold Nanoparticles

	% Injected dose/g	Au (μg)/Tissue Weight (mg)
Injected dose	1.35 g Au/kg	1 g Au/kg
Bio distribution	5 min post injection	24 h post injection
NP Injection	Intravenous	Intravenous
Tumor	4.9 ± 0.6	74.24
Tumor periphery	8.9 ± 3.2	11.55
Liver	2.8 ± 0.1	147
Kidney	132.0 ± 2.7	2.62
Blood	18.6 ± 3.7	1.48
Group	Hainfeld et al. (2004)	Chang et al. (2008)

preferential targeting the cancer cells (Choi et al. 2010). If GNPs can be localized within the tumor, it would lead to a higher dose to the cancerous tissue compared with the dose received by normal tissue during a radiotherapy treatment. The mechanism of action for this effect is likely dependent on intracellular and potentially nuclear localization (Zheng et al. 2008). We believe that cellular uptake producing much higher intracellular concentrations is a prerequisite for radiosensitization (Chithrani and Chan 2007; Chithrani et al. 2006, 2009).

9.4 Mechanisms of Sensitization

A clear understanding of fundamental mechanisms of cancer biology and therapies can lead to improved clinical outcomes (Alberts 2008; Varmus 2006). Understanding the fundamental mechanisms that induce DNA damage and cell death should lead to a clearer picture of the cause of cancers and benefit the development of improved strategies for cancer treatment. Radiation is used in radiotherapy because radiation (x-rays, γ-rays, and fast-moving charged particles such as ions, electrons, and protons) interacts with DNA inside living cells, causing enough damage and that could lead to cell death (von Sonntag 1987). For this reason, such radiation is used in radiotherapy to kill cancer cells.

As illustrated in Figure 9.7, radiation alone produces ions, radicals, and free electrons as they travel through matter (Sanche 2009; Turner et al. 1983; von Sonntag 1987). The electrons, in turn, generate large quantities of a second generation of radicals, ions, and free electrons. Most studies suggest that DNA is damaged indirectly by hydroxyl radicals (Chatterjee and Magee 1985). However, the electrons can also cause damage to DNA, as illustrated in a recent study in which low-energy electrons emitted from metal films were found to cause DNA strand breaks directly (Boudaiffa et al. 2000). This study was performed using dry films. However, it is important to look at the role of electrons in biologically relevant environment, such as water. When electrons are generated in water, they become hydrated and form a complex with several water molecules. It was assumed that these hydrated electrons do not cause much DNA damage as compared to hydroxyl radicals (Zheng and Sanche 2009). Now, the question is whether these hydrated electrons can cause DNA damage. Recently, Wang et al performed an experiment to study the reaction of prehydrated electrons

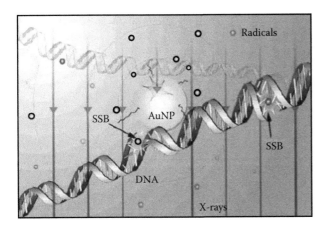

FIGURE 9.7 Schematic diagram of results of a Monte Carlo simulation. Radicals are generated from electrons produced in water and are shown as gray spheres, whereas Auger electrons, secondary electrons, and photoelectrons originating from GNP are shown as black spheres. When radicals are within ~0.5 nm of scDNA, a single-strand break (SSB) occurs with ~25% efficiency. Trajectories of electrons are not shown, and only relative average density of radicals generated from these electrons is displayed. Diameter of GNP shown here is approximately 3 nm. (Reproduced with permission from Carter, J.D. et al., *J. Phys. Chem. B*, 111, 11622–11625, 2007.)

with deoxyribonucleotides, the building blocks of DNA (Yavuz et al. 2009). The authors performed their experiments in water, which provides a good model for cells. They found that significant quantities of single- and double-strand breaks of irradiated aqueous DNA are induced by prehydrated electrons. Based on these recent studies, both electrons and hydroxyl radicals could be responsible for DNA damage in irradiated cells. In the next section, we will discuss the contribution from GNPs to these existing mechanisms of cell damage after exposure to radiation.

Recently, GNPs are being used as sensitizers in radiation therapy (Butterworth et al. 2010; Chen and Zhang 2006; Herold et al. 2000; Kong et al. 2008; Liu et al. 2010; Rahman et al. 2009). As a step toward understanding the mechanism behind enhanced sensitization properties of GNPs, Carter et al. (2007) have performed a Monte Carlo calculation and pointed out that the following effects can be combined to cause this phenomenon: (1) enhanced localized absorption of x-rays by nanostructures, (2) effective release of low-energy electrons from GNPs, and (3) efficient deposition of energy in water in the form of radicals and electrons. When GNPs are present, the electrons released from these NPs could create more radicals as illustrated in Figure 9.7 (Carter et al. 2007). They also confirmed the theoretically predicted nanoscale energy deposition distribution by measuring hydroxyl radical-induced DNA strand breaks. These results provide important information to understand gold-based sensitization mechanisms. However, in these studies, the GNPs were in close proximity to DNA. The exact mechanisms of cell damage when GNPs are localized away from DNA (either when they are in the media or in the cytoplasm of the cell) are not known yet. Hence, more work needs to be done in order to elucidate the mechanism of sensitization due to GNPs.

As discussed earlier, the primary target of radiation is nuclear DNA, with DSB formation being the most lethal DNA damage. The enhancement in DNA damage in the presence of GNPs has been assessed by several research groups (Butterworth et al. 2008; Foley et al. 2005; Zheng et al. 2008; Zheng and Sanche 2009). However, evidence is now immerging to suggest that radiation damage to mitochondria and the cell membrane may also contribute to the cytotoxic effect of radiation (Butterworth et al. 2010). In addition, a recent study showed that GNPs were shown to potentiate the effect of the radiomimetic agent, bleomycin, in the absence of radiation, suggesting that biological interactions of GNPs with cells could be another mechanism by which sensitization occurred (Butterworth et al. 2010). If sensitization is primarily an effect of biological interactions with GNPs, the importance of GNP radical production, hypoxia, and cell signaling pathways needs to be elucidated.

9.5 Future of GNP-Based Therapeutics in Cancer Therapy

In treating cancer, radiation therapy and chemotherapy remain the most widely used treatment options. However, recent developments in cancer research show that the incorporation of gold nanostructures into those protocols has enhanced tumor cell killing. There is evidence of the enhanced radiation sensitization in mice even at megavoltage energies used in conventional radiation practice (6 MVp) and at concentrations feasible for use in humans (see Figure 9.8) (Chang et al. 2008; Zheng et al. 2008). Chang et al. have shown that tumor growth was both retarded in mice receiving either radiation alone or receiving GNP followed by radiation (Figure 9.8a) compared to the controls with no radiation. More importantly, tumor volume in the combination therapy group was significantly smaller compared with that in radiation alone group ($P < 0.05$), whereas administration of GNP or phosphate-buffered saline alone did not exert any antitumor effect on tumor-bearing mice (Figure 9.8a). They have also examined whether apoptosis was associated with the antitumor effects of combination therapy as illustrated in Figure 9.8b (Chang et al. 2008). Apoptotic activity was analyzed by terminal deoxynucleotidyl transferase-mediated deoxyuridine triphosphate nick end labeling (TUNEL) assay. Apoptotic cells were calculated by averaging the number of positive TUNEL signals from eight fields with the highest density of TUNEL signals in each section. Noticeably, the number of apoptotic cells detected was significantly higher in the GNP and radiation combination group than that in the radiation alone group (see Figure 9.8b). These new developments in nanotechnology offer great potential for improvements in the care of cancer patients (Cuenca et al. 2006; Ferrari 2005; Peer et al. 2007). In addition, these nanostructures further provide strategies for improving loading, targeting, and controlling the release of drugs to minimize the side effects of highly toxic anticancer drugs used in chemotherapy and photodynamic therapy. In addition, the heat generation capability of gold nanostructures upon exposure to

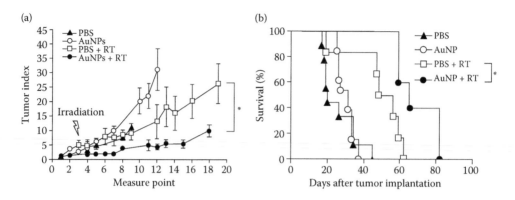

FIGURE 9.8 Antitumor effects of combination treatment of gold nanoparticles (GNPs) and radiotherapy in tumor-bearing mice. C57BL/6 mice were inoculated subcutaneously with B16F10 cells (1×10^6) at day 0. At day 7, tumor-bearing mice were injected intravenously with 200 μL of 200 nM AuNP, or with 200 μL of phosphate-buffered saline (PBS) 24 h before irradiation (25 Gy/mouse). Mice were monitored for tumor growth (a) and survival (b) ($n = 4$–7; *$P < 0.05$). RT, radiotherapy. (Reproduced with permission from Chang, M.-Y. et al., *Cancer Sci.*, 99, 1479–1484, 2008.)

near-infrared light is being used to damage tumor cells locally in photothermal therapy. Hence, gold nanostructures provide a versatile platform to integrate many therapeutic options leading to effective combinational therapy in the fight against cancer.

For example, GNPs have been explored to enhance the damage induced by anticancer drugs as discussed in the next section (Brown et al. 2010; Zheng and Sanche 2009). These results may suggest the clinical potential of GNPs in improving the outcome of radiotherapy and chemotherapy. Recently, it was shown that GNPs can be used to enhance DNA damage caused by platinum-based anticancer drugs, and the enhancement effect of cisplatin by GNP was obtained when DNA was exposed to low energy electrons, as produced by ionizing radiation (Zheng and Sanche 2009). These platinum-based anticancer drugs cisplatin,

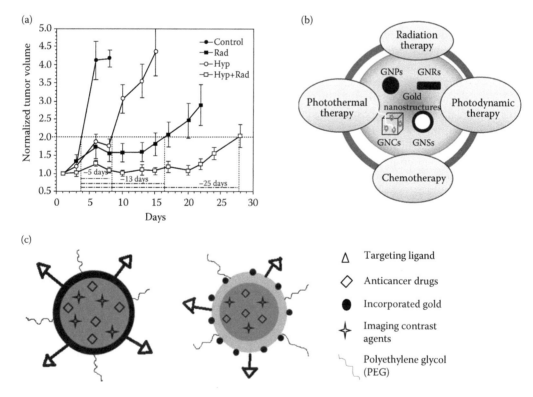

FIGURE 9.9 Gold nanoparticles for combinational therapy in cancer. (a) Modulation of *in vivo* tumor radiation response via gold nanoshell–mediated vascular-focused hyperthermia. Normalized tumor volume plot of control, hyperthermia, radiation, and thermoradiotherapy groups showing mean ± SE values at different time intervals after initiation of each treatment. (b) Combinational therapy using gold nanoparticles. (c) Gold nanoparticles can be incorporated into either polymer or lipid NPs for combinational cancer therapy. (Reproduced with permission from Diagaradjane, P. et al., *Nano Lett.*, 8, 1492–1500, 2008.)

carboplatin, and oxaliplatin are an important component of chemotherapy and have had a major impact, particularly for patients with testicular or ovarian cancers (Kelland 2007). As a proof of concept, Sanche and coworkers have shown that there is an enhancement in the DNA DSBs when anticancer drugs are used in combination with GNPs and ionizing radiation (Zheng and Sanche 2009). They found that radiation-induced DNA DSBs, a highly lethal type of cellular damage, were enhanced by a factor of 7.5 by this combination. In this study, GNPs were in close proximity to DNA. Hence, it would be interesting to carry out further experiments to see the full potential of these anti cancer effects *in vitro* and *in vivo* where NPs are mostly localized in the cytoplasm away from DNA in the nucleus. As a step forward toward *in vitro* and *in vivo* studies, Brown et al. (2010) have tethered the active component of the anticancer drug oxaliplatin to a PEGylated GNP for improved drug delivery, and the *in vitro* study showed that drug-tethered NPs demonstrated as good as, or significantly better, cytoxicity than drug alone in cancer cell lines such as lung epithelial cancer cell line and colon cancer cell lines (HCT 116, HCT15, HT129, and RKO). The larger surface area of NPs facilitates attachment of a large number of drug molecules, and they demonstrated that 280 drug molecules can be attached to a single NP. Based on these new findings, GNPs can be used for improved cancer therapeutics by combining chemotherapy and radiation for a better outcome in future cancer care of patients.

Krishnan and coworkers have reported a method to combine heat generation capability of GNPs with radiation therapy for a better outcome (Diagaradjane et al. 2008). They reported a noninvasive modulation of *in vivo* tumor radiation response using gold nanoshells (see Figure 9.9a). Mild-temperature hyperthermia generated by near-infrared illumination of gold nanoshell-laden tumors caused an early increase in tumor perfusion, reducing the hypoxic fraction of tumors. A subsequent radiation dose induced vascular disruption with extensive tumor necrosis. This novel integrated antihypoxic and *localized* vascular disrupting therapy can potentially be combined with other therapeutic techniques as outlined in Figure 9.9b. Hence, GNPs provide a versatile platform to integrate many therapeutic options leading to effective combinational therapy in the fight against cancer. A multifunctional platform based on GNPs with targeting ligands, therapeutic molecules, and imaging contrast agents will hold the possibility of promising directions in cancer research (Figure 9.9c).

References

Alberts, B. 2008. The promise of cancer research. *Science* 320:19.

Allal, A. S., M. Richter, M. Russo, M. Rouzaud, P. Dulguerov, and J. M. Kurtz. 1998. Dose variation at bone/titanium interfaces using titanium hollow screw osseointegrating reconstructive plates. *International Journal of Radiation Oncology, Biology & Physics* 40:215–219.

Anshup, A., J. S. Venkataraman, C. Subramaniam, R. R. Kumar, S. Priya, T. R. S. Kumar, R. V. Omkumar, A. John, and T. Pradeep. 2005. Growth of gold nanoparticles in human cells. *Langmuir* 21:11562–11567.

Aoyama, Y., T. Kanamori, T. Nakai, T. Sasaki, S. Horiuchi, S. Sando, and T. Niidome. 2003. Artificial viruses and their application to gene delivery. Size-controlled gene coating with glycocluster nanoparticles. *Journal of the American Chemical Society* 125:3455–3457.

Arnida, M. A., and H. Ghandehari. 2009. Cellular uptake and toxicity of gold nanoparticles in prostate cancer cells: A comparative study of rods and spheres. *Journal of Applied Toxicology* 30:212–217.

Bao, G., and X. R. Bao. 2005. Shedding light on the dynamics of endocytosis and viral budding. *Proceedings of the National Academy of Sciences of the United States of America* 102:9997–9998.

Boudaiffa, B., P. Cloutier, D. Hunting, M. A. Huels, and L. Sanche. 2000. Resonant formation of DNA strand breaks by low-energy (3 to 20 eV) Electrons. *Science* 287:1658–1660.

Brown, S. D., P. Nativo, J.-A. Smith, D. Stirling, P. R. Edwards, B. Venugopal, D. J. Flint, J. A. Plumb, D. Graham, and N. J. Wheate. 2010. Gold nanoparticles for the improved anticancer drug delivery of the active component of oxaliplatin. *Journal of the American Chemical Society* 132:4678–4684.

Brun, E., P. Cloutier, C. Sicard-Roselli, M. Fromm, and L. Sanche. 2009. Damage Induced to DNA by low-energy (0–30 eV) electrons under vacuum and atmospheric conditions. *Journal of Physical Chemistry B* 113:10008–10011.

Butterworth, K. T., J. A. Wyer, M. Brennan-Fournet, C. J. Latimer, M. B. Shah, F. J. Currell, and D. G. Hirst. 2008. Variation of strand break yield for plasmid DNA irradiated with high-Z metal nanoparticles. *Radiation Research* 170:381–387.

Butterworth, K. T., J. A. Coulter, S. Jain, J. Forker, S. J. McMahon, G. Schettino, K. M. Prise, F. J. Currell, and D. G. Hirst. 2010. Evaluation of cytotoxicity and radiation enhancement using 1.9 nm gold particles: Potential application for cancer therapy. *Nanotechnology* 21:295101.

Carter, J. D., N. N. Cheng, Y. Qu, G. D. Suarez, and T. Guo. 2007. Nanoscale energy deposition by X-ray absorbing nanostructures. *Journal of Physical Chemistry B* 111:11622–11625.

Cartiera, M. S., K. M. Johnson, V. Rajendran, M. J. Caplan, and W. M. Saltzman. 2009. The uptake and Intracellular fate of PLGA nanoparticles in epithelial cells. *Biomaterials* 30:2790–2798.

Castillo, M. H., T. M. Button, R. Doerr, M. I. Homs, C. W. Pruett, and J. I. Pearce. 1988. Effects of radiotherapy on mandibular reconstruction plates. *American Journal of Surgery* 156:261–263.

Chang, M.-Y., A.-L. Shiau, Y.-H. Chen, C.-J. Chang, H. H.-W. Chen, and C.-L. Wu. 2008. Increased apoptotic potential and dose-enhancing effect of gold nanoparticles in combination with single-dose clinical electron beams on tumor-bearing mice. *Cancer Science* 99:1479–1484.

Chatterjee, A., and J. L. Magee. 1985. Theoretical investigation of the production of strand breaks in DNA by water radicals. *Radiation Protection Dosimetry* 13:137–140.

Chen, W., and J. Zhang. 2006. Using nanoparticles to enable simultaneous radiation and photodynamic therapies for cancer treatment. *Journal of Nanoscience and Nanotechnology* 6:1159–1166.

Chithrani, B. D., and W. C. W. Chan. 2007. Elucidating the mechanism of cellular uptake and removal of protein-coated gold nanoparticles of different sizes and shapes. *Nano Letters* 7:1542–1550.

Chithrani, B. D., A. A. Ghazani, and W. C. W. Chan. 2006. Determining the size and shape dependence of gold nanoparticle uptake into mammalian cells. *Nano Letters* 6: 662–668.

Chithrani, B. D., J. Stewart, C. Allen, and D. A. Jaffray. 2009. Intracellular uptake, transport, and processing of nanostructures in cancer cells. *Nanomedicine: NBM* 5:118–127.

Chithrani, B. D., S. Jelveh, F. Jalali, M. Van Prooijen, C. Allen, R. G. Bristow, R. P. Hill, and D. A. Jaffray. 2010. Gold nanoparticles as a radiation sensitizer in cancer therapy. *Radiation Research* 173:719–728.

Cho, S. H. 2005. Estimation of tumor dose enhancement due to gold nanoparticles during typical radiation treatments: A preliminary Monte Carlo Study. *Physics in Medicine and Biology* 50:N163–N173.

Choi, C. H. J., C. A. Alabi, P. Webster, and M. E. Davis. 2010. Mechanism of active targeting in solid tumors with transferrin-containing gold nanoparticles. *Proceedings of the National Academy of Sciences of the United States of America* 107:1235–1240.

Connor, E. E., J. Mwamuka, A. Gole, C. J. Murphy, and M. D. Wyatt. 2005. Gold nanoparticles are taken up by human cells but do not cause acute cytotoxicity. *Small* 1:325–327.

Corde, S., A. Joubert, J. F. Adam, A. M. Charvet, J. F. Le Bas, F. Esteve, H. Elleaume, and J. Balosso. 2004. Synchrotron radiation based experimental determination of the optimal energy for cell radiotoxicity enhancement following photoelectric effect on stable iodinated compounds. *British Journal of Cancer* 91:544–551.

Cuenca, A. G., H. Jiang, S. N. Hochwald, M. Delano, W. G. Cance, and S. R. Grobmyer. 2006. Emerging implications of nanotechnology on cancer diagnostics and therapeutics. *Cancer* 107:459–466.

Das, I. J. 1997. Forward dose perturbation at high atomic number interfaces in kilovoltage x-ray beams. *Medical Physics* 24:1781–1787.

Das, I. J., and K. L. Chopra. 1995. Backscatter dose perturbation in kilovoltage photon beams at high atomic number interfaces. *Medical Physics* 22:767–773.

Diagaradjane, P., A. Shetty, J. Wang, A. Elliot, J. Schwartz, S. Shentu, H. Park, A. Deorukhkar, R. Stafford, S. Cho, J. Tunnell, J. Hazle, and S. Krishnan. 2008. Modulation of in vivo tumor radiation response via gold nanoshell-mediated vascular-focused hyperthermia: Characterizing an integrated antihypoxic and localized vascular disrupting targeting strategy. *Nano Letters* 8:1492–1500.

Ferrari, M. 2005. Cancer nanotechnology: Opportunities and challenges. *Nature Reviews Cancer* 5:161–171.

Foley, E. A., J. D. Carter, F. Shan, and T. Guo. 2005. Enhanced relaxation of nanoparticle bound supercoiled DNA in x-ray radiation. *Chemical Communications* 25:3192–3194.

Gao, H., W. Shi, and L. B. Freund. 2005. Mechanics of receptor-mediated endocytosis. *Proceedings of the National Academy of Sciences of the United States of America* 102:9469–9474.

Hainfeld, J. F., D. N. Slatkin, and H. M. Smilowitz. 2004. The use of gold nanoparticles to enhance radiotherapy in mice. *Physics in Medicine and Biology* 49:N309–N315.

Hainfeld, J. F., F. A. Dilmanian, Z. Zhong, D. N. Slatkin, J. A. Kalef-Ezra, and H. M. Smilowitz. 2010. Gold nanoparticles enhance the radiation therapy of a murine squamous cell carcinoma *Physics in Medicine and Biology* 55:3045–3059.

Herold, D. M., I. J. Das, C. C. Stobbe, R. V. Iyer, and J. D. Chapman. 2000. Gold microspheres: A selective technique for producing biologically effective dose enhancement. *International Journal of Radiation Biology* 76:1357–1364.

Kelland, L. 2007. The resurgence of platinum-based cancer chemotherapy. *Nature Reviews Cancer* 7:573–584.

Kong, T., J. Zeng, X. Wang, X. Yang, J. Yang, S. McQuarrie, A. McEwan, W. Roa, J. Chen, and J. Z. Xing. 2008. Enhancement of radiation cytotoxicity in breast-cancer cells by localized attachment of gold nanoparticles. *Small* 4:1537–1543.

Leung, M. K. K., J. C. L. Chow, B. D. Chithrani, M. J. L. Lee, B. Oms, and D. A. Jaffray. 2011. Irradiation of gold nanoparticles by x-rays: Monte Carlo simulation of dose enhancements and the spatial properties of the secondary electrons production. *Medical Physics* 38:624–631.

Lewinski, N., V. Colvin, and R. Drezek. 2008. Cytotoxicity of nanoparticles. *Small* 4:26–49.

Li, X. A., J. C. Chu, W. Chen, and T. Zusag. 1999. Dose enhancement by a thin foil of high-*Z* material: A Monte Carlo study. *Medical Physics* 26:1245–1251,

Liu, C.-J., C.-H. Wang, S.-H. Chen, H.-H. Chen, W.-H. Leng, C.-C. Chien, C.-L. Wang, I. M. Kempson, Y. Hwu, T.-C. Lai, M. Hsiao, C.-S. Yang, Y.-J. Chen, and G. Margaritondo. 2010. Enhancement of cell radiation sensitivity by pegylated gold nanoparticles. *Physics in Medicine and Biology* 55:931–945.

Matsudaira, H., A. M. Ueno, and I. Furuno. 1980. Iodine contrast medium sensitizes cultured mammalian cells to x-rays but not to γ rays. *Radiation Research* 84:144–148.

Melian, E., M. Fatyga, P. Lam, M. Steinberg, S. P. Reddy, G. J. Petruzzelli, and G. P. Glasgow. 1999. Effect of metal reconstructive plates on cobalt-60 dose distribution: A predictive formula and clinical implications. *International Journal of Radiation Oncology, Biology & Physics* 44: 725–730

Mello, R. S., H. Callisen, J. Winter, A. R. Kagan, and A. Norman. 1983. Radiation dose enhancement in tumors with iodine. *Medical Physics* 10:75–78.

Montenegro, M., S. N. Nahar, A. K. Pradhan, K. Huang, and Y. Yu. 2009. Monte Carlo simulations and atomic calculations for Auger processes in biomedical nanotheranostics. *Journal of Physical Chemistry A* 113:12364–12369.

Nakai, T., T. Kanamori, S. Sando, and Y. Aoyama. 2003. Remarkably size-regulated cell invasion by artificial viruses. Saccharide-dependent self-aggregation of glycoviruses and its consequences in glycoviral gene delivery. *Journal of the American Chemical Society* 125:8465–8475.

Nath, R., P. Bongiorni, and S. Rockwell. 1990. Iododeoxyuridine radiosensitization by low- and high-energy photons for brachytherapy dose rates. *Radiation Research* 124:249–258.

Niidome, T., K. Nakashima, H. Takahashi, and Y. Niidome. 2004. Preparation of primary amine-modified gold nanoparticles and their transfection ability into cultivated cells. *Chemical Communications* 17:1978–1979.

Norman, A., M. Ingram, R. G. Skillen, D. B. Freshwater, K. S. Iwamoto, and T. Solberg. 1997. X-ray phototherapy for canine brain masses. *Radiation Oncology Investigations* 5:8–14.

Osaki, F., T. Kanamori, S. Sando, and Y. Aoyama. 2004. A quantum dot conjugated sugar ball and its cellular uptake. On the size effects of endocytosis in the subviral region. *Journal of the American Chemical Society* 126:6520–6521.

Pan, Y., S. Neuss, A. Leifert, M. Fischler, F. Wen, U. Simon, G. Schmid, W. Brandau, and W. Jahnen-Dechent. 2007. Size-dependent cytotoxicity of gold nanoparticles. *Small* 3:1941–1949.

Peer, D., J. M. Karp, S. Hong, O. C. Farokhzad, R. Margalit, and R. Langer. 2007. Nanocarriers as an emerging platform for cancer therapy. *Nature Nanotechnology* 2:751–760.

Rahman, W. N., N. Bishara, T. Ackerly, C. F. He, P. Jackson, C. Wong, R. Davidson, and M. Geso. 2009. Enhancement of radiation effects by gold nanoparticles for superficial radiation therapy. *Nanomedicine* 5:136–142.

Regehly, M., K. Greish, F. Rancan, H. Maeda, F. Bohm, and B. Roder. 2007. Water-soluble polymer conjugates of ZnPP for photodynamic tumor therapy. *Bioconjuguate Chemistry* 18:494–499.

Regulla, D. F., L. B. Hieber, and M. Seidenbusch. 1998. Physical and biological interface dose effects in tissue due to x-ray-induced release of secondary radiation from metallic gold surfaces. *Radiation Research* 150:92–100.

Sanche, L. 2009. Beyond radical thinking. *Nature* 461:358–359.

Shi, W., J. Wang, X. Fan, and H. Gao. 2008. Size and shape effects on diffusion and absorption of colloidal particles near a partially absorbing sphere: Implications for uptake of nanoparticles in animal cells. *Physical Review E* 78:061914–061925.

Shukla, R., V. Bansal, M. Chaudhary, A. Basu, R. R. Bhonde, and M. Sastry. 2005. Biocompatibility of gold nanoparticles and their endocytotic fate inside the cellular compartment: A microscopic overview. *Langmuir* 21:10644–10654.

Turner, J. E., J. L. Magee, H. A. Wright, A. Chatterjee, R. N. Hamm, and R. H. Ritchie. 1983. The role of thiols in cellular response to radiation and drugs. *Radiation Research* 96:437–455.

Unezaki, S., K. Maruyama, J.-I. Hosoda, I. Nagae, Y. Koyanagi, and M. Nakata. 1996. Direct measurement of the extravasation of polyethyleneglycol-coated liposomes into solid tumor tissue by in vivo fluorescence microscopy. *International Journal of Pharmaceutics* 144:11–17.

Varmus, H. 2006. The new era in cancer research. *Science* 312:1162–1165.

von Sonntag, C. 1987. *The Chemical Basis for Radiation Biology.* London: Taylor and Francis.

Xu, X.-H. N., W. J. Brownlow, S. V. Kyriacou, Q. Wan, and J. J. Viola. 2004. Real-time probing of membrane transport in living microbial cells using single nanoparticle optics and living cell imaging. *Biochemistry* 43:10400–10413.

Yavuz, M. S., Y. Cheng, J. Chen, C. M. Cobley, Q. Zhang, M. Rycenga, J. Xie, C. Kim, K. H. Song, A. G. Schwartz, L. V. Wang, and Y. Xia. 2009. Gold nanocages covered by smart polymers for controlled release with near-infrared light. *Nature Materials* 8: 935–939.

Zhang, S., J. Li, G. Lykotrafitis, G. Bao, and S. Suresh. 2009. Size-dependent endocytosis of nanoparticles. *Advanced Materials* 21:419–424.

Zheng, Y., and L. Sanche. 2009. Gold nanoparticles enhance DNA damage induced by anticancer drugs and radiation. *Radiation Research* 172:114–119.

Zheng, Y., D. J. Hunting, P. Ayottea, and L. Sanche. 2008. Radiosensitization of DNA by gold nanoparticles irradiated with high-energy electrons. *Radiation Research* 169: 19–27.

10

Quantification of Gold Nanoparticle–Mediated Radiation Dose Enhancement

Sang Hyun Cho
Georgia Institute of Technology

Bernard L. Jones
University of Colorado School of Medicine

10.1 Introduction

The ultimate goal of radiation therapy is to deliver a lethal dose of radiation to a tumor while sparing nearby normal tissues. In theory, the tumor dose during photon-based radiation therapy can be selectively enhanced by loading high atomic number (Z) materials into a tumor. This is because the photoelectric mass attenuation coefficient is approximately proportional to Z^3, thereby resulting in much greater photoelectric absorption within the tumor than in nearby normal tissues. This selective tumor dose enhancement could be used to improve the therapeutic ratio of photon-based radiation therapy beyond the level currently achievable. Earlier attempts to capitalize this idea using high-Z contrast materials (e.g., iodine and gadolinium) and gold microspheres met limited success (Mello et al. 1983; Dawson et al. 1987; Iwamoto et al. 1987; Rose et al. 1994; Mesa et al. 1999; Herold et al. 2000; Robar et al. 2002; Verhaegen et al. 2005; Robar 2006). Meanwhile, more encouraging results were seen with gold nanoparticles (GNPs), for example, showing remarkable tumor regression and long-term survival in mice without any significant toxicity compared to mice irradiated without GNPs (Hainfeld et al. 2004). This dramatic outcome could be attributed to the significant increase in the fluence of photoelectrons and Auger/Coster–Kronig electrons within the tumor (including blood vessels) loaded with high-Z GNPs during x-ray irradiation, resulting in greater physical damage to tumor cells and endothelial cells lining the blood vessels (Hainfeld et al. 2008; Cho et al. 2009).

Many other recent studies have also demonstrated similar GNP-mediated tumor dose enhancement/radiosensitization *in vitro* or *in vivo* using x-ray/gamma-ray sources (Foley et al. 2005; Butterworth et al. 2008; Zhang et al. 2008; Kong et al. 2008; Butterworth et al. 2010; Hainfeld et al. 2010; Chithrani et al. 2010; Jain et al. 2011), electron beams (Chang et al. 2008; Zheng et al. 2008), and proton beams (Kim et al. 2010; Polf et al. 2011). In order to properly account for this intriguing phenomenon, it is necessary to quantify the amount and spatial distribution of dose enhancement associated with a given irradiation scenario. In this chapter, therefore, an overview of various approaches for such quantification is presented, focusing on the current authors' own work in particular. Although the current discussion is limited to GNP-mediated tumor dose enhancement due to photon irradiation, much of it could still be applicable to irradiation scenarios with charged particles (e.g., electrons and protons) capable of producing secondary electrons from GNPs.

10.1.1 Rationale for GNPs

In many aspects (explained below), the concept of GNP-mediated tumor dose enhancement appears to be more attractive than similar approaches with high-Z contrast media and gold microspheres. First, gold ($Z = 79$), the base metal of GNPs, exhibits little toxicity, up to at least 3% by weight, on either the rodent or human tumor cells (Herold et al. 2000). Second, because of the Z^3 dependence of photoelectric absorption probability, the dose to a GNP-loaded tumor can be higher than that to a tumor infused with gadolinium ($Z = 64$) or iodine ($Z = 53$), assuming the same concentration of materials in the tumor and the same radiation quality. Third, GNPs provide a better targeting mechanism than micrometer-sized gold particles (i.e., gold microspheres) in terms of delivering high-Z materials to the tumor. Nanoparticles passively leak into the tumor interstitium from blood vessels feeding the tumor via a phenomenon typically known as "enhanced permeability and retention" (Maeda et al. 2003), because they are smaller by definition (e.g., 1–100 nm) than the typical cutoff size of the pores (e.g., up to 400 nm) in the tumor vasculature (Unezaki et al. 1996). In addition to the above strategy to concentrate GNPs specifically within the tumor, commonly known as "passive targeting," the tumor specificity of GNPs can be further increased through so-called "active targeting." In this approach, GNPs are conjugated with antibodies or peptides directed against tumor markers such as epidermal growth factor receptor (Sokolov et al. 2003; El-Sayed et al. 2005; Qian et al. 2008), human epidermal growth factor receptor-2 (Loo et al. 2005; Hainfeld et al. 2011), and angiogenesis markers such as vascular endothelial growth factor (Mukherjee et al. 2007). Moreover, GNPs can be used to produce plasmonic heating within the tumor for thermal ablation or hyperthermia/thermoradiotherapy (Krishnan et al. 2010).

10.1.2 Clinical Implications of GNP-Mediated Tumor Dose Enhancement

The concept of GNP-mediated tumor dose enhancement can provide a way not only to escalate the overall tumor dose far beyond the current limits but also to enhance the radiation dose in a more tumor-specific manner, while sufficiently sparing normal tissues surrounding the tumor. Thus, any clinical implementation of this concept, such as gold nanoparticle–aided radiation therapy (GNRT) (Cho et al. 2009), would be more effective than conventional radiation therapy, especially in terms of managing radioresistant tumors or the cases subject to a very narrow therapeutic window. For those tumors managed well with current radiation dose regimens, it would provide an option to achieve the prescribed tumor dose using a smaller amount of radiation, thereby reducing the dose to surrounding normal tissues. From a biophysical standpoint, GNP-mediated tumor dose enhancement can be seen as modulating the radiation response of the tumor because GNPs can change the interaction probability of the tumor with radiation and, under active targeting, also control the location of radiation interaction within the tumor. Unlike biological radiosensitizers (chemotherapy drugs, targeted biological agents, vaccine strategies, etc.), whose efficacy varies considerably across tumor types depending on the molecular characteristics of a given tumor, GNPs amass within all vascularized tumor types irrespective of their molecular profiles. Moreover, the physical interactions between high-Z gold and photons are also unaffected by tumor type. Furthermore,

these interactions can occur in a more tumor-specific manner through active targeting as mentioned above.

10.1.3 Approaches for Quantification of GNP-Mediated Radiation Dose Enhancement

Currently, there are several ways to quantify the GNP-mediated dose enhancement. The most conventional method is to compute or measure the average dose enhancement over a volume (e.g., tumor) containing some uniform concentration of GNPs. This approach may be referred to as the *macroscopic estimation* of GNP-mediated dose enhancement, and has several advantages and disadvantages to be explained in detail in Sections 10.2 and 10.3. An alternative approach referred to as the *microscopic estimation* of GNP-mediated dose enhancement has also emerged recently and may be useful to explain biological outcomes seen in *in vitro/in vivo* experiments. This approach enables the estimation of GNP-mediated dose enhancement on a nano-/cellular scale, while taking into account more realistic situations such as the heterogeneous distribution of GNPs and the short-ranged low-energy secondary electrons (e.g., Auger/Coster–Kronig electrons) from GNPs. More details about this approach are presented in Section 10.4.

10.2 Macroscopic Estimation of GNP-Mediated Radiation Dose Enhancement

10.2.1 Uniform Mixture Model

From a physical point of view, radiation dose enhancement mediated by GNPs is essentially the net increase in energy deposition throughout the tissue/tumor region filled with GNPs. This can be quantified by considering the macroscopic (or average) dose enhancement factor (MDEF), defined as the ratio of the average dose in the tissue/tumor region with and without the presence of GNPs after the irradiation of the tissue/tumor. In this approach, each GNP-loaded tissue/tumor is assumed to have a uniform distribution of GNPs. Also, no physical interface between GNPs and tissue/tumor is assumed. These two assumptions constitute the so-called uniform mixture model in which a uniform distribution of GNPs throughout the tissue/tumor is approximated by a uniform distribution gold atoms at a given weight fraction among other tissue elements.

For example, the four-component tissue (i.e., 10.1% hydrogen, 11.1% carbon, 2.6% nitrogen, and 76.2% oxygen) defined by the International Commission on Radiation Units and Measurements (ICRU 1989) can be altered by the given weight fraction of GNPs within the tissue (e.g., 0.7% Au/g tissue). The density of each GNP-loaded tissue may be increased from that of the ICRU tissue (i.e., 1 g/cm^3) to the value reflecting the added weight of gold to the ICRU tissue (e.g., 1.007 g/cm^3 for the tissue loaded with 7 mg Au/g). Note that this type of density scaling is only an approximation for realistic cases but can still be deemed reasonable for the computational/experimental phantom cases.

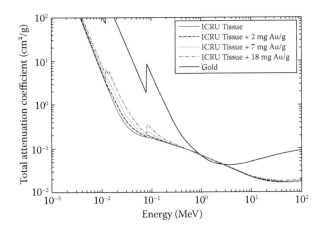

FIGURE 10.1 Total photon interaction cross sections for ICRU tissue, gold-loaded ICRU tissues, and gold. Data are obtained using XCOM software (Berger et al. 2005). Photon absorption edges for gold-loaded tissues become pronounced in this figure as amount of gold within ICRU tissue increases.

Figure 10.1 shows the difference between the four-component tissues with and without GNPs, in terms of their photon interaction cross sections as obtained from XCOM software (Berger et al. 2005). The photoelectric absorption edges for GNP-loaded tissues become pronounced in this figure due to Z^3 dependence as the amount of GNPs within the ICRU tissue increases. Note, as shown in Figure 10.1, that the photon interaction cross sections for GNP-loaded tissues are also slightly increased at high photon energy (e.g., above 10 MeV) because the mass attenuation coefficient for nuclear pair production is roughly proportional to the Z of the medium. However, such a moderate increase in the photon interaction probability might be insufficient to induce significant GNP-mediated dose enhancement. Moreover, there are other issues (e.g., efficiency in low energy secondary electron production, unwanted photonuclear reaction) that appear to make any approach using high energy photon beams less attractive.

10.2.2 Monte Carlo Models

The magnitude of tumor dose enhancement for various irradiation scenarios of GNP-loaded tumors can be quantified by calculating the MDEF or similar quantities using the condensed history Monte Carlo (MC) method and uniform mixture model (Cho 2005; McMahon et al. 2008; Cho et al. 2009; Garnica-Garza 2009; Van den Heuvel et al. 2010). The condensed history MC code systems used for previous investigations on the current topic include EGSnrc/BEAMnrc (Kawrakow and Rogers 2003; Rogers et al. 2002; Walters et al. 2006), MCNP5/MCNPX (X-5 Monte Carlo Team 2003), GEANT4 (Agostinelli et al. 2003), and PENELOPE (Salvat et al. 2003). An overview of previous investigations by the current authors is presented below to exemplify this approach in general.

10.2.2.1 External Beam Radiation Therapy Cases

A previous study (Cho 2005) used several phantom test cases to estimate the level of macroscopic dose enhancement corresponding to typical external beam radiation therapy (EBRT) scenarios with 140-kVp x-rays and megavoltage photon beams (i.e., 4 and 6 MV) from linear accelerators. In each of the cases, GNP-loaded tumors were created based on the ICRU four-component tissue using the uniform mixture model. The composition and density of the tumor were altered by three different concentration levels of gold: 7, 18, and 30 mg Au/g tumor. The first two concentration levels (i.e., 7 and 18 mg Au/g) were based on the animal data for 1.9-nm-diameter GNPs from Hainfeld et al. (2004). The second value (i.e., 18 mg Au/g) was not a gold concentration level inside the tumor during the Hainfeld study but the blood content of gold, 2 min after a mouse was injected with 2.7 g Au/kg body weight. However, it was taken as an upper bound value for a possible gold concentration level within a vascularized tumor at the time of irradiation. The third value was chosen mainly for comparison with previous studies with high-Z contrast media, whereas the tumor gold concentration levels of more than 30 mg/g tumor would be unrealistic, regardless of the particle size and shape, considering the reported tumor gold contents during various animal studies (Herold et al. 2000; Hainfeld et al. 2004; James et al. 2007; Zaman et al. 2007; Qian et al. 2008). The presence of GNPs outside the tumor was assumed for some of the cases studied. Specifically, another type of tissue was created by altering the composition and density of the ICRU four-component tissue for the presence of 2 mg Au/g tissue, which was the concentration level of gold in muscle when the tumor was loaded with 7 mg Au/g tumor (i.e., tumor/muscle gold concentration ratio of 3.5:1), according to the Hainfeld study. This tissue was used to fill the volume outside the tumor region within the phantom during the MC simulation. The geometry used for the EBRT cases involved either a superficial or deep-seated tumor infused with GNPs within a tissue phantom ($30 \times 30 \times 30$ cm³) (see Figure 10.2a). The MC calculations were performed with the BEAMnrc/DOSXYZnrc code (Rogers et al. 2002; Walters et al. 2006). More details about the MC calculations can be found elsewhere (Cho 2005).

Table 10.1 presents the results for the external beam cases, in terms of the values of MDEF. Each of the cases in this table assumed no GNPs outside the tumor, in order to provide a clear relationship between the gold concentration level and beam quality. As shown, the macroscopic (or average) tumor dose enhancement appears to depend on gold concentration within the tumor and the photon beam quality, ranging from several hundred percent for diagnostic x-rays to a few percent for typical megavoltage photon beams. These results also suggest that it would be difficult to achieve clinically meaningful dose enhancement (>10%) with either flattened or unflattened photon beams for the considered phantom test cases. Although not shown here, the loading of GNPs into surrounding normal tissue at 2 mg Au/g for the 7 mg Au/g tumor cases, resulted in an increase in the normal tissue dose, for example, up to 30% for the 140-kVp

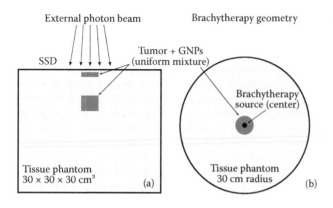

FIGURE 10.2 MC simulation geometry: (a) external beam cases, (b) brachytherapy cases. For external beam cases, size of tumor was $1.0 \times 1.0 \times 0.15$ cm^3 for 140-kVp x-ray case and $2.4 \times 2.4 \times 3.5$ cm^3 for 4- and 6-MV photon beam cases. Center of the tumor was located along central axis of the beam at depths of 0.075 and 6.5 cm for x-rays and photon beams, respectively. For brachytherapy cases, tumor region for ^{125}I and 50 kVp cases was taken as a 1.5-cm-radius sphere centered at origin of spherical phantom excluding the region occupied by source, whereas tumor for ^{169}Yb and ^{192}Ir cases was assumed as a 3.5-cm-radius sphere. Figures are not drawn to scale.

x-ray case, whereas the magnitude of dose enhancement within the tumor was essentially unchanged (Cho 2005). For the 6-MV photon beam cases, there was no significant increase in the normal tissue dose.

A subsequent MC study (Cho et al. 2006) calculated the macroscopic dose enhancement across the tumor volume using actual patient cases containing hypothetical GNP-laden tumors. Unlike the previous study (Cho 2005), megavoltage photon beams further enriched with low energy component (i.e., 2- or 4-MV beams produced with a copper target) were generated by MC simulations and used for patient dose calculations. The results showed that the macroscopic dose enhancement up to 28% and 12% across the tumor volume could be achievable with unflattened 2- and 4-MV photon beams, respectively, at a gold concentration of 1.8% within the tumor (e.g., 18 mg Au/g). These beams were found capable of producing clinically acceptable treatment plans for *GNRT* (Figure 10.3), in spite of their softer photon energy spectra and larger buildup doses, compared to conventional megavoltage beams at the same nominal photon energies. More discussion on this possibility can be found in the last chapter of this book.

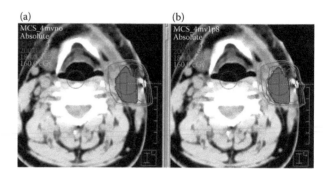

FIGURE 10.3 (See color insert.) Hypothetical GNRT treatment of head and neck tumor using Cu target–produced 4 MV NFF (or FFF) beam: (a) No GNP within tumor, (b) 1.8 wt.% GNP within tumor. 10% enhancement in the target dose (200 cGy) is clearly seen in the panel b. Dose distributions were from MC calculations.

10.2.2.2 Brachytherapy Cases

As shown for the EBRT cases earlier, clinically meaningful macroscopic dose enhancement would be achievable only with low energy photons such as kilovoltage x-rays because such photons will interact with GNPs within the tumor predominantly via the photoelectric effect, which is thought to be the main physical mechanism responsible for the dose enhancement. Because of their limited penetration into condensed media such as human tissue, however, low energy x-rays are generally not suitable for EBRT. Moreover, there are significantly more GNPs in the blood than in the tumor (e.g., tumor/blood ratio of 0.3:1), immediately (e.g., 2 min) after an intravenous injection of GNPs (Hainfeld et al. 2004). Consequently, the endothelial cells lining the vasculature presented along the external beam path would receive significantly higher doses than other tissues, potentially resulting in blood vessel disruption not only inside a tumor but also in surrounding normal tissues. Thus, it is necessary to consider alternative approaches that may help overcome or at least alleviate these difficulties for successful clinical implementation of GNRT. One conceivable approach is apparently to deliver GNRT treatments via brachytherapy. This approach appears to be more promising especially with radioisotopes emitting even lower energy gamma rays than ^{192}Ir such as ^{169}Yb, ^{125}I, and ^{103}Pd or by using miniature x-ray devices producing low energy x-rays (e.g., ~50 kVp). These low-energy gamma-ray/x-ray sources have a higher probability of photoelectric absorption in a GNP-loaded tumor than ^{192}Ir gamma rays and thereby would result in more

TABLE 10.1   Macroscopic Dose Enhancement Factor (MDEF) over Tumor Volume

Concentration (per g tumor)	140 kVp	250 kVp	6 MV FF	6 MV NFF	4 MV FF	4 MV NFF
7 mg Au	2.114	1.632	1.007	1.014	1.009	1.019
18 mg Au	3.811	2.622	1.015	1.032	1.019	1.044
30 mg Au	5.601	3.682	1.025	1.053	1.032	1.074

Note: Results were obtained, assuming no gold presence outside the tumor. FF, flattening filter; NFF, no flattening filter (also known as "flattening-filter-free").

tumor dose enhancement. Note that low-energy gamma-rays/x-rays below the K-edge (i.e., ~80 keV) of gold no longer interact with K-shell electrons of gold but interact predominantly with L-shell electrons during the photoelectric absorption process. This feature has some significant impact on microscopic dose enhancement pattern to be discussed later in this chapter.

In previous studies (Cho 2005; Cho et al. 2009), MC calculations were conducted to determine the typical values of MDEF associated with four different types of brachytherapy sources: [125]I, 50 kVp, [169]Yb, and [192]Ir. The phantom geometry represented a typical geometry used during the MC characterization of brachytherapy sources, namely, a source located at the center of a spherical phantom with a radius of 30 cm (Figure 10.2b). The tumor region for ^{125}I and 50 kVp cases was taken as a 1.5-cm-radius sphere centered at the origin of the spherical phantom excluding the region occupied by the source, whereas the tumor for [169]Yb and ^{192}Ir cases was assumed as a 3.5-cm-radius sphere. The material composition of the tumor and phantom was the same as that for the EBRT cases. The MC calculations for all brachytherapy cases were performed with the MCNP5 code (X-5 Monte Carlo Team 2003). More details about the MC simulation and MCNP5 code can be found elsewhere (Cho et al. 2009; X-5 Monte Carlo Team 2003).

Calculated values of MDEF are plotted in Figures 10.4 and 10.5 as a function of radial distances from the center of the source (or phantom). Similar to the EBRT cases, MDEFs increased with the gold concentration within the tumor. Note, in these figures, the factors beyond the tumor region are not MDEFs but are the dose ratios between the cases with and without GNPs showing the dose reduction behind the tumor loaded with GNPs. As shown in Figures 10.4 and 10.5, macroscopic dose enhancement over a tumor region was estimated to be remarkably large, especially at close radial distances from the center of the source. According to

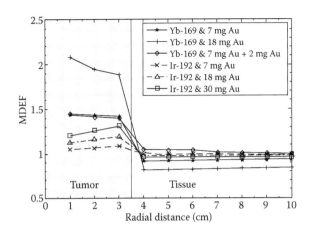

FIGURE 10.5 Calculated macroscopic dose enhancement factor (MDEF) for [169]Yb and ^{192}Ir cases as a function of radial distance along transverse axis of source. If less than unity, factors shown from $r = 4$–10 cm are not MDEFs but show decrease in doses behind the tumor loaded with GNPs. Radius of a spherical tumor centered at origin is 3.5 cm. Amount of gold shown in figure legend is per gram of tumor or tissue.

the results, a significant tumor dose enhancement (e.g., > 40%) could be achievable using ^{125}I, 50 kVp, and [169]Yb sources and realistic concentrations of GNPs (at least in mice). The values of MDEF at 1.0 cm from the center of the source are summarized in Table 10.2.

On a particular note, a comparison of MDEF values between [169]Yb and ^{192}Ir reveals the superiority of [169]Yb to ^{192}Ir, in terms of its effectiveness to induce dose enhancement within a GNP-loaded tumor. In fact, MDEF values for ^{192}Ir estimated at a somewhat unrealistic level of tumor gold concentration (i.e., 30 mg Au/g tumor) were smaller by as much as about 20% than those for [169]Yb estimated at a much lower concentration level of 7 mg Au/g tumor.

As shown in Figures 10.4 and 10.5, MDEFs were also dependent on gamma-/x-ray energy spectra and radial distance. The fall-off of MDEFs through the tumor was more pronounced for ^{125}I and 50 kVp than for [169]Yb and ^{192}Ir, because of increased attenuation of relatively lower energy gamma-rays/x-rays through a high Z GNP-loaded tumor dependent on the tumor gold concentration level. Similar to the EBRT cases shown before, the loading of GNPs into surrounding normal tissue at 2 mg Au/g for the 7 mg Au/g tumor cases, resulted in an increase in the normal tissue dose up to 26%, whereas the tumor dose remained almost the same. Since the tissue dose was already

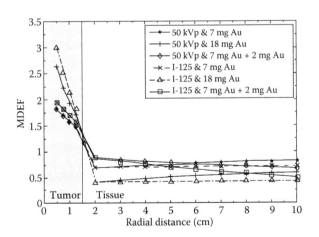

FIGURE 10.4 Calculated macroscopic dose enhancement factor (MDEF) for ^{125}I and 50 kVp cases as a function of radial distance along transverse axis of source. Factors shown from $r = 2$–10 cm are not MDEFs but show decrease in doses behind the tumor loaded with GNPs. Radius of a spherical tumor centered at origin is 1.5 cm. Amount of gold shown in figure legend is per gram of tumor or tissue.

TABLE 10.2 Macroscopic Dose Enhancement Factor (MDEF) at 1.0 cm from Center of Source

Concentration (per g tumor)	[125]I	50 kVp	[169]Yb	[192]Ir
7 mg Au	1.68	1.57	1.44	1.05
18 mg Au	2.16	1.92	2.08	1.13
30 mg Au	–	–	–	1.21

Note: Results were obtained, assuming no gold presence outside the tumor.

reduced significantly owing to the increased photon attenuation through a GNP-loaded tumor, however, the effect of GNPs present in the tissue surrounding the tumor is deemed minimal, at least, for the source-tumor geometry considered here.

10.2.2.3 Non-MC Approach

By performing a non-MC theoretical study based on a systematic analysis of the mass energy absorption coefficients of various mixtures at different photon energies, Roeske et al. (2007) provided the values of MDEF for various photon sources and materials with Z numbers ranging from 25 to 90. The authors reported, despite slight discrepancies, that their external beam results for the gold cases were consistent with previously published MC data by Cho (2005). A similar approach was also used by McMahon et al. (2008). In general, the results obtained from this type of approach would not be identical to those from an MC-based approach, even if the key input data (e.g., source energy spectrum) were perfectly matched between the two approaches. This is because theoretical values are estimated simply by taking the ratios of the photon energy absorption coefficients between water and materials mixed with gold at the initial photon spectra, whereas the MC results are obtained by actually transporting photons through detailed source/tumor geometry to properly take into account the changes in photon spectra throughout the phantom. Nevertheless, a non-MC approach appears to be useful for a quick estimation of the macroscopic dose enhancement under a given irradiation scenario.

10.2.2.4 Validity of Macroscopic Estimation of GNP-Mediated Dose Enhancement

In general, macroscopically estimated GNP-mediated dose enhancement gives a good insight about the net increase in energy deposition within the tumor. Because of the short ranges of 10–100 keV electrons, however, it is likely that the secondary electrons from GNPs may deposit their energies mostly in the vicinity of GNPs themselves, resulting in a nonuniform spatial distribution of dose enhancement within the tumor. Moreover, GNPs are known to be distributed heterogeneously throughout the tumor under both passive and active targeting (Hainfeld et al. 2004; Diagaradjane et al. 2008; Hainfeld et al. 2011). Additionally, they typically aggregate and form clusters within the tumor, and sometimes are taken up by tumor cells (i.e., internalization). Furthermore, significantly more GNPs are found within the tumor vasculature. As a result, GNP-mediated dose enhancement would occur in a very heterogeneous fashion for realistic cases. Accordingly, macroscopic estimation of GNP-mediated dose enhancement would become less meaningful unless it is applied to clinical situations closely approximating a uniform mixture model. In fact, some recent *in vitro* studies have reported a number of findings that cannot be explained well in terms of macroscopically estimated dose enhancement. For example, significant (approximately >10%) radiosensitization effect was reportedly observed in one study in which HeLa cells were treated with GNPs at a very low concentration level (i.e., 0.001 wt.%) (Chithrani et al. 2010). Some studies (Chithrani et al. 2010, Jain et al. 2011) also reported significant radiosensitization effect with 6-MV photon beams, contrary to the prediction based on the concept of MDEF. Appropriate physical accounts for these findings may not be found unless one estimates GNP-mediated dose enhancement microscopically on a nano-/cellular scale (see related discussion in Section 10.4). Moreover, physical models might not be able to fully explain the purely biological effects of GNPs (e.g., possible cytotoxicity). Nevertheless, macroscopic estimation of the dose enhancement may still be useful for some practical situations. For instance, it will be useful for the so-called contrast-enhanced radiation therapy (CERT) with GNPs serving as contrast agents to produce relatively high (approximately on the order of 1%) blood gold content throughout the tumor. It can also be applicable to *in vivo* experiments under passive targeting where no significant internalization of GNPs is expected. Currently, no imaging modality is capable of providing the spatiotemporal distribution of GNPs *in vivo*, which may even vary from one animal to another. Consequently, the tumor gold concentration level, which is often measurable *in vivo* (Zaman et al. 2007; Qian et al. 2008), is probably the only meaningful reproducible information available from *in vivo* studies for computational purpose. A similar argument can be made about the intratumoral and intracellular uptake of GNPs, which are known to be dependent on the size and shape of GNPs but not explicitly taken into account for the macroscopic estimation. Considering all of these, at least as the first approximation, it would still be a reasonable attempt to correlate radiobiological outcome (e.g., tumor shrinkage) with macroscopically estimated dose enhancement through the two globally definable variables across the tumor such as average tumor gold concentration and radiation quality. For example, during an investigation by the current authors and colleagues (Diagaradjane et al. 2008), it was initially predicted by applying an approach based on macroscopic estimation that no significant dose enhancement was to be achieved for mice irradiated with 125 kVp x-rays after the injection of gold nanoshells under a passive targeting scenario, because the tumor gold content due to gold nanoshells was known to be much smaller (e.g., 2 orders of magnitude) than that due to 1.9-nm-diameter GNPs. Later, an animal study performed during the aforementioned investigation (Diagaradjane et al. 2008) confirmed this prediction.

10.3 Measurement of GNP-Mediated Radiation Dose Enhancement by Gel Dosimetry

Although it has been shown well through *in vitro*, *in vivo*, and computational work performed in recent years, GNP-mediated radiation dose enhancement is somewhat difficult to demonstrate by physical measurements, especially over a volume loaded with GNPs, a situation closely mimicking a potential clinical scenario. One of the conceivable ways to demonstrate the dose enhancement over a volume is to use three-dimensional dosimeters such as a gel dosimeter, which can record possible dose enhancement within itself. Once properly fabricated, three-dimensional dosimeters can also be used to validate the MC results mentioned earlier, e.g., MDEFs for various irradiation scenarios. Previously, Fricke solution was used to serve a similar purpose for gold microspheres (Herold et al. 2000). Unlike gold microspheres that can be filtered out after the irradiation, however, GNPs may not be filtered out from the Fricke solution because of their small sizes (e.g., 1–100 nm). Moreover, the overall procedures to perform the Fricke dosimetry appear to be cumbersome in many aspects. Consequently, as previously demonstrated by the current authors and colleagues (Siddiqi et al. 2009), radiosensitive polymer gels can be better suited to measure the macroscopic dose enhancement over a volume loaded with GNPs. A similar approach was also used to show radiation dose enhancement within iodine-loaded gel phantoms (Boudou et al. 2007).

Methacrylic and Ascorbic acid in Gelatin Initiated by Copper (MAGIC) is one of the most widely available radiosensitive normoxic polymer gels. In MAGIC gel, the spin relaxation rate, R_2, is proportional to absorbed dose, producing a linear dose–response curve in the range of 0–30 Gy (Fong et al. 2001). In the aforementioned investigation (Siddiqi et al. 2009), a formaldehyde-containing MAGIC gel following the formula suggested by Fernandes et al. (2008) was used instead of the conventional MAGIC gel, in order to increase the radiosensitivity of MAGIC gel to capture the dose enhancement supposedly occurring within micrometers around GNPs. Specifically, formaldehyde containing MAGIC gel was poured into 2-mL cylindrical plastic containers serving as the phantoms for x-ray irradiation. Four of them had MAGIC gel only, whereas the remaining two were filled with MAGIC gel and commercially available 1.9-nm diameter GNPs (Hainfeld et al. 2004) at a concentration of 1% by weight (1 wt.%). Each gel phantom was irradiated using 110 kVp x-rays. The total dose delivered to the phantom ranged from 0 to 30 Gy. One phantom in each group was not irradiated and taken

as the control. All phantoms were read using magnetic resonance imaging on Bruker 7T Pharmascan. The inverse T_2 relaxation time (i.e., R_2 value) for each phantom was plotted against the delivered dose to obtain the calibration curve.

Figure 10.6 shows the results from the experimental measurements described above. According to these results, addition of GNPs to MAGIC gel did not significantly change the R_2 value, at least at the gold concentration level tested in the investigation. As shown in Figure 10.8, the dose to the gel phantom mixed with GNPs was enhanced more than 100% across the entire volume (i.e., 12 Gy → 27 Gy; 125% enhancement), which agreed reasonably well with the MC estimation. Overall, the results clearly suggest that radiosensitive gels can successfully be used to experimentally show some remarkable GNP-mediated radiation dose enhancement with kilovoltage x-ray sources on a macroscopic scale (or on average). Considering typical experimental uncertainty associated with gel dosimeters (e.g., ~5%), however, radiosensitive gels might not be suitable to detect very small amounts of macroscopic dose enhancement on the order of a few percents such as those predicted for megavoltage sources in Table 10.1.

10.4 Microscopic Estimation of GNP-Mediated Radiation Dose Enhancement

10.4.1 Calculations of Secondary Electron Spectra within a GNP-Loaded Tumor

In previous studies (Cho et al. 2009; Jones 2011), the secondary electron spectra within a GNP-loaded tumor or water due to various photon sources (i.e., [103]Pd, ^{125}I, 50 kVp, [169]Yb, 250 kVp, ^{192}Ir, and 6 MV) were determined using the MC method. The EGSnrc/DOSXYZnrc code (Kawrakow and Rogers 2003; Walters et al. 2006) was modified to output the energy of any

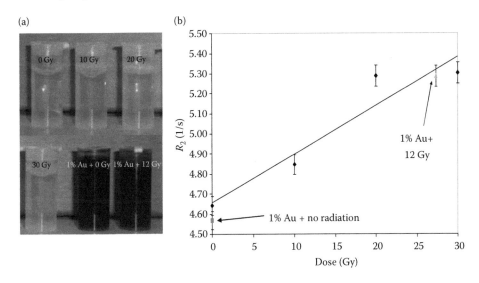

FIGURE 10.6 Results of radiosensitive gel dosimetry measurements: (a) physical appearance of radiosensitive gel with/without GNPs, (b) gel dose calibration curve. Both unirradiated and irradiated gels with GNPs are shown.

electron generated from Compton scattering, photoelectric absorption, and atomic relaxation for a proper binning depending on the electron energy, atom of origin, and interaction type. Two separate simulations for each source were performed to determine electron spectra within a tumor or water with and without GNPs. The ICRU tissue or water was the material for the phantom/tumor and the base material for the tumor loaded with GNPs at 7 mg Au/g. Only one level of tumor gold concentration (i.e., 7 mg/g tumor) was deemed sufficient for the MC calculations, assuming the secondary electron fluence due to each GNP present within ICRU tissue is additive without any significant screening effect by neighboring GNPs. The simulation geometries were also similar to those used for the estimation of MDEF.

Figure 10.7 shows a remarkable change in the photoelectron fluence within the tumor region because of the presence of GNPs during low-energy gamma-ray/x-ray irradiation. As shown in these figures, the photoelectron fluence within a GNP-loaded tumor was significantly larger (e.g., more than 2 orders of magnitude) than that within a tumor without GNPs. Some distinct peaks around each photoelectric absorption edge of gold (i.e., ~81 keV for K-edge and 12–14 keV for L-edge) are also well shown in these figures. Although not shown here, similar trends were also seen for other low-energy gamma-ray sources (i.e., [103]Pd and ^{125}I) and 250 kVp x-ray source. On the other hand, an overall increase in the photoelectron fluence was much less for ^{192}Ir and 6 MV sources, for example, approximately 2 orders of magnitude smaller for 6 MV than that for [169]Yb. More detailed results can be found elsewhere (Cho et al. 2009; Jones 2011).

Figure 10.8 shows the calculated fluence and energy spectra of Auger electrons within the tumor region due to the presence of GNPs during low-energy gamma-ray/x-ray irradiation. Note that Auger electrons below 1 keV are not included in this plot, because the electron cutoff energy of 1 keV was used for this MC study. Nevertheless, these results clearly demonstrate the expected outcome from each irradiation scenario considered. The role of Auger electrons, particularly those with large

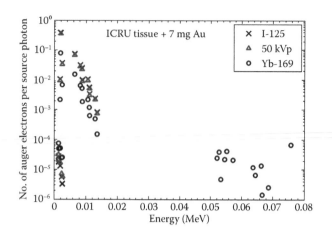

FIGURE 10.8 Auger electron spectra within a 3 × 3 × 3 cm³ tumor irradiated by ^{125}I, 50 kVp, and [169]Yb sources located at center of the tumor. Spectra are shown only for a tumor loaded with GNPs at 7 mg/g, because Auger electrons above 1 keV are not seen for a tumor without GNPs. Note no distinction between Auger and Coster–Kronig electrons is made for these spectra.

abundance because of gold L- and M-shell relaxation processes, would become significant when one considers microscopic dose enhancement on a nano-/cellular scale to find some correlation with radiobiological effects. This is because the spatial variation in physical dose enhancement on a cellular scale is closely related with the energy of secondary electrons, especially photo-/Auger electrons originating from GNPs. For example, despite an almost 2-fold increase in photoelectron fluence as shown in Figure 10.6, an actual increase in cell killing (i.e., radiosensitization) would be due only to those photoelectrons with sufficient energy to reach tumor and endothelial cells from the site of each GNP or those with much less energy but originating from GNPs at close proximity to these cells (or within the cells) capable of reaching nuclear deoxyribonucleic acid (DNA) for single- or

(a)

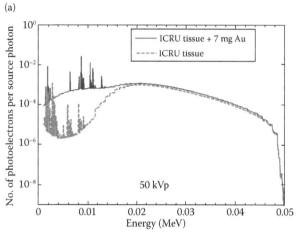

(b)

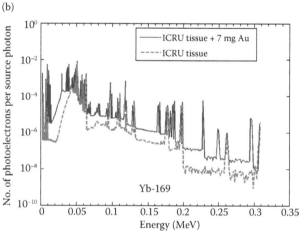

FIGURE 10.7 Photoelectron spectra within a 3 × 3 × 3 cm³ tumor irradiated by a source located at center of the tumor: (a) 50 kVp x-rays and (b) [169]Yb gamma rays. Spectra are shown for a tumor loaded with GNPs at 7 mg/g and for a tumor without GNPs.

double-strand breaks. More detailed discussion on this issue can be found later in this chapter.

10.4.2 Microscopic Estimation of GNP-Mediated Radiation Dose Enhancement

As shown earlier, at least in theory, more than 40% of dose enhancement can easily be achieved macroscopically (or on average across the tumor) during kilovoltage x-ray or low-energy gamma-ray irradiation of a tumor loaded with GNPs at gold concentration levels deemed achievable *in vivo* (e.g., 7 mg/g tumor after an intravenous injection of 1.9-nmdiameter GNPs) (Hainfeld et al. 2004). Because of significant heterogeneity and localization of GNP-mediated dose enhancement (as explained in the preceding section), however, macroscopically estimated dose enhancement alone might be insufficient for a proper correlation with radiobiological outcome (e.g., radiosensitization). A better correlation between GNP-mediated dose enhancement and radiosensitization could be found by considering both the spatial variation and amount of dose enhancement estimated microscopically on a nano-/cellular scale. In previous studies (Cho et al. 2007a; Jones et al. 2010; Jones 2011), therefore, the current authors performed event-by-event (or detailed history) MC calculations to quantify the microscopic dose enhancement around GNPs irradiated by various photon sources. A general overview of these investigations is presented below, along with a brief survey of other approaches developed in recent years.

10.4.2.1 Calculations of Microscopic Dose Point Kernels around GNPs

The first step in the latest investigations by the current authors (Jones et al. 2010; Jones 2011) was to calculate the microscopic dose point kernels around GNPs present within a tumor irradiated by typical photon sources used in radiation therapy such as ^{125}I, [103]Pd, [169]Yb, ^{192}Ir, 50 kVp, 250 kVp, and 6 MV x-rays. The secondary electron spectra required for MC calculations were obtained by the same methodology as described earlier in Section 10.4.1. The secondary electron spectra for gold atoms and water molecules were derived separately to determine the dose point kernels around a GNP and a hypothetical water nanoparticle.

The microscopic dose distribution due to secondary electrons from gold and water nanoparticles was calculated using an event-by-event electron MC code NOREC (Semenenko et al. 2003; Cho et al. 2007b). Each nanoparticle was assumed as a point source of secondary electrons immersed in an infinite medium of water. As a result, self-absorption within relatively large GNPs (e.g., 100 nm in diameter) might be of concern (see Section 10.4.2.4 for related discussion), although this approach is deemed acceptable for smaller GNPs (e.g., 1.9 nm in diameter). In order to allow comparison with the water case under an identical photon irradiation scenario, calculated dose point kernels for gold were scaled by taking into account possible difference in the secondary electron fluence per source photon between the gold and water cases. More details about the scaling of dose

point kernels and the dose point kernels themselves can be found elsewhere (Jones et al. 2010; Jones 2011). For the sake of brevity, the relative dose point kernel is shown in Figure 10.9 for the two selected photon sources only. The dose point kernel in terms of dose per source photon per GNP, or scaled dose point kernel, is also shown in Figure 10.9. The dose point kernel depends strongly on the secondary electron spectrum, as evidenced by the sharp dose fall-off in the lower energy sources (e.g., [103]Pd, ^{125}I, and 50 kVp) as compared to the higher energy sources.

10.4.2.2 Calculations of mDEFs around GNPs

The physical effect of GNPs present within the tumor during radiation therapy (i.e., GNP-mediated radiation dose enhancement) could be quantified by a comparison between gold and water scaled dose point kernels. The ratio of the two dose kernel values at a specific radial distance provides the dose enhancement due to the inclusion of a GNP within the tumor. In other words, this ratio defined as the microscopic dose enhancement factor (mDEF) represents the factor by which the dose would be increased by replacing that point with a GNP. As shown in Figure 10.10, for [103]Pd, ^{125}I, and 50 kVp, remarkable GNP-mediated dose enhancement (e.g., mDEF > 100) would occur mainly within a few microns around a GNP. This could be attributed to the fact that these sources have strong spectral components below the K-edge of gold, resulting in many short-ranged L-shell photoelectrons and Auger electrons. On the other hand, [169]Yb has an average energy very close to the K-edge of gold thereby also capable of producing longer-ranged K-shell photoelectrons very efficiently. As a result, microscopic dose enhancement due to [169]Yb can still be remarkable beyond 100 μm (Jones et al. 2010). For ^{192}Ir and 6 MV x-rays, the amount of microscopic dose enhancement appears to be much less than that for other sources but turns out to be significant enough (e.g., mDEF > 5) over the distance range shown in the figure, mostly because of the action of K-shell photoelectrons.

10.4.2.3 Calculations of Microscopic Dose Enhancement within GNP-Loaded Tissue

To calculate the microscopic dose enhancement due to the presence of GNPs, the dose point kernels were applied to a scanning electron microscopy (SEM) image of a GNP distribution in tissue (Diagaradjane et al. 2008). The microscopic dose enhancement due to GNPs in the sample SEM image (Figure 10.11), defined as the ratio of dose deposited at each point between the gold and water cases, is shown in Figure 10.12 for the three selected sources: [169]Yb, 50 kVp, and 250 kVp. More results can be found elsewhere (Jones et al. 2010; Jones 2011). The 50- and 250-kVp sources, which contained the strongest low-energy component, demonstrated a microscopic dose enhancement as high as 500–2000% within the tumor vasculature. One can also see that the 5% enhancement line extends roughly 10 μm from the nanoparticle clusters. A greater long-range effect was seen in the [169]Yb and 250 kV sources, where the 5% enhancement line extended upward of 30 μm from the nanoparticle clusters, and the dose enhancement inside the vasculature exceeded 200%.

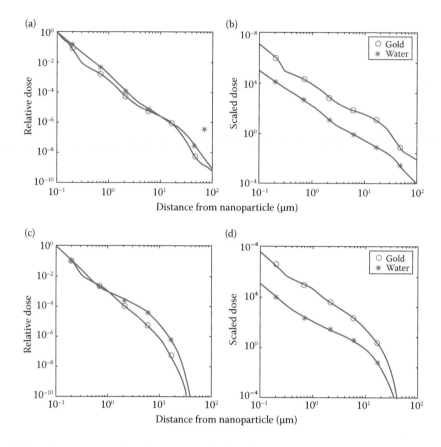

FIGURE 10.9   Dose point kernels in a water medium for a GNP and a hypothetical water nanoparticle under irradiation by [169]Yb and 50 kVp: (a) [169]Yb—relative dose point kernels, (b) [169]Yb—scaled dose point kernels, (c) 50 kVp—relative dose point kernels, (d) 50 kVp—scaled dose point kernels. For scaled dose point kernels, relative dose point kernels are scaled by a factor taking into account differing interaction rates (or secondary electron fluences) between gold and water for varying radiation quality.

10.4.2.4 Other Computational Approaches for Microscopic Estimation

In recent years, various approaches have also been developed to quantify GNP-mediated dose enhancement on a nano-/micro scale. Carter et al. (2007) developed their own MC code to

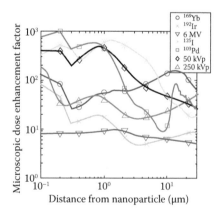

FIGURE 10.10   Spatial variation of microscopic dose enhancement factor (mDEF) around a GNP. mDEF represents factor by which the dose would be increased by replacing that point with a GNP. Results are shown along radial direction from a hypothetical GNP at origin.

simulate interactions between 100-kVp x-rays and 3-nm-diameter GNPs using a simple supercoiled DNA model. They calculated the energy deposition around a GNP far out to 100 nm from the surface of a GNP and claimed to have confirmed their theoretical prediction by measurements. Leung et al. (2011) used the GEANT4 code, a mixed history MC code, to calculate the effective range, deflection angle, dose, energy, and interaction processes of secondary electrons produced from the interactions of photons with a spherical GNP immersed in water. The results were obtained by GEANT4 simulations of a GNP (2, 50, or 100 nm in diameter) irradiated by a GNP-sized photon beam from one of the following

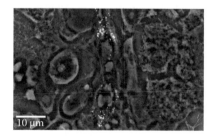

FIGURE 10.11   Scanning electron microscope (SEM) image of tissue containing GNPs (bright white speckles in image) obtained from previous work (Diagaradjane et al. 2008).

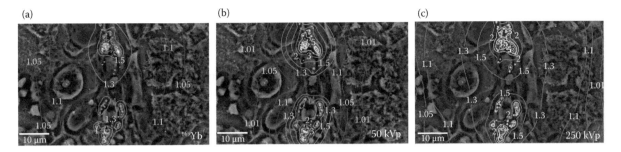

FIGURE 10.12 (See color insert.) Microscopic dose enhancement from three selected photon sources for specific GNP distribution *in vivo*: (a) [169]Yb, (b) 50 kVp, and (c) 250 kVp. Numbers in images correspond to amounts of dose enhancement. For example, 1.05, 1.10, and 2 represent 5%, 10%, and 100% enhancement, respectively.

photon sources: 50 kVp, 250 kVp, ^{60}Co, and 6 MV. One distinct advantage of this type of approach is the ability to take into account the size of a GNP for the quantification of secondary electron spectra and microscopic dose enhancement due to a GNP. For example, the authors reported that, for the 50-kVp irradiation of a 100-nm-diameter GNP, there was about 35% self-absorption within the GNP itself. In spite of a different approach adopted for their investigation, the authors indicated that their conclusion was consistent with that from the current authors' prior work (Jones et al. 2010). Lechtman et al. (2011) used the PELELOPE code, a mixed history MC code, to simulate the production and transport of Auger electrons from GNPs (1.9, 5, 30, and 100 nm in diameter) and subsequent energy deposition around GNPs. PELELOPE simulations were performed with [103]Pd, ^{125}I, [169]Yb, 300 kVp, ^{192}Ir, and 6 MV sources. They reported some similarities in various aspects of their results compared to previous studies including the current authors' work. Besides the MC approaches mentioned above, Ngwa et al. (2010) applied a semiempirical method based on an empirical expression for the electron energy loss in order to calculate microscopic energy deposition within the endothelium in contact with GNPs. The authors considered a model of tumor vascular endothelial cell irradiated by low-energy brachytherapy sources (i.e., [103]Pd, ^{125}I, [169]Yb, and 50 kVp) and reported that they were able to produce the key results comparable to the current authors' published MC data.

10.4.2.5 Validity of Microscopic Estimation of GNP-Mediated Dose Enhancement

In general, computational results from various nano-/microscale studies mentioned above are at least consistent (e.g., on the same order of magnitude) from one another. Although desirable, these nanoscale computational results, especially for a single GNP, appear to be difficult to verify experimentally. Nanoscale radiation detectors or dosimeters based on various nanostructures are conceptually possible and actually under development (Sahare et al. 2007). However, experimental uncertainty associated with such detectors might be comparable to or exceed the uncertainty present in nanoscale computational results. Meanwhile, the validity of nanoscale computational results can often be deduced from experimental findings on a larger scale. For example, the current authors' computational study (Jones et

al. 2010) demonstrated that the microscopic dose around a GNP due to kilovoltage range photon sources could be enhanced by factors up to more than 100. Although performed under somewhat different conditions, a previous experimental study with diagnostic x-rays reported a similar level of dose enhancement (i.e., a factor of 100) on a cellular level because of the presence of gold (i.e., gold foil) (Regulla et al. 1998).

The validity of nanoscale computational results may also be tested through their predicting power for various radiobiological outcomes that are believed to be mainly attributable to physical interactions between cellular structures and radiation. Despite their significant uncertainty (possibly well above the typical experimental uncertainty on a macroscopic scale), nanoscale computational results appear to provide reasonable explanations, at least qualitatively, for some of the puzzling observations from recent *in vitro* experiments. For example, many studies reported significant (approximately >10%) radiosensitization effect in cells treated with very low concentrations (e.g., ~µg Au/g) of GNPs, whereas no significant macroscopic dose enhancement would be expected at such levels of GNP concentration. This observation could become at least less puzzling by noting a remarkable level (e.g., more than a factor 100) of microscopic dose enhancement around GNPs as predicted by nanoscale computational studies. Even at such low concentration levels, there could be more than millions of GNPs, all of which contributed to the microscopic dose enhancement on a cellular scale as individual GNPs or GNP clusters. Moreover, many GNPs were reportedly internalized during *in vitro* experiments. As a result, these internalized GNPs were likely located closer to the cellular nucleus or DNA, possibly resulting in more serious radiation damage to the cell such as the DNA double-strand breaks (DSBs), because of the amplified action of short-ranged secondary electrons from GNPs. Taking these arguments together, one can provide a physically valid explanation of the observed radiobiological outcome. Similar arguments can be made for the observed *in vitro* radiosensitization effect with 6 MV or other megavoltage photon beams. The above physical arguments can also be useful to explain many interesting observations about GNP-mediated radiosensitization from *in vitro* studies, such as its dependence on the GNP size and concentration, the cell type, and the type and energy of radiation. In theory, more quantitative predictions for these effects may be possible using physical models, if cellular- or intracellular-level distribution of GNPs can be accurately known.

10.5 Summary and Future Outlook

In this chapter, we have provided an overview of various approaches for the quantification of GNP-mediated radiation dose enhancement. Much of early quantification effort was given to the macroscopic estimation of GNP-mediated dose enhancement under different irradiation scenarios using computational methods. Strictly speaking, the applicability of macroscopic estimation should be limited to the situations closely matching the assumptions behind the so-called uniform mixture model (e.g., gel phantoms mixed with GNPs). Nevertheless, macroscopic estimation seems to be capable of providing at least first-hand accounts for the average tumor dose enhancement and associated biological consequences such as tumor shrinkage during *in vivo* studies with untargeted GNPs serving as contrast agents (resulting in the blood gold content on the order of ~1%). On the other hand, it does not take into account the microscopic actions (e.g., energy deposition) of secondary electrons from heterogeneously distributed GNPs for realistic cases. Consequently, many important findings from recent radiobiology experiments, especially *in vitro* studies, appear to be unaccountable by macroscopic approaches.

In recent years, significant research effort has been devoted to the microscopic estimation of GNP-mediated dose enhancement on a nano-/cellular scale. In spite of some difficulty in experimental validation, nanoscale computational results from recent investigations have provided better insight into GNP-mediated radiation dose enhancement at the scale relevant to biological processes under both passive and active targeting approaches. As a result, many of puzzling observations from recent *in vitro* studies with GNPs appear to be accountable, at least qualitatively, from a physics point of view. In principle, physical models might be able to reasonably correlate microscopically estimated dose enhancement with certain radiobiological outcomes seen *in vitro* (e.g., increase in the DSB formation), if cellular-/intracellular-level GNP distribution could be accurately identified. However, there are purely biological issues for which physical modeling effort becomes less meaningful. Moreover, physical models developed based on *in vitro* studies might be inapplicable to *in vivo* experiments, because the distribution of GNPs in tissue becomes significantly more heterogeneous—so does the dose enhancement, and many of the effects seen *in vitro* (e.g., internalization of GNPs) might be weakened or even absent *in vivo*. Nevertheless, physical models based on the microscopic estimation of GNP-mediated dose enhancement appear to provide a reasonable ground for a proper correlation between physical dose enhancement and radiobiological outcome.

Based on the findings from numerous investigations on GNP-mediated dose enhancement/radiosensitization, GNRT would likely be implemented either as CERT or cellular-targeted radiation therapy (CTRT) under passive/active targeting (see the last chapter of this book for more discussion). For the CERT implementation, the dose response model of GNRT would primarily be based on the macroscopic estimation of dose enhancement using an overall or average gold concentration level across the tumor based on predetermined GNP biodistribution data or *in vivo* assay methods applicable to humans. On the other hand, for the CTRT implementation, a more comprehensive, possibly semiempirical, model would have to be developed to properly project the treatment outcome from GNRT, taking into account both microscopic dose enhancement pattern specific to each targeting scenario (assuming animal data being extendable to humans) and important biological pathways.

Acknowledgments

This work was supported in part by NIH/NCI grant 1R01CA155446. The authors acknowledge Nivedh Manohar for help in creating some figures in this chapter.

References

Agostinelli, S. et al. 2003. GEANT4 — A simulation toolkit. *Nuclear Instruments and Methods in Physics Research A* 506:250–303.

Berger, M. J., J. H. Hubbell, S. M. Seltzer, J. Chang, J. S. Coursey, R. Sukumar, and D. S. Zucker. 2005. *XCOM: Photon Cross Section Database* (version 1.3). [Online] Available: http://physics.nist.gov/xcom [2008, September 30]. Gaithersburg, MD: National Institute of Standards and Technology.

Boudou, C., I. Troprès, J. Rousseau, L. Lamalle, J. F. Adam, F. Estève, and H. Elleaume. 2007. Polymer gel dosimetry for synchrotron stereotactic radiotherapy and iodine dose-enhancement measurements. *Physics and Medicine Biology* 52:4881–4892.

Butterworth, K. T., J. A. Wyer, M. Brennan-Fournet, C. J. Latimer, M. B. Shah, F. J. Currell, and D. G. Hirst. 2008. Variation of strand break yield for plasmid DNA irradiated with high-Z metal nanoparticles. *Radiation Research* 170:381–387.

Butterworth, K. T., J. A. Coulter, S. Jain et al. 2010. Evaluation of cytotoxicity and radiation enhancement using 1.9 nm gold particles: Potential application for cancer therapy. *Nanotechnology* 21(29):295101.

Carter, J. D., N. N. Cheng, Y. Qu, G.D. Suarez, and T. Guo. 2007. Nanoscale energy deposition by x-ray absorbing nanostructures. *J. Phys. Chem. B* 111:11622–11625.

Chang, M. Y., A. L. Shiau, Y. H. Chen, C. J. Chang, H. H. W. Chen, and C. L.Wu. 2008. Increased apoptotic potential and dose-enhancing effect of gold nanoparticles in combination with single-dose clinical electron beams on tumor-bearing mice. *Cancer Science* 99:1479–1484.

Chithrani, D. B., S. Jelveh, F. Jalali et al. 2010. Gold nanoparticles as radiation sensitizers in cancer therapy. *Radiation Research* 173(6):719–728.

Cho, S. H. 2005. Estimation of tumor dose enhancement due to gold nanoparticles during typical radiation treatments: A preliminary Monte Carlo study. *Physics in Medicine and Biology* 50(15):N163–N173.

Cho, S. H., O. N. Vassiliev, S. Jang, and S. Krishnan. 2006. Modification of megavoltage photon beams for gold nanoparticle-aided radiation therapy (GNRT): A Monte Carlo study. *Medical Physics* 33:2121.

Cho, S. H., O. N. Vassiliev, and S. Krishnan. 2007a. Microscopic estimation of tumor dose enhancement during gold nanoparticle-aided radiation therapy (GNRT) using diagnostic energy range x-rays. *Medical Physics* 34:2468.

Cho, S. H., O. N. Vassiliev, and J. L. Horton. 2007b. Comparison between an event-by-event Monte Carlo code, NOREC, and ETRAN for electron scaled point kernels between 20 keV and 1 MeV. *Radiation and Environmental Biophysics* 46:77–83.

Cho, S. H., B. L. Jones, and Krishnan, S. 2009. Dosimetric feasibility of gold nanoparticle-aided radiation therapy (GNRT) via brachytherapy using low energy gamma-/x-ray sources. *Physics in Medicine Biology* 54(16):4889–4905.

Dawson, P., M. Penhaligon, E. Smith, and J. Saunders. 1987. Iodinated contrast agents as radiosensitizers. *British Journal of Radiology* 60:201–203.

Diagaradjane, P., A. Shetty, J. Wang, J. Schwartz, A. M. Elliot, S. Shentu, H. C. Park, A. Deorukhkar, R. J. Stafford, S. H. Cho, J. W. Tunnell, J. D. Hazle, and S. Krishnan. 2008. Modulation of in vivo tumor radiation response via gold nanoshell mediated vascular-focused hyperthermia: Characterizing an integrated antihypoxic and localized vascular disrupting targeting strategy. *Nano Letters* 8(5):1492–1500.

El-Sayed, I., X. Huang, and M. El-Sayed. 2005. Surface plasmon resonance scattering and absorption of anti-EGFR antibody conjugated gold nanoparticles in cancer diagnostics: Applications in oral cancer. *Nano Letters* 5:829–834.

Fernandes, J. P., B. F. Pastorello, D. B. Araujo, and O. Baffa. 2008. Formaldehyde increases MAGIC gel dosimeter melting point and sensitivity. *Physics in Medicine and Biology* 53:N53–N58.

Foley, E. A., J. D. Carter, F. Shan, and T. Guo. 2005. Enhanced relaxation of nanoparticle-bound supercoiled DNA in x-ray radiation. *Chemical Communications* July 7(25):3192–3194.

Fong, P. M., D. C. Keil, M. D. Does, and J. C. Gore. 2001. Polymer gels for magnetic resonance imaging of radiation dose distributions at normal room atmosphere. *Physics in Medicine and Biology* 46:3105–3113.

Garnica-Garza, H. M. 2009. Contrast-enhanced radiotherapy: Feasibility and characteristics of the physical absorbed dose distribution for deep-seated tumors. *Physics in Medicine and Biology* 54(18):5411–5425.

Hainfeld, J., F. Dilmanian, Z. Zhong, D. Slatkin, J. Kalef-Ezra, and H. Smilowitz. 2010. Gold nanoparticles enhance the radiation therapy of a murine squamous cell carcinoma. *Physics in Medicine and Biology* 55:3045.

Hainfeld, J., M. O'connor, F. Dilmanian, D. Slatkin, D. Adams, and H. Smilowitz. 2011. Micro CT enables microlocalisation and quantification of Her2-targeted gold nanoparticles within tumour regions. *British Journal of Radiology* 84:526.

Hainfeld, J. F., D. N. Slatkin, and H. M. Smilowitz. 2004. The use of gold nanoparticles to enhance radiotherapy in mice. *Physics in Medicine and Biology* 49:N309–N315.

Hainfeld, J. F., F. A. Dilmanian, D. N. Slatkin, and H. M. Smilowitz. 2008. Radiotherapy enhancement with gold nanoparticles. *Journal of Pharmacy and Pharmacology* 60:977–985.

Herold, D. M., I. J. Das, C. C. Stobbe, R. V. Iyer, and J. D. Chapman. 2000. Gold microspheres: A selective technique for producing biologically effective dose enhancement. *International Journal of Radiation Biology* 76:1357–1364.

ICRU. 1989. Tissue substitutes in radiation dosimetry and measurement. *ICRU Report 44* (Oxford, UK).

Iwamoto, K. S., S. T. Cochran, J. Winter, E. Holburt, R. T. Higashida, and A. Norman. 1987. Radiation dose enhancement therapy with iodine in rabbit VX-2 brain tumors. *Radiotherphy and Oncology* 8:161–170.

Jain, S., J. A. Coulter, A. R. Hounsell et al. 2011. Cell-specific radiosensitization by gold nanoparticles at megavoltage radiation energies. *International Journal of Radiation Oncology Biology Physics* 79(2):531–539.

James, W. D., L. R. Hirsch, J. L. West, P. D. O'Neal, and J. D. Payne. 2007. Application of INAA to the build-up and clearance of gold nanoshells in clinical studies in mice. *Journal of Radioanalytical and Nuclear Chemistry* 271(2):455.

Jones, B. L. 2011. Development of dosimetry and imaging techniques for pre-clinical studies of gold nanoparticle-aided radiation therapy. *Ph.D. Dissertation, Georgia Institute of Technology*.

Jones, B. L., S. Krishnan, and S. H. Cho. 2010. Estimation of microscopic dose enhancement factor around gold nanoparticles by Monte Carlo Calculations. *Medical Physics* 37(7):3809–3816.

Kawrakow, I., and D. W. O. Rogers. 2003. The EGSnrc code system. *National Research Council Report PIRS-701* (Ottawa, Canada).

Kim, J.-K., S.-J. Seo, K.-H. Kim et al. 2010. Therapeutic application of metallic nanoparticles combined with particle-induced x-ray emission effect. *Nanotechnology* 21(42):425102.

Kong, T., J. Zeng, X. P. Wang, X. Y. Yang, J. Yang, S. McQuarrie, A. McEwan, W. Roa, J. Chen, and J. Z. Xing. 2008. Enhancement of radiation cytotoxicity in breast-cancer cells by localized attachment of gold nanoparticles. *Small* 4(9):1537–1543.

Krishnan, S., P. Diagaradjane, and S. H. Cho. 2010. Nanoparticle-mediated thermal therapy: Evolving strategies for prostate cancer therapy. *International Journal of Hyperthermia* 26(8):775–789.

Lechtman, E., N. Chattopadhyay, Z. Cai, S. Mashouf, R. Reilly, and J. P. Pignol. 2011. Implications on clinical scenario of gold nanoparticle radiosensitization in regards to photon energy, nanoparticle size, concentration and location. *Physics in Medicine and Biology* 56:4631–4647.

Leung, M. K. K., J. C. L. Chow, D. B. Chithrani et al. 2011. Irradiation of gold nanoparticles by x-rays: Monte Carlo simulation of dose enhancements and the spatial properties of the secondary electrons production. *Medical Physics* 38(2):624–631.

Loo, C., L. Hirsch, M. H. Lee et al. 2005. Gold nanoshell bioconjugates for molecular imaging in living cells. *Optics Letters* 30:1012–1014.

Maeda, H., J. Fang, T. Inutsuka, and Y. Kitamoto. 2003. Vascular permeability enhancement in solid tumor: Various factors, mechanisms involved and its implications. *International Immunopharmacology* 3(3):319–328.

McMahon, S. J., M. H. Mendenhall, S. Jain, and F. Currell. 2008. Radiotherapy in the presence of contrast agents: A general figure of merit and its application to gold nanoparticles. *Physics in Medicine and Biology* 53:5635–5651.

Mello, R. S., H. Callison, J. Winter, A. R. Kagan, and A. Norman. 1983. Radiation dose enhancement in tumors with iodine. *Medical Physics* 10:75–78.

Mesa, A. V., A. Norman, T. D. Solberg, J. J. DeMarco, and J. B. Smathers. 1999. Dose distribution using kilovoltage x-ray and dose enhancement from iodine contrast agents. *Physics in Medicine and Biology* 44:1955–1968.

Mukherjee, P., R. Bhattacharya, N. Bone et al. 2007. Potential therapeutic application of gold nanoparticles in B-chronic lymphocytic leukemia (BCLL): Enhancing apoptosis. *Journal of Nanobiotechnology* 5:4.

Ngwa, W., G. M. Makrigiorgos, and R. I. Berbeco. 2010. Applying gold nanoparticles as tumor-vascular disrupting agents during brachytherapy: Estimation of endothelial dose enhancement tumor dose enhancement using modified megavoltage photon beams and contrast media. *Physics in Medicine and Biology* 47:2433–2449.

Polf, J. C., L. F. Bronk, W. H. P. Driessen, W. Arap, R. Pasqualini, and M. Gillin. 2011. Enhanced relative biological effectiveness of proton radiotherapy in tumor cells with internalized gold nanoparticles. *Applied Physics Letters* 98(19):193702.

Qian, X., X.-H. Peng, D. O. Ansari, Q. Yin-Goen, G. Z. Chen, D. M. Shin, L. Yang, A. N. Young, M. D. Wang, and S. Nie. 2008. In vivo tumor targeting and spectroscopic detection with surface-enhanced Raman nanoparticle tags. *Nature Biotechnology* 26:83–90.

Regulla, D. F., L. B. Hieber, and M. Seidenbusch. 1998. Physical and biological interface dose effects in tissue due to x-ray-induced release of secondary radiation from metallic gold surfaces. *Radiation Research* 150:92–100.

Robar, J. L. 2006. Generation and modeling of megavoltage photon beams for contrast-enhanced radiation therapy. *Physics in Medicine and Biology* 51:5487–5504.

Robar, J. L., S. A. Riccio, and M. A. Martin. 2002. Tumor dose enhancement using modified megavoltage photon beams and contrast media. *Physics in Medicine and Biology* 47:2433–2449.

Roeske, J. C., L. Nunez, M. Hoggarth, E. Labay, and R. R. Weichselbaum. 2007. Characterization of the theoretical radiation dose enhancement from nanoparticles. *Technology in Cancer Research and Treatment* 6:395–402.

Rogers, D. W. O., C.-M. Ma, G. X. Ding, B. R. Walters, D. Sheikh-Bagheri, and G. G. Zhang. 2002. BEAMnrc users manual. *National Research Council Report PIRS-0509(A) rev. G* (Ottawa, Canada).

Rose, J. H., A. Norman, and M. Ingram. 1994. First experience with radiation therapy of small brain tumors delivered by a computerized tomography scanner. *International Journal of Radiation Oncology, Biology, Physics* 30:24–25.

Sahare, P. D. et al. 2007. K3Na(SO4)(2): Eu nanoparticles for high dose of ionizing radiation *Journal of Physics D-Applied Physics* 40(3):759–764.

Salvat, F., J. M. Fernandez-Varea, and J. Sempau. 2003. PENELOPE — A Code System for Monte Carlo Simulation of Electron and Photon Transport. OECD Nuclear Energy Agency, Issy-les-Moulineaux, France.

Semenenko, V. A., J. E. Turner, and T. B. Borak. 2003. NOREC, a Monte Carlo code for simulating electron tracks in liquid water. *Radiation and Environmental Biophysics* 42(3):213–217.

Siddiqi, A., Y. Yang, K. Dextraze, T. Hu, S. Krishnan, and S. Cho. 2009. Experimental demonstration of dose enhancement due to gold nanoparticles and kilovoltage x-rays using radio-sensitive polymer gel dosimeter. *Medical Physics* 36:2511.

Sokolov, K., M. Follen, J. Aaron, I. Pavlova, A. Malpica, R. Lotan, and R. Richards-Kortum. 2003. Real-time vital optical imaging of precancer using anti-epidermal growth factor receptor antibodies conjugated to gold nanoparticles. *Cancer Research* 63:1999–2004.

Unezaki, S., K. Maruyama, J.-I. Hosoda, I. Nagae, Y. Koyanagi, M. Nakata, O. Ishida, M. Iwatsuru, and S. Tsuchiya. 1996. Direct measurement of the extravasation of polyethyl-eneglycol-coated liposomes into solid tumor tissue by in vivo fluorescence microscopy. *International Journal of Pharmaceutics* 144:11–17.

Van den Heuvel, F., J.-P. Locquet, and S. Nuyts. 2010. Beam energy considerations for gold nano-particle enhanced radiation treatment. *Physics in Medicine and Biology* 55(16):4509–4520.

Verhaegen, F., B. Reniers, F. Deblois, S. Devic, J. Seuntjens, and D. Hristov. 2005. Dosimetric and microdosimetric study of contrast-enhanced radiotherapy with kilovolt x-rays. *Physics in Medicine and Biology* 50:3555–3569.

Walters, B. R., I. Kawrakow, and D. W. O. Rogers. 2006. *DOSXYZnrc users manual. National Research Council Report PIRS-794revB* (Ottawa, Canada).

X-5 Monte Carlo Team. 2003. MCNP – A general Monte Carlo N-particle transport code, Version 5. *LA-UR-03-1987* (Los Alamos, NM).

Zaman, R. T., P. Diagaradjane, J. Wang, J. Schwartz, N. Rajaram, K. L. Gill-Sharp, S. H. Cho, H. G. Rylander, III, D. Payne, S. Krishnan, and J. W. Tunnell. 2007. In vivo detection of gold nanoshells in tumors using diffuse optical spectroscopy. *IEEE Journal of Selective Topics in Quantum Electronics* 13(6):1715–1720.

Zhang, X. J., J. Z. Xing, J. Chen, L. Ko, J. Amanie, S. Gulavita, N. Pervez, D. Yee, R. Moore, and W. Roa. 2008. Enhanced radiation sensitivity in prostate cancer by gold-nanoparticles. *Clinical and Investigative Medicine* 31:E160–E167.

Zheng, Y., D. J. Hunting, P. Ayotte, and L. Sanche. 2008. Radiosensitization of DNA by gold nanoparticles irradiated with high-energy electrons. *Radiation Research* 169:19–27.

Semiconductor Nanomaterials for Radiotherapy

Ke Sheng
UCLA School of Medicine

11.1 Radiation Therapy and Need for Radiosensitizers

There has been a long history of successfully using ionizing radiation for cancer therapy. Although much of recent research interest has shifted to more tumor-specific, molecular-targeted treatments, radiation remains as one of the most important forces in cancer management. It has an irreplaceable role in noninvasively killing, debulking, and controlling malignant tumor cells. Because it is unspecific, ionizing radiation is a double-edged sword. It is toxic to both tumor and normal tissue. Unlike chemotherapy, the effectiveness of radiotherapy does not substantially diminish with evolving tumor biology and genetic mutation. High dose of radiation will overcome tumor resistance at a cost of surrounding normal tissue toxicity. In practice, x-rays can penetrate tissues and reach deep-seated tumors for therapy while depositing radiation doses along the beam path. With intensity-modulated radiotherapy, a homogeneous high dose region can be created conforming to the tumor shape with relatively steep dose gradient in the transition region to surrounding normal tissue. Figure 11.1 shows the dose distribution from one of the advanced dose delivery systems, helical tomotherapy (Mackie et al. 1999). In this case, the lesion is covered by a nearly uniformly high dose with excellent spinal cord sparing. The steeper dose gradient compared with conventional

radiotherapy requires a higher setup accuracy, which has been facilitated by image-guided radiotherapy in the treatment room. Because of these improvements, significantly higher tumor dose, and resultant improvement in tumor control probability, has been achieved in numerous dose escalation studies without exceeding patient tolerance to treatment (Hiraoka et al. 2007; Timmerman et al. 2007; Molinelli et al. 2008; Zelefsky et al. 2008). The robustness, low cost, geometrical accuracy, and precise dosimetry are unique to radiotherapy.

On the other hand, advanced treatment planning and delivery cannot change the physics of x-ray transportation. Improvements in the therapeutic ratio have been attributed to redistribution of x-ray doses to less critical and more radioresistant tissues, such as muscle and fat while avoiding functionally important and sensitive organs such as the spinal cord and parotids. Calculation on integral doses versus treatment modality showed that although varying organs-at-risk (OARs) sparing can be achieved by increasing the number of beams and intensity modulation, the integral doses are nearly constant (D'Souza and Rosen 2003; Reese et al. 2009). When the tumor is encompassed by OARs or abutting OARs, dose constraints on tumor and OAR doses become difficult to meet at the same time, and compromises have to be made. This fundamentally limits radiation therapy effectiveness for radioresistant or large tumors. Although particle therapy can be used to improve dosimetry,

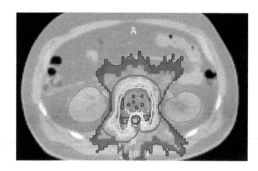

FIGURE 11.1 Radiation dose highly conforming to a spinal lesion form helical tomotherapy but low dose spillage to surrounding normal tissue is still unavoidable.

because of cost constraints, it will not be an option for most patients in the near future. Adjuvant therapy selectively increasing tumor toxicity may shed light on the dilemma.

Chemical radiosensitizers have been developed to increase tumor cell sensitivity to radiation via various biological targets. For example, electrophilic chemicals have been used to reduce hypoxia-associated radioresistance (Adams 1973; Fowler et al. 1976). Tirapazamin is more toxic in a hypoxic environment so it has been used to treat more radioresistant tumor cells (Brown and Wilson 2004). Pyrimidines substituted with bromine or iodine have been incorporated into DNA and enhanced free radical damage (Poggi et al. 2001). Drugs involved in DNA repair have also been evaluated, with mixed results (Eberhardt et al. 2006; Hao et al. 2006). Proteins involved in cell signaling, such as the Ras family, are attractive targets linked to radioresistance (Chinnaiyan et al. 2006; Choudhury et al. 2006). The suppression of radioprotective thiols has also been investigated (Minchinton et al. 1984). Although these applications have shown promise in one or more areas, they are generally toxic to normal tissues, further reducing their tolerance to radiation; they often act via uncertain mechanisms, and sometimes rely on a modulating cellular target, which can change over the time. It has been concluded that clinical gains from these chemical radiosensitizers have been marginal (Tannock 1996; Wardman 2007). Despite differences in pathways, these chemical radiosensitizers share the similarity that the synergistic effect takes place at the cellular level. The sensitizer either renders the cells more vulnerable to ionizing radiation damage or interferes with the repairing process of radiation-damaged cells. The lack of direct interaction between the drug and radiation has led to often unpredictable and unreliable outcomes because of the enormously complex tumor cell biology. There has been research effort to fabricate a very different sensitizer that simply intercepts more radiation energy locally and amplifies the damage to tumor cells. Although high atomic number (Z) nanoparticles used to increase radiation absorption are generally not semiconductor nanoparticles, the energy transfer pathway is relevant to the semiconductor nanoparticle–photosensitizer pair used in combined radiation–photodynamic therapy, which will be introduced in Section 11.3. Therefore, to provide a complete

picture of the physical radiosensitizer, it is essential to include these metal nanoparticles in this chapter. For the same reason, metal oxide nanoparticles are briefly introduced in Section 11.4 as a radiation protector.

11.2 Physical Radiation Enhancer

11.2.1 Concept of Physical Radiation Enhancer

In order to increase x-ray doses for increased tumor cell killing, the physical radiation enhancer must have increased cross section with x-ray photons. X-rays interact with matter and lose energy by one or more of the following mechanisms: the photoelectric effect, Compton scatter, and pair production. Lower energy x-rays react with high-Z materials more efficiently through the photoelectric effect, which allows for higher absorption of energy proportional to the cube of the atomic number (Attix 1986). At the K-edge of gold, the relative absorption coefficient of gold is approximately 1217 ($79^3/7.4^3$) times that of normal tissue with an average atomic number of 7.4. At medium to high energy (100 keV–10 MeV), x-rays lose energy primarily through Compton scatter, whose cross section is proportional to the electron density of the medium and nearly independent of the atomic number. Therefore, the relative absorption coefficient for an electron-dense material such as gold (density 19.3 g/cm^3) is roughly 20 times that of normal human tissue. At higher energy, pair production becomes more important. In pair production, a paired positron and electron is produced by converting photon kinetic energy to rest mass. The mass attenuation coefficient from pair production is proportional to Z^2, but the advantage of high-Z material is not significant until the photon energy is higher than 10 MeV. Because of the shallow penetration, low-energy kV x-rays with prominent photoelectric components are rarely used to treat tumor today. In practice, for deeply seated tumors, high-energy x-rays (6 MV or higher) and isotopes such as Co-60 emitting gamma rays of energy 1.25 MeV are exclusively used in external beam therapy. On the other hand, when the radiation source can be placed near the tumor in brachytherapy (Latin for contact therapy), isotopes emitting lower energy gamma rays are used for superior normal tissue sparing. A commonly used isotope is I-125 emitting 35.5-keV gamma rays, which delivers a very high dose to a distance within several millimeters of the source but the dose drops quickly beyond this range for normal tissue sparing. Clearly, same tissue concentrations of high-Z materials will achieve different dose enhancement for different x-ray energies. For example, a 0.1% mass concentration of gold in the tumor can double the radiation dose for x-rays with energy at the K-edge of gold, but will not have appreciable effects with MV x-rays.

Before proceeding to the following sections, it is necessary to clarify the beam properties and common terminology associated with x-rays generated by x-ray tubes, an accelerator, and gamma ray from isotopes. Different from the nominal energy of gamma rays from radioisotopes, photons with nominal energy of 100 kVp are produced by electrons accelerated to 100 keV before hitting an anode, where they are decelerated and bremsstrahlung x-ray photons are produced. These bremsstrahlung photons are not monoenergetic but rather have a broad spectrum (polychromatic) with 100 keV as the maximum (peak) energy. The x-ray spectrum is determined by thickness and atomic composition of the anode. The bremsstrahlung photons are normally filtered before leaving the x-ray head to harden the x-rays for more penetrating beams but still leaving a large numbers of photons at lower energy interacting with the matter by the photoelectric effect. Similar to the kV x-ray units, MV x-rays from a linear accelerator are polyenergetic. Monte Carlo simulation and beam commissioning are needed to accurately characterize the beam quality of a megavoltage linear accelerator. As a rule of thumb, the average energy of the x-rays from a modern megavoltage clinical accelerator is approximately one-third of the peak energy. Historically, various units have been used to describe x-ray energies. keV and MeV are used to describe the average energy (for polychromatic emission) of x-rays, the exact emission peak (for monochromatic emission) of gamma rays, or monochromatic electron beams. kVp is used to denote the peak energy of kV x-rays produced by electrons hitting anodes. Similar to kVp, MV refers to the peak energy of megavoltage x-rays produced by linear accelerators.

The potential of using a large cross section of kV x-rays with high-Z materials has been exploited in various cell and animal apparatuses for radiation sensitization. Iodine ($Z = 53$) was one of the first elements tested *in vitro* (Santos Mello et al. 1983; Matsudaira et al. 1980), because it can be easily integrated into DNA with agents such as iododeoxyuridine (IUdR). A 3-fold increase in tumor cell killing was observed with radiation after incubation in IUdR solution (Nath et al. 1990). In animal experiments, increased tumor growth delay was observed (Santos Mello et al. 1983; Iwamoto et al. 1987) with iodine contrast medium and 100 kVp x-rays. Improved tumor-bearing dog survival (53%) was observed with the CT (computed tomography) iodine contrast media and orthovoltage (140 kVp) x-rays delivered by a CT scanner (Norman et al. 1997). Instead of typical open beam geometry, the scanner was modified by a collimator to deliver conformal radiation to the tumor and spare the normal tissue surrounding it. Following the animal experiments, a phase I clinical trial was conducted using iodine contrast and the modified CT scanner on eight human patients, each with multiple metastatic brain tumors. In addition to the total brain radiation of 40 Gy, after contrast injection, for the same patient, one of the metastatic tumors was treated by kV x-rays for 15–25 Gy, and the other was spared from the additional irradiation. Two tumors that received the additional dose showed complete response, but no statistical conclusion could be drawn because of the small number of subjects (Rose et al. 1999). One disadvantage associated with iodine as a physical radio-enhancer is that the highest absorbance energy for iodine right above its K-edge of 33.2 keV is not penetrating enough for most deep-seated tumor treatment and only consists of a small percentage of the 140-kVp x-ray. For more practical higher energy x-rays, to achieve high dose enhancement effects *in vivo*, the percentage

of thymine that has to be replaced by iodouracil is prohibitively high (Nath et al. 1990).

Compared with iodine, gold (Au; $Z = 79$) has a higher and more desirable K-edge at 80.7 keV, which is also farther away from the K-edge of the bones and tissue. Gold was first applied in the form of gold foil and showed the ability to enhance cell killing by approximately 100-fold (Regulla et al. 1998). Micron-sized gold particles directly injected into tumors before radiation also reduced the cells' viability (Herold et al. 2000).

11.2.2 Enhanced Radiation Therapy by Gold Nanoparticles

With gold, it is difficult to apply bulk materials such as foil and to achieve uniform dose enhancement, since the range of dose enhancement with the foil is on the order of 50 µm (Regulla et al. 1998). Gold microspheres were not able to infiltrate densely packed tumor cells (Herold et al. 2000). To overcome the difficulties in the enhancement of radiation therapy by gold materials, gold nanoparticles have emerged as an attractive solution. The *in vitro* efficacy was shown by Kong et al. (2008), who compared cell survival after treatment with kV radiation only or kV radiation in the presence of gold nanoparticles. Cells containing gold nanoparticles survived significantly less, and the dark toxicity of the gold nanoparticles was found negligible. Although the degree of radiosensitization is arguably the same as micron-sized gold particles for a given concentration, gold nanoparticles are more versatile and biocompatible. It has been shown that gold particles less than 2 nm (and without surface modification) can effectively evade the immune system and liver retention. They can also exploit the leaky nature of the tumor vascular structure to achieve tumor/liver concentration ratios of 1.6 (Hainfeld et al. 2004). In the study conducted by Hainfeld et al. (2004), 0.01 mL/g of 1.9 ± 0.1 nm gold nanoparticles were injected into the tail vein of mice. Xenograft tumor uptake of gold nanoparticles was observed shortly thereafter. These mice were subsequently irradiated by 250-kVp x-rays and compared with mice without gold nanoparticles. In the group with radiation only, the long-term survival (1 year) was 20%. In the groups irradiated with lower (135 mg Au/kg) and higher (270 mg Au/kg) gold nanoparticle loads, the long-term survival was 50% and 86%, respectively. The 250-kVp x-rays are traditionally referred to as orthovoltage x-ray, which had been widely used to treat human patients, but its utilization virtually has been replaced by MV x-rays with a few exceptions in surface and intraoperative applications (Bachireddy et al. 2010). This study showed that high-Z nanoparticles were able to effectively enhance radiation therapy without significant side effects. Without surface modification and tumor targeting, gold nanoparticle concentrations were slightly higher in tumors than in the liver simply because of size selection. The concentration in tumors is still lower than that in the kidney and blood and similar to that of other tissues, which would have received a higher dose correspondingly because of the interaction of gold with kV x-rays. Therefore, without high specificity to the tumors, the overall effects are arguably equivalent to radiation dose escalation, which will likely achieve similar improvement in animal survival. For physical radio-enhancers, the only way to improve therapeutic ratio is by delivering a higher concentration of particles to the tumor but not to the surrounding normal tissue.

Achieving higher tumor specificity requires surface modifications that increase serum half-life, affinity to a tumor hosting environment, and specific binding to receptors on the tumor cells. Kong et al. showed that gold nanoparticles coated by glucose were selectively internalized by breast cancer cells and that the selectivity could be fine-tuned by modification of the surface charge (Kong et al. 2008). Li et al. (2009) demonstrated four times higher tumor cell uptake of gold nanoparticles functionalized by transferrin compared with normal cells. Similarly, increased uptake was observed on prostate cancer cells as well, but the cell killing enhancement was not proportional to the gold nanoparticle loading, indicating a saturation mechanism in the effectiveness (Zhang et al. 2008). In addition to nonspecific coating molecules, gold nanoparticles have also been conjugated with peptides (Porta et al. 2007; Surujpaul et al. 2008) and antibodies (Pissuwan et al. 2007) for more specific tumor cell targeting.

The advantages of nanoparticle physical radiosensitizers are evident. The interaction between radiation and the high Z nanoparticles is well characterized. On the other hand, it is not without drawbacks in its current form. Most cancer patients today are treated with megavoltage x-rays, which can penetrate deeper into the tissue for skin sparing and higher dose conformity. The efficacy of gold nanoparticles under these conditions is modest owing to the lack of photoelectric interaction. Dramatically increasing therapeutic ratio depends on the specificity of tumor targeting, by itself a daunting task owing to many physiological barriers in penetrating a solid tumor; the very high loading required for radiosensitization (0.5–5%) may saturate cell uptake. Clearly, controllability of the physical radio-enhancer is desired but to dramatically improve its efficacy, increased tumor targeting alone is insufficient. A novel mechanism using the radiation for new cell killing pathways is desired. There has been an increasing interest in simultaneously delivering photodynamic therapy (PDT) with radiation therapy using semiconductor nanoparticles as the energy mediators.

11.3 Radiation Therapy in Combination with PDT Using Semiconductor Nanoparticles as Energy Mediator

11.3.1 Photodynamic Therapy

Radiation therapy and photodynamic therapy (PDT) are similar in many ways. Both obtain energy from an external irradiation. Both cause damage to tumor cells indirectly by ways of secondary molecules such as free radical species or singlet oxygen molecules. Despite these similarities, PDT and radiation therapy are fundamentally different. In PDT, a separate drug referred to as the photosensitizer is needed. The photosensitizer is excited by light. Through intersystem crossing, the excited state becomes

a metastable triplet that can exist for a few microseconds. Photosensitizers in the triplet state react with molecules in the environment and release energy by type I and type II mechanisms. In the type I reaction, through hydrogen-atom abstraction or electron transfer, free radical species such as the superoxide radical anion are generated. In the type II reaction, which is considered the primary reaction in PDT treatment, the triplet state photosensitizer reacts directly with the ground state triplet molecular $^3O_2$ to generate excited singlet $^1O_2$, which is highly reactive and toxic to cell membranes, lysosomes, and mitochondria.

Besides its non-oncologic applications, PDT is used to treat many types of cancer, including skin, head and neck, esophagus, and bladder cancer. When applicable, PDT is effective and potent, with few long-term side effects. However, one major limitation of PDT is that the light required for activation has shallow penetration. For example, the wavelength of the activating light for a Food and Drug Administration–approved photosensitizer, Photofrin, is 620 nm, which has an attenuation coefficient of approximately 1 mm^{-1} in tissue and thus an effective treatment depth of 5 mm before the light intensity drops to less than 1% of the surface intensity. New classes of photosensitizers, such as phthalocyanines (Pcs), were developed to utilize longer wavelengths for activation. With activation in the near-infrared band, the treatment depth can be practically increased from less than 1 cm to several centimeters (Moan and Anholt 1990). To treat deeply seated solid tumors, optic fibers have to be inserted into patients through orifices or incisions, significantly adding challenges to the procedure. Consequently, PDT to nonsuperficial sites is limited to a very small number of institutes specialized in such procedures. In addition to shallow penetration, accurate modeling of the light dosimetry is difficult. The estimate varies with scatter, reflective light, and distribution of oxyhemoglobin, which is a strong absorber of the red light. A combination of crude light dosimetry and variable photosensitizer tissue concentrations has rendered PDT dosimetry more empirical than quantitative. In contrast, the accuracy of 3-D radiation dosimetry is on the order of 2–3%.

It has been an attractive idea combining photosensitizers with radiotherapy and using the highly quantifiable and penetrating of x-rays for excitation (Schaffer et al. 2002; Kulka et al. 2003; Schaffer et al. 2003, 2005; Luksiene et al. 2006a, 2006b). Moderate radiosensitization was observed on several aggressive mouse and human cell lines, both *in vitro* and *in vivo*. The mechanism has not been completely understood, since in theory, photosensitizers such as porphyrins used in these studies have narrow absorption spectra and cannot be excited by x-rays directly to generate singlet oxygen. A theory was proposed and tested by Luksiene et al. (2006a) that ligands of peripheral benzodiazepine receptors, which are overexpressed in aggressive tumor cells, might diminish the cell growth. Dicarboxylic porphyrins are the ligands for such receptors. The hypothesis was supported by the observation that, in these experiments, the effect of the photosensitizers was primarily antiproliferatory rather than causing apoptosis, a more common effect of singlet oxygen. Therefore, the working mechanism of photosensitizers as radiosensitizers is independent of x-rays. To use more penetrating x-rays, however, an energy mediator is needed. The energy mediator needs to satisfy the following properties: (1) It has a high extinction coefficient to high energy x-rays. In other words, materials with higher density and atomic number will intercept more radiation energies. (2) It has a high quantum yield converting absorbed energy to energy quanta matching the absorption peak of the photosensitizer. (3) Its size is small enough to infiltrate cell membranes and preferably, evade the immune system. (4) Its surface is highly modifiable to conjugate with photosensitizers and other moieties for both targeted therapy and imaging. The search quickly narrowed down to fluorescent semiconductor nanoparticles, the most well known of which is the quantum dot (QD).

11.3.2 Quantum Dots

The band gap of a bulk semiconductor material is determined by its chemical composition. However, when the dimension of the semiconductor material is reduced to 1–5 nm, the Bohr radius, the quantum confinement effect emerges and determines its fluorescent characteristics. The nanoscaled semiconductor materials, named quantum dots, behave like a single atom with discrete energy states. When excited, the QDs will return to the ground state by emitting photons with characteristic wavelengths. The energy levels can be solved by the Schrödinger equation. Assuming spherical symmetry, the energy levels of a QD are expressed by

$$E_g(\text{QD}) = E_{g,0} + \frac{h^2 \alpha_{n,l}}{8\pi^2 m_{eh} R^2}, \qquad (11.1)$$

where $\alpha_{n,l}$ are the energy states ($n = 1, 2, 3\dots, l = $ s, p, d, …) similar to a single atom, h is Planck's constant, $E_{g,0}$ is the band gap of the bulk material, R is the radius of the dot, and $m_{eh} = m_e \cdot m_h$ is the effective mass of an electron–hole pair (exciton). It is clear that the energy level is proportional to the inverse square of the radius. Therefore, by controlling the size of the particle during chemical synthesis, a full spectrum of visible light can be obtained. Quantum confinement was first described in theory (Bryant 1988; Stucky and MacDougall 1990; Norris 1994) and experimentally investigated as the QD by Alivosatos (1996) and Bawendi et al. (1990). In addition to the high quantum yield, QDs are not subject to photobleaching that diminishes fluorescent lights. Most importantly, for their role in combined radiotherapy and photodynamic therapy, QDs can be excited by a broad range of photon energies; Figure 11.2 shows the fluorescent light from CdSe QDs when irradiated by 6 MV x-ray from a clinical accelerator. Therefore, they are ideal energy mediators for the combination of radiation with photodynamic therapy.

There are several different methods to synthesize QDs or more generally, semiconductor nanocrystals. With the development of material science, these methods are also evolving for better-quality particles at a lower manufacturing cost. A representative method was introduced by Peng et al. (2000). To prepare cadmium selenide (CdSe) QDs, dimethylcadmium and selenium powder are dissolved in a tri-alkyl phosphine (-butyl or -octyl) before injection into hot (340–360°C) trioctyl phosphine

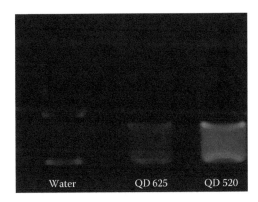

FIGURE 11.2 (See color insert.) Fluorescent light from CdSe QD excited by 6 MV x-rays from a clinical accelerator. Image shows two emission wavelengths.

oxide (TOPO). Nucleation begins shortly after and followed by growth at a slightly lower temperature. Kinetic control is used to select the average particle size and size distribution because the smaller particles grow faster than the larger ones.

11.3.3 Combination of Photodynamic Therapy and Radiotherapy Using QDs

11.3.3.1 Background

Independent of radiotherapy, nanoparticles were initially tested as delivery vehicles for photosensitizers. Most photosensitizers, including porphyrins and Pcs, are not highly water soluble and tend to aggregate in tissue and impair the efficiency of their photochemical activities. Gold nanoparticles coated with Zn-Pc were synthesized as a more efficient hydrophilic PDT delivery system. Biodegradable liposome nanoparticles were also used to facilitate the transportation of photosensitizer molecules to tumor sites (Konan et al. 2002). In addition to delivery vehicles, it has been found that the energy transfer efficiency can be improved when photosensitizers are conjugated to fluorescent semiconductor nanoparticles. Samia et al. (2003) first demonstrated that CdSe QDs can be used to mediate energy from UVA light to a PDT agent via a Förster resonance energy transfer (FRET) mechanism. QDs can be used to excite conjugated Pcs (Samia et al. 2003), but the conjugates were not water-soluble and the quantum yield was also very low (~5%), rendering it impractical for biological application. Shi et al. (2006) synthesized QDs overcoated by phytochelatin-related peptides for improved water solubility. Tsay et al. (2007) covalently bound QD with a similar surface coating to a photosensitizer, Rose Bengal. As a result, not only were excellent colloidal properties observed, but 3–4 times higher singlet oxygen yield from the QD/photosensitizer conjugate was also reported compared with the photosensitizer alone. Based on this platform, the quantum yield can be potentially improved by shortening the link between the QD and photosensitizer to increase the FRET efficiency and increasing the number of photosensitizer molecules on each QD. On the other hand, visible or UV lights were used in these experiments, and the fundamental limitation of PDT treatment depth was not addressed.

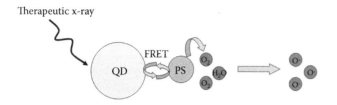

FIGURE 11.3 Energy transfer from x-ray to singlet oxygen mediated by QD and photosensitizer.

To utilize more tissue penetrating near infrared light, much effort was made to convert near-infrared light to visible photons that can be used to excite the conjugated photosensitizer. There are two known mechanisms to convert the energy from two photons with lower energy (longer wavelength) into a photon with higher energy (shorter wavelength). Simultaneous two-photon absorption requires a single nonlinear optical conversion with combined energy sufficient to induce the transition from the ground state to an excited electronic state. This conversion relies on sequential discrete absorption and luminescence steps where at least two metastable energy states are involved, the first serving as a temporary excitation reservoir. The energy in the reservoir is later combined with the second photon, and second higher excitation state can be reached. A higher energy photon can then be emitted from this state. In the first mechanism, a virtual intermediate state is involved from the quantum mechanical view. The two excitation photons have to be coherent. It is only achievable by a laser source with extremely fine temporal resolution (10^{-15} s). In the second mechanism, because of the intermediate metastable state, the demand for temporal resolution from the excitation source is lower; therefore, the efficiency is higher.

The opposite of up-conversion is a more straightforward and efficient process that converts photons with higher energy to the visible range for photosensitizer excitation. The idea was first explored by Samia et al. (2003) using QDs as energy mediators to more efficiently excite conjugated photosensitizers with UV light. UV light is not more penetrating than visible light, but this energy pathway is important in the combination of radiation therapy and PDT.

With its obvious photoluminescence ability and wide absorption spectrum, it is natural to consider the QD as a potential candidate to transfer its energy to chemically bonded photosensitizers. To utilize more penetrating x-ray photons as the excitation source, the innovative idea of combining radiation therapy and PDT to excite nanoparticles in the tissue at depth was first proposed by Chen and Zhang (2006). It was demonstrated that excitation by kV x-rays induced fluorescence and phosphorescence in LaF_3:Eu nanoparticles. Different from classical QDs, the fluorescent emission from LaF_3:Eu is a result of electric–dipole transition rather than quantum confinement (Pi et al. 2005), but the overall energy transfer pathway is similar. Using QDs, Yang et al. (2007, 2008) demonstrated the energy transfer from MV to photosensitizers in a feasibility study.

The energy transfer pathway depicted in Figure 11.3 shows that the QD is first excited by therapeutic x-rays, and its energy is then transferred to the chemically conjugated photosensitizer by a mechanism known as FRET. The excited photosensitizer then releases energy by type I and type II reactions to generate free radicals or singlet oxygen. The conjugation chemistry is shown in Figure 11.4. The carboxylic acid group on the Photofrin is activated by 1-ethyl-3-(3-dimethylaminopropyl)-carbodiimide and then reacts with the amine group on the QD to form a covalent bond. Figure 11.5 shows that the fluorescence emission of

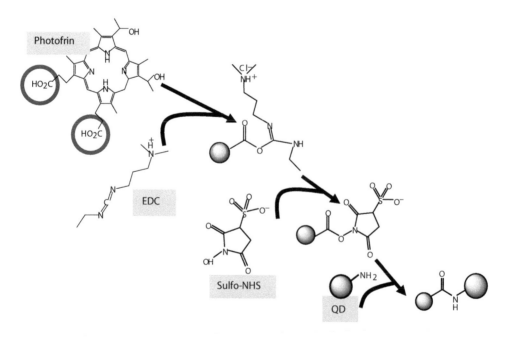

FIGURE 11.4 Conjugation chemistry of QD/photosensitizer synthesis.

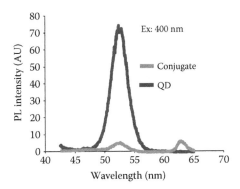

FIGURE 11.5 Photoluminescence from QD and QD/Photofrin conjugates.

QDs is quenched in the conjugate. Instead, energy from the QD is transferred to the Photofrin, which emits characteristic photons with a wavelength of 630 nm.

11.3.3.2 Energy Transfer Calculation and Measurement

FRET is an important pathway for the transfer of energy through dipole–dipole coupling of two conjugated molecules without fluorescence (Andrews 1989). The FRET efficiency (η) can be calculated by the separation distance between the QD (donor) and the photosensitizer molecule (acceptor) using (Andrews 1989; Lakowicz 1999):

$$\eta = \frac{1}{1 + \left(\dfrac{r}{R_0}\right)^6}. \tag{11.2}$$

where r is the actual separation distance and R_0 is the Förster distance. It is obvious that the energy transfer efficiency is 0.5 when the actual separation is the same as the Förster distance, which can be calculated as (Lakowicz 2006)

$$R_0 = (BQ_D I)^{1/6}, \tag{11.3}$$

where Q_D is the quantum yield of the donor, I is the spectral overlap between the QD and Photofrin, and B is a constant that can be expressed as

$$B = \frac{\left[9000\ln(10)\right]k_p^2}{128\pi^5 n_D^4 N_A}. \tag{11.4}$$

In this equation, n_D is the refractive index of the medium, k_p^2 is the orientation factor and varies from 0 (when the dipoles of the donor and the acceptor are perpendicular) to 4 (when they are parallel), and N_A is Avogadro's constant. Equation 11.2 has been traditionally used to describe the energy transfer between two molecules with smaller size than a semiconductor nanocrystal, so initially it was not clear whether it could be applied

to the QD/small organic molecule conjugate, because the size of a PEG-coated QD with a core–shell structure is several orders of magnitude larger than the small molecule conjugated to it. Pons et al. (2007) conducted an elegant experiment using self-assembling CdSe–ZnS core–shell QDs decorated with a series of Cy3-labeled beta-strand peptides of increasing length. The bridging peptides are rigid and have fixed lengths, as confirmed by electron microscopy. Using this system, the FRET efficiency as a function of the separation distance was determined, and a very good agreement between Equation 11.2 and the experimental measurement was observed.

Because of the large gradient introduced by the sixth-order term, the uncertainties in estimating r, and the possibility of multiple conjugation points between the QD and photosensitizer molecules, the efficiency is more commonly determined by quenching experiments. QDs have limited channels through which to release their energy upon excitation. In this case, the energy has to be released by either photon emission, FRET, or other channels such as singlet oxygen emission through the triplet state of the QD. It was demonstrated that the last pathway constitutes less than 5% of the total energy release (Samia et al. 2003), leaving the first two competing against each other for the remaining 95%. Therefore, the efficiency can be expressed as (Biju et al. 2006; Pons et al. 2007):

$$\eta = 1 - \frac{I_{conj}}{I_{QD}}, \tag{11.5}$$

which compares the emission of photons of the QD simply mixed with the acceptor molecule without conjugation (I_{QD}) to the emission of the QD when conjugated (I_{conj}). The percentage of energy transferred to the acceptor by FRET can thus be determined. Using energy quenching, high FRET efficiencies between QDs and conjugated Photofrin were reported by several groups ranging between 58% (Idowu et al. 2008) and 77% (Samia et al. 2003).

FRET efficiency is also influenced by the number of acceptors conjugated to the donor. Because QDs have a large surface area that is usually functionalized with multiple binding sites, more than one acceptor can be attached to a single donor. It was reported that n multiple bound acceptors can increase the FRET efficiency according to the following equation (Sapsford et al. 2007),

$$\eta = \frac{nR_0^6}{nR_0^6 + r^6} \tag{11.6}$$

Yang et al. (2008) showed that the FRET efficiency increased with the number of Photofrins conjugated to the surface of the QD, which was terminated by multiple amine groups (Figure 11.6). When the number of Photofrins per QD in the conjugation chemistry increased to 20, the FRET efficiency approached 100% (Yang et al. 2007), following the curve shown in Figure 11.7.

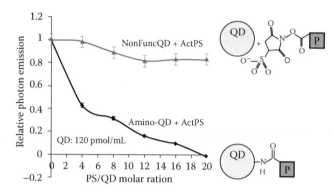

FIGURE 11.6   Quenching as a function of the Photofrin/QD ratio.

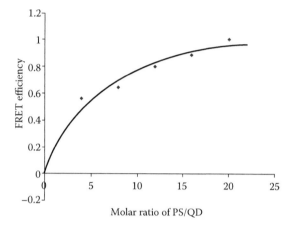

FIGURE 11.7   FRET efficiency vs. the average number of Photofrin on a single QD.

It was also pointed out (Sapsford et al. 2007) that Equation 11.5 is valid only when the number of acceptors bound to a QD is uniform; in practice, the heterogeneity in conjugate valence, that is, the acceptor/donor ratio, can vary significantly. In such cases, the FRET efficiency for multiple acceptors bound to the QD is more precisely described by a Poisson distribution as follows:

$$\eta(N) = \sum_n^N \frac{e^{-N}N^m}{n!}\eta(n), \qquad (11.7)$$

where N is the nominal valence and n is the actual number of acceptors bound to a donor.

With the FRET efficiency determined experimentally, it is possible to estimate the singlet oxygen produced from a given amount of radiation. The number of singlet oxygens produced in a cell was estimated by Morgan et al. (2009) based on LaF$_3$ luminescent nanoparticles, using the following formula:

$$N_{^1O_2} = 3.2DM\nu\Phi_{^1O_2}, \qquad (11.8)$$

where D is the radiation dose in Gy, M is the absorption of the nanoparticle cores relative to that of tissue and is strongly dependent on incident x-ray energy, ν is the concentration of the nanoparticle, and $\Phi_{^1O_2}$ is the energy transfer efficiency. The conversion factor of 3.2 comes from the fact that 1 Gy of radiation deposits 3.2 MeV to a cell with an estimated volume of 0.52 pL. A wide range of nanoparticle cell loading values between 0.1% and 5% was summarized by Morgan et al. (2009). The lower Niedre limit (Hainfeld et al. 2008) of 5.6×10^7 was used as the number of singlet oxygen molecules per cell for effective tumor cell killing. The FRET efficiency was assumed to be 0.75, and the quantum yields of the LaF$_3$ nanoparticle and the photosensitizer were 0.5 and 0.89, respectively. It was also assumed to generate 3.9×10^5 photons per 1 MeV of radiation energy using the excitation wavelength of 480 nm. The resultant singlet oxygen production as a function of radiation dose is plotted in Figure 11.8. Because of the photoelectric effect at lower energy and higher atomic numbers of the LaF$_3$, low energy x-rays generate higher numbers of singlet oxygen molecules than MV x-rays that translate to lower required dose for combined PDT and radiotherapy.

An alternative way to estimate the physical energy transfer is to compare the energy deposition from x-ray therapy and conventional PDT as follows using CdSe QDs as the energy mediator.

Assuming Compton scattering is the dominant effect for 6-MV radiation to interact with the media, the energy transferred to Photofrin per mass can be expressed as

$$E_c = \eta D\rho_m M, \qquad (11.9)$$

where D is the radiation dose, ρ_m is the molar concentration of the conjugates, η is the FRET efficiency, and M is the molar mass of the conjugates. η is assumed to be 0.76 based on the quenching data reported by Yang et al. (2008).

The molar mass of QDs (CdSe core and ZnS shell, PEG coating, amine terminated; Evident, Troy, NY) is distributed between 1×10^5 and 3×10^5 g/mol. The molar mass of Photofrin is 600 g/mol. Both are provided by the manufacturer. Because the excitation

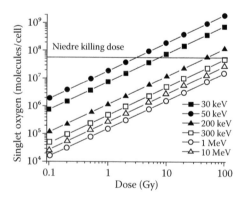

FIGURE 11.8   Theoretical $^1O_2$ production as a function of photon energy. (After Morgan, N. Y. et al., *Radiation Research*, 171(2), 236–244, 2009. With permission.)

efficiency of the Photofrin by 520 nm light is approximately 3 times greater (Bonnett 1995; Dougherty et al. 1998; Macdonald and Dougherty 2001; Allison et al. 2004; Karotki et al. 2006) than the excitation efficiency by the 630-nm light used clinically, the energy transferred to Photofrin per gram of tissue (assuming tissue density 1 g/mL) at depth d in this conventional PDT is

$$E_p = \frac{1}{3}\psi e^{-(d,\kappa)}\frac{\rho t}{0.1 \text{ cm}}, \tag{11.10}$$

where ψ is the photon energy density, converted to energy per mass by the 0.1 cm in the denominator; κ is the attenuation coefficient ~1 mm^{-1} (Whitehurst et al. 1990); and ρ is the clinically achieved tissue concentration ~10^{-6} (Hahn et al. 2006). t is the percentage energy deposited to 1 mm of tissue, as follows:

$$t = \frac{de^{-(a\times\kappa)}}{da}\bigg|_{a=1 \text{ mm}} = 0.37 \tag{11.11}$$

E_c can be derived using Equation 11.9:

$$E_c = \eta D\rho_m M = 0.76 \times 50 \text{ Gy} \times 24 \text{ pmol/g} \times 2 \times 10^{15} \text{ g/mol}$$
$$= 1.8 \times 10^{-7} \text{ J/g} \tag{11.12}$$

As for the conventional PDT, the energy deposited at 0.5-cm depth based on Equation 11.10 is

$$E_p = \frac{1}{3}\omega e^{-(d,\kappa)}\frac{\rho}{0.1 \text{ cm}} \times 0.37 = 0.33 \times 80 \times e^{-5} \times 10^{-6} \times 10 \times 0.37$$
$$= 6.6 \times 10^{-7} \text{ J/g} \tag{11.13}$$

In Equation 11.12, 50 Gy was used as the typical clinical radiotherapy dose; as a result, E_c was 27% of E_p. It is important to note that the energy deposition here is calculated to the Photofrin *only*. After converting to tissue dose, the energy deposition would be 10^6 greater in the calculation for traditional PDT, resulting in 0.66 J/g, which agrees well with previous estimates of 0.3–1 J/g (Lilge et al. 1996; Farrell et al. 1998). This theoretical calculation thus demonstrates that the energy transferred to Photofrin using standard-dose x-rays is comparable to the low dose end of a conventional PDT procedure. The result is consistent to the order of magnitude with the calculation reported by Morgan et al. (2009), confirming that the role of the conjugate in MV radiation is in the category of radiation sensitizer.

Both estimates have made a number of crude assumptions. The energy deposition from x-rays to the nanoparticle is not exactly known. The energy deposition to nanoparticles from radiation may have underestimated the contribution from scatter photons and electrons from surrounding molecules, such as the shell and PEG coating of the nanoparticles. Radiation dose

is defined as the total energy deposited in a finite volume, and it may be different than the dose received by a microscopic particle with high atomic number. The Niedre killing limit, a rough estimate by itself, was derived based on a uniformly distributed $^1O_2$. As this chapter will show later, the intracellular distribution of the nanoparticle is highly heterogeneous, and so are the $^1O_2$. Depending on the locations with higher concentrations of $^1O_2$, effective cell killing can still be achieved if a large number of $^1O_2$ is produced at cell organelles susceptible to $^1O_2$ damage. Therefore, theoretical estimates can only be used as an order of magnitude estimate; the actual $^1O_2$ yield and biological effects will need to be quantified experimentally using *in vivo* and *in vitro* assays.

Experimental quantification of the singlet oxygen production from x-ray and QD–photosensitizer conjugates was performed using Singlet Oxygen Sensor Green (SOSG; Invitrogen, Carlsbad, CA) (Wang et al. 2010). SOSG is highly selective to singlet oxygen and does not show reactivities to hydroxyl radicals and superoxide (Flors et al. 2006). This new singlet oxygen indicator initially exhibits weak blue fluorescence, with excitation peaks at 372 and 393 nm and emission peaks at 395 and 416 nm. In the presence of singlet oxygen, it emits a green fluorescence (excitation/emission maxima ~504/525 nm). The nanoconjugate with a concentration of 48 nM was prepared. SOSG was added to the solution, reaching a concentration of 5 μM. The solution was then irradiated by 6 MV x-rays to doses of 6, 10, 20, and 30 Gy using a dose rate of 6 Gy/min. Because the fluorescence from SOSG remains constant for at least 60 min based on the manufacturer's manual, the fluorescent emission from SOSG was then measured 4 times by a fluorospectrometer 10, 15, 20 and 30 min after irradiation. Stable fluorescence from SOSG was detected 10–30 min after irradiation with little variation between time points. The fluorescent photon counting, with background subtracted from water irradiated by the same dose, increased with higher radiation dose, but the increase was not linear (Figure 11.9) at higher radiation doses. Compared with the $^1O_2$ from 50 J/cm^2 of blue laser light, the amount of singlet

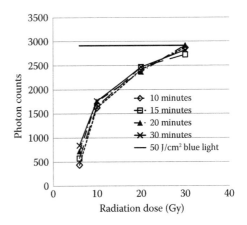

FIGURE 11.9 Fluorescent light from SOSG, which is proportional to the number of $^1O_2$ as a function of the radiation dose.

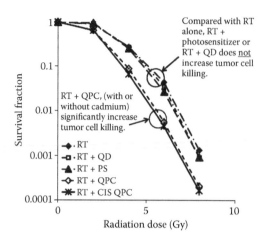

FIGURE 11.10 Cell survival vs. radiation dose for H460 cells incubated in photosensitizer alone, QD alone, and QD/photosensitizer conjugate. Both CdSe and Cd free QD are tested.

oxygen is approximately 1 order of magnitude lower with 6 Gy of radiation.

11.3.3.3 Biological Testing

Biological verification of the efficacy of the conjugate has been demonstrated that with the conjugate, lung carcinoma cell killing from the same radiation dose was increased significantly (Yang et al. 2008), as shown in Figure 11.10. QDs alone, however, did not sensitize the tumor cells at all, confirming that the amount of singlet oxygen production by QD alone when excited by x-rays is insufficient (Samia et al. 2003) to cause any biological effects. It also verified that Photofrin alone is sufficient to sensitize this particular tumor cell line without QDs as the energy mediator.

The cell killing mechanism by the combined PDT and radiotherapy has been another interesting topic. From a biological perspective, the combined therapy not only increases the physical energy deposition but also adds new tumor cell killing pathways

as PDT has very different biological targets than radiation therapy. In PDT, DNA is not the major target (typically photosensitizers localize in/on cell membranes, frequently mitochondrial membranes and lysosomes), and cell death is somatic (i.e., apoptosis) rather than antiproliferative. As a result, tissue responses are rapid and sometimes detectable even before treatment has been completed (Penning 1994; Gomer et al. 1996; Oleinick 1998; Wilson 2008), regardless of tumor radiation sensitivity. Although the cell killing mechanisms of PDT are well understood and highly complementary to those of RT, it should be worth investigating if the same mechanisms apply to combined therapy. In a recent study conducted by Wang et al., a high geometrical coincidence was observed between the lysotracker and the QD (Figure 11.11) under the confocal microscope, validating the intracellular distribution to be primarily in the lysosomes. Also as expected, QD was not significantly presented in the cell nuclei. Additionally, apoptosis as indicated by the TUNEL stain with an emission wavelength of 525 nm was observed in the cells treated by the QD-photosensitizer conjugate (QPC) and radiation but not in control cells treated with radiation or QPC alone (Figure 11.12). Apoptotic activity was also verified by Western blot cleaved caspase-3 assay showing a band at 17 kDa in cells treated by NC and radiation with much more cleaved protein compared to radiation alone and none in the NC alone control. The additional apoptosis is important to kill radioresistant cells that otherwise would not be eliminated by simple dose escalation.

An *in vivo* study using a xenograft tumor model was conducted using H460 NSCLC cells. One hundred microliters of the QD conjugate, at a concentration of 4 nmol/mL, was injected intravenously into mice with xenograft tumors when the average tumor volume reached a palpable size (about 0.2 cm³). These mice were treated 24 h after injection. Significant growth delay was observed with 6 Gy of radiation and 48 nM concentration on a total of eight mice with four in each group (Figure 11.13). Figure 11.14 shows the fluorescent microscope images of the xenograft tumor 24 h after red light-emitting QDs are injected. With a nonspecific targeting folate modification, QD fluorescent lights were observed in most tumor cells.

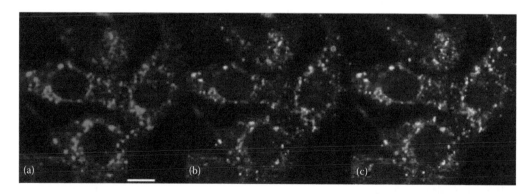

FIGURE 11.11 (See color insert.) Fluorescent microscope images of the QD uptake and subcellular localization. (a) Red fluorescence from lysotracker; (b) green emission from QD decorated by folate at the same view; (c) green emission from QD superimposed on the red emission from the lysotracker results in the orange color.

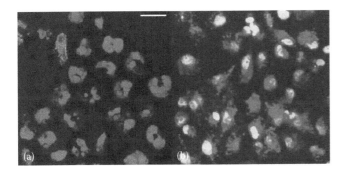

FIGURE 11.12 (See color insert.) Apoptotic study using TUNEL stain. (a) H460 cells treated by 6 Gy only. (b) H460 cells treated by 6 Gy and 48 nM QPC. TUNEL stains in green overlay on the cell nuclei that are shown by red propidium iodide (PI) stains. In (b), scattered dots outside of the cell nuclei are from emission of Photofrin at 650 nm.

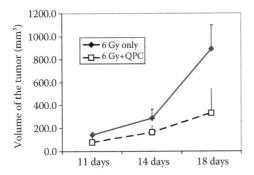

FIGURE 11.13 Tumor growth delay with radiation and QPC, compared with radiation alone. Mice were euthanized after 18 days because of the size of tumor.

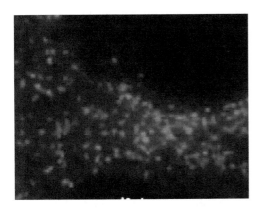

FIGURE 11.14 Fluorescent microscopic image of 650 nm QD distribution in the tumor.

11.3.3.4 Toxicity, Potentials, and Alternative to Cadmium Containing QDs

One major obstacle to the biological application of QDs is their potential toxicity. QDs have a cadmium core, which is normally encapsulated by a bioinert shell structure composed of ZnS. In biological applications, it not uncommon to further coat the QDs with a PEG layer to improve their water solubility and biocompatibility. Although the core of QDs is between 2 and 5 nm in diameter, with the shell, PEG layer, and additional biological functionalization, the hydrodynamic diameter of QDs could approach 25 nm. Particles of this size cannot be cleared by the kidneys. However, long-term circulation of QDs in a biological subject may result in the breakdown of the shell structure, allowing the Cd^{2+} ions to leak into the cytoplasm and cause cytotoxicity. An official conclusion on QD toxicity is not easily drawn (Hardman 2006), largely because the toxicity is affected by many environmental and intrinsic variables. Size, charge, concentration, outer coating bioactivity (capping material and functional groups), and stability toward oxidation, photolysis, and mechanical force can affect QD toxicity (Bouldin et al. 2008). It can be difficult to compare toxicology studies because of these variables and the fact that the quality of QDs varies significantly across manufacturers and laboratories. In an experiment conducted by our group, the toxicity of the conjugate was evaluated in mixed organotypic brain cell cultures, MDCK cells (a kidney cell line), and 3T3 fibroblasts, by measuring propidium iodide uptake after various exposure times. Brain cells and kidney are two critical targets for Cd toxicity. Figure 11.15 shows cell death induced by NC in MDCK cells. Concentrations of QD from 24 fM to 24 pM showed no toxicity as compared to controls for up to 5 days post-delivery. There was an increase in cell death in conjugate-treated cells at 5 and 10 days post-delivery; however, this was less than 0.7% of the total number of cells. No significant toxicity was noted over the 24-fM to 24-nM doses for up to 10 days. Similar results were observed in organotypic brain cell cultures, which are highly sensitive to toxin exposure. In considering the toxicity of nanoparticles containing Cd, one must consider the potential metabolism of coating materials that would reveal the Cd core. To assess this, QDs were exposed to rat liver microsomes for 2 h to allow for metabolic alterations to the conjugates, followed by exposure to MDCK cells and mixed organotypic brain cell cultures, but no increased toxicity in microsome-exposed Photofrin/QD conjugates was observed. In any case, the uncertainty surrounding QD toxicity has been a major roadblock for its further human application, and efforts

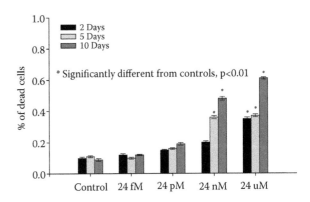

FIGURE 11.15 Effect of Photofrin/QD conjugates on survival of MDCK kidney cells in culture. Results are expressed as percentage of total cells that were dead.

to develop other photoluminescent nanoparticles have been made to circumvent this long-term obstacle. ZnS nanoparticles doped with Mn^{2+} (Chen et al. 2001) or Eu^{2+} (Chen et al. 2000) were fabricated. These particles are both photoluminescent and magnetic, making dual optical and magnetic resonance imaging applications possible. In addition to ZnS semiconductive nanoparticles, silicon nanoparticles (He et al. 2009), carbon dots (Buitelaar et al. 2002; Sun et al. 2006; Cao et al. 2007), SiC-based QDs (Botsoa et al. 2008), InAs/InP/ZnSe QDs (Gao et al. 2010), and CuInS2/ZnS QDs (Pons et al. 2010) have been synthesized. These particles have similar photoluminescent properties and reportedly low or negligible toxicity. Among them, CuInS2/ZnS QDs showed promise for semiconductor mediated simultaneous PDT and radiotherapy.

Synthesis of core $CuInS_2$ QDs is achieved in organic solvent. $InCl_3$ and CuCl, salts are dissolved in octadecene in the presence of trioctylphosphine (TOP) and oleylamine. A sulfur precursor is injected at 190°C in the form of bis(*N*-hexyldithiocarbamate) zinc, $Zn(NHDC)_2$. Core $CuInS_2$ QDs are produced by stirring the mixture for 10 min. Purified QDs are coated by ZnS subsequently. The resultant QDs are highly similar to the CdSe QD in terms of quantum yield (20–30%) and hydrodynamic diameters (20–22 nm). 6-MV clinical x-rays were used to test the excitation of $CuInS_2$ QDs under identical experimental condition as CdSe QDs. A slightly higher emission was observed compared with the emission of CdSe QDs. The cell killing efficacy was tested using clonogenic assay on H460 cells, and the result is shown in Figure 11.10. Therefore, the toxicity of Cd-containing QDs is likely a surmountable obstacle, but further studies are needed on particle delivery, combined dosimetry, and applicable tumor sites.

In vivo applications of these surface modifications and improved tumor control using combined PDT and radiotherapy have not been widely studied. This indicates the difficulty in *in vivo* targeting, namely, that the uptake of the nanoparticles is not determined simply by the affinity between the particles and the tumor cells, as it is in a Petri dish. The particle may be intercepted by the immune system before entering the tumor, as shown in Figure 11.16. QD distribution was measured by inductively coupled plasma mass spectrometry (ICP-MS) for Cd. As shown, QD distribution to the brain appears to be low, possibly due to the lack of penetration of the blood–brain barrier. Similarly, deposition in the heart was also low. The main sites of incorporation appear to be the liver, kidney, spleen, and lung. In addition, tumors can have higher intratumoral pressure as a result of the lack of lymphatic drainage, making it more difficult for the nanoparticles to penetrate into the core of the tumor, especially when the tumor is not well vascularized.

11.4 Radiation Protection with Nanoparticles

11.4.1 Role of Free Radical Scavenger in Radiation Protection

Working from the opposite direction, the therapeutic ratio can be improved by protection of normal tissue more than tumor tissue from radiation damage. Since radiation-induced injury to cells is caused primarily by free radicals generated by excitation and ionization events during the interaction of radiation with the tissue, free radicals have been the primary target of research in radiation protection.

Amifostine is the only approved treatment for radioprotection in patients with head-and-neck cancer (Spencer and Goa 1995). In normal cells, amifostine hydrolyzes by alkaline phosphatase to the active thiol metabolite, WR-1065, which scavenges superoxide radicals generated from ionizing radiation. However, common side effects of amifostine include hypocalcemia, diarrhea, nausea, vomiting, sneezing, somnolence, and hiccups. Serious side effects include hypotension (found in 62% of patients) and erythema multiforme. These side effects have prevented the wider application of amifostine in radiation therapy.

11.4.1.1 CeO₂ Nanoparticles

Recent nanotechnology-based molecular engineering advancements have produced new classes of molecules, such as cerium oxide (CeO_2), which was developed as a potent free radical scavenger (Tarnuzzer et al. 2005). CeO_2 is a rare earth oxide material from the lanthanide series of the periodic table. The typical synthesis process of CeO_2 nanoparticles is by microemulsion process consisting of surfactant sodium bis(2-ethylhexyl) sulfosuccinate (AOT), toluene, and water. AOT is dissolved in 50 mL of toluene, and 2.5 mL of 0.1 M aqueous cerium nitrate solution are added. The mixture of cerium nitrate, AOT, and toluene is stirred for separation into two layers. The upper layer is toluene containing nonagglomerated CeO_2 nanoparticles. Figure 11.17 shows the transmission electron microscopy (TEM) image of the resultant nanoparticles. CeO_2 are highly efficient redox reagents used in various applications such as ultraviolet absorbents, oxygen sensors, and automotive catalytic converters. All of these applications are based on the capability of CeO_2 to reduce oxidation species in a catalytic manner (Rzigalinski et al. 2006). The cerium atom can exist in either the +3 (fully reduced) or +4 (fully oxidized) state. In its oxidative form, CeO_2 also exhibits oxygen vacancies, or defects, in the lattice structure,

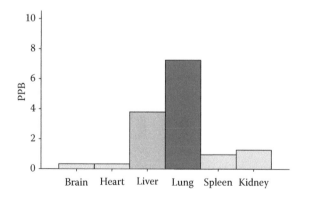

FIGURE 11.16 Biodistribution of the QD as determined by ICP-MS.

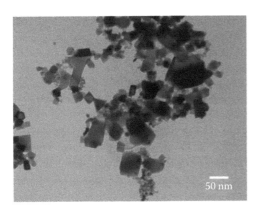

FIGURE 11.17 TEM image of CeO_2 nanoparticles.

through loss of oxygen and/or its electrons, alternating between CeO_2 and CeO_{2-x} during redox reactions. The change in cerium valence during a redox event subsequently alters the structure of the oxide lattice, possibly creating additional oxygen vacancies by lattice expansion (Rzigalinski et al. 2006). This electron translation within the lattice provides reducing power for free radical scavenging. After the scavenging event, the original lattice structure may be regenerated by releasing H_2O while the cerium atom returns to the +3 state. It was also reported that the redox efficiency of CeO_2 was inversely proportional to the CeO_2 nanoparticle size (Rzigalinski et al. 2006). By using nano-sized (10–20 nm) CeO_2 particles, the balance between high antioxidant efficiency and cell penetration is optimized.

Because of their free-radical scavenger ability, CeO_2 nanoparticles have been used to protect cells and animals against ionizing radiation. Investigators have reported that CeO_2 particles were able to protect 90% of normal cells from 10 Gy of radiation with minimal tumor cell protection (Tarnuzzer et al. 2005). The mechanism behind the different protection is not entirely clear. It may arise from the differential uptake of the particles, although this has not been proved. Another possible mechanism was offered by Jonathan et al. (1999), who hypothesized that tumor cells expose more bases of the chromatin structure as targets for free-radical attack. The greater number of vulnerable sites in tumor cells makes radiation protection by CeO_2 more difficult.

CeO_2 can reduce large numbers of free radicals more rapidly than amifostine, rendering it suitable for use as a radioprotector during standard radiation therapy. Since the burst of free radical generation that occurs during radiation is completed within milliseconds of treatment, the longer retention time of the CeO_2 nanoparticle makes fewer infusions required during radiation treatment. In a toxicology study conducted by Rzigalinski (2005) and Rzigalinski et al. (2006), CeO_2 did not exhibit toxicity in neuronal and macrophage cell lines as long as the particle size was less than 20 nm (Rzigalinski 2005). Significantly reduced normal lung fibroblast cell was observed on cells administrated CeO_2 nanoparticles 24 h before the radiation challenge. The same study also showed that CeO_2 nanoparticles can protect mice receiving 12–18 Gy of radiation and reduce the severity of radiation pneumonitis (Colon et al. 2009). In a subsequent study, the same group demonstrated that CeO_2 nanoparticles reduce reactive oxygen species (ROS) levels and protect normal human colon cells from radiation-induced cell death *in vitro* and mice gastrointestinal cells *in vivo* (Colon et al. 2010). These initial studies suggest that the CeO_2 nanoparticle is an attractive candidate for clinical development as a novel radiation protector, but the differential protection of tumor and normal cells is yet to be demonstrated in an animal model.

11.4.1.2 Carboxyfullerene

Carboxyfullerene has been described as a free radical sponge that can absorb multiple radicals to a single nanoparticle. For this reason, researchers have shown strong interests in using it for protection against cell oxidation damage since its discovery (Krusic et al. 1991; Dugan et al. 1996; Monti et al. 2000). A widely used carboxyfullerene for protection against oxidation is (C_3) (Osuna et al. 2010). Although there has been experimental evidence that carboxyfullerenes decrease ROS production, the mechanism was not clearly understood until recently. A computational model was used to reveal the reaction process between C_3 and free radicals (Osuna et al. 2010). The unpaired electron of superoxicide is transferred to C_3, and the free radical becomes neutral. This step has been found to be the rate-limiting factor of the entire reaction. Optimization of this reaction would lead to more effective free radical scavenging. In the second step, C_3 radical anion bearing an extra electron reacts with a second superoxicide. Excessive electrons are transferred to an OO moiety and convert it to a more stable singlet state, which acquires two protons from the COOH link on C_3, resulting in hydrogen peroxide molecule through a number of intermediate steps. However, further metabolism of the hydrogen peroxide molecule with C_3 is not entirely clear.

Effective protection of human keratinocytes against UVB was demonstrated with carboxyfullerene (Fumelli et al. 2000). C_3 has also been tested for radioprotective function. A protection factor, defined as the ratio of survival with and without C_3, of up to 2.38 was demonstrated in normal hematopoietic progenitor cells, but much less protection was observed in mouse and human tumor cell lines (Lin et al. 2001). The antioxidative stress action of C_3 was also seen after C_3 treatment of $Sod2^{-/-}$ mice, which lack expression of mitochondrial manganese superoxide dismutase; their life span increased by 300% (Ali et al. 2004). Further experiments conducted by Yin et al. (2009) showed that ROS, superoxide radical anion, singlet oxygen, and hydroxyl radicals can be effectively inhibited by the fullerenes. This report also revealed that the radical scavenging ability is affected by surface chemistry–induced differences in electron affinity and physical properties, such as degree of aggregation (Yin et al. 2009). On the other hand, there have been reports on the toxicity of carboxyfullerene. Depending on the specific carboxyfullerene derivative, aqueous carboxyfullerene (nC_{60}) has been reported as a generator of ROS. Investigations with nC_{60} in zebrafish reported significant embryo mortality and malformation (Usenko et al. 2007), but the result is highly controversial

(Henry et al. 2011). The second challenge of using carboxyfullerene as radiation protection is the speed of free radical scavenging. C_3 can remove the superoxide anion at a relatively slow speed compared with CeO_2 nanoparticles. It may be sufficient for an environment with slow production of free radicals, but in a scenario of ionizing radiation challenge, the discrepancy in speed may render carboxyfullerene ineffective as a radiation protector as indicated by a recent study (Brown et al. 2010) that carboxyfullerene has modest activity as a radiation protector *in vivo*. The same study also pointed out the third challenge that there was no evidence of differential protection from irradiation to normal versus tumor cells. Modification of the surface chemistry for more efficient ROS scavenging and radiation protection is an ongoing research topic.

11.5 Radiation Dosimeters Using Semiconductor Nanomaterials

11.5.1 General Principles of Macroscopic Dosimeters and Need for Nanodosimetry

Dose is an important quantity in radiotherapy and radiation safety. Radiation dose is defined as energy deposit per unit mass. The international standard unit of radiation dose is Gy = J/kg. The deposited energy will eventually be converted to heat, but as the unit suggests, within the range of radiation dose seen in clinical application, the temperature change is extremely small, making direct measurement of the radiation dose by thermometry impractical. To ensure accurate dose detection, a dosimeter is needed to convert radiation energy to signals that can be more sensitively quantified. Conventional dosimeters such as ion chambers, diodes, thermoluminescent dosimeters (TLDs), and metal-oxide semiconductor field effect transistors (MOSFETs) are used to perform the task. The choice of dosimeter depends on the desired accuracy, convenience, and accessibility. It can be difficult to attain all of these goals at once. For example, when air in the ion chamber is ionized by x-ray, the ion pairs are collected by the positive and negative electric nodes and a current is generated. The total charge is proportional to the radiation dose. After calibration to a known source, such as Co-60, an ion chamber can be used to measure radiation dose accurately and reproducibly. Ion chambers are also minimally affected by energy in a wide range of megavoltage photon and electron measurements. Residual energy dependency can be quantified for fine calibration. Therefore, ion chambers are commonly used as the reference in radiation detection. However, ion chambers require high voltage to operate, are cumbersome, and are too large for *in vivo* measurement. TLDs use crystals with defects in lattice that result in metastable energy traps for electrons excited by x-rays. Once the crystal is heated up, the trapped electron is released and a photon is emitted. Photomultiplier tubes can detect these photons with very high sensitivity. The amount of photons is proportional to the radiation dose in a wide range of doses. TLDs (Cameron et al. 1969) can be made tissue equivalent and very small so they are suitable for *in vivo* measurement,

but the calibration and readout process is not only tedious but also prone to errors. For a diode used in radiation dosimetry, the excess carriers, electrons or holes generated by radiation, would diffuse to the p–n junction. They are then pushed through the junction by the built-in potential and collected by the electrometer. Because of the higher density of the solid semiconductor material compared with air, diodes are much more sensitive than ion chamber and can be made extremely small. Semiconductor diodes are used without bias, but they are still wired, not tissue equivalent, and dependent on the radiation energy and angles. MOSFETs (Soubra et al. 1994) can be made wireless and convenient to apply but in addition to its energy and angle dependency, they are limited by the number of readings before the shift in threshold voltage is saturated irreversibly. The one direction shift of threshold voltage also makes calibration of these detectors irreproducible. A recent addition to the inventory is optical stimulated luminescent dosimeter (OSL dosimeter) (Pradhan et al. 2008). Similar to TLD, crystals such as MgS, CaS, and SrS doped with rare earth elements exhibit metastable electron traps that can be filled by electrons excited by ionizing radiation. Instead of releasing these electrons by heat, light is used to stimulate the material and release trapped electrons. In the process, light is emitted and detected. The light emission is easily differentiated from the excitation light by wavelength. Compared with TLD, OSL can be read repeatedly after exposure and offers higher resolution. On the other hand, OSL is not strictly reuseable. Not all trapped electrons can be released from optical stimulation, and the detector will saturate after many uses. These dosimeters do not provide two-dimensional (2-D) or three-dimensional (3-D) dose distribution unless hundreds of them are built into a bulky array. Radiographic film is widely available for 2-D dose measurement, but it does not provide absolute readout, is not reuseable, and is cumbersome to process. With the phasing out of conventional film processors in many hospitals with digital films and the gradual deterioration of existing processors, radiographic films are gradually being replaced by radiochromic film (Niroomand-Rad et al. 1998), which does not require chemical processing. Instead, Ionizing radiation induces an instantaneous polymerization and changes in color/opacity. The convenience of exposing radiochromic film is somewhat offset by the more intricate digitization and calibration process that is affected by light scatter, film inhomogeneity, and orientation. 3-D gel dosimeters such as BANG polymer gel (Maryanski et al. 1997; Oldham et al. 1998) polymerize upon radiation and can be used for 3-D dose measurement, but they are expensive and not reusable. In addition, a dedicated optical CT scanner is needed to read the dose.

Clearly, there is an arsenal of tools for the physicist to measure radiation dose with varying accuracy, convenience, resolution, and cost. Nanotechnology may improve the performance of these areas and handle new challenges such as nanodosimetry, where the energy deposited on a scale comparable to the size of a DNA molecule needs to be measured by a device that is in the same scale (De Nardo et al. 2002a, 2002b; Grosswendt 2005, 2006; Schulte et al. 2008). The kind of dosimetry is important when macroscopic dose is not an accurate predictor of the DNA damage. For example, Auger electrons with a very high linear energy transfer efficiency have the potential to kill tumor cells effectively. However, because of the short range of Auger electrons, the magnitude of DNA damage can be very different for the same macroscopic dose. Nanodosimetry, on the other hand, will be able to more accurately describe these processes.

11.5.2 Semiconductor Nanoparticle-Based Dosimeters

The principles used in conventional dosimetry can be adapted to making nanodosimeters for radiation measurement. For example, carbon nanotubes are effective in converting photon energy to electronic signals. The conversion is referred to as optoelectronics (Stewart and Leonard 2005). When a carbon nanotube is irradiated by visible light and a bias voltage is applied to it, an electrical current is generated and part of the energy from the light can be converted to electricity. At present, the efficiency of energy conversion is around 10%. This has made carbon nanotubes attractive for application in solar panels and photodiodes. The same capacity may be applied to radiation dosimetry, where x-ray energy can be converted to electricity. More interestingly, short circuit photocurrents can be theoretically generated (Stewart and Leonard 2004) without a bias voltage on a nanotube, making wireless operation of the nanodosimeter possible.

Nanometer-sized phosphorescence (afterglow) crystals such as $LiF:Mg$ and $MgF_2:Eu$ can be manufactured for this purpose (Salah et al. 2006; Chen et al. 2008). Compared with macro-sized TLD, nano-TLDs are more stable, with high luminescent intensity, low photobleaching, and a large Stokes shift (Yi et al. 2001). The dynamic range of TLDs was improved with the nanometer-scaled phosphors. Sahare el al. synthesized a $K_3Na(SO_4)_2:Eu$ nanocrystalline powder that is four times more sensitive than a $LiF:Mg,Ti$ (TLD-100) phosphor and has a near-linear response up to a radiation dose of 70,000 Gy (Yi et al. 2001).

MOSFETs can also be made on the nanometer scale. A schematic presentation of a typical MOSFET is shown in Figure 11.18. When a positive voltage is applied on the gate, the electric field causes the holes to be repelled from the interface, creating a depletion region containing negatively charged acceptor ions. A further increase in the gate voltage eventually causes a sufficient number of electrons to appear at the inversion layer, and the MOSFET becomes a conductor. The voltage that turns on the MOSFET is referred to as the threshold voltage. When a MOSFET device is irradiated, trapped charges are built up in the oxide layer, the number of interface traps increases, and the number of bulk oxide traps increases. With the excitation from radiation, electron–hole pairs are generated. Electrons quickly move out of the gate electrode, whereas holes move slowly in toward the Si/SiO_2 interface where they become trapped, causing a negative threshold voltage shift that can be measured. The voltage shift is proportional to the amount of radiation received (Gladstone et al. 1994; Soubra et al. 1994). Very small MOSFETs suitable for *in vivo* measurements can be manufactured. Compared with ion chambers, they do not need high voltage to operate and can be used wirelessly. Compared with diodes, MOSFETs are less dependent on the radiation energy and angle. The readout of a MOSFET is instantaneous and much less elaborate than that of a TLD. As opposed to the traditional bulk MOSFET material, field effect transistors (FET) based on single-wall carbon nanotubes (SWNT) have been fabricated for molecular, chemical, and biological sensing (Kong et al. 2000; Pengfei et al. 2003). Using the same platform, Sahare et al. (2007) synthesized SWNT-FETs on a SiO_2/Si substrate using patterned chemical vapor deposition on top of W/Pt electrodes. The structure of the SWCN-FET is similar to the conventional MOSFET except that the p-type Si is replaced by the nanotubes. The authors demonstrated that the SWNT-FET is stable up to doses to 1 Gy and is about 2 orders of magnitude more sensitive than a conventional MOSFET. In addition to these novel applications, there is one particularly interesting dosimeter based on semiconductor nanoparticle that is not feasible with conventional material.

11.5.3 A Case for Nanoparticle-Based *In Vivo* Dosimetry

Unlike surgery, or magnetic resonance–guided focused ultrasound, the result of radiation delivery is not directly visible *in vivo*. A large number of complex processes and devices are involved in radiation delivery. Further compounding factors are added from patient inter- and intrafractional motion. Although the 3-D dose can be precisely computed, the actual delivered dose may deviate from expected levels. Most small deviations are not consequential, but large deviations can lead to local control failure, patient injury, or even death. An ultimate goal in the physics of radiation therapy is thus to directly measure the actual 3-D dose delivered to a patient. However, currently available technology is not able to reach this goal. There are several approaches to measure *in vivo* radiation dose. Dosimeters with small form factors, such as diodes and TLDs, are used to measure point doses on the surface. An implantable dosimeter was used to measure *in vivo* dose delivered to the prostate and breast (Scarantino et al. 2008). Indirectly, 3-D dose distribution can be

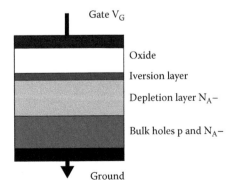

FIGURE 11.18 Illustration of a MOSFET.

reconstructed using exit dose (Cheung et al. 2009), but the accuracy is limited by complicated modeling of scattering photons, uncertain impact from patient interfractional and intrafractional variation, and nonlinearity of the detector. Clearly, none of these approaches meets all criteria of 3-D *in vivo* dosimetry, which are direct measurement, continuous imaging in the spatial domain, and being noninvasive.

Photomagnetism is a process where a nonparamagnetic material becomes paramagnetic under the stimulation of light. Photomagnetism has been discovered for more than a decade (Sato et al. 1996); however, because of the spin interchange with molecules in the liquid environment, the magnetic properties of the illuminated sample relaxed to the initial state. When the temperature of the sample was increased to 150 K, the magnetism is reversed. The temperature is much lower than room temperature, ruling out its clinical application. In order to achieve room temperature light-induced magnetization, magnetic molecules need to be isolated. Photomagnetism of Mn-doped QDs under room temperature was recently discovered (Beaulac et al. 2009). Previously, the formation of excitonic magnetic polaron was hampered by rapid energy transfer to Mn^{2+} from excited nanocrystals. The energy transfer is faster than Mn^{2+} reorientation, preventing the formation of photomagnetism. The problem was solved by modifying the synthesis of the particle so that the excitonic energy levels are lower than the electronic excited states and photon-induced magnetization can be induced. These room temperature spin effects have been attributed to strong zero-dimensional exciton confinement achieved in colloidal doped nanocrystals. Original interest in this technology was for quantum information processing, but the same particles can be readily adopted for radiation dosimetry. It has been demonstrated in previous sections that QDs can be excited by megavoltage x-rays (Yang et al. 2008) used to treat prostate cancer. With x-rays, the excited nanoparticles will then become paramagnetic, resulting in reduced transversal relaxation time of proton (T2) in the vicinity. The change would be detectable by an MR scanner. Radiation dose can then be calculated from digital subtraction. With emerging MR-guided therapy machines, the dose image may directly be measured in real time (Cervino et al. 2011; Crijns et al. 2011; Raaymakers et al. 2011).

11.6 Conclusions

Although the application of semiconductor nanomaterials in radiation therapy is relatively new, the early results are encouraging. Because of their prominent photofluorescent properties, semiconductor nanomaterials were used as an energy reservoir that absorbs a wide range of x-rays and converts them to visible light with specific wavelength tuned to the absorption peak of the photosensitizer, which generates cytotoxic singlet oxygen molecules for enhanced tumor cell killing. This application, compared with the simple energy sink with high-*Z* materials, may activate new biological pathways for tumor cell death and overcome radioresistance. Future research directions, other than tumor-specific targeting, include particle modification for larger

cross section with high energy x-rays and utilization of photosensitizers with higher singlet oxygen yields. Many nanoparticles have high redox ability that enables them to be free radical scavengers for radioprotection, but their differential protection of normal tissue and tumor is yet to be proved. Nanoengineering of these semiconductor materials has produced a new generation of dosimeters that are more nimble and sensitive. On the other hand, the fact that nanomaterials are not routinely used in patient treatment indicates that this field is still in the very early stage of research and development. To realize the full potential of nanotechnology in radiation therapy, technological breakthroughs are needed in terms of effectiveness, toxicity, pharmacokinetics, and cost.

Many of these problems are long-standing in the biological application of nanomaterials. Remarkable progress has been made, but there are no well-established general protocols. Success is usually achieved on a case-by-case basis. Moreover, it is difficult—if not impossible—to fabricate a nanodevice that will carry all functions or satisfy all requirements. It is important to prioritize these properties for a specific application. For radiotherapy, long-term toxicity and biological clearance may arguably have a lower priority than effectiveness, particularly for patients with terminal-stage cancer.

References

Adams, G. E. 1973. Chemical radiosensitization of hypoxic cells. *British Medical Bulletin* 29(1):48–53.

Ali, S. S., J. I. Hardt et al. 2004. A biologically effective fullerene (C-60) derivative with superoxide dismutase mimetic properties. *Free Radical Biology and Medicine* 37(8):1191–1202.

Alivisatos, A. P. 1996. Perspectives on the physical chemistry of semiconductor nanocrystals. *Journal of Physical Chemistry* 100(31):13226–13239.

Allison, R. R., G. H. Downie et al. 2004. Photosensitizer in clinical PDT. *Photodiagnosis and Photodynamic Therapy* 1:27.

Andrews, D. L. 1989. A unified theory of radiative and radiationless molecular-energy transfer. *Chemical Physics* 135(2):195–201.

Attix, F. 1986. *Introduction to Radiological Physics and Radiation Dosimetry*. New York: Wiley-Interscience.

Bachireddy, P., D. Tseng et al. 2010. Orthovoltage intraoperative radiation therapy for pancreatic adenocarcinoma. *Radiation Oncology* 5:105.

Bawendi, M. G., M. L. Steigerwald et al. 1990. The quantum-mechanics of larger semiconductor clusters (quantum dots). *Annual Review of Physical Chemistry* 41:477–496.

Beaulac, R., L. Schneider et al. 2009. Light-induced spontaneous magnetization in doped colloidal quantum dots. *Science* 325(5943):973–976.

Biju, V., T. Itoh et al. 2006. Quenching of photoluminescence in conjugates of quantum dots and single-walled carbon nanotube. *Journal of Physical Chemistry B* 110(51):26068–26074.

Bonnett, R. 1995. Photosensitizers of the porphyrin and pthalocynanie series for photodynamic therapy. *Chemical Society Reviews* 24:19.

Botsoa, J., V. Lysenko et al. 2008. Application of 3C-SiC quantum dots for living cell imaging. *Applied Physics Letters* 92(17): 173902–173903.

Bouldin, J. L., T. M. Ingle et al. 2008. Aqueous toxicity and food chain transfer of Quantum DOTs in freshwater algae and *Ceriodaphnia dubia*. *Environmental Toxicology and Chemistry* 27(9):1958–1963.

Brown, A. P., E. J. Chung et al. 2010. Evaluation of the fullerene compound DF-1 as a radiation protector. *Radiation Oncology* 5:34.

Brown, J. M., and W. R. Wilson. 2004. Exploiting tumour hypoxia in cancer treatment. *Nature Reviews Cancer* 4(6):437–447.

Bryant, G. W. 1988. Excitons in quantum boxes: Correlation effects and quantum confinement. *Physical Review B Condensed Matter* 37(15):8763–8772.

Buitelaar, M. R., A. Bachtold et al. 2002. Multiwall carbon nanotubes as quantum dots. *Physical Review Letters* 88(15): 156801.

Cameron, J. R., M. G. Ort et al. 1969. A TLD measurement of x-ray quality and output simultaneously. *Physics in Medicine and Biology* 14(2):338.

Cao, L., X. Wang et al. 2007. Carbon dots for multiphoton bioimaging. *Journal of the American Chemical Society* 129(37):11318–11319.

Cervino, L. I., J. Du et al. 2011. MRI-guided tumor tracking in lung cancer radiotherapy. *Physics in Medicine and Biology* 56(13):3773–3785.

Chen, W., and J. Zhang 2006. Using nanoparticles to enable simultaneous radiation and photodynamic therapies for cancer treatment. *Journal of Nanoscience and Nanotechnology* 6(4):1159–1166.

Chen, W., J. O. Malm et al. 2000. Energy structure and fluorescence of Eu^{2+} in ZnS:Eu nanoparticles. *Physical Review B* 61(16):11021–11024.

Chen, W., A. G. Joly et al. 2001. Up-conversion luminescence of Mn^{2+} in $ZnS:Mn^{2+}$ nanoparticles. *Physical Review B* 6404(4): 041202–041202-4.

Chen, W., S. L. Westcott et al. 2008. Dose dependent x-ray luminescence in MgF_2: Eu^{2+}, Mn^{2+} phosphors. *Journal of Applied Physics* 103(11): 113103–113103-5.

Cheung, J., J. F. Aubry et al. 2009. Dose recalculation and the Dose-Guided Radiation Therapy (DGRT) process using megavoltage cone-beam CT. *International Journal of Radiation Oncology Biology Physics* 74(2):583–592.

Chinnaiyan, P., G. W. Allen et al. 2006. Radiation and new molecular agents: Part II. Targeting HDAC, HSP90, IGF-1R, PI3K, and Ras. *Seminars in Radiation Oncology* 16(1):59–64.

Choudhury, A., A. Cuddihy et al. 2006. Radiation and new molecular agents: Part I. Targeting ATM-ATR checkpoints, DNA repair, and the proteasome. *Seminars in Radiation Oncology* 16(1):51–58.

Colon, J., L. Herrera et al. 2009. Protection from radiation-induced pneumonitis using cerium oxide nanoparticles. *Nanomedicine* 5(2):225–231.

Colon, J., N. Hsieh et al. 2010. Cerium oxide nanoparticles protect gastrointestinal epithelium from radiation-induced damage by reduction of reactive oxygen species and upregulation of superoxide dismutase 2. *Nanomedicine* 6(5):698–705.

Crijns, S. P., J. G. Kok et al. 2011. Towards MRI-guided linear accelerator control: Gating on an MRI accelerator. *Physics in Medicine and Biology* 56(15):4815–4825.

D'Souza, W. D., and Rosen, I. I. 2003. Nontumor integral dose variation in conventional radiotherapy treatment planning. *Medical Physics* 30(8):2065–2071.

De Nardo, L., A. Alkaa et al. 2002a. A detector for track-nano-dosimetry. *Nuclear Instruments & Methods in Physics Research Section A-Accelerators Spectrometers Detectors and Associated Equipment* 484(1–3):312–326.

De Nardo, L., P. Colautti et al. 2002b. Track nanodosimetry of an alpha particle. *Radiation Protection Dosimetry* 99(1–4):355–358.

Dougherty, T. J., C. J. Gomer et al. 1998. Photodynamic therapy. *Journal of the National Cancer Institute* 90:889–905.

Dugan, L. L., J. K. Gabrielsen et al. 1996. Buckminsterfullerenol free radical scavengers reduce excitotoxic and apoptotic death of cultured cortical neurons. *Neurobiology of Disease* 3(2):129–135.

Eberhardt, W., C. Pottgen et al. 2006. Chemoradiation paradigm for the treatment of lung cancer. *Nature Clinical Practice Oncology* 3(4):188–199.

Farrell, T. J., B. C. Wilson et al. 1998. Comparison of the in vivo photodynamic threshold dose for photofrin, mono- and tetrasulfonated aluminum phthalocyanine using a rat liver model. *Photochemistry and Photobiology* 68(3):394–399.

Flors, C., M. J. Fryer et al. 2006. Imaging the production of singlet oxygen in vivo using a new fluorescent sensor, Singlet Oxygen Sensor Green. *Journal of Experimental Botany* 57(8):1725–1734.

Fowler, J. F., G. E. Adams et al. 1976. Radiosensitizers of hypoxic cells in solid tumors. *Cancer Treatment Reviews* 3(4): 227–256.

Fumelli, C., A. Marconi et al. 2000. Carboxyfullerenes protect human keratinocytes from ultraviolet-B-induced apoptosis. *Journal of Investigative Dermatology* 115(5):835–841.

Gao, J., K. Chen et al. 2010. Ultrasmall near-infrared non-cadmium quantum dots for in vivo tumor imaging. *Small* 6(2): 256–261.

Gladstone, D. J., X. Q. Lu et al. 1994. A miniature mosfet radiation dosimeter probe. *Medical Physics* 21(11):1721–1728.

Gomer, C. J., M. Luna, et al. 1996. Cellular targets and molecular responses associated with photodynamic therapy. *J Clin Laser Med Surg* 14(5):315–321.

Grosswendt, B. 2005. Nanodosimetry, from radiation physics to radiation biology. *Radiation Protection Dosimetry* 115(1–4):1–9.

Grosswendt, B. 2006. Nanodosimetry, the metrological tool for connecting radiation physics with radiation biology. *Radiation Protection Dosimetry* 122(1–4):404–414.

Hahn, S. M., M. E. Putt et al. 2006. Photofrin uptake in the tumor and normal tissues of patients receiving intraperitoneal photodynamic therapy. *Clinical Cancer Research* 12(18): 5464–5470.

Hainfeld, J. F., D. N. Slatkin et al. 2004. The use of gold nanoparticles to enhance radiotherapy in mice. *Physics in Medicine and Biology* 49(18):N309–N315.

Hainfeld, J. F., F. A. Dilmanian et al. 2008. Radiotherapy enhancement with gold nanoparticles. *Journal of Pharmacy and Pharmacology* 60(8):977–985.

Hao, D., M. A. Ritter et al. 2006. Platinum-based concurrent chemoradiotherapy for tumors of the head and neck and the esophagus. *Seminars in Radiation Oncology* 16(1):10–19.

Hardman, R. 2006. A toxicologic review of quantum dots: Toxicity depends on physicochemical and environmental factors. *Environmental Health Perspectives* 114(2):165–172.

He, Y., Y. Su et al. 2009. Photo and pH stable, highly-luminescent silicon nanospheres and their bioconjugates for immunofluorescent cell imaging. *Journal of the American Chemical Society* 131(12):4434–4438.

Henry, T. B., E. J. Petersen et al. 2011. Aqueous fullerene aggregates (nC60) generate minimal reactive oxygen species and are of low toxicity in fish: A revision of previous reports. *Current Opinion in Biotechnology* 22(4):533–537.

Herold, D. M., I. J. Das et al. 2000. Gold microspheres: A selective technique for producing biologically effective dose enhancement. *International Journal of Radiation Biology* 76(10):1357–1364.

Hiraoka, M., Y. Matsuo et al. 2007. Stereotactic body radiation therapy (SBRT) for early-stage lung cancer. *Cancer/Radiotherapie* 11(1–2):32–35.

Idowu, M., J. Y. Chen et al. 2008. Photoinduced energy transfer between watersoluble CdTe quantum dots and aluminium tetrasulfonated phthalocyanine. *New Journal of Chemistry* 32:290–296.

Iwamoto, K. S., S. T. Cochran et al. 1987. Radiation dose enhancement therapy with iodine in rabbit VX-2 brain tumors. *Radiotherapy & Oncology* 8(2):161–170.

Jonathan, E. C., E. J. Bernhard et al. 1999. How does radiation kill cells? *Current Opinion in Chemical Biology* 3(1):77–83.

Karotki, A., M. Khurana et al. 2006. Simultaneous two-photon excitation of photofrin in relation to photodynamic therapy. *Photochemistry and Photobiology* 82(2):443–452.

Konan, Y. N., R. Gurny et al. 2002. State of the art in the delivery of photosensitizers for photodynamic therapy. *Journal of Photochemistry and Photobiology B* 66(2):89–106.

Kong, J., N. R. Franklin et al. 2000. Nanotube molecular wires as chemical sensors. *Science* 287(5453):622–625.

Kong, T., J. Zeng et al. 2008. Enhancement of radiation cytotoxicity in breast-cancer cells by localized attachment of gold nanoparticles. *Small* 4(9):1537–1543.

Krusic, P. J., E. Wasserman et al. 1991. Radical reactions of C60. *Science* 254(5035):1183–1185.

Kulka, U., M. Schaffer et al. 2003. Photofrin as a radiosensitizer in an in vitro cell survival assay. *Biochemistry and Biophysical Research Communications* 311(1):98–103.

Lakowicz, J. 2006. *Principles of Fluorescence Spectroscopy*. New York: Spinger.

Lakowicz, J. R. 1999. *Principles of Fluorescence Spectroscopy*. 2nd ed. New York: Plenum Publishing Corporation.

Li, J. L., L. Wang et al. 2009. In vitro cancer cell imaging and therapy using transferrin-conjugated gold nanoparticles. *Cancer Letters* 274(2):319–326.

Lilge, L., M. C. Olivo et al. 1996. The sensitivity of normal brain and intracranially implanted VX2 tumour to interstitial photodynamic therapy. *British Journal of Cancer* 73(3): 332–343.

Lin, H. S., T. S. Lin et al. 2001. Fullerenes as a new class of radioprotectors. *International Journal of Radiation Biology* 77(2): 235–239.

Luksiene, Z., P. Juzenas et al. 2006a. Radiosensitization of tumours by porphyrins. *Cancer Letters* 235(1):40–47.

Luksiene, Z., D. Labeikyte et al. 2006b. Mechanism of radiosensitization by porphyrins. *Journal of Environmental Pathology, Toxicology and Oncology* 25(1–2):293–306.

Macdonald, I., and Dougherty T. 2001. Basic principles of photodynamic therapy. *Journal of Porphyrins and Pthalocyanines* 5:105.

Mackie, T. R., J. Balog et al. 1999. Tomotherapy. *Seminars in Radiation Oncology* 9(1):108–117.

Maryanski, M. J., C. Audet et al. 1997. Effects of crosslinking and temperature on the dose response of a BANG polymer gel dosimeter. *Physics in Medicine and Biology* 42(2):303–311.

Matsudaira, H., A. M. Ueno et al. 1980. Iodine contrast medium sensitizes cultured mammalian cells to X rays but not to gamma rays. *Radiation Research* 84(1):144–148.

Minchinton, A. I., A. Rojas et al. 1984. Glutathione depletion in tissues after administration of buthionine sulphoximine. *International Journal of Radiation Oncology Biology Physics* 10(8):1261–1264.

Moan, J., and H. Anholt 1990. Phthalocyanine fluorescence in tumors during PDT. *Photochemistry and Photobiology* 51(3):379–381.

Molinelli, S., J. de Pooter et al. 2008. Simultaneous tumour dose escalation and liver sparing in Stereotactic Body Radiation Therapy (SBRT) for liver tumours due to CTV-to-PTV margin reduction. *Radiotherapy & Oncology* 87(3):432–438.

Monti, D., L. Moretti et al. 2000. C60 carboxyfullerene exerts a protective activity against oxidative stress-induced apoptosis in human peripheral blood mononuclear cells. *Biochemistry and Biophysical Research Communications* 277(3):711–717.

Morgan, N. Y., G. Kramer-Marek et al. 2009. Nanoscintillator conjugates as photodynamic therapy-based radiosensitizers: Calculation of required physical parameters. *Radiation Research* 171(2):236–244.

Nath, R., P. Bongiorni et al. 1990. Iododeoxyuridine radiosensitization by low- and high-energy photons for brachytherapy dose rates. *Radiation Research* 124(3):249–258.

Niroomand-Rad, A., C. R. Blackwell et al. 1998. Radiochromic film dosimetry: Recommendations of AAPM Radiation Therapy Committee Task Group 55. American Association of Physicists in Medicine. *Medical Physics* 25(11):2093–2115.

Norman, A., M. Ingram et al. 1997. X-ray phototherapy for canine brain masses. *Radiation Oncology Investigations* 5(1):8–14.

Norris, D. J. 1994. Electronic structurein semiconductor nanocrystals. In: *Semiconductor and Metal Nanocrystals: Synthesis and Electronic and Optical Properties*, ed. V. I. Klimov, 65–102. New York: Marcel Dekker.

Oldham, M., I. Baustert et al. 1998. An investigation into the dosimetry of a nine-field tomotherapy irradiation using BANG-gel dosimetry. *Physics in Medicine and Biology* 43(5):1113–1132.

Oleinick, N. L. and H. H. Evans. 1998. The photobiology of photodynamic therapy: cellular targets and mechanisms. *Radiat Res* 150(5 Suppl):S146–156.

Osuna, S., M. Swart et al. 2010. On the mechanism of action of fullerene derivatives in superoxide dismutation. *Chemistry* 16(10):3207–3214.

Peng, X., L. Manna et al. 2000. Shape control of CdSe nanocrystals. *Nature* 404(6773):59–61.

Pengfei, Q. F., O. Vermesh et al. 2003. Toward large arrays of multiplex functionalized carbon nanotube sensors for highly sensitive and selective molecular detection. *Nano Letters* 3(3):347–351.

Penning, L. C. and T. M. Dubbelman. 1994. Fundamentals of photodynamic therapy: cellular and biochemical aspects. *Anticancer Drugs* 5(2):139–146.

Pi, D., F. Wang et al. 2005. Luminescence behavior of Eu^{3+} doped LaF_3 nanoparticles. *Spectrochimica Acta Part A Molecular and Biomolecular Spectroscopy* 61(11–12):2455–2459.

Pissuwan, D., C. H. Cortie et al. 2007. Gold nanosphere–antibody conjugates for hyperthermal therapeutic applications. *Gold Bulletin* 40(2):121–129.

Poggi, M. M., C. N. Coleman et al. 2001. Sensitizers and protectors of radiation and chemotherapy. *Current Problems in Cancer* 25(6):334–411.

Pons, T., I. L. Medintz et al. 2007. On the quenching of semiconductor quantum dot photoluminescence by proximal gold nanoparticles. *Nano Letters* 7(10):3157–3164.

Pons, T., E. Pic et al. 2010. Cadmium-free CuInS2/ZnS quantum dots for sentinel lymph node imaging with reduced toxicity. *ACS Nano* 4(5):2531–2538.

Porta, F., G. Speranza et al. 2007. Gold nanoparticles capped by peptides. *Materials Science and Engineering B-Solid State Materials for Advanced Technology* 140(3):187–194.

Pradhan, A. S., J. I. Lee et al. 2008. Recent developments of optically stimulated luminescence materials and techniques for radiation dosimetry and clinical applications. *Journal of Medical Physics* 33(3):85–99.

Raaymakers, B. W., J. C. de Boer et al. 2011. Integrated megavoltage portal imaging with a 1.5 T MRI linac. *Physics in Medicine and Biology* 56(19):N207–N214.

Reese, A. S., S. K. Das et al. 2009. Integral dose conservation in radiotherapy. *Medical Physics* 36(3):734–740.

Regulla, D. F., L. B. Hieber et al. 1998. Physical and biological interface dose effects in tissue due to X-ray-induced release of secondary radiation from metallic gold surfaces. *Radiation Research* 150(1):92–100.

Rose, J. H., A. Norman et al. 1999. First radiotherapy of human metastatic brain tumors delivered by a computerized tomography scanner (CTRx). *International Journal of Radiation Oncology Biology Physics* 45(5):1127–1132.

Rzigalinski, B. A. 2005. Nanoparticles and cell longevity. *Technology in Cancer Research and Treatment* 4(6):651–659.

Rzigalinski, B. A., K. Meehan et al. 2006. Radical nanomedicine. *Nanomedicine (London)* 1(4):399–412.

Sahare, P. D., R. Ranjan et al. 2007. $K_3Na(SO_4)(2)$: Eu nanoparticles for high dose of ionizing radiation. *Journal of Physics D Applied Physics* 40(3):759–764.

Salah, N., P. D. Sahare et al. 2006. TL and PL studies on $CaSO_4$: Dy nanoparticles. *Radiation Measurements* 41(1):40–47.

Samia, A. C. S., X. B. Chen et al. 2003. Semiconductor quantum dots for photodynamic therapy. *Journal of the American Chemical Society* 125(51):15736–15737.

Santos Mello, R., H. Callisen et al. 1983. Radiation dose enhancement in tumors with iodine. *Medical Physics* 10(1):75–78.

Sapsford, K. E., T. Pons et al. 2007. Kinetics of metal-affinity driven self-assembly between proteins or peptides and CdSe–ZnS quantum dots. *Journal of Physical Chemistry C* 111(31):11528–11538.

Sato, O., T. Iyoda et al. 1996. Photoinduced magnetization of a cobalt–iron cyanide. *Science* 272(5262):704–705.

Scarantino, C. W., B. R. Prestidge et al. 2008. The observed variance between predicted and measured radiation dose in breast and prostate patients utilizing an in vivo dosimeter. *International Journal of Radiation Oncology Biology Physics* 72(2):597–604.

Schaffer, M., P. M. Schaffer et al. 2002. Application of Photofrin II as a specific radiosensitising agent in patients with bladder cancer— a report of two cases. *Photochemical & Photobiological Sciences* 1(9):686–689.

Schaffer, M., B. Ertl-Wagner et al. 2003. Porphyrins as radiosensitizing agents for solid neoplasms. *Current Pharmaceutical Design* 9(25):2024–2035.

Schaffer, M., B. Ertl-Wagner et al. 2005. The application of Photofrin II as a sensitizing agent for ionizing radiation— a new approach in tumor therapy? *Current Medical Chemistry* 12(10):1209–1215.

Schulte, R. W., A. J. Wroe et al. 2008. Nanodosimetry-based quality factors for radiation protection in space. *Zeitschrift Fur Medizinische Physik* 18(4):286–296.

Shi, L., B. Hernandez et al. 2006. Singlet oxygen generation from water-soluble quantum dot–organic dye nanocomposites. *Journal of the American Chemical Society* 128(19):6278–6279.

Soubra, M., J. Cygler et al. 1994. Evaluation of a dual bias dual metal-oxide–silicon semiconductor field-effect transistor detector as radiation dosimeter. *Medical Physics* 21(4):567–572.

Spencer, C. M., and K. L. Goa. 1995. Amifostine. A review of its pharmacodynamic and pharmacokinetic properties, and therapeutic potential as a radioprotector and cytotoxic chemoprotector. *Drugs* 50(6):1001–1031.

Stewart, D. A., and F. Leonard. 2004. Photocurrents in nanotube junctions. *Physical Review Letters* 93(10):107401.

Stewart, D. A., and F. Leonard. 2005. Energy conversion efficiency in nanotube optoelectronics. *Nano Letters* 5(2):219–222.

Stucky, G. D., and J. E. Mac Dougall. 1990. Quantum confinement and host/guest chemistry: Probing a new dimension. *Science* 247(4943):669–678.

Sun, Y. P., B. Zhou et al. 2006. Quantum-sized carbon dots for bright and colorful photoluminescence. *Journal of the American Chemical Society* 128(24):7756–7757.

Surujpaul, P. P., C. Gutierrez-Wing et al. 2008. Gold nanoparticles conjugated to [Tyr(3)]Octreotide peptide. *Biophysical Chemistry* 138(3):83–90.

Tannock, I. F. 1996. Treatment of cancer with radiation and drugs. *Journal of Clinical Oncology* 14(12):3156–3174.

Tarnuzzer, R. W., J. Colon et al. 2005. Vacancy engineered ceria nanostructures for protection from radiation-induced cellular damage. *Nano Letters* 5(12):2573–2577.

Timmerman, R. D., B. D. Kavanagh et al. 2007. Stereotactic body radiation therapy in multiple organ sites. *Journal of Clinical Oncology* 25(8):947–952.

Tsay, J. M., M. Trzoss et al. 2007. Singlet oxygen production by peptide-coated quantum dot-photosensitizer conjugates. *Journal of the American Chemical Society* 129(21): 6865–6871.

Usenko, C. Y., S. L. Harper et al. 2007. In vivo evaluation of carbon fullerene toxicity using embryonic zebrafish. *Carbon N Y* 45(9):1891–1898.

Wang, L., W. Yang et al. 2010. Tumor cell apoptosis induced by nanoparticle conjugate in combination with radiation therapy. *Nanotechnology* 21(47):475103.

Wardman, P. 2007. Chemical radiosensitizers for use in radiotherapy. *Clinical Oncology (Royal College of Radioliogists)* 19(6):397–417.

Whitehurst, C., M. L. P., J. V. Moore, T. A. King, and N. J. Blacklock. 1990. In vivo to post-mortem change in tissue penetration of red light. *Lasers in Medical Science* 5(4):395–398.

Wilson, B. C. and M. S. Patterson. 2008. The physics, biophysics and technology of photodynamic therapy. *Phys Med Biol* 53(9):R61–109.

Yang, W., P. W. Read et al. 2007. Novel FRET-based radiosensitization using quantum dot–photosensitizer conjugates. *Signals, Systems and Computers ACSSC* 2007:1861–1865.

Yang, W., P. W. Read et al. 2008. Semiconductor nanoparticles as energy mediators for photosensitizer-enhanced radiotherapy. *International Journal of Radiation Oncology Biology Physics* 72(3):633–635.

Yi, G. S., B. Q. Sun et al. 2001. Bionic synthesis of ZnS:Mn nanocrystals and their optical properties. *Journal of Materials Chemistry* 11(12):2928–2929.

Yin, J. J., F. Lao et al. 2009. The scavenging of reactive oxygen species and the potential for cell protection by functionalized fullerene materials. *Biomaterials* 30(4):611–621.

Zelefsky, M. J., Y. Yamada et al. 2008. Long-term results of conformal radiotherapy for prostate cancer: Impact of dose escalation on biochemical tumor control and distant metastases-free survival outcomes. *International Journal of Radiation Oncology Biology Physics* 71(4):1028–1033.

Zhang, X. J., J. Z. Xing et al. 2008. Enhanced radiation sensitivity in prostate cancer by gold-nanoparticles. *Clinical and Investigative Medicine* 31(3):E160–E167.

12

Radioactive Gold Nanoparticles for Tumor Therapy

Raghuraman Kannan
University of Missouri–Columbia

Kattesh V. Katti
University of Missouri–Columbia

Cathy Cutler
University of Missouri–Columbia

12.1 Introduction

The aim of this chapter is to review recent advances in the development of radioactive gold nanoparticles (NPs) as imaging and therapeutic agents in the diagnosis and treatment of cancer. The field of nanomedicine utilizes particulate matter with sizes that are the same as or smaller than that of cellular components allowing for NPs to cross multiple biological barriers to enhance delivery and retention for optimal treatment (Lammers et al. 2011; Nystrom and Woley 2011). The size similarity to cellular components makes NPs attractive candidates for curing functional abnormalities that instigate disease at the cellular level (Lammers et al. 2010). A major challenge in cancer therapy has been delivery and retention (Satija et al. 2007). Current approaches result in serious adverse side effects that significantly limit the number of therapeutic molecules that can be delivered to the tumor site, resulting in low efficacy. For effective tumor treatment, it is vital to increase the therapeutic payload to destroy cancer cells (Kievet and Zhang 2011). Control over the supply of therapeutic payload enables oncologists to deliver an optimized effective treatment for cancer patients. In this regard, it is important to note that nanoparticulates containing radioactive isotopes provide an opportunity to tune the radioactive therapeutic dose delivered to tumor cells (Katti et al. 2006).

NP-based radiopharmaceuticals can be broadly classified into two types: type A and type B (Figure 12.1). In type A nanoradiopharmaceuticals, radioisotopes are incorporated inside biodegradable NPs of organic matrices (Hamoudeh et al. 2007, 2008). For example, Hamoudeh and coworkers (2007, 2008) have shown that dirhenium decacarbonyl [Re$_2$(CO)$_{10}$] can be incorporated into biocompatible poly-L-lactide (PLLA) NPs. Rhenium-loaded PLLA NPs, upon exposure to neutron irradiation, result in the generation of radioactive rhenium, which subsequently can be used for cancer therapy (Hamoudeh et al. 2007, 2008). The group

has shown that the structural integrity of rhenium carbonyl is maintained after exposure to neutron irradiation. Type B NPs are derived from direct doping of radioisotopes into an inorganic NP matrix. In this type, radioisotopes become an integral part of the nanoparticulate matter. Type B nanoparticulates consist of radioisotopes generated in an inorganic nanomatrix, and that can, subsequently, be surface-conjugated to antibodies, peptides, or proteins (Woodward et al. 2011). For example, Actinium-225 can be incorporated into a LaPO$_4$ NP matrix. Decay of ^{225}Ac results in stable ^{209}Bi, with release of four α-particles. The study further shows a retention of ~50% of daughter nuclides within the La(^{225}Ac)PO$_4$ NPs over a period of 1 month (Woodward et al. 2011). In a similar fashion, therapeutic ^{198}Au NPs can be incorporated within the nonradioactive gold nanoparticle (AuNP) matrix to yield [^{198}AuAuNP] NPs (Kannan et al. 2006; Chanda et al. 2010a). These radioactive gold NPs have shown excellent therapeutic properties in animal models (Chanda et al. 2010b). In fact, recent publications from around the globe offer a wealth of information on how radioactive gold NPs can aid in the therapy of cancer (Chanda et al. 2010b; Kannan et al. 2012). This chapter will summarize the recent advances in radioactive gold NP-based cancer therapy, the inherent advantages of radioactive gold, and the future prospects of NPs in clinical applications. Recent advances in the synthesis of radioactive gold NPs and their applications in cancer therapy are discussed below.

12.2 Radioactive Gold

Gold has two medically useful radioactive isotopes, ^{198}Au and ^{199}Au, for locally irradiating and killing tumor cells. The radionuclidic properties of these isotopes are presented in Table 12.1. The half-life of ^{198}Au is 2.7 days, with the radiation of 90% β radiation with a mean energy of 312 KeV and a maximum energy of 961 KeV. ^{198}Au is a short-range β emitter, with a maximum

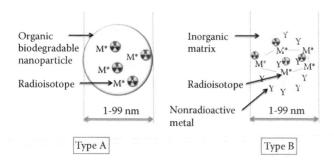

FIGURE 12.1 Type A and type B nanoradiopharmaceuticals.

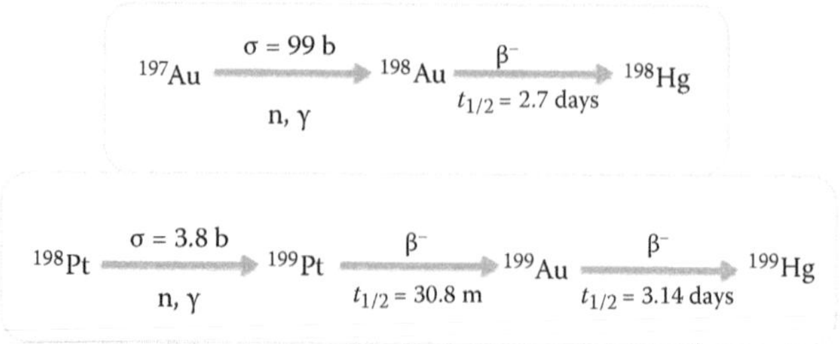

FIGURE 12.2 Production of Au-198 from natural gold and production of Au-199 from enriched platinum.

penetration in soft tissue of 3.8 mm. Fifty percent of β emissions penetrate only 0.38 mm and are the primary source of the radiating effect in localized areas. The therapeutic (or destructive) effect of ^{198}Au is localized in the area immediately around the source. Each ^{198}Au nucleus decays with subsequent transformation into ^{198}Hg. The excited state ^{198}Hg emits a 0.41-MeV γ-ray and transition to the stable state. Even though ^{198}Au decays to mercury, the nano- to picomolar amounts of mercury generated *in vivo* upon administration of therapeutic doses of gold is far below chemical toxicity levels. The half-life of ^{198}Au of 2.7 days is long enough to accommodate transportation and chemical synthesis for clinical applications. Both ^{198}Au and ^{199}Au emit γ-rays suitable for single-photon emission computed tomography (SPECT) imaging. However, the γ-ray of ^{198}Au is not optimal for SPECT imaging studies as the high energy is not easily collimated. The gamma energy of ^{199}Au is very close to that of technetium-99m, the most commonly used radioisotope for imaging, and is ideal for SPECT imaging. Imaging studies with ^{199}Au can be used to assess an individual's uptake, biodistribution and excretion of the drug, thus allowing for calculation of a tailored personalized dose and dosimetry assessment of the ^{198}Au analog. This type of imaging results in the delivery of a more effective dose as radiotherapy without imaging has been shown to be either over or under prescribed dose levels 25–60% of the time. Imaging with ^{199}Au can be used to obtain the exact patient dosimetry before the delivery of the therapeutic dose of ^{198}Au-198 for treatment. Gold provides excellent opportunities for the development of imaging-therapy "matched pair" agents. Development of such dual diagnostic-therapy (theranostic) and matched pairs would provide tremendous consistency in the follow-up of therapy studies and will also minimize regulatory steps leading to final approval by the Food and Drug Administration. Only a few elements in the periodic table possess the magic pair of both therapeutic β and imaging γ emission possibilities; ^{198}Au and ^{199}Au isotopes are among them.

^{198}Au is produced in high specific activity in research reactors by neutron bombardment of natural gold (Au-197) (see Figure 12.2 for details). Natural gold possesses a high capture cross section for neutrons (100 barns); thus, the production of ^{198}Au is fairly efficient. Irradiation times used are typically a few hours to days at a neutron flux of 8×10^4 neutrons/cm^2 s. Average sample size for irradiation is mgs of natural gold, resulting in ~100 mCi/g of radioactive gold. Typically, one in 3000 gold atoms is converted to ^{198}Au, with the number varying depending on irradiation time and flux. Usually, 90% of the radiogold produced is ^{198}Au and the remaining 10% is ^{199}Au. However, this ratio varies depending on the irradiation time and neutron flux used.

^{199}Au is produced carrier free (~200 mCi/μg) by irradiating enriched ^{198}Pt (see Figure 12.2 for details). Only a small fraction of ^{198}Pt atoms are converted into ^{199}Pt, which beta decays to produce ^{199}Au upon irradiation, requiring subsequent purification to yield ^{199}Au free from platinum. ^{199}Au decays by β decay with a half-life of 3.1 days yielding 159- and 208-keV photons with relative intensities of 100 and 23.42, respectively. ^{199}Au is used in tumor imaging applications. In this chapter, ^{198}Au and its therapeutic properties will be presented.

12.3 Colloidal Radioactive Gold

Radioactive colloidal gold has been used for treating tumors in human patients. Colloidal gold is a watery liquid with a characteristic intense cherry red color enabling easy identification of contamination. It is readily available in sterile, nonpyrogenic form. Colloidal gold is not produced with a definite size. One report indicates that it has two discrete ranges of size, as measured by electron microscopy, the larger size particles range from 40 to 90 nm and the smaller particles are in the range of 1–1.5 nm (Kerr et al. 1957). Other publications report the size of these colloidal particles in the range of ~60 nm, with each mL containing ~2×10^{12} particles (Wheeler et al. 1952). Results obtained from animal and human studies are summarized below:

- Colloidal gold has been used for interstitial irradiation in patients with prostatic cancer. Flocks and coworkers (1952, 1954; Bulkey and O'Connor 1959) have treated more than 1000 patients mainly with prostate cancer. The colloidal particles were directly injected into the tumor or into the tumor bed. The study reported 5- and 10-year survival rates of 54% and 38%, respectively.

TABLE 12.1 Radionuclidic Properties of Au-198 and Au-199

Radionuclide	$t_{1/2}$	β (keV)	γ (keV)
Au-198	2.7	312 (avg.); 961 (max)	412 (95%)
Au-199	3.1	86 (avg.); 453 (max)	159 (37); 208 (22)

- Several studies have been reported on the intraperitoneal application of radioactive gold for the treatment of ovarian carcinoma (Muller 1963; Keettel and Elkins 1956; Aure et al. 1971). Muller and coworkers (1963; Keettel and Elkins 1956; Aure et al. 1971) have established the life-saving role of colloidal gold. Other studies have reported higher overall survival rates in primary ovarian cancer patients treated with gold. Some stage I ovarian carcinomas with intact capsules shed abnormal cells that result in a late recurrence of the cancer (Muller 1963; Keettel and Elkins 1956; Aure et al. 1971). Utilization of colloidal gold led to the destruction of the harmful cells, resulting in a significant increase in overall survival rates.

- In another study, 165 patients with ovarian cancer initially treated by surgery and subsequently administered intraperitoneal ^{198}Au demonstrated stage-specific survival rates comparable to that of survival rate data in the literature. For stage I, the 5-year tumor-free survival was 70%; for stage II and stage III, the 5-year survival rates were 24% and 18%, respectively (Patyanik et al. 2002).

- A combination of ^{198}Au and external beam radiation has also been used to treat human patients with carcinoma (Hodgkinson et al. 1956). This combination treatment offers both local high-dose treatment for the primary tumor and lower dose treatment to remove microscopic metastatic spread in the tumor region. A comparative study between ovarian carcinoma patients treated with external beam therapy and radioactive ^{198}Au colloid therapy has shown that patients treated with colloidal ^{198}Au exhibited survival rates 14% to 20% higher than those given external radiation alone. This trend looks similar for patients in stages I, II, and III. The combination of ^{198}Au and external beam therapy results in the effective destruction of microscopic metastases in the peritoneal cavity (Hodgkinson et al. 1956).

- Toxicity and adverse effects of colloidal gold treatments were investigated in several studies (Muller 1963; Keettel and Elkins 1956; Aure et al. 1971; Patyanik et al. 2002; Hodgkinson et al. 1956). All studies have shown that colloidal gold did not induce any toxicity to the liver, spleen, or marrow in the quantities administered for therapeutic treatment. The studies also showed no radioactivity was observed in the human patients' urine or feces (Muller 1963; Keettel and Elkins 1956; Aure et al. 1971; Patyanik et al. 2002; Hodgkinson et al. 1956).

The studies discussed above clearly show there are definite benefits to using colloidal radioactive gold for tumor therapy (Muller 1963; Keettel and Elkins 1956; Aure et al. 1971; Patyanik et al. 2002; Hodgkinson et al. 1956). However, a few disadvantages are widely noted. One of the major disadvantages is the nonuniform distribution of therapeutic ^{198}Au within the tumor region. The size of the colloidal gold particles plays a crucial role in determining the diffusion and distribution within the organ. The size irregularity observed in colloidal gold NPs poses a significant problem in distribution and further utilization for humans. It is important to recognize that homogenous size distribution can be achieved by developing uniform-sized ^{198}Au NPs. The synthesis and current status of nanoparticulate ^{198}Au for therapeutic applications is presented in the next section.

12.4 Nanosized Radioactive Gold

Recently, there has been widespread interest in designing and developing well-defined ^{198}Au NPs for tumor therapy applications. Two different synthetic methodologies have been developed, and the therapeutic efficacies of these NPs in animal models have been published. A major advantage of nanosized radioactive particles is their potential to contain several radioactive atoms within a single NP. Delivery of a high therapeutic payload to tumor can be achieved by this method. Details on the nanosized ^{198}Au NPs are presented below.

Balogh and coworkers (2003) and Bielinska et al. (2002) have used a nanocomposite device (NCD) for encapsulation of radioisotopes, providing a nanoparticle with defined size and surface properties. Radioactive ^{198}Au incorporated NCDs can serve to stabilize and prevent agglomeration; in addition, they can serve as vehicles to transport radioactive AuNPs to tumor sites. By controlling the size of the NCD, the amount of radioactivity delivered to a tumor site can be tailored to meet the specific dose requirements for cell death. Using this method, the number of radioactive gold atoms can be increased without destroying the targeting ability of the NCD. Gold NCDs are synthesized as monodisperse hybrid NPs composed of radioactive guests immobilized by dendritic polymer hosts. In order to generate NPs, commercially available polymers including, poly(amidoamine) PAMAM dendrimers and tecto dendrimers are used as nanocomposites. The synthesis of ^{198}Au NPs by this method involves encapsulation of ^{198}Au within PAMAM dendrimers. Encapsulation was achieved by mixing dilute solutions of PAMAM dendrimer with an aqueous solution of $HAuCl_4$. Salt formation between the tetrachloroaurate anions and the dendrimer nitrogens ensure effective encapsulation of gold within the dendrimer matrix. Upon encapsulation, elemental gold was converted into ^{198}Au within the dendrimer matrix. The conversion of ^{197}Au to ^{198}Au in NCD was carried out both in solid and solution phase by direct neutron irradiation. Biodistribution studies using 5-nm-sized tritium-labeled PAMAM dendrimers in tumor models including mouse B16 melanoma, human prostate DU 145, and human KB squamous cell carcinoma mouse xenografts have been performed. Uptake of tritium activity in tumor tissue and also retention of activity for several weeks provided proof of principle that the radioactive isotope encapsulated nanodevices (such as PAMAMs with β emitting ^{198}Au) could be very useful in tumor therapy. A recent study showed that intratumoral injections of 74 μCi of poly{198Au} with a diameter of 22 nm, in a mouse model, resulted in a 45% reduction in tumor volume when compared with untreated mice.

Researchers at the University of Missouri have developed a method for direct generation of nanoparticulate ^{198}Au, using aqueous based nontoxic reducing agents, in biomolecular matrices (Kannan et al. 2006; Chanda et al. 2010a, 2010b; Kannan et al. 2012). Generation of [198]Au involves irradiation of natural gold ([197]Au) foil in a neutron flux of 8×10^{13} n cm^{-2} s^{-1}. Subsequently, irradiated gold foil is dissolved in *aqua regia* and finally reconstituted in dilute hydrochloric acid to form H[198]AuCl$_4$. Traditional hydridic and carboxylate-based reduction strategies cannot be extended to the production of [198]AuNPs. For example, reduction with NaBH$_4$ will not proceed in acidic medium; likewise, citric acid tends to protonate itself under acidic conditions, making it less desirable for reducing [198]AuCl$_4^-$ to produce [198]AuNPs. Therefore, more effective reducing agents that work under acidic pH with favorable kinetics for rapid reduction of radiometals at low concentrations (~10^{-8} M) are needed for the production of radioactive [198]AuNPs. In order to circumvent existing problems associated with the production of [198]AuNPs, trimeric alanine-based phosphine, P(CH$_2$NHCH(CH$_3$)COOH)$_3$ (THPAL), is used as a reducing agent (Figure 12.3). THPAL reduces [198]Au^{3+} ions under acidic conditions, in the presence of biocompatible gum arabic (GA) matrix to create GA-[198]AuNPs with excellent yields. It is believed that simple mixing of THPAL with [198]Au^{3+} ions initiates the phosphane-mediated reduction process with simultaneous oxidation into phosphanoxide. Subsequently, amino acid carboxylic groups in THPAL further reduce to yield uniform-sized GA-[198]AuNPs. GA serves as an excellent backbone for the stabilization of [198]AuNPs. GA is a plant extract approved by the Food and Drug Administration for use as a food additive in a variety of foods including yogurts, chocolates, soup mixes, and candies. The UV–visible overlay of GA-AuNP and GA[198]AuNP spectra is shown in Figure 12.4. The following studies were conducted to establish the therapeutic efficacy of GA[198]AuNPs and the results are presented in the subsequent section.

1. Tumor retention: In this study, GA[198]AuNPs were injected intratumorally in prostate tumor–bearing mice and analyzed for the retention of radioactivity.
2. Tumor ablation: Therapeutic efficacy of GA-[198]AuNPs in prostate tumor–bearing mice.

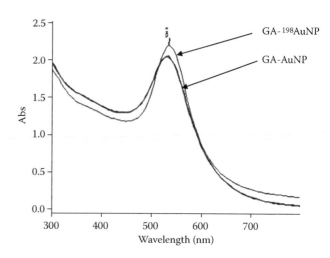

FIGURE 12.4 UV–visible overlay of GA-AuNP and GA[198]AuNP spectra.

12.4.1 Tumor Retention GA-[198]AuNPs

Retention of radioactive NPs within the tumor site, after intratumoral injection, is dependent on matching the size of the NPs with the tumor vasculature. In order to evaluate the retention of NPs in the tumor, GA-[198]AuNPs (3.5 µCi/tumor) were injected intratumorally in prostate tumor–bearing SCID mice. The distribution of radioactivity in various organs at different time points was analyzed (Figure 12.5). More than 75% of the injected dose was retained within the tumor site after the 24-h post injection period. The pore size for tumor vasculature ranges from 150 to 300 nm. GA-[198]AuNPs possess 12–18 nm core diameter and 85 nm hydrodynamic diameter. It is believed the approximate size match between the NPs and tumor vasculature results in the NP being retained within the tumor region. In a separate study, nonradioactive NPs were injected intratumorally in prostate tumor–bearing mice, and x-ray CT images were recorded. Twenty-four hours after administration, the NPs were observed to be distributed homogeneously within the tumor region (Figure 12.6).

12.4.2 Therapeutic Efficacy GA-[198]AuNPs

Therapeutic efficacy of GA[198]AuNPs was investigated by intratumoral administration of GA-[198]AuNP in prostate tumor-bearing SCID mouse models (Kannan et al. 2006; Chanda et al. 2010a, 2010b; Kannan et al. 2012). A single dose of GA-[198]AuNP (408 µC$_i$) was injected directly into the tumor. The tumor volumes were monitored for a period of 30 days post injection. The tumor volume reduction compared to untreated mice at the end of 30 days reached 82% (Figure 12.7). Any radioactivity not retained in the tumor site was cleared through the renal pathway. The reduction in tumor volume, as shown by GA-[198]AuNP in prostate tumor–bearing SCID mice, is an important clinical development demonstrating the potential of this agent for reducing the size of tumors before surgical resection, and perhaps to reduce or eliminate the need for surgical resection under certain circumstances.

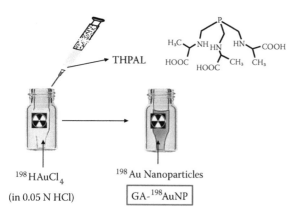

FIGURE 12.3 Scheme for generation of GA-[198]AuNP.

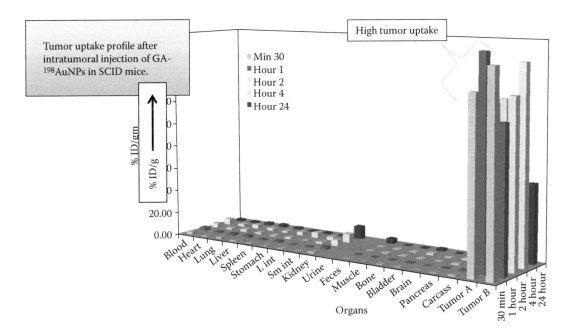

FIGURE 12.5 Retention of more than 80% radioactivity in tumor after 24 h in prostate tumor–bearing SCID mice model.

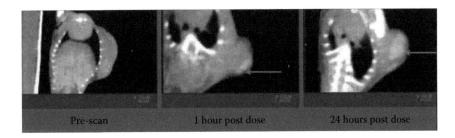

FIGURE 12.6 X-ray CT images of GA-AuNPs after intratumoral injection in prostate tumor–bearing SCID mice.

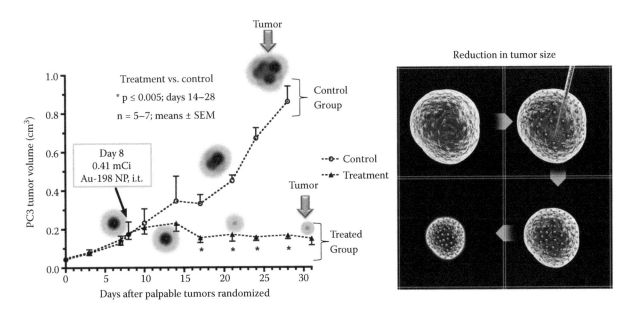

FIGURE 12.7 Decrease in prostate tumor volume after 30 days of intratumoral injection of GA-^{198}AuNPs in SCID mice model.

References

Aure, J. C., K. Hoeg, and P. Kolstad. 1971. Radioactive colloidal gold in the treatment of ovarian carcinoma. *Acta Radiologica: Therapy, Physics, Biology* 10(4):399–407.

Balogh, L. P., S. S. Nigavekar, A. C. Cook, L. Minc, and M. K. Khan. 2003. Development of dendrimer–gold radioactive nanocomposites to treat cancer microvasculature. *PharmaChem* 2(4):94–99.

Bielinska, A., J. D. Eichman, I. Lee, J. R. Baker, Jr., and L. Balogh. 2002. Imaging {Au0-PAMAM} gold–dendrimer nanocomposites in cells. *Journal of Nanoparticle Research* 4(5):395–403.

Bulkey, G. J., and V. J. O'Conor, Sr. 1959. Treatment of carcinoma of the prostate by interstitial irradiation with radioactive gold. Experimental and clinical studies. *Transactions of the American Association of Genito-Urinary Surgeons* 51:126–133.

Chanda, N., P. Kan, L. D. Watkinson, R. Shukla, A. Zambre, T. L. Carmack, H. Engelbrecht, J. R. Lever, K. Katti, G. M. Fent et al. 2010a. Radioactive gold nanoparticles in cancer therapy: therapeutic efficacy studies of GA-198AuNP nanoconstruct in prostate tumor-bearing mice. *Nanomedicine (Philadelphia, PA, United States)* 6(2):201–209.

Chanda, N., V. Kattumuri, R. Shukla, A. Zambre, K. Katti, A. Upendran, R. R. Kulkarni, P. Kan, G. M. Fent, S. W. Casteel et al. 2010b. Bombesin functionalized gold nanoparticles show in vitro and in vivo cancer receptor specificity. *Proceedings of the National Academy of Sciences of the United States of America* 107(19):8760–8765.

Flocks, R. H., H. D. Kerr, H. B. Elkins, and D. Culp. 1952. Treatment of carcinoma of the prostate by interstitial radiation with radio-active gold (Au 198): A preliminary report. *The Journal of Urology* 68(2):510–522.

Flocks, R. H., H. D. Kerr, H. B. Elkins, and D. Culp. 1954. The treatment of carcinoma of the prostate by interstitial radiation with radioactive gold (Au198): A follow-up report. *The Journal of Urology* 71(5):628–633.

Hamoudeh, M., H. Salim, D. Barbos, C. Paunoiu, and H. Fessi. 2007. Preparation and characterization of radioactive dirhenium decacarbonyl-loaded PLLA nanoparticles for radionuclide intra-tumoral therapy. *European Journal of Pharmaceutics and Biopharmaceutics* 67(3):597–611.

Hamoudeh, M., H. Fessi, H. Mehier, A. Al Faraj, and E. Canet-Soulas. 2008. Dirhenium decacarbonyl-loaded PLLA nanoparticles: Influence of neutron irradiation and preliminary in vivo administration by the TMT technique. *International Journal of Pharmaceutics* 348(1–2):125–136.

Hodgkinson, C. P., L. E. Preuss, L. A. Swinehart, and A. H. Velduhuis. 1956. Experimental studies on interstitial injection of radioactive colloidal gold (Au198). *Surgery, Gynecology & Obstetrics* 103(5):539–551.

Kannan, R., V. Rahing, C. Cutler, R. Pandrapragada, K. K. Katti, V. Kattumuri, J. D. Robertson, S. J. Casteel, S. Jurisson, C. Smith et al. 2006. Nanocompatible chemistry toward fabrication of target-specific gold nanoparticles. *Journal of the American Chemical Society* 128(35):11342–11343.

Kannan, R., A. Zambre, N. Chanda, R. Kulkarni, R. Shukla, K. Katti, A. Upendran, C. Cutler, E. Boote, and K. V. Katti. 2012. Functionalized radioactive gold nanoparticles in tumor therapy, Wiley interdisciplinary reviews. *Nanomedicine and Nanobiotechnology* 4(1):42–51.

Katti, K. V., R. Kannan, K. Katti, V. Kattumori, R. Pandrapraganda, V. Rahing, C. Cutler, E. J. Boote, S. W. Casteel, C. J. Smith et al. 2006. Hybrid gold nanoparticles in molecular imaging and radiotherapy. *Czechoslovak Journal of Physics* 56(Suppl. D):d23–d34.

Keettel, W. C., and H. G. Elkins. 1956. Experience with radioactive colloidal gold in the treatment of ovarian carcinoma. *American Journal of Obstetrics and Gynecology* 71(3):553–568.

Kerr, W. K., J. C. F. MacDonald, C. A. Smythe, and V. Comet. 1957. Experimental studies on radioactive colloidal gold in the prostate. *Canadian Journal of Surgery* 1:63–68.

Kievit, F. M., and M. Zhang. 2011. Cancer nanotheranostics: Improving imaging and therapy by targeted delivery across biological barriers. *Advanced Materials (Weinheim, Germany)* 23(36):H217–H247.

Lammers, T., F. Kiessling, W. E. Hennink, and G. Storm. 2010. Nanotheranostics and image-guided drug delivery: Current concepts and future directions. *Molecular Pharmaceutics* 7(6):1899–1912.

Lammers, T., S. Aime, W. E. Hennick, S. Gert, and F. Kiessling. 2011. Theranostic nanomedicine. *Accounts of Chemical Research* 44(10):1029–1038.

Muller, J. H. 1963. The treatment of cancer of the ovaries. Its improvement by intraperitoneal administration of radioactive colloidal gold. *Revue Francaise de Gynecologie et d'obstetrique* 58:197–214.

Nystrom, A. M., and K. L. Wooley. 2011. The importance of chemistry in creating well-defined nanoscopic embedded therapeutics: Devices capable of the dual functions of imaging and therapy. *Accounts of Chemical Research* 44(10):969–978.

Patyanik, M., A. Mayer, and I. Polgar. 2002. Results of ovary tumor treatment with abdominally administered ^{198}Au evaluated on the basis of long term follow up. *Pathology Oncology Research* 8:54–57.

Satija, J., U. Gupta, and N. K. Jain. 2007. Pharmaceutical and biomedical potential of surface engineered dendrimers. *Critical Reviews in Therapeutic Drug Carrier Systems* 24(3):257–306.

Wheeler, H. B., J. H. Rubenstein, M. D. Coleman, and T. W. Botsford. 1952. Techniques and radiation precaution for intratumor injections with radioactive colloidal gold. *A.M.A. Archives of Surgery* 65(2):283–289.

Woodward, J., S. J. Kennel, A. Stuckey, D. Osborne, J. Wall, A. J. Rondinone, R. F. Standaert, and S. Mirzadeh. 2011. LaPO$_4$ nanoparticles doped with actinium-225 that partially sequester daughter radionuclides. *Bioconjugate Chemistry* 22(4):766–776.

Nanoparticles for Selective Radioprotection of Normal Tissues

Jacob M. Berlin
City of Hope

Kathy A. Mason
M. D. Anderson Cancer Center

13.1 Introduction

Since the dawn of the nuclear era, radiation scientists have recognized the importance of protecting normal tissues—but not tumors—during radiotherapy and the need to apply this technology to related scenarios such as space travel for astronauts, radiation exposure from nuclear accidents, and as countermeasures for radiation terrorism (Weiss and Landauer 2009). Radiation therapy is widely used to treat cancer, with more than half of all patients receiving some form of radiotherapy as a component of their multidisciplinary care. Although the search for such agents has now lasted more than 60 years, only a single chemical agent, amifostine, has been translated to limited clinical use. The search continues for radiation protectors and mitigators with a renewed sense of urgency (Williams et al. 2010) largely as a response to world political pressures such as the 9/11/2001 terrorist attack on the World Trade Center in New York, the development of non-peaceful radiation capabilities by rogue nations, and environmental catastrophes such as the 2011 earthquake and subsequent tsunami in Japan that destroyed most of a nuclear power facility.

The general desirable characteristics of a radiation protector or mitigator that would be useful in combination with cancer radiotherapy include the following: (1) should ameliorate the toxic effects of radiotherapy to a clinically meaningful level; (2) should not inhibit or negatively modify the antitumor efficacy of radiotherapy; (3) should have an acceptable toxicity profile itself (Ryan et al. 2011); and (4) should demonstrate a favorable cost/benefit ratio. Although identification of such ideal agents has been slow, the current pipeline of potentially useful agents or combination of agents is expanding to include thiols, nitroxides, inhibitors of angiotensin-converting enzyme, metallothioneins, protease inhibitors, antioxidant vitamins, metalloelements, calcium antagonists, adenosine analogues, methylxanthines, superoxide dismutase, Chinese herbal medicines, antibiotics, cytokines,

immunomodulators (Murray and McBride 1996), and most recently, nanoparticles.

Within cells, radiation induces the formation of reactive oxygen species (ROS) that, in turn, react with DNA and RNA, resulting in permanent damage and eventual cell death. Despite marked advances in focusing the applied radiation on the tumor by methods such as proton therapy or intensity modulated radiotherapy, damage to surrounding healthy tissue remains the primary limitation of this treatment modality, resulting in many of the side effects commonly associated with radiation (gastrointestinal symptoms of nausea, vomiting, diarrhea, and dry mouth; acute and delayed inflammatory effects such as mucositis and pneumonitis; late tissue injuries including kidney failure, central nervous system demyelination, and dermal fibrosis and telangectasia; behavioral disturbances manifested as fatigue, somnolence, changes in appetite and taste). Moreover, these side effects can often be dose limiting, preventing complete tumor eradication.

Thus, selective protection of normal tissues during radiotherapy is an as yet unfulfilled goal in cancer therapy that has not been achieved in more than six decades of research. The snail's pace of progress was recently documented in excellent reviews of the history (Weiss and Landauer 2009) and development of radiation-protective agents (Murray and McBride 1996). There is only one chemical agent, amifostine, currently Food and Drug Administration (FDA)–approved for clinical use to prevent radiation-induced normal tissue injury (xerostomia). The selectivity exhibited by amifostine was not achieved by design, as amifostine emerged from the thousands of compounds evaluated by the U.S. military for their goal of protecting troops and civilians in the event of a nuclear attack (Alberts 1996; Spencer and Goa 1995). The endpoint of that research was a compound that would protect normal tissue, with no regard for avoiding protection of cancerous tissue. Amifostine is a small molecule phosphorothioate prodrug that must be dephosphorylated into

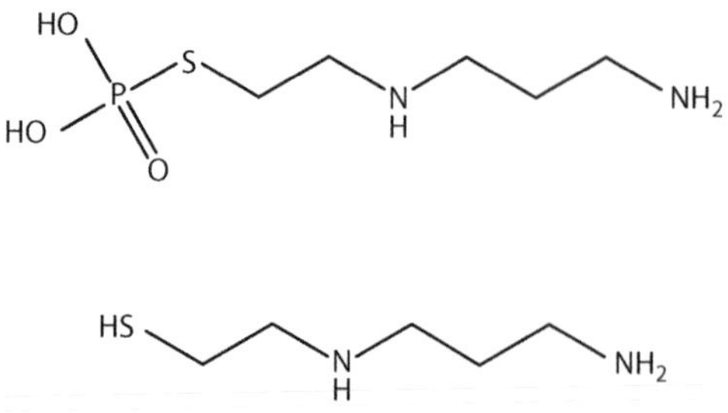

FIGURE 13.1 Structures of amifostine and its active metabolite, WR-1065, that results from dephosphorylation.

its active thiol form, which acts as an antioxidant to reduce cell damage due to ROS (Figure 13.1). Amifostine's selective protection of normal tissue is attributable to preferential accumulation of the active thiol form in normal cells as compared to cancer cells. This is ascribed to two factors: decreased activity of alkaline phosphatase in the tumor microenvironment and reduced cellular uptake of the active thiol because of the lower pH of the tumor microenvironment.

There are several major limitations to the clinical use of amifostine. Most importantly, it has significant side effects itself. Common side effects of amifostine include hypocalcemia, diarrhea, nausea, vomiting, sneezing, somnolence, and hiccups. In addition, although it is effective when administered orally to small animals such as mice, it is most clinically effective when administered intravenously to large animals, such as dogs or humans (Seeney 1979). Despite decades of work since amifostine's discovery, no other small molecule antioxidant has proven effective and selective enough to reach the clinic. There is thus a compelling need to find and evaluate new classes of biologically compatible antioxidants.

The definition of nanotechnology has been debated, but as it pertains to the development of materials for use in medicine, the most salient opinion is likely the recent guidance issued by the U.S. FDA, indicating that when considering whether an FDA-regulated product contains nanomaterials or otherwise involves the application of nanotechnology, the FDA will ask:

1. Whether an engineered material or end product has at least one dimension in the nanoscale range (approximately 1 to 100 nm).
2. Whether an engineered material or end product exhibits properties or phenomena, including physical or chemical properties or biological effects that are attributable to its dimension, even if these dimensions fall outside the nanoscale range, up to 1 μm.

Such nanomaterials are of particular interest for evaluation of biologically compatible antioxidants because nanoparticles have markedly different biodistribution and metabolic profiles than small molecules or enzymes and may afford improved *in vivo* efficacy (Riehemann et al. 2009). This may allow them to be directed to sites of interest for antioxidant therapy and to

persist in those locations for much longer times than are possible for small molecules or enzymes. Intriguingly, a variety of nanoparticles have proven to be biologically compatible antioxidants. Fullerene derivatives (Ali et al. 2008, 2004; Dugan et al. 1997, 2001; Quick et al. 2008), hydrophilic carbon clusters (Lucente-Schultz et al. 2009), nanoparticles composed of cerium oxide (nanoceria) (Das et al. 2007; Hirst et al. 2009; Schubert et al. 2006) and other metals (Martin et al. 2010) have been shown to be antioxidants and neuroprotective. However, although nanoparticles have shown promising antioxidant strength, little selectivity has been demonstrated for protecting normal cells or tissue in preference to cancer cells.

13.2 The Problem of Biodistribution

Preferential accumulation of nanoparticles in normal tissue is a significant challenge. In order to achieve selective distribution to normal tissue two major problems must be overcome: clearance by the reticuloendothelial system (RES) and an inherent bias toward accumulation in tumor tissue. The RES is composed of the liver, spleen, lymph nodes, bone marrow, and associated circulating cells, including macrophages. The RES is specifically designed to phagocytose objects on the size scale of nanoparticles, i.e., viruses and bacteria. Thus, it is no surprise that when nanoparticles are administered intravenously, the vast majority are distributed to the liver and spleen. A very small amount is also generally found in the lymph nodes and bone marrow. Doxil® can be considered as the exception that proves the rule. Doxil is a liposomal formulation of doxorubicin that is FDA-approved and used in cancer chemotherapy. It is thus an exception as most compounds never reach the clinic, and this suggests that Doxil likely has a better biodistribution profile than most nanoparticles. Yet, animal models indicate that only 5% of the injected dose reaches the tumor with >37% residing in the liver and spleen (Harrington et al. 2000). So, if the goal is to localize nanoparticles in normal tissue surrounding a tumor that is to be irradiated, the first challenge is to get enough of the injected dose into that local area to prove effective.

However, in many ways, the challenge becomes even tougher in the local area of the tumor as one of the main reasons nanoparticles have received so much attention for cancer therapy is that they preferentially accumulate in the tumor microenvironment relative to adjacent normal tissue because the enhanced permeability and retention (EPR) effect (Matsumura and Maeda 1986). Since the perivascular cells and the basement membrane, or the smooth-muscle layer, are frequently absent or abnormal in the vascular wall in malignant tissue, the blood vessels in malignant tissue are more disorderly, dilated, leaky, or defective than those in normal tissue. Thus, larger structures, such as nanoparticles, preferentially extravasate from the vasculature at the tumor site. In addition, malignant tissues have poor lymphatic drainage, resulting in prolonged retention of the nanoparticles in the tumor. Taken together, RES clearance and the EPR effect pose a significant hurdle to preferential distribution of nanoparticles to normal tissue.

13.3 Emerging Strategies for Selective Radioprotection

There are two strategies for realizing selective protection of normal tissue: nanoparticles can be designed to preferentially accumulate in normal tissue or they can be engineered to be more potent antioxidants in the normal tissue microenvironmental milieu (Figure 13.2). As described above, preferential accumulation in normal tissue is difficult to achieve, particularly outside of the RES system. Indicative of this, this strategy has thus far mostly been used in magnetic resonance imaging (MRI) of RES organs. For example, ferucarbotran (Resovist®) is a superparamagnetic iron oxide (SPIO) nanoparticle clinically approved in Europe for liver-specific MRI contrast enhancement (Reimer and Balzer 2003). Within minutes of injection of ferucarbotran, 80% of the injected dose is found in the liver and 5–10% is found in the spleen (Hamm et al. 1994; McLachlan et al. 1994; Weissleder et al. 1989). SPIO particles function as MRI contrast agents by decreasing the intensity of one component of the MRI signal (T2). After the phagocytic cells have accumulated the ferucarbotran, the MRI signal is decreased in healthy liver and spleen tissue. However, malignant tumors generally lack large numbers of phagocytic cells, so they appear as hyperintense/bright lesions contrasted against the hypointense/black liver. The same strategy has been extended in clinical studies to lymph-node metastases in prostate cancer (Harisinghani et al. 2003). Thus, there is suggestive evidence that nanoparticles can be distributed to normal RES cells and avoid tumor accumulation. At this time, we are not aware of any reports making use of this strategy for treating RES tumors with the combination of radiation therapy and radioprotective nanoparticles. We expect that such studies will be pursued, and we are hopeful that they will improve the efficacy of treatment for these tumor types.

In order to apply a strategy of preferential accumulation outside of the RES system, a different approach must be taken. The primary strategy for altering the biodistribution of nanoparticles is to modify the surface with poly(ethylene glycol) to prolong blood circulation and limit RES clearance and additionally conjugate to the particles an antibody that binds to the tissue of interest. To date, the vast majority of research in this area has sought to use antibodies that bind to cells in the tumor microenvironment. There is a shortage of antibodies known to bind to normal tissue and not to cancerous tissue. However, in the imaging literature, there is an example of using this strategy to image pancreatic cancer (Montet et al. 2006). It was found that bombesin peptide binding receptors were highly expressed on normal pancreatic tissue but were depleted on pancreatic ductal adenocarcinoma. Nanoparticles that are MRI contrast agents were then functionalized with bombesin so that they would be preferentially bound by the normal pancreatic tissue. This resulted in a marked enhancement in detection of pancreatic cancer in a mouse model. To the best of our knowledge, this strategy has not been used for many other tumor types and, in the case of pancreatic cancer, it has not yet been tested for treatment with the combination of radiation therapy and radioprotective nanoparticles targeted to normal tissue.

The final strategy for selective radioprotection is to make use of nanoparticles that are inherently more potent antioxidants in normal tissue than cancerous tissue. In this manner, it does not matter if the problem of selective biodistribution cannot be solved. The nanoparticles that reside in normal tissue will simply be so much more active than those in cancerous tissue that selective protection will occur. There are several reports in the literature on this emerging strategy, although the mechanisms behind the enhanced potency remain uncertain or unknown. In one example, silica nanoparticles were coated

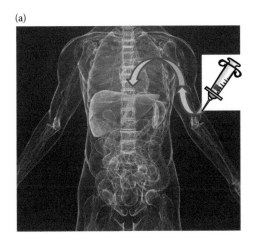
(a)

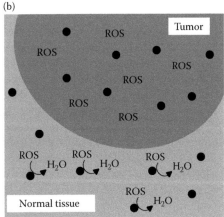
(b)

FIGURE 13.2 Strategies for radioprotection of normal tissue by nanoparticles are envisioned as: (a) uptake of nanoparticles by liver and spleen is exploited for protection of those organs during radiotherapy, as tumors in these organs generally have fewer macrophages and uptake fewer nanoparticles than normal tissue, and (b) nanoparticles are designed to be more potent antioxidants when in normal tissue than when in the tumor microenvironment.

with melanin for the protection of bone marrow during radiation therapy in a mouse model of melanoma (Schweitzer et al. 2010). Melanin was selected for the coating as the authors had found that melanin could control the dissipation of high-energy recoil electrons, preventing secondary ionizations and the generation of ROS (Schweitzer et al. 2009). Only 0.3% of the administered nanoparticles were found in the bone marrow 3 h after the injection. Unfortunately, distribution to the tumor was not reported. Nonetheless, the melanin-coated nanoparticles seemed to reduce the impact of radioimmunotherapy (RIT) on nude mice bearing A2058 human metastatic melanoma tumors on their flanks. When these mice were treated with 1 mCi of [188]Relabeled melanin-binding monoclonal antibody 6D2 alone or were pretreated with melanin-coated nanoparticles and then given the RIT, there was no difference in antitumor efficacy. The melanin-coated nanoparticle-treated group had a significantly smaller reduction in white blood cells at day 3 after therapy. However, 28 days after therapy, both groups had an equivalent number of white blood cells. Although this is a pilot study, the results suggest that the nanoparticles may be more effective antioxidants in the bone marrow as compared to the tumor. The fact that the amount of nanoparticles in the tumor was not measured makes it impossible to evaluate if this is truly the case, since it is possible that there are simply fewer nanoparticles in the tumor. Further studies are certainly required to evaluate if the difference in white blood cell numbers is attributable to radioprotection of the bone marrow, if the nanoparticles are more potent antioxidants in the bone marrow as compared to the tumor and, if this is the case, what is the mechanism for the difference in potency.

On the other hand, studies have identified nanoceria as more active antioxidants in normal tissue as compared to the tumor microenvironment, but no study has yet reported whether antitumor efficacy is maintained when radiotherapy is combined with nanoceria. In an intriguing finding, it was reported that in tissue culture nanoceria can protect a normal breast cancer cell line (CRL8798) from irradiation, but that the nanoceria do not confer the same protection on a breast cancer cell line (MCF-7) (Tarnuzzer et al. 2005). The nanoceria used in this study were cerium oxide nanoparticles 3–5 nm in diameter that contained mixed valence states of cerium (Ce^{3+} and Ce^{4+}). For the cell studies, CRL8798 and MCF-7 cells were plated in 96-well plates and exposed to 10 Gy of radiation. This resulted in 40–50% cell death for both CRL8798 and MCF-7 cells alone. However, when the cells were pretreated with 10 nM cerium oxide nanoparticles 24 h before irradiation, CRL8798 cells were protected almost 100%, whereas no effect was seen for MCF-7 cells, which experienced equivalent cell death as when untreated. At the time of this study, it was unclear what mechanism was responsible for the difference in protection.

In a follow-up study, nanoceria were observed to be taken up by normal human dermal fibroblasts and a squamous carcinoma cell line (SCL-1) in a similar manner (Alili et al. 2011). In both cell lines, the nanoceria were broadly distributed in the cytoplasm and were aggregated. The intracellular size for observed

Fenton reaction:	$2 H_2O_2 \rightarrow HO^\bullet + OH^- + HOO^\bullet + H^+$
Superoxide dismutase reaction:	$O_2^- + 2 H^+ \rightarrow H_2O_2$
Catalase reaction:	$2 H_2O_2 \rightarrow 2 H_2O + O_2$

FIGURE 13.3 Nanoceria are capable of at least the three different redox reactions shown. It may be the balance between these reactions that confers different activity on nanoceria when they are in different biological environments.

nanoparticles was ≥50 nm, whereas the administered nanoparticles were 3–5 nm. The most significant point is that no difference was observed for uptake between the two cell lines, so a difference in intracellular concentration of nanoceria is unlikely to explain the dramatically different radioprotection afforded the normal breast cells as compared to the cancerous cells in the previous study. In another study, the same authors demonstrated that nanoceria are capable of carrying out a Fenton-like reaction that produces hydroxyl and peroxide radicals from hydrogen peroxide (Heckert et al. 2008). This is in addition to the catalase activity (conversion of hydrogen peroxide to water and oxygen) and superoxide dismutase (SOD) activity (conversion of superoxide into oxygen and hydrogen peroxide) that the nanoceria possess (Figure 13.3). It is possible that the mechanism for selective protection is a trade-off between these three modes of action. For example, the higher level of ROS present in cancer cells may lead to the nanoceria favoring a Fenton-like reaction over the other pathways. This would lead to a further increase in ROS in the cancer cells, whereas in normal cells catalase and SOD activity could be favored, leading to a decrease in ROS. Further studies are necessary to better understand the mechanism as the selectivity demonstrated by the nanoceria in the pilot study is remarkable. Studies are also required to investigate if this selectivity is maintained *in vivo*.

13.4 Conclusion

Nanoparticles are poised to make an impact on the field of selective radioprotection of normal tissue during cancer radiotherapy and nonmedically related radiation exposures. Their unique modes of action and biopersistence offer possibilities unimagined for small molecules and enzymes. Numerous reports have described antioxidant properties for a variety of nanomaterials and corresponding protection of normal cells from radiation induced injury. However, there are limited reports on selective protection of normal cells in preference to cancer cells and still fewer reports on selective protection *in vivo* of normal tissue in preference to tumor tissue. Significant hurdles will have to be overcome to achieve selective protection in animal models. Nanoparticles are inherently biased toward accumulation in the RES organs, and the particles that do reach the area near the tumor generally accumulate in the tumor. As such, the most developed selective accumulation in normal tissue has been observed in the RES organs. Thus far, this property has only been exploited for imaging, but we expect this will be the area with the lowest barrier to implementing nanoparticles as selective radioprotectors. Beyond this application, particles must be developed with radioprotective activity that is modulated by the local tissue microenvironment such that they behave differently in normal tissue and cancerous tissue. There are intriguing reports of this behavior, but the mechanisms remain unclear. Future work will hopefully illuminate these mechanisms and lead to the refinement of particles with enough selectivity to progress toward the clinic. Selective radioprotection is an unmet clinical need that in the near future may be addressed with novel nanoparticles.

References

Alberts, D. S. 1996. Introduction: Applications of amifostine in cancer treatment. *Seminars in Oncology* 23:1.

Ali, S. S., J. I. Hardt, K. L. Quick, J. S. Kim-Han, B. F. Erlanger, T. T. Huang et al. 2004. A biologically effective fullerene (C60) derivative with superoxide dismutase mimetic properties. *Free Radical Biology and Medicine* 37:1191–1202.

Ali, S. S., J. I. Hardt, and L. L. Dugan. 2008. SOD activity of carboxyfullerenes predicts their neuroprotective efficacy: A structure–activity study. *Nanomedicine* 4:283–294.

Alili, L., M. Sack, A. S. Karakoti, S. Teuber, K. Puschmann, S. M. Hirst et al. 2011. Combined cytotoxic and anti-invasive properties of redox-active nanoparticles in tumor–stroma interactions. *Biomaterials* 32:2918–2929.

Das, M., S. Patil, N. Bhargava, J. F. Kang, L. M. Riedel, S. Seal et al. 2007. Auto-catalytic ceria nanoparticles offer neuroprotection to adult rat spinal cord neurons. *Biomaterials* 28:1918–1925.

Dugan, L. L., D. M. Turetsky, C. Du, D. Lobner, M. Wheeler, C. R. Almli et al. 1997. Carboxyfullerenes as neuroprotective agents. *Proceedings of the National Academy of Sciences of the United States of America* 94:9434–9439.

Dugan, L. L., E. G. Lovett, K. L. Quick, J. Lotharius, T. T. Lin, and K. L. O'Malley. 2001. Fullerene-based antioxidants and neurodegenerative disorders. *Parkinsonism and Related Disorders* 7:243–246.

Hamm, B., T. Staks, M. Taupitz, R. Maibauer, A. Speidel, A. Huppertz et al. 1994. Contrast-enhanced MR imaging of liver and spleen: First experience in humans with a new superparamagnetic iron oxide. *Journal of Magnetic Resonance Imaging* 4:659–668.

Harisinghani, M. G., J. Barentsz, P. F. Hahn, W. M. Deserno, S. Tabatabaei, C. H. van de Kaa et al. 2003. Noninvasive detection of clinically occult lymph-node metastases in prostate cancer. *New England Journal of Medicine* 348:2491–2499.

Harrington, K. J., G. Rowlinson-Busza, K. N. Syrigos, P. S. Uster, R. M. Abra, and J. S. Stewart. 2000. Biodistribution and pharmacokinetics of 111In-DTPA-labelled pegylated liposomes in a human tumour xenograft model: Implications for novel targeting strategies. *British Journal of Cancer* 83:232–238.

Heckert, E. G., S. Seal, and W. T. Self. 2008. Fenton-like reaction catalyzed by the rare earth inner transition metal cerium. *Environmental Science and Technology* 42:5014–5019.

Hirst, S. M., A. S. Karakoti, R. D. Tyler, N. Sriranganathan, S. Seal, and C. M. Reilly. 2009. Anti-inflammatory properties of cerium oxide nanoparticles. *Small* 5:2848–2856.

Lucente-Schultz, R. M., V. C. Moore, A. D. Leonard, B. K. Price, D. V. Kosynkin, M. Lu et al. 2009. Antioxidant single-walled carbon nanotubes. *Journal of the American Chemical Society* 131:3934–3941.

Martin, R., C. Menchon, N. Apostolova, V. M. Victor, M. Alvaro, J. R. Herance et al. 2010. Nano-jewels in biology. Gold and platinum on diamond nanoparticles as antioxidant systems against cellular oxidative stress. *ACS Nano* 4:6957–6965.

Matsumura, Y., and H. Maeda. 1986. A new concept for macromolecular therapeutics in cancer chemotherapy: Mechanism of tumoritropic accumulation of proteins and the antitumor agent smancs. *Cancer Research* 46:6387–6392.

McLachlan, S. J., M. R. Morris, M. A. Lucas, R. A. Fisco, M. N. Eakins, D. R. Fowler et al. 1994. Phase I clinical evaluation of a new iron oxide MR contrast agent. *Journal of Magnetic Resonance Imaging* 4:301–307.

Montet, X., R. Weissleder, and L. Josephson. 2006. Imaging pancreatic cancer with a peptide–nanoparticle conjugate targeted to normal pancreas. *Bioconjugate Chemistry* 17:905–911.

Murray, D., and W. H. McBride. 1996. Radioprotective agents. In: *Kirk-OthmerEncyclopedia of Chemical Technology*, edited by J. I. Kroschwitz, and M. Howe-Grant, 963–1006. New York: John Wiley & Sons.

Quick, K. L., S. S. Ali, R. Arch, C. Xiong, D. Wozniak, and L. L. Dugan. 2008. A carboxyfullerene SOD mimetic improves cognition and extends the lifespan of mice. *Neurobiology of Aging* 29:117–128.

Reimer, P., and T. Balzer. 2003. Ferucarbotran (Resovist): A new clinically approved RES-specific contrast agent for contrast-enhanced MRI of the liver: Properties, clinical development, and applications. *European Radiology* 13:1266–1276.

Riehemann, K., S. W. Schneider, T. A. Luger, B. Godin, M. Ferrari, and H. Fuchs. 2009. Nanomedicine—challenge and perspectives. *Angewandte Chemie International Edition in English* 48:872–897.

Ryan, J. L., S. Krishnan, B. Movsas, C. N. Coleman, B. Vikram, and S. S. Yoo. 2011. Decreasing the adverse effects of cancer therapy: An NCI workshop on the preclinical development of radiation injury mitigators/protectors. *Radiation Research* 176:688–691.

Schubert, D., R. Dargusch, J. Raitano, and S. W. Chan. 2006. Cerium and yttrium oxide nanoparticles are neuroprotective. *Biochemical and Biophysical Research Communications* 342:86–91.

Schweitzer, A. D., R. C. Howell, Z. Jiang, R. A. Bryan, G. Gerfen, C. C. Chen et al. 2009. Physico-chemical evaluation of rationally designed melanins as novel nature-inspired radioprotectors. *PLoS One* 4:e7229.

Schweitzer, A. D., E. Revskaya, P. Chu, V. Pazo, M. Friedman, J. D. Nosanchuk et al. 2010. Melanin-covered nanoparticles for protection of bone marrow during radiation therapy of cancer. *International Journal of Radiation Oncology, Biology, Physics* 78:1494–1502.

Seeney, T. R. 1979. A Survey of Compounds from the Antiradiation Drug Development Program of the U.S. Army Medical Research Development Command. Walter Reed Army Inst. of Research. U.S. Army Medical Research Development Command, Washington, D.C.

Spencer, C. M., and K. L. Goa. 1995. Amifostine. A review of its pharmacodynamic and pharmacokinetic properties, and therapeutic potential as a radioprotector and cytotoxic chemoprotector. *Drugs* 50:1001–1031.

Tarnuzzer, R. W., J. Colon, S. Patil, and S. Seal. 2005. Vacancy engineered ceria nanostructures for protection from radiation-induced cellular damage. *Nano Letters* 5:2573–2577.

Weiss, J. F., and M. R. Landauer. 2009. History and development of radiation-protective agents. *International Journal of Radiation Biology* 85:539–573.

Weissleder, R., D. D. Stark, B. L. Engelstad, B. R. Bacon, C. C. Compton, D. L. White et al. 1989. Superparamagnetic iron oxide: Pharmacokinetics and toxicity. *American Journal of Roentgenology* 152:167–173.

Williams, J. P., S. L. Brown, G. E. Georges, M. Hauer-Jensen, R. P. Hill, A. K. Huser et al. 2010. Animal models for medical countermeasures to radiation exposure. *Radiation Research* 173:557–578.

IV

Nanomaterials for Hyperthermia and Thermal Therapy

Gold Nanoparticle–Mediated Hyperthermia in Cancer Therapy

Dev K. Chatterjee
M. D. Anderson Cancer Center

Sunil Krishnan
M. D. Anderson Cancer Center

14.1 Introduction

The interaction of heat with the human body in health and disease has long fascinated physicians and given rise to the question—can heat aid healing? Over many centuries, experience has led to the tentative conclusion that under suitable conditions heat can indeed play a role in the treatment of several diseases, not the least of which is cancer. Heat can be used as a means of directly killing tumors cells, akin to surgical removal of tumors. It can result in a spectrum of effects ranging from a protective response via heat shock proteins to cellular death largely through necrosis mediated by irreparable coagulation of proteins and other biological macromolecules. However, like all therapies, lethal levels of untargeted heat can cause significant collateral damage to normal tissues. Alternatively, a milder rise in temperature, by itself not lethal, can aid and abet other treatment modalities, increasing their therapeutic efficiency. Therefore, to begin with, we need to define some terms to unambiguously understand which form of heat we are discussing. In medical literature, hyperthermia is generally understood to be a rise in temperature of body tissues, globally or locally, sometimes induced with a therapeutic intent. When the intent is destruction of cancerous tissue solely through the agency of heat, we define the therapeutic modality to be *thermoablation*. However, as mentioned above heat can also be used at levels less than that required for ablation. This is done to make the cancerous tissue more vulnerable to other forms of treatment, such as ionizing radiation or chemotherapy. Often, this form of subablative heating is simply called *hyperthermia*, usually understood to be distinct from thermoablation in this context. In this review, we will use the terms thermoablation and hyperthermia to be mutually exclusive—separated in meaning by the intent of the treatment rather than in terms of a definite temperature.

The threshold temperature for ablation depends on several factors, with time of exposure being possibly the most important of them. However, for practical purposes, a temperature above 45°C is generally considered ablative, although typically much higher temperatures (>50°C) are obtained focally for necrosis of tumors. Hyperthermia treatment generally aims to restrict temperature rises to 41–45°C.

Although burning off tumors with ablative temperatures is an attractive option, lower temperatures of about 41–42°C (mild temperature hyperthermia) are efficacious as adjuncts to radiation therapy and chemotherapy. This is largely driven by an *increase in blood flow* (often sustained for 1 to 2 days) and oxygen delivery and a decrease in oxygen demand (due to hyperthermia-induced cell death and metabolic suppression resulting in a shift toward anaerobic metabolism) that converge to *increase tumor tissue oxygenation*. Hyperthermia can also *activate immunological responses*. The molecular mechanisms of these effects of hyperthermia are being unraveled, and we now have a greater understanding of the subcellular events that render cells susceptible to various forms of damage (Fuller et al. 1994; Harmon et al. 1991). It is now known that there is no basic difference among tumor cells, tumor vascular endothelial cells, and normal cells in their sensitivity to heat-induced cytotoxicity. However, inefficient blood flow and oxygen transport through disordered tumor neovasculature results in an acidotic and nutrient-deprived environment within the tumor that makes them more thermosensitive (Bass et al. 1978). The greater sensitivity of hypoxic areas to heat allows for synergy with radiation therapy since hypoperfused areas within the tumor core are less sensitive to radiation-induced cytotoxicity, which depends on the generation of oxygen free radicals within well perfused regions. Increased perfusion also improves the delivery of chemotherapeutic drugs to the poorly vascularized tumor cores.

Accordingly, numerous clinical and preclinical studies have documented significant improvements in outcome when hyperthermia is combined with chemotherapy or radiation therapy, particularly for tumors of the prostate, breast, bladder, brain, cervix, head and neck, lung, rectum, and esophagus. Typically, this combined treatment regimen does not increase treatment toxicity but improves local control, cure, and/or palliation (Franckena et al. 2010; Huilgol et al. 2010a, 2010b; Hurwitz et al. 2011; Moros et al. 2010; Van den Berg et al. 2006; Vasanthan et al. 2005; Zagar et al. 2010).

All these point to significant advantages in adding hyperthermia to the cancer caregiver's portfolio of therapeutic choices. Yet, despite knowledge of the therapeutic effect of hyperthermia on cancer being available to us for more than a century, there is very sparse clinical adoption of this treatment modality. The reason lies not in the *effectiveness* of hyperthermia, but in the difficulties in *effecting* hyperthermia in a controlled and specific manner. This is not a trivial challenge for clinicians, and the solution of this problem might well result in a universal adoption of this effective remedy.

The earliest methods of induction of hyperthermia in cancer came via serendipitous observations of the effect of fever on tumors. A correlation between erysipelas (a streptococcal skin infection) and tumor regression had been observed for over a century before Dr. William Coley first documented evidence of a relationship between infection and cancer regression in sarcoma patients in 1891 (Coley 1891). His attempts to recreate this phenomenon for the treatment of cancers culminated in the generation of cocktails of bacteria (Coley's toxin) that intentionally induced a fever to effect an antitumor response. Although this probably represents one of the first instances of the clinical use of hyperthermia for cancer therapy, it was also among the first demonstrations of the efficacy of immunotherapy. Since then, more localized and relatively safer methods of hyperthermia—either singly or in combination with conventional therapy—have been used by many investigators to treat cancer (Doss and McCabe 1976; Friedenthal et al. 1981; Irish et al. 1986; Kim et al. 1982; Lele 1980; Luk et al. 1984; Magin and Johnson 1979;

Seegenschmiedt et al. 1993; Stewart and Gibbs 1984; Thrall 1980). The fascinating history and progress of tumor management with hyperthermia has been covered in detail elsewhere (Chen et al. 2011; Day et al. 2009; DeNardo and DeNardo 2008; Everts 2007; Kennedy et al. 2011; Krishnan et al. 2010; Rao et al. 2010). A brief overview of the more modern techniques for achieving hyperthermia is presented below.

14.2 Hyperthermia Techniques

The ideal means of achieving hyperthermia would be the one that could result in *specific, controlled, uniform* hyperthermia through a *clinically robust* technique. Often, there is a payoff between these needs. Three types of hyperthermia are traditionally used in clinical practice—whole body, regional, and local hyperthermia (Figure 14.1). Whole body hyperthermia, as the name indicates, raises body temperature as a whole. Dr. Foley's attempts as therapeutic induction of fever can be considered as an early example of this method. Currently, it is achieved by such methods as hot water blankets and thermal chambers. Although this method obviously exposes normal tissues to the rigors of higher temperature, it can be advantageous when dealing with metastatic cancer where focal hyperthermia would be ineffective in controlling the cancer. Nonetheless, whole body hyperthermia techniques are rarely used in clinical practice. However, recent unpublished data seem to indicate that tumor-bearing mice kept in housing facilities with a higher ambient temperature seem to have significantly longer tumor doubling time than control groups kept in regular air-conditioned rooms. Although more research needs to be done to corroborate these findings and extrapolate to clinical scenarios, it appears to support the hypothesis that mild temperature hyperthermia has a positive effect on the body's immune response to cancer.

Regional hyperthermia is generally done by perfusing the body's cancer-bearing region with heated liquids. Two of the more popular techniques are the perfusion of a part of the patient's blood—taken out and warmed *ex vivo*—into an artery supplying the limb containing the tumor, and perfusing

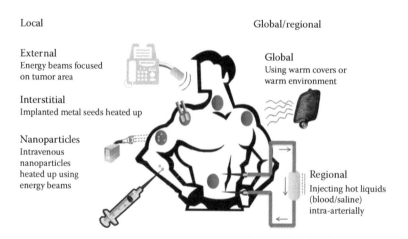

Local

External
Energy beams focused on tumor area

Interstitial
Implanted metal seeds heated up

Nanoparticles
Intravenous nanoparticles heated up using energy beams

Global/regional

Global
Using warm covers or warm environment

Regional
Injecting hot liquids (blood/saline) intra-arterially

FIGURE 14.1 Methods of achieving tumor hyperthermia.

the peritoneum with a heated solution of anticancer drugs for peritoneal cancers such as mesothelioma (Chang et al. 2001; Feldman et al. 2003). In general, whole body and regional hyperthermia techniques are not very tumor-specific in their generation of heat, but they are also not very toxic. Nevertheless, patients treated with whole body hyperthermia occasionally develop gastrointestinal symptoms such nausea, vomiting, and diarrhea, and may rarely have serious cardiovascular side effects such as myocardial ischemia, thrombosis, and cardiac failure. The positive attributes of regional hyperthermia have to be weighed against the potential morbidities associated with the invasiveness of the procedure and the consequent requirements of a dedicated facility that involves nontrivial challenges of setup cost and expert manpower.

Local hyperthermia, on the other hand, has the advantage of being tumor-focused. The three major methods of generation of local hyperthermia, in order of increasing invasiveness of the required interventions, are external, luminal, and interstitial. *Luminal hyperthermia* uses special metallic probes placed as close to the tumor as possible, as in the vagina for treatment of cervical and vaginal cancers, whereas *interstitial hyperthermia* uses an array of probes placed within tumor parenchyma to achieve a more uniform tumor heating. Heat sources are inserted into these probes or the probes are heated by external heating sources. In general, these methods tend to be relatively invasive, and the resulting thermal dosimetric profile is heterogeneous with heat being most intense along the probe and falling off exponentially as one moves away from the probe. To some extent, this nonuniformity can be countered by applying multiple interstitial probes closer together, but this increases the invasiveness of the approach. Alternatively, the invasiveness of this form of interstitial hyperthermia can be reduced by replacing the probe array with smaller metal antennas ("seeds") in the tumor parenchyma before exposure to an external energy source that activates these seeds noninvasively. Typically, these seeds are made of ferromagnetic material (e.g., iron) that heat up in an alternating magnetic field to generate hyperthermia in the form of heat that emanates from *inside the tumor* rather than filtering through from outside. The placement of seeds within the tumor is fairly invasive and potentially challenging for deep-seated tumors. However, the advantages are that the hyperthermia session is temporally spaced from the invasive procedure, and the procedure is noninvasive and fully controllable by regulating the strength of the magnetic field. Administering this treatment incurs some capital investment to comply with federal regulations and construct special electromagnetically shielded rooms. *External hyperthermia* can be achieved either with electromagnetic radiation (e.g., microwave, laser, and radiofrequency) or high-intensity focused ultrasound, all of which transduce energy from an external source to the tumor. Tumor specificity is typically achieved via image guidance and focusing of energy upon tumor visualized by imaging techniques. There is potential for such techniques to deposit some energy within normal tissues along the path of the beam, similar to the heating of tumors by placing something warm (like a hot water bottle) against the body surface overlying the tumor. In the latter case, the insulation provided by the subcutaneous fat layer and plexus of blood vessels necessitates higher surface temperatures to achieve target tumor temperatures along a temperature gradient from the skin to the tumor—higher surface temperatures increase the probability of skin erythema and desquamation.

The foregoing discussion highlights some of the prospects and perils of traditional methods of generating hyperthermia. Clearly, hyperthermia is a treatment modality that holds a lot of promise, but the methods of attaining, maintaining, monitoring, and modeling it suffer from many inadequacies. Therefore, there remains a continuing need for newer methods of generating hyperthermia. Preferably, this would be tumor-focused, minimally invasive, and more uniform in the hyperthermia generated. This quest has fueled the investigation of a new method of achieving hyperthermia, namely, via the use of very tiny bits of matter made up of a few tens of thousands of atoms—nanoparticles. *Nanoparticles*, as outlined in preceding chapters, are materials with their longest dimension less than 100 nm although particles up to 1 μm in size are also often lumped within this definition. Most components of the cellular machinery providing structure, signaling function, interactions, and control operate at this molecular or nano-scale. Controlling and manipulating these processes often requires scaling down interventions to this level. Nanoparticles, therefore, are of great interest since they can potentially control and manipulate nanoscale interactions at the molecular or supramolecular level via a degree of precision and/or design unsurpassed by other materials. Moreover, matter at the nanoscale often demonstrates surprising properties that can be exploited for therapeutic gain.

The most popular nanoparticles under investigation for hyperthermia include various forms of gold nanoparticles, superparamagnetic iron oxide nanoparticles (SPIONs), and carbon nanotubes (CNTs). These "nanotransducers" have different mechanisms of action, but all depend on quantum phenomenon to trap radiant energy from an external source and deposit it in tissues. Apart from the three major nanoparticles, there are reports of other nanoparticles that can potentially be exploited for hyperthermia. For example, fluorescent quantum dots (Glazer and Curley 2010), silver and zinc nanoparticles, and lanthanum manganite particles with impregnated silver ions (Melnikov et al. 2009) have been explored as potential hyperthermic agents. Across all these particles are some unique advantages to the utilization of nanoparticles for generating hyperthermia over conventional hyperthermia methods.

14.3 Nanoparticles for Local Hyperthermia

Nanoparticles, once sequestered within tumors, can be energized extrinsically to generate heat. The use of nanoparticles for induction of tumor-specific mild temperature hyperthermia envisages the following steps—intravenous inoculation of the nanoparticles that have been designed to accumulate preferentially in tumors or interstitial injection of nanoparticles into tumors, followed after an appropriate period by exposure of the tumor-bearing part with incident energy. The nanoparticles trap the incident energy and convert it to heat. Hyperthermia achieved in this manner has some unique properties and advantages over other forms of hyperthermia.

(a) Location of nanoparticles—A few well-established principles have guided the use of nanoparticles in the realms of oncology. First, it is widely recognized that intravenously administered nanoparticles accumulate passively within tumors because of extravasation from leaky, immature, and chaotic vasculature of tumors with large fenestrations, more porous basement membranes, and less efficient lymphatic draining than surrounding normal tissues (Maeda 2001; Maeda et al. 2000). This is often referred to as the enhanced permeability and retention (EPR) effect. Second, it is also known that for most nanoparticle geometries, the reticuloendothelial system is the most common sieve that traps systemically administered nanoparticles and reduces their circulation time. Evasion of this capture often requires either scaling down the size of nanoparticles to about 5 nm (Choi et al. 2007) (whereby they are cleared by the kidneys) or rendering these nanoparticles with "stealth" properties by cloaking them within a shell of biocompatible coatings such as polyethylene glycol (a process known as PEGylation, or pegylation), dextran, chitosan, or other such molecules (Kah et al. 2009; Wang et al. 2010). By evading capture by the reticuloendothelial system and increasing circulation time, these stealth nanoparticles can accumulate more efficiently within tumors via the EPR effect. Third, additional tumor-specific accumulation of nanoparticles may be achieved via decoration of nanoparticles with targeting molecules that allow them to home specifically and preferentially to tumor cells and/or tumor vasculature (El-Sayed et al. 2006; Hosta-Rigau et al. 2010; Patra et al. 2010; Waldman et al. 2006). Although this increases tumor accumulation on a global scale, it also often alters the location of nanoparticles at the cellular and/or tissue level as well (Huang et al. 2010), which may be advantageous for specific applications. From a hyperthermia perspective, tumor-specific accumulation of nanoparticles allows confinement of hyperthermia to the tumor without the use of invasive techniques.

(b) "Inside-out" hyperthermia—With most other forms of hyperthermia, it is commonly recognized that the interface between tumor parenchyma and blood vessels is often an area of low temperature or "cold spots." Blood flowing through the vessel at body temperature cools down the immediately adjacent area of tumor parenchyma (the "heat sink effect")—the inadequate heating of these perivascular zones leads to suboptimal therapeutic effects especially in thermoablation scenarios. However, with nanoparticle-mediated hyperthermia, it is not uncommon

for the nanoparticle to stay sequestered in the perivascular zone and therefore concentrate heat within close proximity of the blood vessel. Furthermore, as with all heat sources, there is a rapid fall-off in temperature gradient as one moves away from the heat source (the perivascularly sequestered nanoparticle). Therefore, achieving a tumor parenchymal temperature of say 42°C requires perivascular temperatures that far exceed this amount—largely reversing the heat sink effect noted with most other forms of hyperthermia. We distinguish this form of hyperthermia by referring to it as "inside-out" hyperthermia as opposed to "outside-in" hyperthermia when an extrinsic heating source delivers energy transduced from the outside to within the tumor. Given that many of the nanoparticles that generate hyperthermia are metallic with excellent thermal conductivities, this internal heat source within the tumor quickly couples and instantly transmits the heat generated to the surrounding tumor tissue.

(c) Vascular-focused hyperthermia—As noted above, a unique feature of nanoparticle-mediated hyperthermia is the location of nanoparticles in close proximity to tumor vasculature (Diagaradjane et al. 2008). This has the advantage of not only thermally sensitizing tumor cells but also vascular endothelia, both for direct antitumor effects and for priming them for cytotoxicity following subsequent radiation therapy. The presence of many tumor stem cell niches in the perivascular zone of tumors also provides some rationale for the preferential sensitization of these stem cells to direct or combined cytotoxic effects of hyperthermia with or without additional conventional therapies.

(d) Theranostics—This refers to the simultaneous use of an agent for therapeutic and diagnostic purposes. Nanoparticles have the potential to be served as a unified platform for simultaneous imaging and treatment of tumors since they preferentially accumulate within tumors allowing sensing, imaging, and targeted image-guided hyperthermic therapy of tumors.

(e) Combination with other therapies—In addition to hyperthermic sensitization of tumors to conventional therapeutic modalities such as chemotherapy and radiotherapy, there is the added potential for nanoparticles to serve as platforms for integration of other functionalities via custom engineering of their surface or interiors. For instance, nanoparticles can be laden with or decorated with drugs (or oligonucleotides) and designed for triggered payload release such that chemotherapy (and gene therapy) and hyperthermia are spatially and temporally controlled and synchronized. Similarly, linking nanoparticles to radioactive tracers offers the possibility of combining selective internal radiation therapy with hyperthermia.

The aforementioned unique properties of nanoparticles provide an overview of the potential advantages of using nanoparticles to create hyperthermia. Outlined below are some of the characteristics of individual classes of nanoparticles: gold nanoparticles, iron oxide nanoparticles, and CNTs.

14.4 Hyperthermia Using Gold Nanoparticles

A unique property of metallic nanoparticles, as opposed to the bulk form of the same metal, is the *optical resonances of their surface plasmons*—when exposed to light of a characteristic wavelength, they strongly absorb and scatter the incident light and convert the resonant energy to heat. This phenomenon occurs because metal atoms readily lose their outer shell electrons. These delocalized electrons—responsible for such a metallic phenomenon like electrical conductivity—exist as an electron cloud. When the free electrons oscillate, a quantum of this "electron wave" is called a *plasmon* (similar to a *photon* for light waves). Plasmons may be categorized as quasiparticles just as photons may be considered elementary particles. When a photon (a light wave particle) and a plasmon (an electron wave particle) interact, they can couple to form another quasiparticle, called a *polariton*. We can think of this as two waves interacting to produce a resonant effect. Most bulk metals have their resonant frequency in the ultraviolet wavelength. Hence, when visible light falls on metals, there is no resonance, and the light is reflected back, making metals look shiny. Some metals, such as gold and copper, have resonance frequencies at visible wavelengths, giving them their distinctive colors.

Plasmons are also found at the surface of metals (which have a negative dielectric constant) exposed to a medium with a positive dielectric constant (like air). These are called *surface plasmons* (SP), and are of lower energy than the electron oscillations occurring in the bulk of the metal. Also, unlike the bulk plasmons, the SP waves propagate parallel to the surface. When a photon interacts with an SP at the resonant frequency, a surface plasmon polariton (SPP) is born, and the phenomenon is called *surface plasmon resonance* (SPR). The SPP propagates along the surface until it loses its energy by adsorption into the metal (primary) or radiation (minor, due to scattering from inhomogeneities of the surface). The intensity of the decay of SPP energy varies exponentially with distance traveled, and hence an SPP discharges its energy quickly into the metal.

Spherical metal nanoparticles, by virtue of the small surface dimensions, can be made resonant to light at particular wavelengths. The SPR frequency, for a particular metal–medium pair, shows a positive correlation with the size of the nanoparticle—larger particles and particle aggregates push the resonant frequency from the yellow to the more red regions of the electromagnetic spectrum. Solid spherical gold nanoparticles, for example, show SPR at visible wavelengths (about 520 nm), which increase slightly for larger particle diameters (Figure 14.2a–c). When fabricated in certain geometries, these plasmon resonances of gold can be tuned to near-infrared (NIR) wavelengths, where light penetrates deepest within human tissues because minimal absorbance by native tissue chromophores (Hirsch et al. 2006).

When the shape of the nanoparticles is changed from spheres to more oblong rodlike forms (called gold nanorods or GNRs; Figure 14.2d–f), the resonant frequency splits into two absorption bands. One band corresponds to the shorter dimension of the nanorods and is called the transverse mode, whereas the band corresponding to the longer axis is called the longitudinal mode. The longitudinal mode is of lower energy (longer wavelength) and usually lies in the red to NIR region of the spectrum, whereas the transverse mode resonates at about 520 nm. The relative strengths of the two peaks and the resonant frequencies have a positive correlation with the aspect ratio of the nanorods. Thus, GNRs made of solid gold with dissimilar dimensions are *optically tunable*, in contrast to solid gold nanospheres, where the optical absorption maximum is in the region of 540 nm and can only be minimally tuned to other wavelengths. Hence, the spherical formulation is of limited use clinically; however, the GNRs, with large length/diameter aspect ratios, can be tuned to the NIR region with potential clinical applications. Typical GNR sizes are in the range of around 45 nm in the longer dimension, usually with an aspect ratio of about 3. The most common synthetic pathway for GNRs involves the use of a strongly charged surfactant such as cetyl trimethylammonium bromide (CTAB) to facilitate the anisotropic elongation of a sphere to a cylindrical structure during seed-mediated chemical synthesis. Although CTAB also prevents GNR aggregation in solution, it can be cytotoxic and is, therefore, often removed by serial centrifugation or dialysis, processes that incur considerable expense and reduce the yield of GNRs. In addition to removal of CTAB, another technique often used to render GNRs more biocompatible is the cloaking of the GNR surface with a layer of PEG or polysaccharides or block copolymers (Choi et al. 2011).

A variation on this tunable plasmonic nanoparticle is the gold nanoshell (GNS), which has a dielectric core of silica and a thin coating of colloidal gold on its surface. Although this particle is spherical like the gold nanoparticle, the resonant frequency can be shifted to the NIR region by reducing the thickness of the gold layer, or by increasing the diameter of the silica core, or both. This is readily achieved by starting with the appropriate size of the silica core and then layering the right thickness of gold on its surface. Fortuitously, silica nanoparticles can be readily obtained as highly monodisperse uniform-sized particles with sizes ranging from nanometers to more than a micrometer, and the epilayer of gold is created by adsorbing gold colloid to the amine groups on the surface of the silica core. Treatment with chloroauric acid reduces additional gold onto the adsorbed colloid, which acts as a nucleation site and catalyzes growth and coalescence of gold colloid with neighboring gold colloid to form a complete shell. This reaction is controllable and dictates the thickness of the shell. The thin gold shell not only enhances the SP response by trapping incident photons and generating heat efficiently but also reduces the amount of gold required and decreases the potential cost of therapy. GNSs are usually close to 50–150 nm in diameter and are generally moderately stable in solution, especially if stored at low temperatures (Figure 14.2g–i). Silica-GNSs that are activable by NIR light tend to be

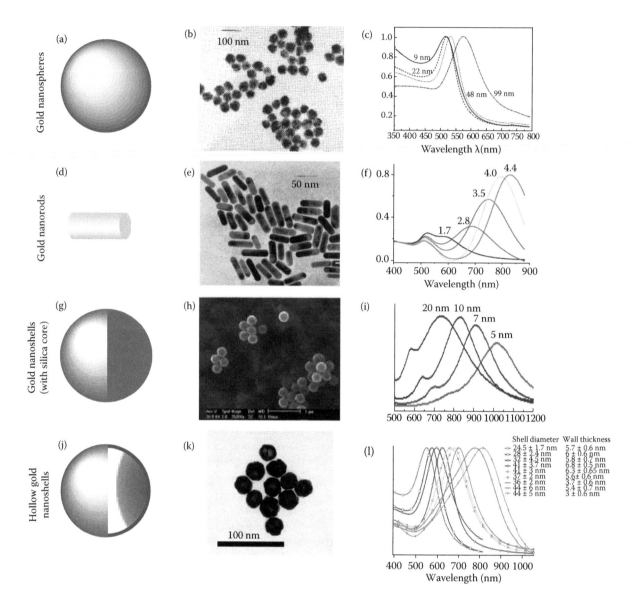

FIGURE 14.2 Comparison of structure (first column), electron microscopic images (second column) and UV–Vis absorption spectra showing SPR frequency (third column) of 48-nm solid gold nanospheres (a–c), gold nanorods (d–f), silica–gold nanoshells (g–i), and 28 ± 2.3 nm hollow gold nanoshells (j–l). UV–Vis absorption spectra of 9-, 22-, 48-, and 99-nm gold nanospheres in water having absorption maxima at 517, 521, 533, and 575 nm, respectively (c). (Reproduced with permission from Link, S., El-Sayed, M.A., *J. Phys. Chem. B*, 103, 4212–4217, 1999.) Absorption spectra of gold nanorods with different aspect ratios (f). (Reproduced with permission from Dickerson, E.B. et al., *Cancer Lett.*, 269, 57–66, 2008.) Absorption spectra of silica–gold core–shell nanoshells as a function of their gold shell thickness (i). (Reproduced with permission from Loo, C. et al., *Technol. Cancer Res. Treat.*, 3, 33–40, 2004.) Absorption spectra of hollow gold nanoshells with varying diameters and wall thicknesses (l). (Reproduced with permission from Schwartzberg, A.M. et al., *J. Phys. Chem. B*, 110, 19935–19944, 2006.)

roughly 150 nm in diameter, with a 120-nm-diameter silica core. Extensive investigations into the safety and tolerability of GNSs suggest that these particles are biocompatible and have paved the way for human clinical trials with investigational device exemption. GNSs are currently in clinical trials for head and neck cancer and prostate cancer using interstitial illumination with NIR lasers for thermoablation applications.

A variant of the core/shell gold nanostructure is one where the core has been dissolved away, leaving a hollow GNS filled with water and salts (Figure 14.2j–l). These novel gold nanoparticles

have a strong tunable plasmon resonance extending deep in the NIR region (~950 nm) with good photothermal heating. Since the ability to fabricate uniform hollow nanoshells also depends on the homogeneity of the core, creating highly monodisperse uniform hollow GNSs ranging in diameter from 30 to 50 nm is eminently feasible. Unlike GNRs, hollow GNSs do not need a cytotoxic surfactant to stabilize them in solution. Earlier reports of hollow GNSs involved using a latex bead as the sacrificial core. Current reports mostly use cobalt nanoparticles as the core. After the growth of the gold shell, exposure of the core/

shell structure to air results in the cobalt core being completely oxidized, leaving only dissolved salts in the core. As with CTAB on GNRs, it is unclear whether all the cobalt can be removed from the hollow core of these particles.

In whichever incarnation, the SPR phenomenon aids to "trap" incident resonant photons to the surface of gold nanostructures. The absorption cross section of gold nanoparticles is typically 4 to 5 orders of magnitude stronger than the strongest absorbing Rhodamine 6G dye molecules. Gold nanoparticles excited by NIR light have absorption cross sections up to a thousand-fold greater than the FDA-approved dye indocyanine green. The transfer of photon energy to the metal nanoparticle, studied using femtosecond transient absorption spectroscopy, is very rapid, on the order of a single picosecond. The metal, in turn, loses its energy to the surrounding media, over a period of about 100 ps.

14.5 Gold Nanoparticle–Mediated Hyperthermia for Cancer Therapy

The photothermal activation of gold nanoparticles, combined with their accumulation within tumors via the EPR effect, has been exploited for thermal ablation of tumors. *In vitro* studies suggest that local temperatures greater than 50°C are required for coagulative cell death (Goldberg et al. 1996). Furthermore, in cultured cells, irreversible cellular damage can be predicted as a composite function of temperature and time, especially in the temperature range of 40–47°C (Roti Roti 2008). Upon translation the *in vivo* setting, identifying a threshold temperature for thermal cytotoxicity is confounded by variability in response depending on the type of tissue, the duration of temperature rise, the uniformity of temperature rise, the vascular perfusion of the tissue (i.e., the heat sink effect), and perhaps the nature of delivery of heat. Nevertheless, a temperature in the 45–50°C range is generally accepted as a bare minimum threshold to be crossed for ablation to be effective, especially when heating is restricted to a few minutes. The effect of laser power on local temperature rise around a 40-nm gold nanoparticle was experimentally determined using a thin film of AlGaN embedded with Er^{3+} ions as an optical temperature sensor with a photoluminescence that is a surrogate for temperature (Carlson et al. 2011) (Figure 14.3). The temperature distribution around a heated nanoparticle conforms to a Gaussian shape with the average temperature change for excitation with a 532-nm laser at 3.8×10^{10} W/m^2 being about 90 K. However, this system depends on heat dissipation through air and partly through the film, instead of water. The calculated thermal interface conductance is only 10 MW/m^2 K, and this reduces the rate of heat dissipation from the nanoparticle to the surrounding matrix. Gold nanoparticles in aqueous solutions have a thermal interface conductance in the range of 100–130 MW/m^2 K (depending on the hydrophilicity of the surface—ranging from 50 for hydrophobic surfaces to 200 for hydrophilic surfaces) (Ge et al. 2006). From these studies, it is evident that laser illumination of tissue loaded with a sufficient number of gold nanoparticles can easily result in ablative temperatures.

Indeed, temperature elevations in this predicted range have been documented *in vivo*. In our experience, laser illumination of subcutaneous human colorectal tumors in nude mice 24 h after intravenous administration of 8×10^8 nanoshells/g results in a steep temperature rise that plateaus after about 5 min. A 0.8-W laser output raised temperature by about 14°C above baseline (about 31°C), whereas 0.6- and 0.4-W laser outputs resulted in about 10°C and 5°C temperature rises, respectively. These temperature elevations can be measured using thermocouples inserted into the tumor or via noninvasive magnetic resonance thermal imaging that relies on documenting the shift in proton resonance frequency as the temperature rises (Diagaradjane et al. 2008).

This experience is similar to that reported with other gold nanoparticles. In one report, pegylated GNRs were injected intravenously (9.6 mg/kg, optical density = 120) in mice bearing sarcomas. Twenty-four hours later, tumors were illuminated for 10 min with an 808-nm laser at 1.2 and 1.6 W/cm^2. Temperature monitoring with a needle thermocouple implanted into the tumor showed that the average equilibrium temperature within the tumor plateaus at 43.6°C and 46.3°C, respectively. In a similar study, mice bearing human glioblastoma tumors, were each injected with 2.5×10^{11} hollow GNSs (either targeted or untargeted), and NIR laser illumination (16 W/cm^2, 3 min, 808 nm) 24 h later resulted in a higher maximum temperature of 57.75 ± 0.46°C in the targeted hollow GNS group than the untargeted group.

In a landmark publication that heralded the use of gold nanoparticles for thermal applications in cancer, Hirsch et al. (2003) demonstrated that subcutaneous tumors directly injected with GNSs can be readily ablated by NIR laser illumination. This was followed by a paper where mice with subcutaneous CT26/wt murine colorectal cancers were divided into three treatment groups—a control group with no intravenous injection or laser treatment, a sham treatment group with intravenous injection of 0.9% sterile saline followed 6 h later with laser treatment (808 nm diode laser, 800 mW, at 4 W/cm^2 for 3 min), and a treatment group with 2.4×10^{10} nanoshells injected intravenously and followed by similar laser treatment. About 90% of mice in the GNS group survived for more than 8 weeks, whereas all mice in the sham treatment and control groups died within about 3 weeks (O'Neal et al. 2004). Since then, studies have demonstrated that active targeting of gold nanoparticles to tumors rather than passive accumulation within tumors results in improved treatment outcomes (Bernardi et al. 2008; Cheng et al. 2009; Dickerson et al. 2008; Skrabalak et al. 2007). These pivotal early experiences highlight some of the potential uses of gold nanoparticles for cancer treatment.

In contrast to stand-alone thermoablation of tumors using gold nanoparticles, there is a growing interest in the use of nanoparticles for hyperthermia (i.e., non-ablative temperatures) in combination with other modalities such as radiation therapy. In principle, these temperatures are less prone to causing

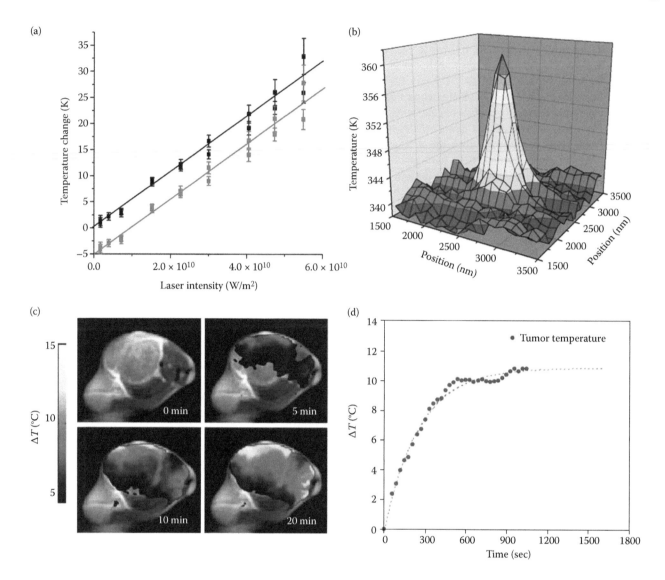

FIGURE 14.3 Gold nanoparticle-mediated hyperthermia. (a) Raw (gray) and corrected (black) temperature change from a 40-nm gold NP as a function of laser intensity. (b) Temperature profile of the NP created from the experimental photoluminescence intensities. (Reproduced with permission from Carlson, M.T. et al., *Nano Lett.*, 11, 1061–1069, 2011.) Low-power laser heating of subcutaneous tumors loaded with gold nanoshells: (c) magnetic resonance thermal imaging (MRTI) of tumor tissues at various time periods following laser illumination, and (d) temperature profile in tumor tissue estimated from the MRTI at various time points. The gray dotted line in (d) represents the best-fit line. (Reproduced with permission from Diagaradjane, P. et al., *Nano Lett.*, 8, 1492–1500, 2008.)

collateral damage of tissues adjacent to the tumor that may also be ablated. In our laboratory, we have investigated the use of GNS-mediated mild temperature hyperthermia (about 41°C for 20 min) to enhance the efficacy of radiation therapy. When such hyperthermia was followed by radiation therapy (single dose of 10 Gy), the time to doubling of tumor volume was nearly twice that with radiation alone (Diagaradjane et al. 2008). This enhancement of radiation response *in vivo* was attributed to an early increase in vascular perfusion of tumors following hyperthermia, as documented by an increase in contrast enhancement in the center of the tumor visualized by dynamic contrast-enhanced magnetic resonance imaging, and a subsequent increase in vascular disruption, as documented by a decrease in

vascular density and an increase in hypoxic and necrotic zones within the tumor (Figure 14.4). This vascular collapse with resultant downstream necrosis was attributed to the sequestration of relatively large GNSs in the perivascular space because of their inability to penetrate deep inside tumor parenchyma. In a separate study, it was demonstrated that cancer stem cells, the putative tumor-initiating cells hypothesized to be the primary reason for treatment failure and metastatic spread, are also sensitized to radiation therapy by GNS-mediated hyperthermia. Whereas radiation of breast cancer xenografts results in tumor volume reduction, the residual tumor has a higher proportion of stem cells. In contrast, GNS-mediated hyperthermia coupled with radiation results not only in a greater reduction of tumor

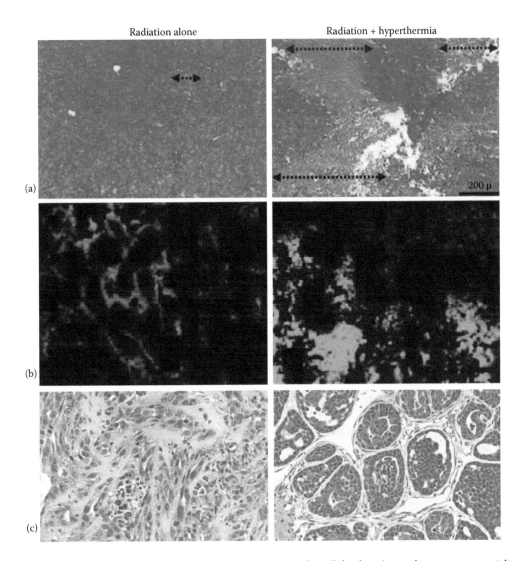

FIGURE 14.4 **(See color insert.)** Comparison of tumors treated with radiation alone (left column) vs. radiation + nanoparticle-mediated hyperthermia (right column). (a) Hematoxylin and eosin–stained slides of the core of human colorectal cancers in mice showing minimal necrosis with radiation alone but significantly more necrosis when hyperthermia is combined with radiation. (Reproduced with permission from Diagaradjane, P. et al., *Nano Lett.*, 8, 1492–1500, 2008.) (b) The tumor core of mice treated with radiation alone has classical tissue architecture with central vascular channels surrounded by orderly layers of cells with decreasing levels of perfusion and increasing hypoxia with increasing distance from the vasculature—hypoxic areas are stained green with pimonidazole and perfused areas are stained blue with Hoechst 33342 in this immunofluorescence image. However, mice treated with combined hyperthermia and radiation (right) have tumor cores with complete disruption of normal stromal structure, suggestive of vascular collapse. (Reproduced with permission from Diagaradjane, P. et al., *Nano Lett.*, 8, 1492–1500, 2008.) (c) Forty-eight hours after treatment with a single dose of radiation with or without hyperthermia, tumors (breast cancer) were digested and reimplantated in syngeneic mice in limiting dilutions. The combined treatment group required more cells reimplanted in mice than the radiation alone group to regrow tumors, suggesting that combined treatment selectively depletes cancer stem cells. In the few mice that tumors regrew upon limiting dilution transplantation following combined treatment, the tumors had a less aggressive phenotype than those that regrew upon limiting dilution transplantation following radiation alone—these better-differentiated tumors would presumably be more readily treated (Reproduced with permission from Atkinson, R.L. et al., *Science Transl. Med.*, 2:55ra79, 2010.)

volume but also a reduction of the proportion of stem cells in the residual tumor. Further investigation of these provocative findings via limiting dilution transplantation of the cancer cells from residual tumors confirmed that more cells needed to be transplanted into mice to regenerate tumors in the combined treatment group compared to the radiation group or the control group (Atkinson et al. 2010).

14.6 Conclusions and Future Directions

Hyperthermia results in increased vascular perfusion of tumors, resulting in decreased hypoxia within tumor cores and a better response to radiotherapy and chemotherapy. However, beyond these traditional effects, nanoparticle-mediated hyperthermia plays additional important roles in cancer—from disruption of microvasculature to sensitization of recalcitrant cancer stem cells to radiation. To some extent, this results from the vascular-focused inside-out hyperthermia that is unique to this form of generating heat within tumors. What makes this form of hyperthermia even more appealing is the ability to achieve this hyperthermia noninvasively and to control the conformality of hyperthermia via both the tumor-specific accumulation of nanoparticles and the collimation of the laser beam to precisely the contours of the tumor. These attributes of nanoparticle-mediated hyperthermia provide the impetus for potential translation of this modality into clinical practice.

The advent of nanoparticles such as SPIONs, GNSs, and CNTs into the realm of cancer hyperthermia has opened up a promising avenue for enhancing the efficacy of traditional therapies such as radiotherapy and chemotherapy. Among these activatable nanoparticles, photothermally activatable gold nanoparticles have several distinct advantages for clinical applications. First, gold is a noble metal that is biologically and chemically inert and molecularly and thermally stable, suggesting that it might be relatively nontoxic and safe. Indeed, there is extensive clinical experience with the use of gold for the treatment of conditions such as rheumatoid arthritis with a long history of safety when administered in small quantities. Second, given that average vascular fenestrations within tumors are 60–400 nm in size, NIR-activatable GNSs measuring about 150 nm and GNRs measuring about 45 nm are well within the size regimes that permit tumor-specific accumulation via the EPR effect. Third, in addition to passive accumulation in tumors via the EPR effect, the gold surface can be readily coupled, via thiol linkages, to PEG or other biomolecules for evasion of the reticuloendothelial system or peptides/antibodies for specific homing to the tumors or tumor vasculature. Lastly, gold nanoparticles have the potential for faster and less expensive FDA clearance since they can probably be classified as devices rather than as drugs, thus saving some time along the road to clinical translation.

Nonetheless, clinical translation faces some of the obstacles that have confounded traditional hyperthermia for many decades. These include the inability to report temperature dose–time attributes accurately, monitor hyperthermia in real time, and model thermal dose distributions within the tumor. Advances such as magnetic resonance thermal imaging may allow us to monitor hyperthermia noninvasively. Newly developed algorithms and tools for predicting thermal dose based on gold nanoparticle concentrations within tumor may facilitate *a priori* treatment planning based on thermal dosimetry similar to radiation dosimetry. The challenge of thermal dosimetry being a reflection of physical heat generation and physiological heat generation plus dissipation does make it more complex than radiation dosimetry, where only physical considerations need to be taken into account. The biggest advantage that nanoparticle-mediated hyperthermia has is the relative ease of generating hyperthermia noninvasively and without the need for expensive equipment. The unique translational issue with this form of hyperthermia is the added requirement to establish the safety and tolerability of the nanoparticle formulation. In addition, adequacy and uniformity of accumulation of nanoparticles at the tumor site needs to be determined in individual treatment scenarios. Even with very small nanoparticles, uniform temperature throughout the core and periphery of tumor is still difficult to obtain especially because of inadequate penetration of nanoparticles into poorly vascularized tumor cores. Another challenge with the use of light (even NIR light) for activation of gold nanoparticles is the limited penetration of light within tissues and the consequent restriction of the utility of photothermal therapy to scenarios where the target is superficial (skin, chestwall, intraoperative tumor bed, etc.), accessible to an endoscope, or implantable with interstitial catheters (which partly negates the advantage of being noninvasive). Lastly, similar to other treatment modalities such as chemotherapy and radiation, hyperthermia is most effective when confined to the tumor. Although the EPR effect allows passive accumulation within tumors, there is also concurrent accumulation in some other tissues, most notably the liver. This makes use of this technique less effective when dealing with tumors of the liver and surrounding areas. Accumulation at other areas, although less prominent, argues for thorough investigation of the pharmacokinetics and biodistribution of each particle proposed for clinical use.

In conclusion, nanoparticles hold promise as a novel means of generating hyperthermia with distinct advantages over traditional methods. Comprehensive toxicity evaluations, optimized methods to ensure uniform, adequate, and specific intratumoral delivery and curtailing nonspecific accumulation in the liver are the key challenges that face effective clinical exploitation of nanoparticle-mediated hyperthermia in cancer treatment.

References

Atkinson, R. L., M. Zhang, P. Diagaradjane et al. 2010. Thermal enhancement with optically activated gold nanoshells sensitizes breast cancer stem cells to radiation therapy. *Science Translational Medicine* 2:55ra79.

Bass, H., J. L. Moore, and W. T. Coakley. 1978. Lethality in mammalian cells due to hyperthermia under oxic and hypoxic conditions. *International Journal of Radiation Biology and Related Studies in Physics, Chemistry & Medicine* 33:57–67.

Bernardi, R. J., A. R. Lowery, P. A. Thompson, S. M. Blaney, and J. L. West. 2008. Immunonanoshells for targeted photothermal ablation in medulloblastoma and glioma: An in vitro evaluation using human cell lines. *Journal of Neuro-oncology* 86:165–172.

Carlson, M. T., A. Khan, and H. H. Richardson. 2011. Local temperature determination of optically excited nanoparticles and nanodots. *Nano Letters* 11:1061–1069.

Chang, E., H. R. Alexander, S. K. Libutti et al. 2001. Laparoscopic continuous hyperthermic peritoneal perfusion. *Journal of the American College of Surgeons* 193:225–229.

Chen, B., W. Wu, and X. Wang. 2011. Magnetic iron oxide nanoparticles for tumor-targeted therapy. *Current Cancer Drug Targets* 11:184–189.

Cheng, F. Y., C. T. Chen, and C. S. Yeh. 2009. Comparative efficiencies of photothermal destruction of malignant cells using antibody-coated silica@Au nanoshells, hollow Au/Ag nanospheres and Au nanorods. *Nanotechnology* 20:425104.

Choi, H. S., W. Liu, P. Misra et al. 2007. Renal clearance of quantum dots. *Nature Biotechnology* 25:1165–1170.

Choi, W. I., J. Y. Kim, C. Kang, C. C. Byeon, Y. H. Kim, and G. Tae. 2011. Tumor regression in vivo by photothermal therapy based on gold-nanorod-loaded, functional nanocarriers. *ACS Nano* 5:1995–2003.

Coley, W. B. 1891. II. Contribution to the knowledge of sarcoma. *Annals of Surgery* 14:199–220.

Day, E. S., J. G. Morton, and J. L. West. 2009. Nanoparticles for thermal cancer therapy. *Journal of Biomechanical Engineering* 131:074001.

DeNardo, G. L., and S. J. DeNardo. 2008. Update: Turning the heat on cancer. *Cancer Biotherapy & Radiopharmaceuticals* 23:671–680.

Diagaradjane, P., A. Shetty, J. C. Wang et al. 2008. Modulation of in vivo tumor radiation response via gold nanoshell-mediated vascular-focused hyperthermia: Characterizing an integrated antihypoxic and localized vascular disrupting targeting strategy. *Nano Letters* 8:1492–1500.

Dickerson, E. B., E. C. Dreaden, X. Huang et al. 2008. Gold nanorod assisted near-infrared plasmonic photothermal therapy (PPTT) of squamous cell carcinoma in mice. *Cancer Letters* 269:57–66.

Doss, J. D., and C. W. McCabe. 1976. A technique for localized heating in tissue: An adjunct to tumor therapy. *Medical Instrumentation* 10:16–21.

El-Sayed, I. H., X. Huang, and M. A. El-Sayed. 2006. Selective laser photo-thermal therapy of epithelial carcinoma using anti-EGFR antibody conjugated gold nanoparticles. *Cancer Letters* 239:129–135.

Everts, M. 2007. Thermal scalpel to target cancer. *Expert Review of Medical Devices* 4:131–136.

Feldman, A. L., S. K. Libutti, J. F. Pingpank et al. 2003. Analysis of factors associated with outcome in patients with malignant peritoneal mesothelioma undergoing surgical debulking and intraperitoneal chemotherapy. *Journal of Clinical Oncology* 21:4560–4567.

Franckena, M., and J. van der Zee. 2010. Use of combined radiation and hyperthermia for gynecological cancer. *Current Opinion in Obstetrics & Gynecology* 22:9–14.

Friedenthal, E., J. Mendecki, C. Botstein, F. Sterzer, M. Nowogrodzki, and R. Paglione. 1981. Some practical considerations for the use of localized hyperthermia in the treatment of cancer. *Journal of Microwave Power* 16:199–204.

Fuller, K. J., R. D. Issels, D. O. Slosman, J. G. Guillet, T. Soussi, and B. S. Polla. 1994. Cancer and the heat shock response. *European Journal of Cancer* 30A:1884–1891.

Ge, Z. B., D. G. Cahill, and P. V. Braun. 2006. Thermal conductance of hydrophilic and hydrophobic interfaces. *Physical Review Letters* 96:186101.

Glazer, E. S., and S. A. Curley. 2010. Radiofrequency field-induced thermal cytotoxicity in cancer cells treated with fluorescent nanoparticles. *Cancer* 116:3285–3293.

Goldberg, S. N., G. S. Gazelle, E. F. Halpern, W. J. Rittman, P. R. Mueller, and D. I. Rosenthal. 1996. Radiofrequency tissue ablation: Importance of local temperature along the electrode tip exposure in determining lesion shape and size. *Academic Radiology* 3:212–218.

Harmon, B. V., Y. S. Takano, C. M. Winterford, and G. C. Gobe. 1991. The role of apoptosis in the response of cells and tumours to mild hyperthermia. *International Journal of Radiation Biology* 59:489–501.

Hirsch, L. R., R. J. Stafford, J. A. Bankson et al. 2003. Nanoshell-mediated near-infrared thermal therapy of tumors under magnetic resonance guidance. *Proceedings of the National Academy of Sciences of the United States of America* 100:13549–13554.

Hirsch, L. R., A. M. Gobin, A. R. Lowery et al. 2006. Metal nanoshells. *Annals of Biomedical Engineering* 34:15–22.

Hosta-Rigau, L., I. Olmedo, J. Arbiol, L. J. Cruz, M. J. Kogan, and F. Albericio. 2010. Multifunctionalized gold nanoparticles with peptides targeted to gastrin-releasing peptide receptor of a tumor cell line. *Bioconjugate Chemistry* 21:1070–1078.

Huang, X., X. Peng, Y. Wang, D. M. Shin, M. A. El-Sayed, and S. Nie. 2010. A reexamination of active and passive tumor targeting by using rod-shaped gold nanocrystals and covalently conjugated peptide ligands. *ACS Nano* 4:5887–5896.

Huilgol, N. G., S. Gupta, and R. Dixit. 2010a. Chemoradiation with hyperthermia in the treatment of head and neck cancer. *International Journal of Hyperthermia* 26:21–25.

Huilgol, N. G., S. Gupta, and C. R. Sridhar. 2010b. Hyperthermia with radiation in the treatment of locally advanced head and neck cancer: A report of randomized trial. *Journal of Cancer Research and Therapeutics* 6:492–496.

Hurwitz, M. D., J. L. Hansen, S. Prokopios-Davos et al. 2011. Hyperthermia combined with radiation for the treatment of locally advanced prostate cancer: Long-term results from Dana–Farber Cancer Institute study 94–153. *Cancer* 117:510–516.

Irish, C. E., J. Brown, W. P. Galen et al. 1986. Thermoradiotherapy for persistent cancer in previously irradiated fields. *Cancer* 57:2275–2279.

Kah, J. C., K. Y. Wong, K. G. Neoh et al. 2009. Critical parameters in the pegylation of gold nanoshells for biomedical applications: An in vitro macrophage study. *Journal of Drug Targeting* 17:181–193.

Kennedy, L. C., L. R. Bickford, N. A. Lewinski et al. 2011. A new era for cancer treatment: Gold-nanoparticle-mediated thermal therapies. *Small* 7:169–183.

Kim, J. H., E. W. Hahn, and S. A. Ahmed. 1982. Combination hyperthermia and radiation therapy for malignant melanoma. *Cancer* 50:478–482.

Krishnan, S., P. Diagaradjane, and S. H. Cho. 2010. Nanoparticle-mediated thermal therapy: Evolving strategies for prostate cancer therapy. *International Journal of Hyperthermia* 26:775–789.

Lele, P. P. 1980. Induction of deep, local hyperthermia by ultrasound and electromagnetic fields: Problems and choices. *Radiation and Environmental Biophysics* 17:205–217.

Link, S., and M. A. El-Sayed. 1999. Size and temperature dependence of the plasmon absorption of colloidal gold nanoparticles. *Journal of Physical Chemistry B* 103:4212–4217.

Loo, C., A. Lin, L. Hirsch et al. 2004. Nanoshell-enabled photonics-based imaging and therapy of cancer. *Technology in Cancer Research & Treatment* 3:33–40.

Luk, K. H., M. E. Francis, C. A. Perez, and R. J. Johnson. 1984. Combined radiation and hyperthermia: Comparison of two treatment schedules based on data from a registry established by the Radiation Therapy Oncology Group (RTOG). *International Journal of Radiation Oncology, Biology, Physics* 10:801–809.

Maeda, H. 2001. The enhanced permeability and retention (EPR) effect in tumor vasculature: The key role of tumor-selective macromolecular drug targeting. *Advances in Enzyme Regulation* 41:189–207.

Maeda, H., J. Wu, T. Sawa, Y. Matsumura, and K. Hori. 2000. Tumor vascular permeability and the EPR effect in macromolecular therapeutics: A review. *Journal of Controlled Release* 65:271–284.

Magin, R. L., and R. K. Johnson. 1979. Effects of local tumor hyperthermia on the growth of solid mouse tumors. *Cancer Research* 39:4534–4539.

Melnikov, O. V., O. Y. Gorbenko, M. N. Markelova et al. 2009. Ag-doped manganite nanoparticles: New materials for temperature-controlled medical hyperthermia. *Journal of Biomedical Materials Research A* 91:1048–1055.

Moros, E. G., J. Penagaricano, P. Novak, W. L. Straube, and R. J. Myerson. 2010. Present and future technology for simultaneous superficial thermoradiotherapy of breast cancer. *International Journal of Hyperthermia* 26:699–709.

O'Neal, D. P., L. R. Hirsch, N. J. Halas, J. D. Payne, and J. L. West. 2004. Photo-thermal tumor ablation in mice using near infrared-absorbing nanoparticles. *Cancer Letters* 209:171–176.

Patra, C. R., R. Bhattacharya, D. Mukhopadhyay, and P. Mukherjee. 2010. Fabrication of gold nanoparticles for targeted therapy in pancreatic cancer. *Advanced Drug Delivery Reviews* 62:346–361.

Rao, W., Z. S. Deng, and J. Liu. 2010. A review of hyperthermia combined with radiotherapy/chemotherapy on malignant tumors. *Critical Reviews in Biomedical Engineering* 38:101–116.

Roti Roti, J. L. 2008. Cellular responses to hyperthermia (40–46 degrees C): Cell killing and molecular events. *International Journal of Hyperthermia* 24:3–15.

Schwartzberg, A. M., T. Y. Olson, C. E. Talley, and J. Z. Zhang. 2006. Synthesis, characterization, and tunable optical properties of hollow gold nanospheres. *Journal of Physical Chemistry B* 110:19935–19944.

Seegenschmiedt, M. H., R. Sauer, C. Miyamoto, J. A. Chalal, and L. W. Brady. 1993. Clinical experience with interstitial thermoradiotherapy for localized implantable pelvic tumors. *American Journal of Clinical Oncology* 16:210–222.

Skrabalak, S. E., L. Au, X. Lu, X. Li, and Y. Xia. 2007. Gold nanocages for cancer detection and treatment. *Nanomedicine (London)* 2:657–668.

Stewart, J. R., and F. A. Gibbs, Jr. 1984. Hyperthermia in the treatment of cancer. Perspectives on its promise and its problems. *Cancer* 54:2823–2830.

Thrall, D. E. 1980. Clinical requirements for localized hyperthermia in the patient. *Radiation and Environmental Biophysics* 17:229–232.

Van den Berg, C. A., J. B. Van de Kamer, A. A. De Leeuw et al. 2006. Towards patient specific thermal modelling of the prostate. *Physics in Medicine & Biology* 51:809–825.

Vasanthan, A., M. Mitsumori, J. H. Park et al. 2005. Regional hyperthermia combined with radiotherapy for uterine cervical cancers: A multi-institutional prospective randomized trial of the international atomic energy agency. *International Journal of Radiation Oncology, Biology, Physics* 61:145–153.

Waldman, S. A., P. Fortina, S. Surrey, T. Hyslop, L. J. Kricka, and D. J. Graves. 2006. Opportunities for near-infrared thermal ablation of colorectal metastases by guanylyl cyclase C-targeted gold nanoshells. *Future Oncology* 2:705–716.

Wang, J., M. Sui, and W. Fan. 2010. Nanoparticles for tumor targeted therapies and their pharmacokinetics. *Current Drug Metabolism* 11:129–141.

Zagar, T. M., J. R. Oleson, Z. Vujaskovic et al. 2010. Hyperthermia combined with radiation therapy for superficial breast cancer and chest wall recurrence: A review of the randomised data. *International Journal of Hyperthermia* 26:612–617.

15

Magnetic Resonance Temperature Imaging for Gold Nanoparticle-Mediated Thermal Therapy

R. Jason Stafford
M. D. Anderson Cancer Center

John D. Hazle
M. D. Anderson Cancer Center

15.1 Introduction

Targeted modulation of tissue temperature for therapeutic purposes is commonly referred to as "thermal therapy" and encompasses a wide array of applications from modulation of the local tissue microenvironment to the ablation of tissue. Extremely cold temperatures below –20°C (cryoablation) or extremely hot temperatures above 50°C (thermal ablation) can be used to rapidly damage tissue. Moderate decreases in temperature (hypothermia) or increases in temperature (hyperthermia) can be used to modulate the biochemistry and physiology to impact endogenous processes such as vascular and cellular permeability, protein and enzyme function, cellular metabolism, and sensitivity to chemo- or radiotherapies, as well as modulate the properties of exogenous agents such as drugs, contrast agents, or micro-/nano-scale particles.

Minimally invasive image-guided thermal ablation of tissue for treatment of soft-tissue diseases, such as cancer, is developing into a viable alternative to conventional surgical interventions for patients who are not good candidates for surgery or could potentially benefit substantially from a less invasive procedure, such as for the treatment of an early stage, locally confined disease (Ahmed et al. 2011; Requart 2011; Thumar et al. 2010; Pua et al. 2010; Sharma et al. 2011; Kurup and Callstrom 2010; Flanders and Gervais 2010; McWilliams et al. 2010; Colen and Jolesz 2010; Goldberg et al. 2009; Ahmed and Goldberg 2005; Callstrom et al. 2009; Rybak 2009; Kunkle and Uzzo 2008; Gillams 2008; Beland et al. 2007). Such procedures often require only minimal anesthesia and are generally associated with reduced complications, blood loss, and normal tissue morbidity, and so have the strong potential to reduce the impact of the intervention on the patient as well as reduce the overall procedure cost. In this manner, minimally invasive thermal therapy procedures, which generally do not rely on ionizing radiation, facilitate low-impact, repeatable procedures ideally suited for the nonsurgical or palliative patient, but are increasingly being investigated in the management of early detected local disease as well. These procedures often rely on a loco-regional deposition of energy, which results in a rapid rise in tissue temperatures, leading to irreversible damage to the target tissue. A variety of energy sources for heating, such as radiofrequency (Hong and Georgiades 2010), microwave (Lubner et al. 2010), high-intensity ultrasound (Jolesz 2009; Pauly et al. 2006), and laser (Stafford et al. 2010), may be used, each being associated with its own advantages and disadvantages.

Laser-induced thermal therapy (LITT) is an interstitial tissue ablation technique (Figure 15.1) that utilizes high-power lasers placed interstitially in the target tissue, often a tumor, to deliver therapy to a variety of sites, such as the brain, prostate, liver, kidney, bone, breast, head, and neck. Multiple laser fibers can be placed into the treatment volume and can be fired individually or simultaneously to rapidly heat a target volume. Modern high power (≥15 W) laser systems utilize compact solid-state diode power supplies along with actively cooled treatment applicators that house the laser fiber and aid in keeping tissue from charring when high powers are used to increase the treatment volume. Laser ablation for tissue destruction primarily relies on photothermal interactions where certain photon-absorbing molecules (chromophores) transduce photon energy into heat via entering a higher vibrational state and exchanging this energy with the surrounding environment. This optical–thermal response to the laser source is typically understood as a bioheat transfer process with the thermal source provided by the radiation transport equation. For applications involving temperatures at which coagulative necrosis of tissue is achievable within a minimal time (>54°C), the exposure can be in the hundreds of W/cm^2 applied for seconds to minutes using light that is generally in the

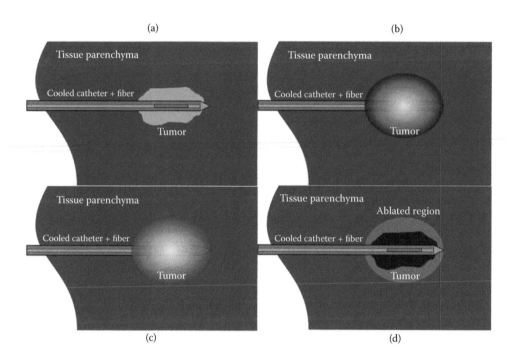

FIGURE 15.1 Laser-induced thermal therapy is often delivered interstitially. The fiber, or actively cooled applicator, is placed directly into the target tissue, usually under image guidance (a). The active element of the fiber is where light energy is emitted into the tissue. Absorption of photons by the tissue results in heating near the fiber and the region of heating grows via conduction to envelope the treatment zone (b). During the exposure, the optical and thermal properties of the tissue in the heated region change as proteins denature and perfusion begins to shut down (c). After exposure, a region is left where heat mediated damage to the tissue results in tissue fixation, loss of perfusion and eventually cell death over the entire region (d). The lesion can be made longer by pulling the fiber back within the applicator. Additional applicators can be placed to generate larger lesions. Nanoparticles hold the promise of creating a heat distribution more conformal to the tumor.

near-infrared (NIR) portion of the spectrum, which provides an excellent window for penetration into tissue.

Incidentally, metal nanoparticles can be tuned to absorb strongly in the NIR part of the spectrum used for laser ablation. This can effectively move the source of energy transduction to primarily come from the distribution of nanoparticles and less from tissue, effectively generating a highly conformal approach to energy delivery when these particles can be delivered into the target tissue. Unlike traditional dyes, these particles do not photobleach, use bioinert gold, and can be easily coated for longer circulation times or targeted for increased retention in target tissue.

To this end, an increasing number of laser-activated metal-based nanoparticles, of varying sizes and conformations, are being investigated for therapeutic uses. The nanoparticle property being exploited is the plasmon resonance of the conduction band electrons that substantially increase absorption near the optical resonance wavelength (Hirsch et al. 2006). Particles can be constructed in a specific manner to "tune" this absorption wavelength to occur in the NIR part of the spectrum, coinciding with the wavelength of lasers currently used for interstitial ablation (Hirsch et al. 2003; Schwartz et al. 2009). The plasmon resonance enhanced absorption results in substantial absorption and subsequent heating at an applied power well below that needed for laser ablation of tissue, thus providing a mechanism for targeted heating with the potential for sublethal heating of adjoining normal tissue not containing the nanoparticles.

Because of their small size, these nanoparticles can accumulate preferentially in tumor tissue passively via enhanced permeability and retention or, for smaller particles, by functionalizing the surface chemistry to bind to specific moieties and enhance retention in targeted tissue. This tends to turn the tumor vasculature into the primary heat source as opposed, or in addition, to the tissue immediately adjacent to the laser source. Thus, nanoparticles may provide a heating profile much more conformal to the tumor tissue.

In this chapter, we provide an overview of how magnetic resonance (MR) guidance and temperature imaging may be incorporated into investigations of the potential for using gold nanoparticles (AuNP) as mediators for delivering highly conformal LITT of soft-tissue tumors using phantoms as well as *in vivo* small and large animal models. MR temperature imaging (MRTI) provides a means to demonstrate the feasibility of these therapies in a variety of scenarios as well as monitor and control therapy delivery. In addition, temperature feedback can be used to validate computational models of AuNP mediated therapy as well. In this manner, MR imaging (MRI) is used to both validate and refine the approach to these therapies.

15.2 Magnetic Resonance Temperature Imaging

Image guidance plays a critical role in minimally invasive thermal therapy delivery. Imaging can increase the safety and efficacy of these procedures by integrating into any of the primary steps in treatment delivery: treatment planning, targeting and localization, treatment monitoring, and immediate and longitudinal post-treatment verification (Ahmed et al. 2011). Treatment planning often encompasses the process of identifying the target volume to be treated, important surrounding critical structures, and a potential approach to therapy that may involve simulation. Treatment targeting and localization may include the use of imaging at the time of the intervention for the purpose of coordinating the device coordinate system with the anatomical coordinate system (e.g., stereotaxy) or real-time visualization of the device location with respect to the anatomy or target tissue. Imaging for treatment monitoring includes a periodic or real-time update of treatment progress, either directly or via a surrogate marker. Examples include visualization of flow or injected contrast agent to assess vascular occlusion or perfusion losses, intensity changes associated with iceball formation during cryotherapy, or estimated tissue temperature changes during thermal therapies. Post-treatment verification is the use of imaging to aid in the assessment of the delivery of therapy and may include imaging for visualizing damage (e.g., anatomy or physiological properties like perfusion) to the target as well as normal tissue (e.g., hemorrhage).

Traditionally, computed tomography (CT) and ultrasound (US) have been the workhorses for guidance of percutaneous interventions involving thermal ablation. Each has been used successfully to place LITT applicators and also to provide limited monitoring during the delivery of therapy. CT guidance has the advantage of generating fast, high-resolution axial images of soft-tissue versus bone anatomy as well as perfusion-weighted contrast-enhanced imaging. CT guidance has traditionally been used primarily for applications in interventional radiology environments, such as lung, bone, liver, and kidney. CT provides reasonable guidance with reasonable access to the patient, but limited soft-tissue contrast capabilities for visualizing pathologies, such as cancer. Additionally, the metal hardware used during interventions can cause streaking in the images that interferes with anatomical visualization. Although laser applicators themselves do not cause streaking, the guide needles to place them do. Also, because of the radiation dose to the patient, in-room staff, and the radiologist, CT is used only sparingly during the localization and monitoring processes.

Ultrasound has also been used for guidance of soft-tissue thermal therapies in interventional radiology and surgical settings, including liver and kidney, as well as being a workhorse for applications such as prostate, breast, and head and neck. Ultrasound is comparatively the least expensive, fastest, and most user-friendly modality for real-time guidance. Weaknesses include the poor contrast of many pathologies against normal anatomy as well as limitations associated with acoustic propagation (e.g., across bone–tissue or tissue–air interfaces).

For soft-tissue disease, MRI is a nonionizing and noninvasive modality that provides incomparable contrast compared to CT or ultrasound. This contrast can often be complemented by multiple physiological functional imaging and metabolic imaging capabilities. Additionally, MRI can acquire in any arbitrary orientation with respect to the anatomy, often at near real-time rates, for guidance. Therefore, despite the potential barriers posed by working in the magnetic field environment, MRI is increasingly being investigated and marketed as a solution for certain image-guided interventions, particularly for those in which there is no comparable imaging modality, such as in brain, prostate, and breast, or where external heating techniques are used, such as focused ultrasound (Jolesz et al. 1988; Hynynen et al. 1993; Cline et al. 1992). When these imaging capabilities are coupled with the inherent qualitative and quantitative temperature imaging capabilities of MRI in conjunction with the compatibility of laser applicators with the high-field MR environment, MR guidance of laser-induced thermal therapies, of which nanoparticle-mediated heating is likely to appear as a subset, has the potential to be a very efficient "one-stop shop" for these procedures (Stafford et al. 2010).

These advantages have opened the door for a new generation of Food and Drug Administration–cleared laser ablation systems that use MRI guidance to treat with lasers safely at high powers in sensitive locations with the potential for increased efficacy over previous approaches. Using compact, solid-state lasers operating in the NIR regime along with actively cooled catheters that facilitate the use of higher powers and larger lesions, interstitial laser ablation of deep-seated lesions in areas such as the brain, prostate, liver, kidney, and bone are being investigated in animal models with extensions into humans. The MRTI feedback during therapy (Rieke and Butts Pauly 2008a; McDannold 2005) is used to help estimate the extent of tissue damage for efficacy and aid in the enforcement of temperature limits of involved critical structures as well as tissue adjacent to the laser fiber in order to prevent potential charring associated with the use of higher powers, thus minimizing complications and enhancing the overall safety of the procedure (Raz et al. 2010; Carpentier et al. 2008; Kickhefel 2011).

Therefore, in addition to MRI providing superlative soft-tissue contrast mechanisms for planning, targeting, and verifying therapy delivery, MRTI adds the potential for quantitatively monitoring the delivery of thermal energy to tissue in real time. This provides several advantages in the research and clinical application of heating using nanoparticles. First, many of the physiological effects of heat, such as tissue damage, are governed by exposure as opposed to a simple temperature threshold (Dewhirst et al. 2003). The spatiotemporal temperature distribution estimated by MRTI, when correlated with post-treatment imaging or histopathology, can be useful in validating or establishing these biological models of activation in living systems in a relatively noninvasive manner (Stafford et al. 2011; Yung et al. 2010; Diagaradjane et al. 2008).

Additionally, MRTI temperature feedback provides information that can aid in the development and validation of physical models of tissue heating under various phantom, *ex vivo*, and *in vivo* scenarios (Cheong et al. 2009; Fuentes et al. 2009, 2010, 2011; Feng et al. 2009; Elliott et al. 2007, 2008). These models can be incorporated into prospective treatment planning or incorporated into the real-time feedback loop with MRTI information to aid in controlling treatment delivery (Fuentes et al. 2009; Oden et al. 2007).

MRTI is a means for qualitative or quantitative mapping of tissue temperature changes over time (McDannold 2005; Raz et al. 2010). Clinical MRI relies primarily on the MR properties of water protons, most of which unsurprisingly demonstrate temperature dependence. However, not all parameters demonstrate linear dependence with temperature over the ranges needed for thermal ablation, nor are all parameters amenable to the fast, high-resolution image acquisition techniques needed for monitoring high-temperature thermal ablations. Because of their sensitivity and relative ease of measurement, the most studied and used temperature-sensitive MR parameters have traditionally been the apparent diffusion coefficient of water (ADC), the spin-lattice relaxation time (T1), and the water proton resonance frequency (PRF). Unfortunately, both the T1 and ADC techniques have a tissue-dependent temperature sensitivity coefficient, making them difficult to use as a sole means of temperature monitoring and requiring the user to calibrate the sequences in different tissue for useful thermometry. Additionally, as the tissue undergoes irreversible changes at high temperatures, both the T1 and the ADC deviate from the presumed linear response (Young et al. 1994; Michael et al. 2002; Graham et al. 1999). Therefore, such techniques may be better suited for lower temperature applications or for providing background temperature estimates over time. Qualitative T1-weighted MRTI, being relatively insensitive to motion between acquisitions, has demonstrated a reasonable means for guiding heating in the liver (Matsumoto et al. 1992), where it has been used as feedback in clinical liver ablation (Vogl et al. 1995, 2002). Additionally, T1-weighted approaches may be the best option for monitoring tissue temperature changes in adipose tissue, which is not suitable for PRF or diffusion techniques (Kuroda et al. 2000; Taylor et al. 2011a). Because of this, hybrid multiparametric approaches to MRTI are being investigated (Taylor et al. 2011b; Todd et al. 2012).

Techniques based on the linear shift in the PRF (McDannold 2005; Schneider et al. 1958; Hindman 1966) with temperature are most often used for quantitative monitoring of rapid, high-temperature ablations, where knowledge of high temperatures at the laser catheter–tissue interface is useful for both safety and efficacy purposes. The PRF shift technique has the primary advantage of being quantitative, with a temperature sensitivity that is relatively independent of tissue type or state of tissue denaturation (Graham et al. 1999; Taylor et al. 2011b; Kuroda et al. 1998; Peters et al. 1998).

This temperature dependence is primarily attributed to the relatively weak hydrogen bonding between water protons and oxygen. Kinetic energy increases with temperature, resulting in a longer hydrogen bond and shorter covalent bond between the hydrogen and parent oxygen, increasing the shielding factor of the hydrogen from the magnetic moment of the electron cloud of the oxygen. The impact of the electron cloud results in a chemical shift (σ) induced deviation from the expected resonance frequency, f, expressed as

$$f = \gamma B_0 (1 - \sigma) \tag{15.1}$$

where γ is the gyromagnetic ratio (42.58 MHz/T for hydrogen atoms) and B_0 is the applied magnetic flux density. Since the PRF

shift arises primarily because of alterations in the mean hydrogen bonding length with temperature, slight deviations between various in vivo tissues are expected as a function of regional ion content due to tissue pH effects on H-bonding, electrical conductivity effects (Peters and Henkelman 2000), or tissue susceptibility (De Poorter 1995). Susceptibility from equipment (i.e., titanium applicators) or heating of tissue itself can also influence measurements (Peters et al. 1999; Boss et al. 2005). In vivo measurements of temperature sensitivity are difficult to execute, and reports in the early literature tend to have more variance than comparable ex vivo measurements, indicating a need for more carefully executed in vivo measurements. When corrected for errors, ex vivo measurements tend to consistently be reported between –0.01 ± 0.001 ppm/°C in a variety of tissues under a variety of heating circumstances (McDannnold 2005; Peters et al. 1998).

Covalently bonded lipid protons do not exhibit the same temperature sensitivity as the water protons and only contribute to the temperature-dependent shifts via susceptibility contributions. Although this can cause errors in temperature measurement when lipid signal is not separated or suppressed (de Zwart et al. 1999; Rieke and Butts Pauly 2008b), it also means lipid tissue can also provide an "internal reference" for self-correcting PRF techniques (Kuroda 2005; Taylor et al. 2008).

The temperature-dependent PRF shift can be measured using chemical shift imaging techniques to directly measure the frequency shift (Kuroda 2005; Taylor et al. 2008, 2009; Kuroda et al. 1996; McDannold et al. 2001; Kuroda et al. 2000), but more often, high spatiotemporal resolution estimates of temperature change (ΔT) are based on indirect measurement of the PRF shift via relating the difference in phase (Φ) between subsequent images using fast gradient-echo acquisitions using the equation (Ishihara et al. 1995; Kuroda et al. 1997; Chung et al. 1996)

$$\Delta T = \frac{\Phi - \Phi_{\text{ref}}}{2\pi \cdot \alpha \cdot \gamma \cdot B_0 \cdot \text{TE}} \tag{15.2}$$

where α is the temperature sensitivity coefficient (ppm/°C) and TE is the echo time (ms). The uncertainty in the phase-difference measurement is dependent on the magnitude of the image signal-to-noise ratio (SNR) (Conturo and Smith 1990)

$$\sigma_{\Delta\phi} \cong \frac{\sigma}{A}\sqrt{2} = \frac{\sqrt{2}}{\text{SNR}_A} \tag{15.3}$$

This expression can be coupled with the TE dependence on the phase difference to obtain a contrast-to-noise ratio term for the phase-difference ($\text{CNR}_{\Delta\phi}$) given by

$$\text{CNR}_{\Delta\phi} \propto \Delta\phi \cdot \text{SNR}_A \propto \text{TE} \cdot e^{-\text{TE}/T2^*} \cdot \sin(\theta) \cdot \frac{1 - e^{-\text{TR}/T1}}{1 - e^{-\text{TR}/T1} \cdot \cos(\theta)} \tag{15.4}$$

for a spoiled gradient-recalled echo approach that results in an optimal parameter selection of TE = T2* (transverse relaxation time) and flip angle (θ) that maximizes signal for the given pulse repetition time (TR) (Conturo and Smith 1990).

Because of the potential for acquisition or device-dependent effects on MRTI, PRF temperature imaging techniques should always be validated, at least in phantom, to verify that measurements are as anticipated. Errors on the order of ±0.001 ppm/°C in knowledge of the true sensitivity coefficient translate into errors of several degrees for very large temperature changes. Uncertainty in the phase-difference calculation from which temperature is estimated varies inversely with magnitude SNR, so large uncertainties in temperature may be expected at large temperatures owing to loss of signal.

15.2.1 MRTI Guidance of Nanoparticle-Mediated Thermal Therapies

Investigations of nanoparticle-mediated heating in phantoms, such as agarose, are useful in that they provide highly controlled conditions under which to assess the heating properties of nanoparticles. Although heterogeneous distributions of particles can be created to mimic treatment scenarios for proof of principle studies (Elliott et al. 2010) (Figure 15.2), often homogeneous distributions of particles are amenable to assessing certain aspects of the particle optical properties (Figure 15.3). As discussed in another chapter, several investigations have compared the impact of different optical fluence models for heating, such as directly modeling the photothermal effect (Cheong et al. 2009), or using traditional approximations to the radiation transport problem, such as the diffusion approximation (Feng et al. 2009; Elliott et al. 2007), delta-P1 (Elliott et al. 2009), or analytical approaches (Elliott et al. 2009). Obtaining real-time spatial temperature maps facilitates the evaluation of heating models using noninvasive methods so as to minimize the errors at boundaries. For invasive laser applicators, it provides a chance to test dosimetry and treatment approaches.

Elliott et al. (2009) used MRTI to quantitatively assess the difference between using the optical diffusion approximation

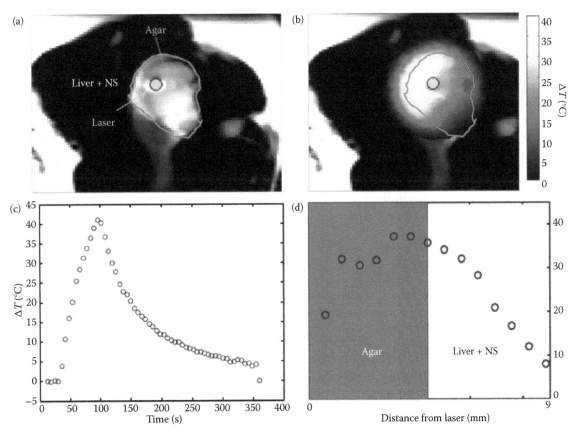

FIGURE 15.2 MRTI can be used for *ex vivo* proof of principle investigations. Elliott et al. (2010) investigated the feasibility of using the nanoparticle uptake in the liver as a beam stop. A T2-weighted MRI of an excised canine liver 24 h after infusion of gold–silica nanoshells is shown in (a) with an artificial agar occlusion representing a tumor, with relatively little nanoparticle uptake, and a laser applicator in the center. The maximum extent of heating is illustrated by magnetic resonance temperature imaging (MRTI) in (b), showing increased heating due to the surrounding nanoshell laden liver even at moderate power. The MRTI temporal profile shows the maximal extent of heating was approximately 40°C above background (c). A spatial profile demonstrates the heating is coming more from the tumor-liver boundary (d). Such a technique could potentially be useful for laser treatment of liver tumors at much higher powers by using the nanoshell laden liver parenchyma to help enhance heating within the tumor during LITT.

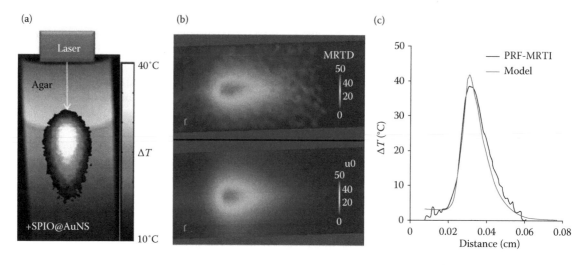

FIGURE 15.3 MRTI of a phantom consisting of homogenous distribution of agar interfaced with a homogeneous distribution of superparamagnetic iron oxides (SPIO)-labeled hollow gold nanoshells (a). From the spatiotemporal temperature distribution, estimates of optical parameters were made by solving the inverse problem on a finite element grid. The finite element solution on bottom in (b) can be compared directly against the MRTI results on top. As can be seen by the axial temperature profile (c), the MRTI developed model has reasonable agreement with measurement.

(ODA) versus a delta-P1 approximation using homogeneous gold–silica nanoshell phantoms similar to that shown in Figure 15.3. By analyzing the root mean squared error between the model and MRTI on a pixel-by-pixel basis, they were able to demonstrate that although the two techniques remain comparable at lower concentrations of nanoshells, the delta-P1 model demonstrated a statistically significant increase in accuracy at higher concentrations of nanoshells. Although it is well known that the delta-P1 approximation should work better in situations where the diffusion approximation is suboptimal, this research provided data and analysis showing quantitatively when, where, and how much error was generated as a function of concentration (Elliott et al. 2009). Such information is a vital first step in designing approaches to simulating therapy for the development of new approaches or treatment planning.

Small animal models are valuable in assessing many aspects of therapy. Orthotopic or, more often, tumors engrafted subcutaneously ("xenografts") can be used to evaluate the feasibility of specific mechanisms, such as the biological effects of heating on the tissue and organs, uptake of nanoparticles in tumor and other organs ("biodistribution"), the impact of synergistic drug adjuvants, as well as a host of physiological responses such as modulation of membrane or vascular permeability (Figure 15.4).

However, xenografts are not useful for evaluating complete tumor destruction via the mechanism of heating as an endpoint. The primary problem here is that nanoparticle-mediated photothermal therapies rely heavily on the interface between physiology, physics, and biology. Many studies focus on demonstrating a particular dose of nanoparticles coupled with a particular exposure is effective for treating cancer. A problem with this approach is that the means of delivery as well as the response to partial treatment of a subcutaneous tumor xenograft is not readily translatable to any real treatment scenario.

This is primarily because the mechanism of cell death in the majority of nanoparticle-induced heating approaches is thermal. When this is the case, the most important information that can be obtained with respect to the nanoparticle therapy is the amount and distribution of uptake of nanoparticles in the

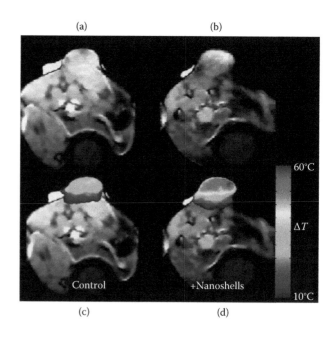

(a)   (b)

(c)   (d)

FIGURE 15.4 (See color insert.) *In vivo* investigation in a Balb-C mouse model of colorectal cancer of gold–silica nanoshell (NS) mediated heating 24 h post-infusion using a fixed laser exposure (4 W/cm² for 3 min at 808 nm) using MRTI. Both control (a) and +NS (b) tumors are of similar size and shape, but the presence of nanoshells in the tumor results in substantially higher heating in the +NS tumor (d) versus control (c).

tumor as well as the spatiotemporal heating response for a given exposure. When validated against histopathology and microscopy, this information can be useful both for demonstrating feasibility and aiding in the development of an approach to therapy in humans. Stafford et al. (2011) used MRTI to investigate the spatiotemporal distribution of temperature in PC-3 xenografts 24 h after an intravenous infusion of gold–silica nanoshells using an external laser with a fixed exposure (4 W/cm^2 for 3 min at 808 nm). MRTI measurements demonstrated a statistically significant ($p < 0.001$) increase in maximum temperature in the tumor cortex (mean = 21 ± 7°C) in +AuNS tumors versus control tumors, but that this heating fell off dramatically with distance away from the cortex. The depth of damage was significantly deeper than the nominal damage that sometimes occurred at the surface of the control tumors as well. Although the research demonstrated that passive uptake of nanoparticles resulted in ablative temperatures being able to be reached at that which did not result in significant damage to control tumor tissue, the depth of penetration indicated that it would be unlikely that approaches to ablation of tumors from the outside the tumor are unlikely to be successful except in the case of smaller tumors.

At lower temperatures (hyperthermia) over short durations, heat can increase the permeability of cells and vasculature. A group led by Sunil Krishnan investigated mild-temperature hyperthermia generated by NIR illumination of gold nanoshell-laden tumors noninvasively quantified by MRTI. They demonstrated an early increase in tumor perfusion related to the low heating, which was visualized using dynamic contrast enhancement before and after therapy. This increased perfusion reduced the hypoxic tumor fraction and enhanced the sensitivity to radiotherapy (Diagaradjane et al. 2008).

Rylander et al. (2011) investigated the activation of Hsp70 and Hsp27 in PC-3 xenografts in the peri-ablational region containing sublethal damage with MRTI correlation. Mice were infused intravenously with nanoparticles 24 h before exposure to an external laser beam (5 W/cm^2 for 3 min at 810 nm) under MRTI monitoring. Tumors were sectioned 16 h after laser treatment and stained for Hsp27 and Hsp70 (Rylander et al. 2011). The investigation is illustrated in Figure 15.5. Biological models for cell death and Hsp expression from *in vitro* studies are derived from the MRTI data to demonstrate how Hsp expression is elevated in the peri-ablational region.

Small animal work will continue to factor prominently in future research of nanomedicine, and MRI methods are useful for demonstrating efficacy and proof of principle. To increase the uptake of particles in tumor, creating smaller nanoparticles is of interest to increase uptake. However, smaller particles tend to wash out at a higher rate as well. To this end, investigators are researching the impact of using functional nanoparticles that target specific ligands on the tumor or the tumor stroma. Melancon et al. (2008) have used MRTI in their *in vivo* assessment of the

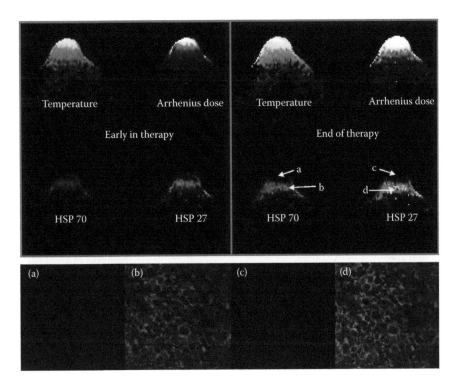

FIGURE 15.5 (See color insert.) Use of MRTI (1.5-T clinical scanner, 4 × 2 cm FOV, 5 s per image) and a model of HSP expression kinetics to dynamically image the expression HSP70 and HSP27 during laser thermal ablative therapy of a PC3 xenograph. Treatment post i.v. delivery of gold–silica nanoshells to human prostate carcinoma xenographs implanted on the backs of SCID mice. Laser exposure was 4 W/cm^2 for 3 min at 808 nm. Zones where maximum damage occurred resulted in no HSP expression (a and c), whereas zones where there was less damage demonstrated expression (b and d). Immunofluorescent staining for HSP expression appears to correlate with expression model (unpublished results).

impact of targeting small, hollow gold nanoshells (hAuNS). In this research, they used temperature distribution histograms generated from a region of interest within their tumors to demonstrate a substantial increase in heating between the targeted nanoparticles versus the untargeted particles and control groups.

It should be noted that the concentrations in tumor found after a typical intravenous injection preclude visualization of most gold nanoshells with CT, much less MRI. Although radionuclide tagging would be the most sensitive technique, the signal would fade with time. Superparamagnetic iron oxides (SPIO) perturb the local magnetic field well beyond the dipole–dipole interaction length that governs paramagnetic T1 relaxation decreases. The result is a reduction in the spin–spin (T2) or transverse (T2*) relaxation rates that result in dark signal. Some groups are combining SPIO with gold nanoshells (SPIO@AuNS) to investigate MR–visible photothermal therapy agents (Ji et al. 2007; Melancon et al. 2009). Melancon et al. (2011a, 2011b) have investigated an epidermal growth factor receptor targeted version of these particles using both phantom and *in vivo* mouse xenograft investigations with MRI to demonstrate the linear relationship between R2* changes and concentration, that SPIO paramagnetism does not interfere with the PRF temperature sensitivity coefficient at biologically relevant concentrations, and that targeting was successful *in vivo*.

Despite the usefulness of small animal models for investigating the feasibility of significant nanoparticle uptake for therapy and resulting heating as well as modulation of biological and physiological parameters, in order to more aptly demonstrate the potential to safely and effectively treat larger tumors and further develop and refine the physical and biological models needed for planning and guiding these treatments, some investigations into larger animals is often required. Within these models, realistically sized tumors can often be grafted in the organ of interest for therapy. Although these models do not often display the same relevant biological targets that one would be interested in during testing chemotherapy, they present an experimental arena where realistic considerations can be made with respect to treatment of larger tumors, such as realistic parameters for parenchymal and tumor perfusion, optical parameters, and implications regarding nearby critical structures. Furthermore, human size treatment applicators and approaches can be tested in a manner more closely resembling the actual approach to treatment.

An excellent example of an application that requires such consideration is nanoparticle-mediated thermal ablation in the brain (Figure 15.6). Schwartz et al. (2009) reported on a pilot study demonstrating a proof of concept for the passive delivery of AuNS to an orthotopic model—canine transmissible venereal tumor—in the brain of immunosuppressed dogs. AuNS (~150 nm) were infused intravenously and allowed to passively accumulate in the intracranial tumors over a period of 24 h. NIR was delivered interstitially through the skull using a cooled catheter with a 1-cm-long diffusion tip laser fiber with a fixed exposure (3.5 W, 3 min at 808 nm). Multiplanar echo-planar MRTI demonstrated the temperature of +AuNS tumor tissue to be 65.8 ±

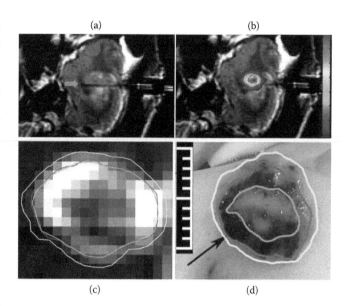

(a)   (b)

(c)   (d)

FIGURE 15.6   (See color insert.) MRTI spatiotemporal analysis of nanoshell-mediated tumor heating and damage. (a) The tumor was outlined (magenta) from a FLAIR sequence (TR/TE=1 s/145 ms, FOV = 20 × 20 cm, receiver BW = ±25 kHz, thickness = 4 mm). The laser catheter (green) is posterior to tumor. (b) Real-time multiplanar MRTI overlay at end of exposure (3.5 W for 180 s et 808 nm) demonstrates confinement of heating to tumor region (color bar ΔT range is 5–40°C). (c) Post-treatment contrast-enhanced T1-weighted imaging (TR/TE = 800 ms/9.2 ms, FOV = 20 × 20 cm, receiver BW: ±25 kHz) demonstrates a region of damage surrounded by edema. The Arrhenius estimate for coagulation (red) and 54C isotherm (cyan) show excellent correlation, whereas an Arrhenius model for vascular stasis (yellow) appears to match well with the very edge of the edema and is approximately. A distribution of damage similar to that predicted by the MRTI and modeling was observed by the pathologist where the inner region (green) represents the tumor surrounded by coagulated tissue (red) and a ring of edema (yellow) about 1 mm out (similar to MRTI predictions). Note observations are not necessarily exactly in same scale nor same location and orientation in brain.

4.1°C versus 48.6 ± 1.1°C in normal brain. An Arrhenius model of the thermal damage demonstrated damage in the +AuNS animals that agreed well with histopathology and further demonstrated that MRTI guidance would be essential for monitoring such therapies in the brain to assure safety and efficacy.

15.3 Summary

NIR activated nanoparticles present an exciting opportunity for enhancing current approaches to laser ablative therapies by enhancing heating in a manner that leads to a more conformal therapy delivery. MRTI has already played a critical role in the development and clinical implementation of high-temperature ablative therapies such as focused ultrasound and laser ablation. MRTI feedback from carefully constructed phantom, *ex vivo*, and *in vivo* investigations can be used to demonstrate proof of concept as well as gather information for modeling and simulation of therapy. MRTI is an invaluable tool for facilitating the clinical translation research needed to fully evaluate the implementation of this minimally invasive thermal ablative therapy. Additionally, for many sensitive areas of application, such as brain, MRTI may prove to be a facilitating technology that can be effectively leveraged to help ensure both the safety and efficacy of thermal therapy procedures using metal nanoparticles.

References

Ahmed, M., and S. Goldberg. 2005. Image-guided tumor ablation: Basic science. *Tumor Ablation: Principles and Practice*, van Sonnenberg, E., McMullen, W., and Solbiati, L., eds., 23. Springer: Heidelberg.

Ahmed, M. et al. 2011. Principles of and advances in percutaneous ablation. *Radiology* 258(2):351–269.

Beland, M., P. R. Mueller, and D. A. Gervais. 2007. Thermal ablation in interventional oncology. *Seminars in Roentgenology* 42(3):175–190.

Boss, A. et al. 2005. Magnetic susceptibility effects on the accuracy of MR temperature monitoring by the proton resonance frequency method. *Journal of Magnetic Resonance Imaging* 22(6):813–820.

Callstrom, M. R. et al. 2009. Research reporting standards for image-guided ablation of bone and soft tissue tumors. *Journal of Vascular and Interventional Radiology* 20(12):1527–1540.

Carpentier, A. et al. 2008. Real-time magnetic resonance-guided laser thermal therapy for focal metastatic brain tumors. *Neurosurgery* 63(1 Suppl 1):ONS21–ONS28; discussion ONS28–ONS29.

Cheong, S. K., S. Krishnan, and S. H. Cho. 2009. Modeling of plasmonic heating from individual gold nanoshells for near-infrared laser-induced thermal therapy. *Medical Physics* 36(10):4664–4671.

Chung, A. H. et al. 1996. Optimization of spoiled gradient-echo phase imaging for in vivo localization of a focused ultrasound beam. *Magnetic Resonance in Medicine* 36(5): 745–752.

Cline, H. E. et al. 1992. MR-guided focused ultrasound surgery. *Journal of Computer Assisted Tomography* 16(6):956–965.

Colen, R. R., and F. A. Jolesz. 2010. Future potential of MRI-guided focused ultrasound brain surgery. *Neuroimaging Clinics of North America* 20(3):355–366.

Conturo, T. E., and G. D. Smith. 1990. Signal-to-noise in phase angle reconstruction: Dynamic range extension using phase reference offsets. *Magnetic Resonance in Medicine* 15(3):420–437.

De Poorter, J. 1995. Noninvasive MRI thermometry with the proton resonance frequency method: Study of susceptibility effects. *Magnetic Resonance in Medicine* 34(3):359–367.

de Zwart, J. A. et al. 1999. Fast lipid-suppressed MR temperature mapping with echo-shifted gradient-echo imaging and spectral–spatial excitation. *Magnetic Resonance in Medicine* 42(1):53–59.

Dewhirst, M. W. et al. 2003. Basic principles of thermal dosimetry and thermal thresholds for tissue damage from hyperthermia. *International Journal of Hyperthermia* 19(3):267–294.

Diagaradjane, P. et al. 2008. Modulation of in vivo tumor radiation response via gold nanoshell-mediated vascular-focused hyperthermia: Characterizing an integrated antihypoxic and localized vascular disrupting targeting strategy. *Nano Letters* 8(5):1492–1500.

Elliott, A. M. et al. 2007. Laser-induced thermal response and characterization of nanoparticles for cancer treatment using magnetic resonance thermal imaging. *Medical Physics* 34(7):3102–3108.

Elliott, A. et al. 2008. Analytical solution to heat equation with magnetic resonance experimental verification for nanoshell enhanced thermal therapy. *Lasers in Surgery and Medicine* 40(9):660–665.

Elliott, A. M. et al. 2009. Quantitative comparison of delta P1 versus optical diffusion approximations for modeling near-infrared gold nanoshell heating. *Medical Physics* 36(4):1351–1358.

Elliott, A. M. et al. 2010. Use of gold nanoshells to constrain and enhance laser thermal therapy of metastatic liver tumours. *International Journal of Hyperthermia* 26(5):434–440.

Feng, Y. et al. 2009. Nanoshell-mediated laser surgery simulation for prostate cancer treatment. *Engineering Computations* 25(1):3–13.

Flanders, V. L., and D. A. Gervais. 2010. Ablation of liver metastases: Current status. *Journal of Vascular and Interventional Radiology* 21(8 Suppl):S214–S222.

Fuentes, D. et al. 2009. Computational modeling and real-time control of patient-specific laser treatment of cancer. *Annals of Biomedical Engineering* 37(4):763–782.

Fuentes, D. et al. 2010. Adaptive real-time bioheat transfer models for computer-driven MR-guided laser induced thermal therapy. *IEEE Transactions on Biomedical Engineering* 57(5):1024–1030.

Fuentes, D. et al. 2011. Magnetic resonance temperature imaging validation of a bioheat transfer model for laser-induced thermal therapy. *International Journal of Hyperthermia* 27(5):453–464.

Gillams, A. 2008. Tumour ablation: Current role in the liver, kidney, lung and bone. *Cancer Imaging* 8(Spec No A):S1–S5.

Goldberg, S. N. et al. 2009. Image-guided tumor ablation: Standardization of terminology and reporting criteria. *Journal of Vascular and Interventional Radiology* 20(7 Suppl):S377–S390.

Graham, S. J. et al. 1999. Quantifying tissue damage due to focused ultrasound heating observed by MRI. *Magnetic Resonance in Medicine* 41(2):321–328.

Hindman, J. C. 1966. Proton resonance shift of water in the gas and liquid states. *Journal of Chemical Physics* 44(12):4582–4592.

Hirsch, L. R. et al. 2003. Nanoshell-mediated near-infrared thermal therapy of tumors under magnetic resonance guidance. *Proceedings of the National Academy of Sciences of the United States of America* 100(23):13549–13554.

Hirsch, L. R. et al. 2006. Metal nanoshells. *Annals of Biomedical Engineering* 34(1):15–22.

Hong, K., and C. Georgiades. 2010. Radiofrequency ablation: Mechanism of action and devices. *Journal of Vascular and Interventional Radiology* 21(8 Suppl):S179–S186.

Hynynen, K. et al. 1993. MRI-guided noninvasive ultrasound surgery. *Medical Physics* 20(1):107–115.

Ishihara, Y. et al. 1995. A precise and fast temperature mapping using water proton chemical shift. *Magnetic Resonance in Medicine* 34(6):814–823.

Ji, X. et al. 2007. Bifunctional gold nanoshells with a superparamagnetic iron oxide–silica core suitable for both MR imaging and photothermal therapy. *Journal of Physical Chemistry C Nanomaterials and Interfaces* 111(17):6245.

Jolesz, F. A. 2009. MRI-guided focused ultrasound surgery. *Annual Review of Medicine* 60:417–430.

Jolesz, F. A. et al. 1988. MR imaging of laser-tissue interactions. *Radiology* 168(1):249–253.

Kickhefel, A. et al. 2011. Clinical evaluation of MR temperature monitoring of laser induced thermotherapy in human liver using the proton resonance frequency method and predictive models of cell death. *Journal of Magnetic Resonance Imaging* 33(3):704–709.

Kunkle, D. A., and R. G. Uzzo. 2008. Cryoablation or radiofrequency ablation of the small renal mass: A meta-analysis. *Cancer* 113(10):2671–2680.

Kuroda, K. 2005. Non-invasive MR thermography using the water proton chemical shift. *International Journal of Hyperthermia* 21(6):547–560.

Kuroda, K. et al. 1996. Temperature mapping using water proton chemical shift obtained with 3D-MRSI: Feasibility in vivo. *Magnetic Resonance in Medicine* 35(1):20–29.

Kuroda, K. et al. 1997. Temperature mapping using the water proton chemical shift: A chemical shift selective phase mapping method. *Magnetic Resonance in Medicine* 38(5):845–851.

Kuroda, K. et al. 1998. Calibration of water proton chemical shift with temperature for noninvasive temperature imaging during focused ultrasound surgery. *Journal of Magnetic Resonance Imaging* 8(1):175–181.

Kuroda, K. et al. 2000. Temperature mapping using the water proton chemical shift: Self-referenced method with echo-planar spectroscopic imaging. *Magnetic Resonance in Medicine* 43(2):220–225.

Kurup, A. N., and M. R. Callstrom. 2010. Ablation of skeletal metastases: Current status. *Journal of Vascular and Interventional Radiology* 21(8 Suppl):S242–S250.

Lubner, M. G. et al. 2010. Microwave tumor ablation: Mechanism of action, clinical results, and devices. *Journal of Vascular and Interventional Radiology* 21(8 Suppl):S192–S203.

Matsumoto, R., K. Oshio, and F. A. Jolesz. 1992. Monitoring of laser and freezing-induced ablation in the liver with T1-weighted MR imaging. *Journal of Magnetic Resonance Imaging* 2(5):555–562.

McDannold, N. 2005. Quantitative MRI-based temperature mapping based on the proton resonant frequency shift: Review of validation studies. *International Journal of Hyperthermia* 21(6):533–546.

McDannold, N. et al. 2001. Temperature monitoring with line scan echo planar spectroscopic imaging. *Medical Physics* 28(3):346–355.

McWilliams, J. P. et al. 2010. Percutaneous ablation of hepatocellular carcinoma: Current status. *Journal of Vascular and Interventional Radiology* 21(8 Suppl):S204–S213.

Melancon, M. P. et al. 2008. in vitro and in vivo targeting of hollow gold nanoshells directed at epidermal growth factor receptor for photothermal ablation therapy. *Molecular Cancer Therapeutics* 7(6):1730–1739.

Melancon, M., W. Lu, and C. Li. 2009. Gold-based magneto/optical nanostructures: Challenges for in vivo applications in cancer diagnostics and therapy. *Materials Research Bulletin* 34(6):415.

Melancon, M. P. et al. 2011a. Targeted multifunctional gold-based nanoshells for magnetic resonance-guided laser ablation of head and neck cancer. *Biomaterials* 32(30):7600–7608.

Melancon, M. P. et al. 2011b. Theranostics with multifunctional magnetic gold nanoshells: Photothermal therapy and t2* magnetic resonance imaging. *Investigative Radiology* 46(2):132.

Michael, P. et al. 2002. T1 relaxation time at 0.2 Tesla for monitoring regional hyperthermia: Feasibility study in muscle and adipose tissue. *Magnetic Resonance in Medicine* 47(6):1194–1201.

Oden, J. T. et al. 2007. Dynamic data-driven finite element models for laser treatment of cancer. *Numerical Methods for Partial Differential Equations* 23(4):904–922.

Pauly, K. B. et al. 2006. Magnetic resonance-guided high-intensity ultrasound ablation of the prostate. *Topics in Magnetic Resonance Imaging* 17(3):195–207.

Peters, R. D., and R. M. Henkelman. 2000. Proton-resonance frequency shift MR thermometry is affected by changes in the electrical conductivity of tissue. *Magnetic Resonance in Medicine* 43(1):62–71.

Peters, R. D., R. S. Hinks, and R. M. Henkelman. 1998. Ex vivo tissue-type independence in proton-resonance frequency shift MR thermometry. *Magnetic Resonance in Medicine* 40(3):454–459.

Peters, R. D., R. S. Hinks, and R. M. Henkelman. 1999. Heat-source orientation and geometry dependence in proton-resonance frequency shift magnetic resonance thermometry. *Magnetic Resonance in Medicine* 41(5):909–918.

Pua, B. B., R. H. Thornton, and S. B. Solomon. 2010. Ablation of pulmonary malignancy: Current status. *Journal of Vascular and Interventional Radiology* 21(8 Suppl):S223–S232.

Raz, O. et al. 2010. Real-time magnetic resonance imaging-guided focal laser therapy in patients with low-risk prostate cancer. *European Urology* 58(1):173–177.

Requarth, J. 2011. Image-guided palliative care procedures. *Surgical Clinics of North America* 91(2):367–402.

Rieke, V., and K. Butts Pauly. 2008a. Echo combination to reduce proton resonance frequency (PRF) thermometry errors from fat. *Journal of Magnetic Resonance Imaging* 27(3):673–677.

Rieke, V., and K. Butts Pauly. 2008b. MR thermometry. *Journal of Magnetic Resonance Imaging* 27(2):376–390.

Rybak, L. D. 2009. Fire and ice: Thermal ablation of musculoskeletal tumors. *Radiologic Clinics of North America* 47(3):455–469.

Rylander, M. N. et al. 2011. Heat shock protein expression and temperature distribution in prostate tumours treated with laser irradiation and nanoshells. *International Journal of Hyperthermia* 27(8):791–801.

Schneider, W. G., H. J. Bernstein, and J. A. Pople. 1958. Proton magnetic resonance chemical shift of free (gaseous) and associated (liquid) hydride molecules. *Journal of Chemical Physics* 284:601.

Schwartz, J. A. et al. 2009. Feasibility study of particle-assisted laser ablation of brain tumors in orthotopic canine model. *Cancer Research* 69(4):1659–1667.

Sharma, R., J. L. Wagner, and R. F. Hwang. 2011. Ablative therapies of the breast. *Surgical Oncology Clinics of North America* 20(2):317–339.

Stafford, R. J. et al. 2010. Laser-induced thermal therapy for tumor ablation. *Critical Reviews in Biomedical Engineering* 38(1):79–100.

Stafford, R. J. et al. 2011. MR temperature imaging of nanoshell mediated laser ablation. *International Journal of Hyperthermia* 27(8):782–790.

Taylor, B. A. et al. 2008. Dynamic chemical shift imaging for image-guided thermal therapy: Analysis of feasibility and potential. *Medical Physics* 35(2):793–803.

Taylor, B. A. et al. 2009. Autoregressive moving average modeling for spectral parameter estimation from a multigradient echo chemical shift acquisition. *Medical Physics* 36(3): 753–764.

Taylor, B. A. et al. 2011a. Correlation between the temperature dependence of intrinsic MR parameters and thermal dose measured by a rapid chemical shift imaging technique. *NMR in Biomedicine* 24(10):1414–1421.

Taylor, B. A. et al. 2011b. Measurement of temperature dependent changes in bone marrow using a rapid chemical shift imaging technique. *Journal of Magnetic Resonance Imaging* 33(5):1128–1135.

Thumar, A. B. et al. 2010. Thermal ablation of renal cell carcinoma: Triage, treatment, and follow-up. *Journal of Vascular and Interventional Radiology* 21(8 Suppl):S233–S241.

Todd, N. et al. 2012. Hybrid proton resonance frequency/T(1) technique for simultaneous temperature monitoring in adipose and aqueous tissues. *Magnetic Resonance in Medicine.*

Vogl, T. J. et al. 1995. Malignant liver tumors treated with MR imaging-guided laser-induced thermotherapy: Technique and prospective results. *Radiology* 196(1):257–265.

Vogl, T. J. et al. 2002. Magnetic resonance (MR)-guided percutaneous laser-induced interstitial thermotherapy (LITT) for malignant liver tumors. *Surgical Technology International* 10:89–98.

Young, I. R. et al. 1994. Modeling and observation of temperature changes in vivo using MRI. *Magnetic Resonance in Medicine* 32(3):358–369.

Yung, J. P. et al. 2010. Quantitative comparison of thermal dose models in normal canine brain. *Medical Physics* 37(10):5313–5321.

16

Modeling of Plasmonic Heat Generation from Gold Nanoparticles

Francisco Reynoso
Georgia Institute of Technology

Sang Hyun Cho
Georgia Institute of Technology

16.1 Background

The remarkable power of heat as a therapeutic tool has been well documented since ancient times. Reports of hot irons and stones used for medicinal purposes have existed as early as 3000 BC in the Egyptian Edwin Smith surgical papyrus (Breasted 1991) and later in the Greek Hippocratic Corpus around 450–350 BC (Lloyd et al. 1983). Contemporary heating methods include techniques that use radiofrequency, microwaves, or ultrasound waves to elevate the temperature of the clinical target. Therapeutic modalities aimed at treating a local disease via elevated temperatures is now referred to as hyperthermia. Clinical hyperthermia refers to treatment of tumors by heating the lesions between 40°C and 45°C on the order of tens of minutes. Selective treatment of tumors is possible and is a result of the decreased ability of tumors to dissipate heat because of their highly disorganized system of blood vessels. However, routine clinical applications of hyperthermia are still not optimal and major improvements are needed. The temperature distributions achievable from conventional approaches are far from satisfactory and, as a result, improved temperature control and monitoring are still in need of further development for appropriate clinical translation.

Light amplification by stimulated emission of radiation (LASER) yields beams of light that are monochromatic and easily manipulated. The ability to produce highly coherent, monochromatic, collimated beams of light has revolutionized many disciplines, including medicine. There are multiple ways that laser light can interact with tissue and these mechanisms lead to a variety of medical procedures that range from tissue cutting and welding in surgical procedures, to photodynamic therapy in oncology (Boulnois 1986). Laser light can easily be converted to an intense beam of light that can penetrate deep into tissues, which has enabled a procedure termed laser-induced thermal therapy (LITT). High-intensity laser beams can be delivered to deep-seated tumors via optical fibers inserted directly into tumors with great precision. A major drawback of LITT is the lack of selective damage; cancerous and healthy tissues are both equally susceptible to damage. Tissue penetration, inability to selectively heat the target, and a lack of predictive heat control have prevented its widespread clinical use.

Recently, gold nanoparticles (GNPs) have been proposed to enhance the treatment efficacy of LITT. This type of nanoparticle exhibits a plasmon resonance that can be easily tuned to absorb strongly in the near-infrared (NIR) region. In this region, absorption by hemoglobin and water molecules, the strongest chromophores in tissue, is at a minimum, therefore making it ideal for LITT (Weissleder 2001). Gold nanorods (Huang et al. 2006), gold nanoshells (Hirsch et al. 2003), and gold nanocages (Chen et al. 2005) have shown to be promising candidates for NIR-LITT.

16.1.1 Surface Plasmon Resonance

The interesting optical properties of noble metal nanoparticles and GNPs in particular have been unknowingly in use for hundreds of years. Their bright and intense colors have been seen in stained cathedral windows and other forms of artwork for centuries (Jain et al. 2007; El-Sayed 2001; Kelly et al. 2003). Michael Faraday, a well-known physicist, was the first to show that the unique properties of these intense dyes were due to colloidal gold present in the solution (Faraday 1857). Furthermore, in 1908,

Gustav Mie presented the first solution to Maxwell's equations to describe the extinction spectra due to scattering and absorption of spherical nanoparticles of any size (Mie 1908).

The physical basis for the aforementioned phenomena is called surface plasmon resonance (SPR). In a bulk metal, free electrons or quasi-free electrons move freely by the action of external fields and, as a result, conduction occurs. In the Drude model, these free and quasi-free electrons are assumed to behave like a gas of free charge carriers or plasma. The Drude model assumes that the electrons are accelerated by the external fields and move in a straight line until they collide with one another in average time τ, the relaxation time (Ashcroft and Mermin 1976). The ease of motion of electrons through the metal is characterized by the frequency of collisions through the metal. If the frequency of the applied field is comparable to the collision frequency, electrons may accelerate and decelerate between collisions, and the model treating the metal as a single gas loses the validity (Jackson 1998). At higher frequencies, the electrons and their positive ion counterparts are accelerated in opposite directions, and charge density oscillations occur as a result of Coulomb attraction. These longitudinal charge density fluctuations are referred to as plasma oscillations, and its quantum of energy is termed as a plasmon. It can be shown that the frequency of oscillation of the charge density is the plasma frequency ω_p (Jackson 1998):

$$\omega_p = \frac{n_o e^2}{m \varepsilon_0} \qquad (16.1)$$

where n_o is the charge density, e is the electron charge, m is the mass of the electron, and ε_0 is the permittivity of free space. Plasmons can be excited in a metal by passing an electron through a thin film or by reflecting an electron or photon from the surface of a metal, which distinguishes surface plasmons from bulk plasmons. Surface plasmons are bound to the interface between the plasma and a dielectric such as water or air. Surface plasmons exist in a variety of metals at a wide range of electromagnetic frequencies.

At the nanometer scale, the distinction between bulk plasmons and surface plasmons disappears as the size of metallic structures drop below the penetration depth of electromagnetic fields into the metal. The illumination of a metallic nanoparticle by light causes the conduction electrons to respond to the electric field of the incident light. These electrons are collectively displaced, and as a result of the Coulomb attraction from the nucleus, the electron cloud collectively oscillates. This phenomenon is referred to as SPR and is the basis for the enhanced absorption cross section of metallic nanoparticles. Mie scattering theory is an exact analytical description of SPR of metal nanoparticles, and the scattering and absorption of light by small particles. The theory is a solution of Maxwell's equations by expansion of the electromagnetic fields into vector spherical harmonic functions. The solution is expressed as a multipole expansion via spherical harmonics with each term including

finer angular features. The extinction cross section and scattering cross section for a spherical nanoparticle can be derived using the Mie scattering theory as (Kreibig and Vollmer 1995; Bohren and Huffman 1983):

$$\sigma_{sca} = \frac{2\pi}{k^2} \sum_{l=1}^{\infty} (2l+1)\left(\left|a_l\right|^2 + \left|b_l\right|^2\right) \qquad (16.2)$$

$$\sigma_{ext} = \frac{2\pi}{k^2} \sum_{l=1}^{\infty} (2l+1) Re\left\{a_l + b_l\right\} \qquad (16.3)$$

where k is the wave-vector and the absorption cross section is $\sigma_{abs} = \sigma_{ext} - \sigma_{sca}$. The scattering coefficients a_l and b_l are problem-dependent, and for a sphere of permeability μ_l embedded in a medium of permeability μ_m they are

$$a_l = \frac{\mu_m m^2 j_l(mx)[xj_l(x)]' - \mu_1 j_l(x)[mxj_l(mx)]'}{\mu_m m^2 j_l(mx)[xh_l^{(1)}(x)]' - \mu_1 h_l^{(1)}(x)[mxj_l(mx)]'} \qquad (16.4)$$

$$b_l = \frac{\mu_1 j_l(mx)[xj_l(x)]' - \mu_m j_l(x)[mxj_l(mx)]'}{\mu_1 j_l(mx)[xh_l^{(1)}(x)]' - \mu_m h_l^{(1)}(x)[mxj_l(mx)]'} \qquad (16.5)$$

where x is the size parameter related to the particle radius as $x = kr$, m is the relative refractive index $m = n_1/n_m$, j_l and $h_l^{(1)}$ are the spherical Bessel functions and spherical Hankel functions, respectively. The solutions show the direct relationship between the scattering and absorption properties of the nanoparticles with the frequency of incident light via the frequency dependence of the permeability of the sphere and surrounding material. It can be further shown that for particles small compared to the wavelength of light ($x \gg 1$), the scattering and absorption cross sections reduce to (Kreibig and Vollmer 1995; Bohren and Huffman 1983):

$$\sigma_{abs} = 4\pi k r^3 Im\left\{\frac{\varepsilon_1 - \varepsilon_m}{\varepsilon_1 + 2\varepsilon_m}\right\} \qquad (16.6)$$

$$\sigma_{sca} = \frac{8}{3}\pi k^4 r^6 \left|\frac{\varepsilon_1 - \varepsilon_m}{\varepsilon_1 + 2\varepsilon_m}\right|^2 \qquad (16.7)$$

which demonstrates a strong resonance condition at the electromagnetic frequency when $\varepsilon_1 = -2\varepsilon_m$, defining the SPR frequency. The SPR frequency depends on the size, shape, and type of metal, and the dielectric properties of the surrounding media, yielding remarkable ability to tune the optical properties of metal nanoparticles.

16.1.2 Plasmonic Heating

The unique optical properties of noble metal nanoparticles have made them unique tools in biomedical imaging and therapeutics. An attractive feature of this class of nanoparticles for the purpose of LITT is the efficient conversion of absorbed energy to heat. The thermalization process in biological tissue is a nonradiative, deexcitation pathway that occurs when a molecule absorbs a photon and collides with other molecules during its excited state. This efficient heating mechanism may help overcome the difficulty of conventional LITT to selectively heat tumor regions. GNPs can be specifically accumulated within a tumor via the enhanced permeability and retention effect (Maeda 2000), and target molecules to produce highly conformal heat distribution when used in conjunction with LITT.

The heating mechanism starts with the absorption of electromagnetic energy by electrons via SPR, leading to a temperature differential between the electron gas and the lattice. The added energy disturbs the Fermi–Dirac distribution of the electrons, and electron–electron scattering within the electron gas occurs until the system has thermalized. Electron–phonon interactions then transfer the energy between valence electrons and phonons within the lattice. The absorption of a photon by a molecule hinges on the availability of accessible vibrational states, which are numerous for most biomolecules. This makes the absorption process highly efficient in biological tissues (Boulnois 1986). The vibrational energy in the excited molecule is transferred to other molecules as translational kinetic energy, which macroscopically manifests itself as a temperature increase. This process of thermal relaxation occurs on the order of picoseconds but a macroscopic increase in temperature is only evident on a much larger timescale on the order of seconds (Link and El-Sayed 2000).

16.1.3 Optically Tunable Nanomaterials

The dependence of SPR frequency on size and shape of nanoparticle is better appreciated by deriving the polarizability of metal nanoparticles. Using electrostatic approximation, it can be shown that the polarizability of an ellipsoidal metal nanoparticle along one of its principal axes is (Bohren and Huffman 1983):

$$\alpha_j = 4\pi abc \frac{\varepsilon_1 - \varepsilon_m}{3\varepsilon_m + 3L_j\left(\varepsilon_1 - \varepsilon_m\right)} \quad j = 1, 2, 3 \quad (16.8)$$

where ε_1 and ε_m are the permittivities of the nanoparticle and surrounding media, respectively. The parameters a, b, and c are the length of the semiprincipal axes of the ellipsoidal nanoparticle, with $a = b = c$ for a sphere. L_j's are geometrical factors that dictate the dependence of polarizability on nanoparticle shape, with $L_1 + L_2 + L_3 = 1$ and $L_1 = L_2 = L_3 = 1/3$ for spherical nanoparticles. This expression for polarizability highlights the strong dependence of the SPR frequency on size and shape of the nanoparticle and surrounding dielectric. The expression also

confirms the resonance condition of $\varepsilon_1 = -2\varepsilon_m$ for a spherical nanoparticle.

The dependence of SPR frequency on the type of metal, size, and shape of the nanoparticle, and permittivity of surrounding media imparts unique tunability to metal nanostructures. The polarizability expression also reveals the strong dependence on shape that gives rise to multiple resonance bands in asymmetric nanoparticles, as is the case with gold nanorods. For gold nanorods, the absorption spectrum reveals two resonance bands: a short wavelength resonance band corresponding to transverse electron oscillation and a stronger, long wavelength resonance band corresponding to longitudinal oscillation of electrons. An increase in the aspect ratio of gold nanorods therefore redshifts the maximum SPR wavelengths as the longitudinal oscillation band increases.

Gold nanoshells are another type of gold nanostructure that has been demonstrated to have unique tuning capabilities (Oldenburg et al. 1998). Gold nanoshells contain a silica core surrounded by a thin gold shell. The SPR frequency of these structures can be tuned by adjusting the ratio of gold shell thickness to silica core diameter (Figure 16.1). The maximum SPR wavelength can be red-shifted by decreasing the ratio of gold shell thickness to silica core diameter.

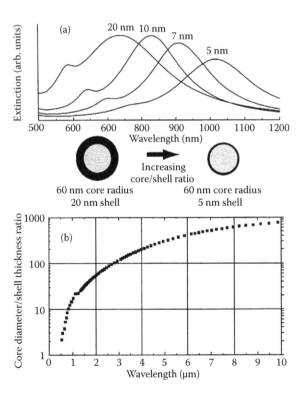

FIGURE 16.1 (a) Theoretically calculated surface plasmon resonance of metal nanoshells for various core radius/shell thickness ratios. (b) Resonance wavelength versus core radius to shell thickness ratio. (Reproduced with permission from Oldenburg, S.J. et al., *Chem. Phys. Lett.*, 288, 243–247, 1998.)

16.2 Modeling of Plasmonic Heat Generation in Tissue

Thermal effects in tissue are unique as there is no specific reaction pathway required to achieve damage; heat can be absorbed by any biomolecule and lead to tissue damage. Thermal effects are also unique in the sense that tissue damage is only dependent on the temperature that is reached and the duration at which it remains at that temperature (McKenzie 1990). This means that, in order to characterize and correctly predict tissue damage from thermal therapy, only the spatial and temporal distribution of heat is needed. This can be obtained in three major steps: determine the optical and thermal parameters of the tissue and GNPs, determine photon absorption via photon transport modeling, and determine the temperature distribution by solving the bioheat diffusion equation.

The heat distribution within the tissue is determined by Pennes' bioheat equation (Pennes 1948):

$$\rho C \frac{\partial T(\vec{r},t)}{\partial t} = \nabla \cdot \left[\kappa \nabla T(\vec{r},t) \right] + q_m - q_b + Q \qquad (16.9)$$

where ρ (kg/m^3) is the density, C (J/kg K) is the specific heat, κ (W/m K) is the thermal conductivity of the medium, Q (W/m^3) is the heat source term, and q_m is the heat generated through metabolic activity. The last term q_b is the heat removed through blood perfusion and defined as

$$q_b = \omega \rho_b c_b (T_a - T) \qquad (16.10)$$

where ω (m^3/s) is the blood perfusion, ρ_b (kg/m^3) is the density of blood, c_b (J/kg K) is the specific heat capacity of blood, and T_a (K) is the temperature of arterial blood. At the boundaries, heat transfer can be modeled via Neumann boundary condition:

$$\kappa \frac{\partial T(\vec{r},t)}{\partial t} = h(T_b - T_\infty) \qquad (16.11)$$

where h is the heat convection constant, T_b is the temperature at the boundary, and T_∞ is the ambient temperature. The heat loss due to blood perfusion q_b can be quite significant for cases in which LITT is used for mild hyperthermia (40–45°C). However, it should be pointed out that the efficacy of heat loss due to blood perfusion decreases the longer the tissue is kept at a high temperature. Tissues kept at an elevated temperature for long enough become necrotic and blood flow ceases. If the exposure duration is long enough, blood perfusion does not significantly affect the temperature rise at the end of laser exposure (McKenzie 1990).

In recent years, many different computational approaches have been developed to solve the plasmonic heating problem in tissue. Among them, the two distinct approaches are discussed here. The first approach attempts to estimate the bulk heating of GNP-laden issue from a macroscopic point of view (Elliott et al. 2007, 2008, 2009) by solving photon transport and bioheat equations taking into account the reduced optical properties of GNP-laden tissue. On the other hand, the second approach aims to account for heat generation due to laser and individual GNPs separately (Cheong et al. 2009) by performing photon and heat transport based on the optical properties of tissue and the heat absorption properties of GNPs. Although not covered in detail here, there are other approaches worth mentioning, one assuming complete transparency of the tissue in the NIR region (Bayazitoglu and Tjahjono 2008) and another solving for photon transport considering the optical properties of the tissue and GNPs separately (Xu et al. 2011).

16.2.1 Bulk Heating Models

In 2007, Elliot et al. developed a three-dimensional (3-D) finite element modeling (FEM) technique to calculate the laser fluence and temperature distribution in water-based gel phantoms embedded with gold nanoshells. This computational model treated the nanoshell-laden gel phantom as a new medium with enhanced absorption properties. It also applied the optical diffusion approximation (ODA) to calculate the laser power density in the medium, and the 3-D FEM method to solve the bioheat equation.

An approximation can be made when scattering processes dominate absorption processes, as is the case in biological tissue ($\mu_a \ll \mu_s$). This is referred to as the ODA and described by

$$-D\nabla^2 \varphi(\vec{r}) + \mu_a \varphi(\vec{r}) = s(\vec{r}) \qquad (16.12)$$

where $\varphi(\vec{r})$ (W/m^2) is the light fluence rate, $s(\vec{r})$ (W/m^3) is the light source term, and D (m) is the diffusion coefficient defined as

$$D = \frac{1}{3(\mu_s' + \mu_a)} \qquad (16.13)$$

where $\mu_s' = (1-g)\mu_s$ is the reduced scattering coefficient and g is the anisotropy factor. The diffusion approximation is appropriate and studies (Heusmann et al. 1996; Niemz 2007) have shown values of μ_s (scattering coefficient) are 2 to 3 orders of magnitude larger than μ_a (absorption coefficient) for biological tissues in the NIR region. Furthermore, this approximation should hold true for water and water-based gel phantoms. Although there are no published absorption and scattering coefficients in the NIR region for such phantoms, it is reasonable to assume that the optical properties of gel phantoms approximate those of water.

The solution to the light diffusion approximation for a continuous wave laser beam at the origin is given by

$$\phi(\vec{r}) = \frac{P_o \exp(-\mu_{eff} \vec{r} \cdot \hat{n})}{4\pi D r} \qquad (16.14)$$

where $\mu_{eff} = \sqrt{3\mu_a(\mu_a + \mu_s')}$ is the effective attenuation coefficient, $\hat{n}$ is the direction of beam travel, and P_o is the laser power.

The model described above was applied to various two-layer phantoms of multiple optical densities (OD) and validated by magnetic resonance temperature imaging (MRTI). The model was applied to a 0.55-OD and a 0.695-OD phantom for various laser powers. Spatial and temporal distributions showed good agreement between FEM calculations and MRTI experiments (Elliott et al. 2007). However, the model becomes less accurate in the 0.695-OD phantom than in the 0.55-OD phantom because of the inherent limitations in the ODA. The higher the concentration of nanoshells, the closer one gets to the limit of applicability of the diffusion approximation ($\mu_a \ll \mu_s$), and the model becomes less relevant.

In 2008, Elliot et al. developed an analytical solution to the problem in order to improve calculation speed. The analytical solution allows the laser power to be changed without the need to recalculate the model. The method was tested for its validity using the same phantoms previously used (Elliott et al. 2007). The model hinges on the use of a Green's function to construct the solution to heat equation. The solution is given by

$$T(q_i, t) = \frac{\alpha}{\kappa} \int_{\tau=0}^{t} \left(\int_{q_i} G(q_i, t | q_i', \tau) Q(q_i', \tau) dq_i' \right) d\tau \qquad (16.15)$$

where T represents temperature, q_i and q_i' represent the cylindrical coordinates (ρ, z). The laser term Q is given by the fluence rate obtained from diffusion approximation times the absorption coefficient of the media. The terms t and τ both represent time, and $\alpha = \kappa/\rho c$ is the thermal diffusivity. This analytical expression shows good agreement with experimental MRTI spatiotemporal temperature distributions for multiple gold nanoshell concentrations and laser powers. The advantage of this model is that it allows for the solution to be readily obtained for a different laser power without recalculating the model. However, the model still suffers from the inaccuracies raised by the ODA.

In order to improve their computational model, Elliot et al. (2009) developed a technique in which the so-called $\delta - P_1$ approximation was used to obtain the optical power density throughout the media. The basis of the approximation is the decomposition of the light source into a collimated primary beam and a secondary scattered beam with the use of a Dirac delta function. The core of the $\delta - P_1$ approximation lies with the Boltzmann transport equation that describes angular photon flux (radiance) $L(\vec{r}, \hat{\Omega})$, which represents the rate of photon arrival at $\vec{r}$ traveling in direction $\hat{\Omega}$ per unit area per unit solid angle. The radiance function is decomposed into a primary forward-scattered component and a diffuse component:

$$L(\vec{r}, \hat{\Omega}) = L_f(\vec{r}, \hat{\Omega}) + L_d(\vec{r}, \hat{\Omega}) \qquad (16.16)$$

where $L_f(\vec{r}, \hat{\Omega})$ is the forward scattered component and $L_d(\vec{r}, \hat{\Omega})$ is the diffuse component. The phase function of the $\delta - P_1$ approximation is given by (Hayakawa et al. 2004)

$$p_{\delta-P_1}(\hat{\Omega} \cdot \hat{\Omega}') = \frac{1}{4\pi} \left\{ 2f\delta[1 - (\hat{\Omega} \cdot \hat{\Omega}')] + (1-f)[1 + 3g * \hat{\Omega} \cdot \hat{\Omega}'] \right\}, \qquad (16.17)$$

where f represents the forward scattered fraction of light photons, $\hat{\Omega}$ and $\hat{\Omega}'$ are the unit vectors of incident and scattered light, respectively. The second term in the expression represents the diffusively scattered fraction of light photons as given by the P_1 approximation to the phase function and $g*$ is the scattering asymmetry coefficient.

The forward-scattered component of the radiance $L_f(\vec{r}, \hat{\Omega})$ is computed by multiplying the spatial distribution of the primary laser source $E(\vec{r}, \hat{\Omega})$ by a Dirac delta function that filters the fraction scattered into the forward direction $\hat{z}$:

$$L_f(\vec{r}, \hat{\Omega}) = \frac{1}{2\pi} E(\vec{r}, \hat{\Omega}) \delta(1 - \hat{\Omega} \cdot \hat{z}) \qquad (16.18)$$

The diffusely scattered component $L_d(\vec{r}, \hat{\Omega})$, on the other hand, can be obtained via the standard ODA. The optical power density obtained via the $\delta - P_1$ approximation is then used to solve the bioheat equation via FEM computation.

The results of the model were compared to experimental MRTI spatiotemporal temperature distributions and the results obtained through the ODA model (Elliott et al. 2009). The $\delta - P_1$ and ODA both showed good agreement with each other and the MRTI data for phantoms with relatively low optical density of up to 0.66 OD. However, as the optical density was increased, the $\delta - P_1$ method demonstrated improved agreement with MRTI spatiotemporal distributions over the ODA approach. Figure 16.2 compares the results obtained through ODA and $\delta - P_1$ model with those measured using MRTI for a 0.66-OD phantom and a 1.49-OD phantom illuminated with a 1.2-W laser for 3 min. A pixel-by-pixel subtraction demonstrates the improved accuracy of the $\delta - P_1$ model and its superiority to the ODA model for phantoms with a higher optical density.

The computational model demonstrates that the $\delta - P_1$ approximation is an accurate technique to determine spatiotemporal temperature distributions in high nanoshell concentration regions (Elliott et al. 2009). The ability to correctly model high concentrations regions is imperative for the model to have any clinical relevance as gold nanostructures are usually distributed heterogeneously throughout living tissues. These high concentration regions and the nonuniformity of optical parameters in tissue present the biggest challenge for GNP-mediated LITT modeling *in vivo*.

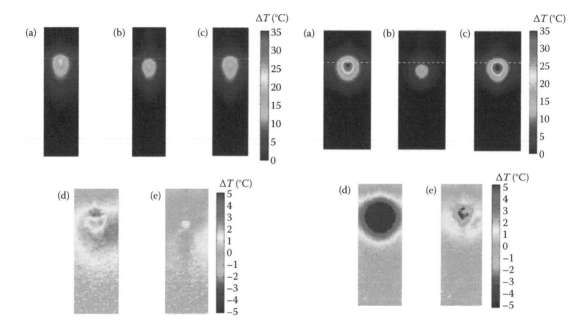

FIGURE 16.2 **(See color insert.)** Comparison of experimental MRTI data with ODA solution and $\delta-P_1$ solution for a 0.659-OD phantom (left) and a 1.49-OD phantom (right) after 3 min of exposure to a 1.2-W laser. Dashed line indicates boundary between portion of the phantom containing nanoshells and that without any nanoshells (top). (a) MRTI, (b) ODA solution, (c) $\delta-P_1$ solution, (d) pixel-by-pixel subtraction of the MRTI data and the ODA solution e) pixel-by-pixel subtraction of the MRTI data and the $\delta-P_1$ solution. (Reproduced with permission from Elliot, A.M. et al., *Med. Phys.*, 36, 1351–1358, 2009.)

16.2.2 Linear Superposition Model

Cheong et al. (2009) developed a computational model taking into account for the heat generated from individual GNPs and the laser source separately. In principle, this model would allow for the prediction of local heat distributions for any arbitrary distribution of GNPs within the target site such as a tumor. Unlike the approaches described in the preceding section, it does not require prior knowledge of the altered optical properties of any GNP-filled tissue, which might not be easily obtained during routine clinical applications. Instead, this model can be applied to any GNP-filled tissue by taking into account the absorption and scattering properties of the tissue and GNPs separately. The absorption and scattering properties of tissue and GNPs can be measured independently and the distribution of GNPs can be deduced from imaging studies. Additionally, this approach can provide the information about the contribution of each heat source (i.e., laser or GNPs) to the overall temperature rise, which might help facilitate the treatment planning of actual clinical applications.

The computational model hinges on the use of multiple heat source terms in the heat diffusion equation, a laser heat source, and one independent heat source for each individual nanoparticle:

$$Q = Q_{\text{laser}} + Q_{\text{NP}} \tag{16.19}$$

$$Q = Q_{\text{laser}} + \sum_{i=1}^{N} Q_i \tag{16.20}$$

where Q_{laser} is the heat source due to laser light alone, Q_{NP} is the heat source due to an ensemble of GNPs, Q_i is the heat source due to an individual GNP, and N is the total number of GNPs.

The heat generated due to laser light at $\vec{r}$ at time t is denoted by

$$Q_{\text{laser}}(\vec{r},t) = \mu_a \varphi(\vec{r},t) \tag{16.21}$$

where μ_a is the tissue absorption coefficient and $\varphi(\vec{r},t)$ (W/m²) is the laser light fluence rate. The heat generated by the ith nanoparticle at $\vec{r}_i$ at time t is described by

$$Q_i(\vec{r},t) = \sigma_a \varphi(\vec{r}_i,t)\delta(\vec{r}-\vec{r}_i) \tag{16.22}$$

where σ_a is the absorption cross section of the GNPs and $\delta(\vec{r}-\vec{r}_i)$ is the Dirac delta function. The Dirac delta function is used to describe the spatial distribution of the heat generated by each nanoparticle.

Since there is no analytical solution to the bioheat equation with the heat source presented above, Cheong et al. (2009) proposed a multistep method to calculate the rise in temperature due to individual GNPs first and then combined all contributions to obtain the full heat distribution. The time-independent temperature rise at point $\vec{r}$, due to constant NIR laser illumination of a single GNP at $\vec{r}_i$ in a homogeneous medium, is derived from the equation of heat conduction as presented previously (Carslaw and Jaeger 1993) and given by

$$\Delta T_i(\vec{r}) = \frac{\sigma_a \varphi(\vec{r}_i)}{4\pi\kappa|\vec{r}-\vec{r}_i|} \quad (16.23)$$

The time-dependent temperature rise in a homogeneous medium with no perfusion can be obtained using a Green's function as shown on a model presented before (Nyborg 1988):

$$\Delta T_i(\vec{r}) = \text{erfc}\left(|\vec{r}-\vec{r}_i|\sqrt{\frac{C\rho}{4\kappa t}}\right)\frac{\sigma_a \varphi(\vec{r}_i)}{4\pi\kappa|\vec{r}-\vec{r}_i|} \quad (16.24)$$

for $t \le t_0$, where t_0 is the time at which the laser is turned off. When the laser is turned off at $t \le t_0$ the heat equation becomes homogeneous and the solution is simply given by

$$\Delta T_i(\vec{r},t) = \Delta T_{0i}\exp\left(-\frac{t}{\tau}\right) \quad (16.25)$$

where ΔT_{0i} is the temperature rise at $t \le t_0$ and τ is the thermal diffusion time constant of the medium.

The temperature increase due to the laser alone, ΔT_{laser}, can be obtained via FEM. The method is therefore a simple superposition of the solution of the heat equation due to each individual heat source, and referred to as the linear superposition method here. The total change in temperature at $\vec{r}$ at time t is therefore:

$$\Delta T(\vec{r},t) = \Delta T_{\text{laser}} + \Delta T_N \quad (16.26)$$

$$\Delta T_N = \sum_{i=1}^{N}\Delta T_i \quad (16.27)$$

This computational model was tested for its validity using the same two-layer tissue phantom used in a previous study (Elliott et al. 2007). In order to reduce computation time, the number of GNP was minimized and a corrective multiplication factor was used:

$$\Delta T_N = M\sum_{i=1}^{n_p}\Delta T_i \quad (16.28)$$

where $M = N/n_p$ is the multiplication factor, and n_p is the reduced number of GNPs. Figure 16.3 shows the convergence of the solution to within 0.5% as the number of reduced nanoparticles reaches $n_p = 1 \times 10^4$ nanoshells/mL (Cheong et al. 2009).

The model was tested for its validity using the MRTI phantom results from Elliot et al. (2007). The test showed the capability of the superposition model to produce qualitatively reasonable results, but it also indicated that a number of key assumptions deduced from Elliott et al.'s phantom experiment were less likely to be applicable to the phantom calculations. As a result, some model parameters were adjusted to properly reflect the changes in the assumptions, matching the experimental results to within 10%. Temperature profiles as a function of depth and time for a phantom illuminated with 1.5-W NIR laser for 3 min are shown in Figure 16.4. These profiles show the temperature rise due to

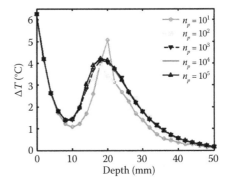

FIGURE 16.3 Depth versus temperature distribution after 3 min of exposure along the central axis of the phantom. Curves are shown for different number of nanoshells/mL and appropriate multiplication factor (M). (Reproduced with permission from Cheong, S.-K. et al., *Med. Phys.*, 36, 2662–4671, 2009.)

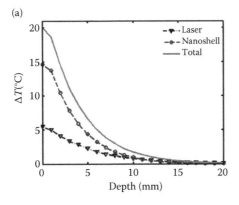

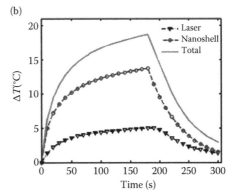

FIGURE 16.4 Temperature profiles as a function of depth and time after 3 min of exposure to a 1.5 W laser. (a) Depth versus temperature distribution at $t = 3$ min. (b) Temperature rise as a function of time at depth of 1 mm from the surface. Both plots show changes in temperature based on heat generated by laser source alone (broken line with reverse triangle) and due to individual nanoshells separately (broken line with circle). (Reproduced with permission from Cheong, S.-K. et al., *Med. Phys.*, 36, 2662–4671, 2009.)

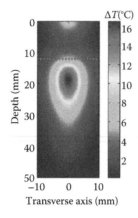

FIGURE 16.5 Cross-sectional view of the phantom showing temperature distribution after 3 min of exposure to a 1.5-W laser. Dashed line indicates boundary between the portion of the phantom containing nanoshells and that without any nanoshells (top). (Reproduced with permission from Cheong, S.-K. et al., *Med. Phys.*, 36, 2662–4671, 2009.)

laser and nanoshells separately, illustrating the GNP-mediated plasmonic heating within the host medium. Figure 16.5 shows a cross-sectional view of a similar phantom after 3 min of illumination, matching the characteristics of plasmonic heating as previously shown by Elliot et al.'s experiment.

16.3 Conclusions

The modeling of plasmonic heating in tissues is a major challenge for the successful clinical translation of GNP-mediated LITT. The currently available computational models have demonstrated great ability to predict heat distribution in phantoms under certain assumptions and conditions. In order for any plasmonic heating model to be clinically useful, it should be able to predict temperature distributions for a wide range of conditions associated with actual clinical cases. For example, the model's ability to handle a highly heterogeneous distribution of GNPs is a key requirement for clinical applications. The distribution of GNPs within a tumor is usually highly heterogeneous and, consequently, an assumption of homogenously distributed GNPs within a tumor dramatically decreases the accuracy of a computational model. As demonstrated in some models, therefore, the ability to account for heterogeneous distributions of GNPs is imperative for any computational model to be clinically useful. Finally, more effort in validating computational models under clinically relevant scenarios should be made. In order to facilitate experimental validation of any computational model, further studies are also necessary for a more accurate determination of the GNP distribution in tissue and the basic optical parameters for various gel phantom materials, especially in the NIR range.

References

Ashcroft, N. W., and N. D. Mermin. 1976. *Solid State Physics*. Philadelphia: Brooks Cole.

Bayazitoglu, Y., and I. K. Tjahjono. 2008. Near-infrared light heating of a slab by embedded nanoparticles. *International Journal of Heat and Mass Transfer* 51:1505–1515.

Bohren, C. F., and D. R. Huffman. 1983. *Absorption and Scattering of Light by Small Particles*. New York: Wiley-International Publication.

Boulnois, J.-L. 1986. Photophysical processes in recent medical laser developments: A review. *Lasers in Medical Science* 1:47–66.

Breasted, J. H. 1991. *The Edwin Smith Surgical Papyrus*. Published in facsimile and hieroglyphic transliteration with translation and commentary in two volumes. Chicago: University of Chicago Press.

Carslaw, H. S., and J. C. Jaeger. 1993. *Conduction of Heat in Solids*. Oxford: Clarendon Press.

Chen, J., F. Saeki, B. Wiley, H. Cang, M. J. Cobb, Z.-Y. Li, L. Au, H. Zhang, M. B. Kimmey, X. Li, and Y. Xia. 2005. Gold nanocages: Bioconjugation and their potential use as optical imaging contrast agents. *Nano Letters* 5:473–477.

Cheong, S.-K., S. Krishnan, and S. H. Cho. 2009. Modeling of plasmonic heating from individual gold nanoshells for near-infrared laser-induced thermal therapy. *Medical Physics* 36:4664–4671.

Elliot, A. M., J. Stafford, J. Schwartz, J. Wang, A. M. Shetty, C. Bourgoyne, P. O'Neal, and J. D. Hazle. 2007. Laser-induced thermal response and characterization of nanoparticles for cancer treatment using magnetic resonance thermal imaging. *Medical Physics* 34:3102–3108.

Elliot, A. M., J. Schwartz, J. Wang, A. M. Shetty, J. D. Hazle, and J. Stafford. 2008. Analytical solution to heat equation with magnetic resonance experimental verification for nanoshell enhanced thermal therapy. *Lasers in Surgery and Medicine* 40:660–665.

Elliot, A. M., J. Schwartz, J. Wang, A. M. Shetty, C. Bourgoyne, P. O'Neal, J. D. Hazle, and J. Stafford. 2009. Quantitative comparison of δ-P1 versus optical diffusion approximations for modeling near-infrared gold nanoshell heating. *Medical Physics* 36:1351–1358.

El-Sayed, M. A. 2001. Some interesting properties of metals confined in time and nanometer space of different shapes. *Accounts of Chemical Research* 34:257–264.

Faraday, M. 1857. Eperimental relations of gold (and other metals) to light. *Philosophical Transactions of the Royal Society London* 147:145–181.

Hayakawa, C. K., B. Y. Hill, J. S. You, F. Bevilacqua, J. Spanier, and V. Venugopalan. 2004. Use of the δ-P1 approximation for recovery of optical absorption, scattering, adn asymmetry coefficients in turbid media. *Applied Optics* 43: 4677–4684.

Heusmann, H., J. Kölzer, and G. Mitic. 1996. Characterization of female breasts in-vivo by time resolved and spectroscopic measurements in near infrared spectroscopy. *Journal of Biomedical Optics* 1:425–434.

Hirsch, L. R., R. J. Stafford, J. A. Bankson, S. R. Sershen, B. Rivera, R. E. Price, J. D. Hazle, N. J. Halas, and J. L. West. 2003.

Nanoshell-mediated near-infrared thermal therapy of tumors under magnetic resonance guidance. *Proceedings of the National Academy of Sciences* 100:13549–13554.

Huang, X., I. H. El-Sayed, W. Qian, and M. A. El-Sayed. 2006. Cancer cell imaging and photothermal therapy in the near-infrared region by using gold nanorods. *Journal of the American Chemical Society* 128:2115–2120.

Jackson, J. D. 1998. *Classical Electrodynamics*. New York: Wiley.

Jain, P. K., X. Huang, I. H. El-Sayed, and M. A. El-Sayed. 2007. Review of some interesting surface plasmon resonance-enhanced properties of noble metal nanoparticels and their application to biosystems. *Plasmonics* 2:107–118.

Kelly, K. L., E. Coronado, L. L. Zhao, and G. C. Schatz. 2003. The optical properties of metal nanoparticles: The influence of size, shape, and dielectric environment. *The Journal of Physical Chemistry B* 107:668–677.

Kreibig, U., and M. Vollmer. 1995. *Optical Properties of Metal Clusters*. Berlin: Springer-Verlag.

Link, S., and M. A. El-Sayed. 2000. Shape and size dependence of radiative, non-radiative and photothermal properties of gold nanocrystals. *International Reviews in Physical Chemistry* 19:409–453.

Lloyd, G. E. R., J. Chadwick, and W. N. Mann. 1983. *Hippocratic Writings*. London: Penguin.

Maeda, H. 2000. Tumor vascular permeability and the EPR effect in macromolecular therapeutics: A review. *Journal of Controlled Release* 65:271–284.

McKenzie, A. L. 1990. Physics of thermal processes in laser–tissue interactions. *Physics in Medicine and Biology* 35:1175–1209.

Mie, G. 1908. Contributions to the optics of turbid media, particularly of colloidal metal solutions. *Annals of Physics* 25: 377–445.

Niemz, M. H. 2007. *Laser–Tissue Interactions: Fundamentals and Application*. Berlin: Springer.

Nyborg, W. L. 1988. Solutions of the bio-heat transfer equation. *Physics in Medicine and Biology* 33:785–792.

Oldenburg, S. J., R. D. Averitt, S. L. Westcott, and N. J. Halas. 1998. Nanoengineering of optical resonances. *Chemical Physics Letters* 288:243–247.

Pennes, H. H. 1948. Analysis of tissue and arterial blood temperatures in the resting human forearm. *Journal of Applied Physiology* 1:93–122.

Xu, X., A. Meade, and Y. Bayazitoglu. 2011. Numerical investigation of nanoparticle-assisted laser-induced interstitial thermotherapy toward tumor and cancer treatments. *Lasers in Medical Science* 26:213–222.

Weissleder, R. 2001. A clearer vision for in vivo imaging. *Nature Biotechnology* 19:316–317.

Carbon Nanotubes for Thermal Therapy

Ravi N. Singh*
Wake Forest School of Medicine

Peter Alexander*
Wake Forest School of Medicine

Andrew R. Burke
Wake Forest School of Medicine

Frank M. Torti
*University of Connecticut
School of Medicine*

Suzy V. Torti
*University of Connecticut
School of Medicine*

17.1 Introduction

Carbon nanotubes (CNTs) have generated enormous interest for a wealth of applications including field emission (Saito et al. 2002; Milne et al. 2004), energy storage (Patchkovskii et al. 2005), molecular electronics (Javey et al. 2003; Keren et al. 2003; Weisman 2003), and atomic force microscopy (Wong et al. 1998). These nanoparticles exhibit a number of novel properties including extraordinary strength, unique electrical properties, and a specific heat and thermal conductivity that are among the highest known for any material (Iijima 1991; Dresselhaus et al. 1995; Berber et al. 2000).

CNTs were first characterized by Iijima (1991) more than two decades ago. Initially, they were described as cylindrical molecules of pure carbon ranging from 1.4 nm in diameter for a single-walled nanotube (SWNT) to 30–50 nm for concentrically arranged, multiwalled nanotubes (MWNT) and possessed widely variable lengths on the order of several microns. Advances in the synthesis and processing of nanotubes have now made it possible to produce nanotubes with both well-defined wall numbers and narrow length distributions ranging from tens of nanometers (ultrashort nanotubes) to several microns. Methods to synthesize CNTs include arc discharge (Ebbesen and Ajayan 1992), laser vaporization of graphite (Puretzky et al. 2000), magnetic field synthesis (Doherty et al. 2006), chemical vapor deposition using gaseous metal catalysts such as cobalt, nickel,

molybdenum, or iron (Cassell et al. 1999; Nobuhito et al. 2007), or water-assisted chemical vapor deposition (Hata et al. 2004).

CNTs possess high aspect ratios (the ratio of the longer dimension to the shorter dimension). The large surface area allows conjugation of multiple moieties including peptides, proteins, nucleic acids, radionuclides, other nanoparticles, and drugs to the surface of these nanoparticles (Pantarotto et al. 2003; Pantarotto et al. 2004a, 2004b; Kam et al. 2005; Singh et al. 2005; Kam and Dai 2005; Lacerda et al. 2006) and permits a strong amplification of signals generated from these agents (Ajayan et al. 2002; Patri et al. 2004; Talanov et al. 2006). Because of these exploitable properties, it is not surprising that CNTs have been studied for applications to enhance the treatment of human malignancies (Kim 2007). Over the past decade, investigations into the use of CNTs for biomedical applications have greatly increased. Of particular interest is their ability to act as delivery vehicles for anticancer agents, including chemotherapeutic agents (Feazell et al. 2007; Liu et al. 2008), radionuclides (McDevitt et al. 2007), and nucleic acids (Wang et al. 2008). Furthermore, they have been shown to be effective as antitumor vaccine delivery systems (Meng et al. 2008) and as cancer imaging and diagnostic agents in animal models (Yu et al. 2006). The attractiveness of CNTs for use in these applications derives from their ability to easily cross cell membranes (Shi Kam et al. 2004; Kostarelos et al. 2007; Selvi et al. 2008) as well as their ability to be functionalized with target-specific molecules that preferentially target cancer cells (Liu et al. 2007; Chen et al. 2008; Kostarelos et al. 2009).

---

* These authors contributed equally to this chapter.

Another feature of CNTs that renders them particularly well suited for anticancer therapy is their ability to act as high-efficiency absorbers of near-infrared radiation (NIR) to promote the generation of therapeutic heat in tumors (Torti et al. 2007; Chakravarty et al. 2008; Klingeler et al. 2008; Biris et al. 2009; Burke et al. 2009; Day et al. 2009; Marches et al. 2009; Zhou 2009; Boldor et al. 2010; Burlaka et al. 2010; Fisher et al. 2010; Picou et al. 2010). CNTs are amenable to stimulation by a range of energy sources including NIR, microwave (MW), and radio-frequency (RF) radiation emitters directed at the site of the CNTs from outside the body. Following exposure to electromagnetic radiation, CNTs release vibrational energy and deliver substantial heat to the tumor site. Because of their potential to deliver multiple rounds of heat therapy in a noninvasive manner and their imaging capabilities, which permits more precise localization of heat delivery (Ding et al. 2011), CNTs have potential to improve the thermal treatment of cancer. Although other nanomaterials share some of these properties, CNTs offer perhaps the best combination of attributes for the development of a noninvasive, thermal therapy.

In this chapter, we will examine the rationale for using CNTs for photothermal cancer therapy. The differences in heating properties between carbon nanomaterials and various sources of radiation will be discussed below. We will consider current knowledge that could be used to optimize the design of CNTs for heat generation and localization, and analyze the current state of CNT-based anticancer photothermal therapy (PTT). Finally, we will examine at the prospects of translating CNT-based photothermal therapeutics to the clinic.

17.2 Photothermal Properties of CNTs

The interaction of light with CNTs is important for a number of applications, including biomedical use in PTT. For PTT, a photothermal sensitizer delivered to cancer cells is excited by a specific wavelength of light, generally in the NIR region of the electromagnetic spectrum (700–1100 nm), which in turn causes the photosensitizer to enter an excited state and release vibrational energy that is transformed into heat, leading to cell death (Kam et al. 2005; Torti et al. 2007). Because biological systems largely lack chromophores that absorb in the NIR region, transmission of NIR light through the body is poorly attenuated (Konig 2000; Weissleder 2001). As noted earlier, CNTs possess an extremely broad electromagnetic absorbance spectrum, covering not only the NIR window, but both the RF and MW bands as well (Gannon et al. 2007). This suggests that, in conjunction with an appropriate energy source, CNTs can be used to treat deeply seated lesions without the need for direct access to the tumor site. Although the photophysical effects governing the interaction of CNTs with electromagnetic energy have been studied through Raman scattering, fluorescence, and nonlinear optical analysis (Avouris et al. 2008; Biris et al. 2011; Kanemitsu 2011; Lehman et al. 2011; Saito et al. 2011; Yin et al. 2011), more research is needed to develop a clear understanding of these properties. It appears that CNTs act as ballistic conductors due

to lack of energy dissipation through electron movement and quantized resistance, which are related to low dimensionality and quantum confinement of electrons within the carbon lattice of the nanotube wall (Brigger et al. 2002; Schonenberger and Forro 2000). Because of this, the specific structure of the nanotube, especially the number of walls, has a great effect on the efficiency of this material for use in photothermal applications. In the following section, we will examine the role nanotube structure plays in determining the photothermal heat transduction efficiency of CNTs.

17.2.1 Effect of Nanotube Structure on Photothermal Properties

After excitation by electromagnetic energy, CNTs exhibit a wide variety of vibrational (phonon) modes created by phonon–phonon and phonon–electron interactions (Kempa 2002; Hagen et al. 2004; Liu et al. 2011). These phonon interactions are dominant in determining the thermal properties of CNTs. In general, more phonon modes appear as the nanotube diameter and the size of the unit cell increases, meaning that the heat capacity and ability to generate and transport heat are unique to each nanotube structure. In one sense, phonons can be considered as quanta of heat. Thus, a basic understanding of the specific thermal conductivity properties of different types of nanotubes is essential to understanding the photothermal heating behavior of these materials. Here, we will briefly examine the thermal conductivity and the interaction of electromagnetic radiation with SWCNTs, MWCNTs, and related nanostructures, focusing on exposure to NIR.

SWCNTs. Measurements of bulk samples of SWCNTs indicate a room-temperature thermal conductivity over 200 W/(m K) (Hone et al. 2000). This is far less than measurements of the room temperature thermal conductivity of an individual SWCNT along its axis which range from 2200 W/(m K) (Hone 2000) to about 3500 W/(m K) (Pop et al. 2006) because of the disorganized orientation of nanotubes in bulk samples. The phonon thermal conductivity displays a peak around 100 K and decreases with increasing temperature (Hone et al. 2000). At higher temperatures, thermal conductivity is predicted to decrease in a generally linear fashion because of the increased phonon–phonon and electron–phonon scattering interactions (known as the Umklapp processes) (Hone 2000; Osman and Srivastava 2001).

SWNHs. SWCNTs can be modified by sealing one end. These nanoparticles are called single-wall nanohorns (SWNHs) and form aggregate structures with typical diameters from 50 to 100 nm. SWNHs are produced without the use of metal catalysts by laser ablation of pure graphite (Whitney et al. 2011). The diameter can be adjusted by modifying the laser pulse length during production, and the size can be adjusted by modifying growth time (Geohegan et al. 2007). Nanohorns have been studied less extensively than either SWCNTs or MWCNTs for photo thermal applications. However, Whitney et al. found a linear relationship between the optical attenuation coefficient and concentration of

SWNHs. They also demonstrated that the attenuation coefficient increased with shorter wavelengths, indicating that SWNHs likely will heat most efficiently when exposed to shorter wavelength NIR (Zhang et al. 2008; Whitney et al. 2011). Because tissue penetrance is significantly better at longer NIR wavelength (Konig 2000), the use of SWNH for photothermal applications may be limited to the treatment of superficial diseases. The few studies to examine the use of SWNH for PTT have required significantly higher SWNH concentrations and longer NIR exposure times than needed for SWCNTs or MWCNTs to generate enough heat to be effective for photothermal ablation of cancer (Zhang et al. 2008; Whitney et al. 2011).

MWCNTs. Similar to SWCNTs, the large number of phonon modes in MWCNTs indicates that these structures also are exceptional heat conductors. The thermal conductivity of a single MWCNT along its axis at room temperature appears to be greater than that of comparable SWCNTs, with measured conductivities in the range of 3000 W/(m K) (Kim, Shi et al. 2001) to 6600 W/(m K) (Berber et al. 2000). However, other research groups have reported somewhat lower thermal conductivity with a range from 1500 to 3500 W/(m K) depending on the diameter of the tube (Fujii, Zhang et al. 2005). It is not clear if these differences are due to variations in the methods used to measure conductance, differences in the MWCNT preparations used, or some other cause. A study using bundles of bulk MWCNTs found a far lower room temperature thermal conductivity of only 20 W/(m K) (Yi, Lu et al. 1999). This lower value may be due to defects in the CNTs, but also suggests a potential inhibition of heating due to aggregation-induced resistive thermal junctions. As expected, the thermal conductivity of MWCNTs increases as diameter decreases, suggesting that interactions of photons and electrons between the walls affect conductivity (Osman and Srivastava 2001).

MWCNTs possess a broader absorption spectra compared to SWCNTs and other plasmonic nanoparticles including nanoshells (Dresselhaus 2004; Torti et al. 2007; Burke et al. 2009). In contrast to SWCNTs, the larger number of electrons available in the MWCNTs for transport, together with a smaller electronic bandgap or metallic behavior, suggests that the most efficient optical coupling of light and CNTs occurs when the nanotube length is comparable to half that of the wavelength of the incident radiation (Hanson 2005), which is consistent with the classic behavior of dipole antennae (Wang 2004; Hanson 2005). Accordingly, in response to illumination with a 1064-nm laser, MWCNTs with lengths of approximately one (1100 nm) or one-half (700 nm) that of the laser wavelength readily heated (Torti et al. 2007). In contrast, MWCNTs with a length of one-third that of the laser (330 nm) failed to generate appreciable heat. As a further indication of this antenna effect, MWCNTs also demonstrate polarization effects, meaning that the antenna response of the CNTs is suppressed when the electric field of the incoming radiation is polarized perpendicular to the dipole axis of the CNTs (Wang 2004).

In practice, MWCNTs appear to be far more efficient at heat production than SWCNTs following exposure to electromagnetic radiation, possibly due to increased numbers of electrons

(carriers) available for photon interactions and a greater mass per particle (Burke et al. 2009; Fisher et al. 2010). Furthermore, most SWCNTs act as semiconductors, and only a fraction of the as-produced tubes exhibit the metallic behavior shown in MWCNTs (Burlaka et al. 2010). This means that the electrons in MWNCTs can more easily become excited than those in most SWCNTs and begin releasing excess energy in the form of heat. Experimental studies comparing the NIR absorbance at 1064 nm of aqueous dispersions of MWCNTs to similarly dispersed SWCNTs indicate that MWCNTs absorb approximately three times more light per particle than SWCNTs (Fisher et al. 2010). Interestingly, the heat-generating capacity of MWCNTs following exposure to an equivalent dose of NIR has been shown to be up to 20-fold higher than that of SWCNTs (Burke et al. 2009; Levi-Polyachenko et al. 2009). This effect is far greater than the optical absorbance data would indicate, and the mechanism by which this occurs remains to be explained. However, the difference in heating between SWCNTs and MWCNTs may be an important feature because a smaller dose of MWCNTs could be used to achieve an equivalent temperature rise following NIR exposure. Thus, MWCNTs could potentially achieve equivalent clinical responses at doses less likely to engender systemic toxicity and off-target effects than SWCNTs.

17.2.2 Effect of Doping on Photothermal Properties of CNTs

Doping of non-carbon atoms into CNTs represents a method to control the photoelectronic properties of the tubes by chemistry rather than through alterations in specific geometry (Esfarjani et al. 1999). Doping can be used to alter the electronic band structure to increase the overall conductivity (hence the antenna behavior) of the nanotubes. Numerous types of dopants can be introduced into CNT walls through methods such as intercalation of electron donors such as alkali metals or acceptors such as halogens, substitutional doping, encapsulation in the interior space, coating on the surface, molecular absorption, and covalent sidewall functionalization (Terrones et al. 2008; Stoyanov et al. 2009; Ayala et al. 2010; Kong et al. 2010).

Doping CNTs with other atoms can have a dramatic effect on their photothermal properties, potentially enhancing their optical absorption and heat transductance capability. Substitutional doping (replacing carbon in the lattice with a non-carbon atom) of the tubes alters the Fermi level of the valence band: the greater the doping, the stronger is the shift of the Fermi level. For pure CNTs, the valence and conduction bands appear to be symmetric about the Fermi level. By comparison, nitrogen doping introduces an impurity located 0.27 eV below the bottom of the conduction bands and boron doping induces a level that is 0.16 eV above the top of the valence bands found in undoped CNTs (Schonenberger and Forro 2000). The lowering of the Fermi level by boron dopants increases the number of conduction channels without introducing strong carrier scattering (Dai 2002). Thus, boron-doped nanotubes show metallic behavior with weak electron–phonon coupling. In contrast to undoped CNTs, which even in idealized conditions show a small bandgap (semiconducting or semimetallic behavior), the valence band of boron-doped MWNTs is filled with a prominent acceptor-like peak near the Fermi level. Although there is only limited experimental evidence, one effect of doping appears to be an increased optical coupling of MWNTs to NIR due to increasing the number of free carriers available, leading to the generation of higher temperatures following exposure to NIR when compared to equivalent undoped tubes (Liu and Fan 2005; Liu and Gao 2005; Torti et al. 2007).

The effect of incorporation of non-carbon atoms into the interior of CNTs on photothermal properties is less explored. Experimental studies have demonstrated that MWCNTs produced with a high concentration of the iron-based catalyst ferrocene in their lumen appear to heat more efficiently than iron-free MWCNTs, achieving temperatures of up to 5–7°C greater following exposure to an equivalent dose of NIR (Levi-Polyachenko et al. 2009). On the other hand, others have found that increasing concentrations of ferrocene in the lumen of MWCNTs has no effect on heating properties (Ding et al. 2011). Although the mechanism accounting for this difference remains poorly understood, it should be noted that the enhanced heating effect observed by Levi-Polyachenko occurred when heating ferrocene-containing MWCNTs at nanotube concentrations greater than 100 µg/mL, whereas Ding et al. only investigated the heating of 100 µg/mL MWCNT samples.

17.3 RF and MW Heating of CNTs

In addition to efficiently absorbing NIR, CNTs are also capable of generating heat upon irradiation with MW or RF radiation. The MW spectrum ranges from 300 MHz to 300 GHz, whereas the broader RF spectrum overlaps and extends from 3 kHz to 300 GHz. MW heating of CNTs causes polarization, producing a similar antenna effect as seen by NIR heating (Wang 2004). Experiments performed with CNTs in viscous dense environments found decreased heating, attributed to reduced photon–photon and photon–electron interactions due to inhibited vibrations (Ye et al. 2006). CNTs irradiated by MWs can induce heat by conduction and dipolar polarization, enabling localized heating (Vazquez and Prato 2009). The modes of CNT heat generation using MW/RF are similar to NIR, as MW irradiation transforms electromagnetic energy into mechanical vibrations and ultimately heat. Residual metals in CNTs may also donate free charges that help expedite MW coupling. Because of the overlap in wavelengths, the heating properties of MW and RF irradiation of CNTs will be discussed together.

Based on theoretical modeling, Dumitrica et al. (2004) found that SWCNTs blended in polycarbonate should absorb MW radiation at a 6- to 20-GHz range. In this study, capped SWCNTs treated with ~100-fs pulses are predicted to remain intact with 8% of valence electrons promoted to antibonding states. Irradiation would also result in opening of the caps without damaging the cylindrical structure. In contrast, heating of bulk (noncapped) SWCNTs is predicted to promote 10% of valence electrons and result in fragmentation to the particles. With both types of NTs, the maximum temperature predicted was 800 K, which stabilized to 300 K. In a separate study, SWCNTs heated with 700 W at 2.45 GHz were observed to spread to twice their original volume during heating before contracting again. Many of the SWCNTs fused after heating and formed junctions (Imholt et al. 2003). A temperature of at least 1500°C must be reached for this phenomenon to occur, suggesting that tremendous temperature increases were achieved in this experiment (Ajayan et al. 2002). Another factor that can influence heating of CNTs is their purity. Unpurified SWCNTs (containing Fe catalyst impurities) reached a temperature of 1850°C upon heating, whereas purified SWCNTs only reached 650°C (Wadhawan 2003). Both types of SWCNTs had diameters of 1.1 nm and were irradiated with 1000-W MW radiation at 2.45 GHz. This suggests that residual metals may play an important role in CNT heating.

Reulet et al. found that RF irradiation at 100 MHz–10 GHz of a single SWCNT (1 nm diameter and several different lengths) caused electron heating by dissipation of mechanical energy. No change was seen in the resonance spectrum upon heating, suggesting that the electrostatic forces on the tube and Coulomb force produced by the RF field are responsible for excitation and vibrations. Different resonant frequencies were seen depending on the length of the NT (Reulet et al. 2000). SWCNTs irradiated with 800 W by a 13.56-MHz RF field produced a temperature increase of 30–40°C (1.6 K/s rate). The heating rate is higher than predicted, and may be due to spontaneous self-assembly of SWCNTs into longer antennae. The total thermal power deposition was found to be 130,000 W/g, with over half specifically from the NTs (75,000 W/g). RF heating of CNTs does not appear

TABLE 17.1 Thermal Ablation Using Radiofrequency Radiation to Heat Carbon Nanomaterials

Nanomaterial/ Functionalization	Power Input	NT Dose	Temperature Change	Cancer Model and Therapeutic Efficacy	Ref.
SWCNT; coated in Kentera polymer	13.56 MHz RF field (400–1000 W)	50–500 µg/mL	33–45°C linear increase with power, exponential increase of heat at fixed power with increasing conc. Enhanced bulk heating at 5 µg/mL	Hep3B, HepG2 and Panc-1 human liver and pancreatic cancer cells; 60–70% cell death with 50 µg/mL, ~90% with 100 µg/mL and 100% with 500 µg/mL. Treated for 2 min 800 W.	Gannon et al. (2007)
SWCNT; coated in Kentera polymer	13.56 MHz RF field (600 W)	500 µg/mL	Not reported	Rabbits with VX2 liver xenografts, intratumoral injection. 2 min RF treatment. Complete thermal necrosis	Gannon et al. (2007)

to be due to excitement of electronic transitions or resonance because of the long wavelengths (Gannon et al. 2007).

Table 17.1 summarizes cellular effects of stimulation of nanotubes with RF radiation. RF-mediated heating of hepatocellular and pancreatic cancer cell lines *in vitro* using 250–500 µg/mL concentration of SWCNTs killed almost all the treated cells, with dose-dependent increases in cell death observed. About 25% of cells that were heated in media alone (without SWCNTs) were killed, suggesting the potential impact of nonspecific ion stimulation in heat generation upon treatment (Gannon et al. 2007). *In vivo* treatment of a VX2 hepatocellular carcinoma xenograft in rabbits resulted in complete thermal necrosis of the tumor. No toxicity was seen, but there was a 2- to 5-mm zone of thermal injury to the surrounding liver (Gannon et al. 2007). This demonstration indicates that SWNTs may be capable of noninvasively treating tumors in any part of the body, a capability currently not shared by NIR laser-based treatments. However, little research has been conducted on the efficacy of this type of therapy both *in vitro* and *in vivo*, possibly because of the risk of significant off-target heating. Therefore, we will focus only on NIR-mediated therapies for the remainder of this chapter.

17.4 Strategies to Localize Heat Distribution and Monitor Photothermal Heating of CNTs

Control of the spatial and temporal distribution of heat used for thermal ablation is essential for localizing heat to a target and reducing collateral damage to normal cells and tissues (Picou et al. 2010).

Simply viewed, heat delivery for photothermal applications is dependent on the total laser energy incident upon the target and the efficiency of the CNT target at converting that energy into heat. Thus, heat generations is limited only by the maximum laser output and achievable nanomaterial concentration. Conversely, heat dissipation away from the CNTs is a complex process that is dependent on tube environment, tube proximity to heat absorbers, solvent, and the substrate into which the tubes are dispersed. As discussed in more detail below, it is unlikely that continuous heating of nanoscale sources can produce a significant temperature increase adjacent to the surface of a nanoparticle, nanowire, or nanotube because of heat transfer away from the site of irradiated nanomaterials, unless the heating power is extremely large (Keblinski et al. 2006).

Several studies have reported that nanoscale temperature localization following NIR heating of isolated nanoparticles can be achieved through use of high-powered, rapidly (femto- to nanosecond time scales) pulsed lasers (Plech et al. 2003; Hartland et al. 2004; Pustovalov and Babenko 2004; Ge et al. 2005). Ultrashort laser pulses of approximately 100 fs are believed to immediately promote electrons in CNTs to antibonding states, whereas pulses that are greater than a picosecond transfer energy from the promoted electrons to atomic thermal motion, resulting in potentially uncontrolled structural

changes in the material (Dumitrica et al. 2004). Nanosecond NIR pulses have been reported to produce temperature increases of 150–300°C in samples containing gold nanoparticles because there was insufficient time for heat dissipation from the several micrometer heated area (Zharov et al. 2005). Similar strategies using nanosecond pulsed lasers have been successfully applied for the treatment of scattered cancer cells following uptake of CNTs (Zharov et al. 2005; Biris et al. 2009; Vitetta et al. 2011).

Unfortunately, nanosecond or picosecond pulsed lasers are not commonly available in clinical environments, and as noted, temperature increases over larger volumes tend to be very small following such brief exposures. However, the use of somewhat longer (millisecond to tens of seconds) NIR pulses to irradiated CNTs may offer a few opportunities for cancer therapy that are not dependent on macroscale temperature increases (Panchapakesan et al. 2005; Kang et al. 2009). In one study, cancer cells that had taken up bundles of SWCNTs were exposed to low intensity NIR (800 nm; 50–200 mW/cm^2) for 60 s. An insignificant temperature rise was measured following treatment. Instead, water molecules entrapped inside and between the bundled SWCNTs boiled; as the water molecules evaporated, extreme pressures developed in SWCNT bundles causing them to explode and kill nearby cancer cells (Panchapakesan et al. 2005). The key to this strategy is the use of bundles of SWCNTs, as the "nanobomb" effect is not observed for well-dispersed samples. In a different study, Kang et al. (2009) heated cancer cells that had taken up CNTs with a millisecond pulsed laser (1064 nm; 200 mW/cm^2) for 20 s, resulting in the death of 85% of the treated cancer cells. Significantly, almost no temperature change was detected, and it is believed that the mode of cell killing was a photoacoustic explosion induced by photon–electron interactions that generated a shockwave, physically disrupting the cells' membranes.

Although there may be some benefit to other treatment strategies as noted above, in general, continuous NIR treatment may be best suited for treating bulky tumors, as the generated heat can effectively spread throughout the tumor (Biris et al. 2009). Moreover, sustained heating of a large number of nanoparticles dispersed across a tumor volume under conditions typically used for *in vivo* thermal ablation produces a global temperature rise that is far larger than the localized temperature rise near each particle, allowing for substantial heating across the entire volume to be treated, which is necessary for anticancer therapy (Keblinski et al. 2006; Picou et al. 2010). The problem is that for long heating times (several seconds or greater), heat transfer away from the target area is significant and not only reduces the effectiveness of the therapy, but may result in collateral damage to healthy tissue surrounding the treatment area.

To reduce the spread of heat from the tumor target to the surrounding tissue, the total electromagnetic energy deposited into the tissue should be minimized such that only the amount of heat needed for treatment is delivered to the targeted area. This requires real-time monitoring of spatiotemporal changes in temperature resulting from NIR irradiation of CNTs. There are several ways to monitor the temperature distribution in a tumor volume following CNT delivery and NIR irradiation, including the use of infrared cameras (Huang et al. 2010; Picou et al. 2010) and magnetic resonance imaging (MRI)-based methods (Burke et al. 2009; Ding et al. 2011). Infrared cameras have proven useful for optimizing both CNT concentration and NIR irradiation parameters needed to generate specific temperature profiles in model tissue (Picou et al. 2010) and in tumor-bearing mice (Huang et al. 2010). However, NIR cameras do not offer the possibility of noninvasively imaging temperature changes deep within tissue.

A noninvasive method of temperature mapping that also allows for superposition of temperature information over high-resolution anatomical images taken at any depth is an MRI-based thermometry method known as proton resonance frequency (PRF) MR temperature mapping (reviewed by Rieke and Pauly 2008). This technique is based on the principle that when the temperature rises, hydrogen bonds break between water molecules in tissue and this causes a PRF shift that varies linearly with temperature changes. Clinically, PRF MR temperature mapping is used to provide control over the treatment outcome by relating the treatment temperature to actual thermal tissue damage. With regard to CNT enhanced photothermal cancer therapy, Burke et al. (2009) demonstrated that this technique could be used to both locate the tumor target in a mouse model of kidney cancer and to calculate the delivered thermal dose to that same tissue following NIR exposure. PRF MRI thermometry showed that a maximum temperature of 76°C was achieved in the tumors injected with 100 µg of MWCNTs following a 30-s NIR exposure (1064 nm; 3 W/cm²). In the absence of CNTs, the maximum temperature rose to only 46°C after the same NIR treatment.

The demonstration that CNTs are compatible with PRF MR temperature mapping is a key step toward future clinical applications, as this technique helps to monitor whether thermal ablative temperatures are reached, and also aids in reducing the risk of collateral damage to neighboring normal tissue. However, this technique could further be refined; optimally, CNTs used for thermal therapy should also be capable of MR contrast enhancement, which would allow for accurate monitoring of nanomaterial distribution in the tumor and image guided placement of the NIR source (Salvador-Morales et al. 2009). Several studies have shown that elements that enhance magnetic resonance (MR) contrast, such as iron (Ding et al. 2011) or gadolinium (Gd) (Sitharaman et al. 2005; Hartman et al. 2008; Richard et al. 2008; Ananta et al. 2010; Zhang et al. 2010), can be incorporated into CNTs to enable their detection by noninvasive imaging.

As a step toward this goal, MWCNTs containing iron were studied for their potential as dual-modality agents for both MR contrast enhancement and photothermal energy transduction (Ding et al. 2011). In this study, MR imaging provided an accurate picture of the distribution of iron-containing MWCNTs inside the tumor, which is essential information for pretreatment planning and determination of laser positioning for MR image guided PTT of tumors in mice. The contrast and heating

properties of such MWCNTs did not change upon multiple rounds of NIR exposure, even after reaching thermal ablative temperatures. Thus, the distribution of the MWCNTs could be monitored over time, and multiple or fractionated laser treatments could be targeted to the tumor as necessary without the need for additional injections.

However, a potential limitation of iron-containing MWCNTs is their propensity to attenuate MR signals. Although this enables iron-containing MWCNTs to act as highly effective T2 contrast agents, extensive MR signal attenuation can potentially interfere with temperature mapping by PRF MR thermometry. Ideally, a CNT specifically engineered for the clinical application of nanoparticle-assisted photothermal cancer therapy will both be compatible with an imaging modality such as MRI to spatially define the margins of the target lesion and assess the distribution of injected nanoparticles within the tumor, and also compatible with a noninvasive temperature mapping technique to ensure that the appropriate thermal dose was achieved in the target area. The optimization of such a material will be a critical step toward the realization of the full potential of nanoparticle enhanced PTT.

17.5 Anticancer Efficacy of CNT-Enhanced PTT

In vitro and *in vivo* tests of the antitumor efficacy of CNTs have been highly encouraging. The first study describing the use of SWCNTs for PTT of cancer cells was published in 2005 by Kam et al. (2005); subsequently, MWCNTs were also shown to be effective (Torti et al. 2007). Although there are many variables that will be discussed in detail below, the general therapeutic approach involves exposing adherent cancer cells, cancer cells in suspension, or tumors grown in mice to CNTs followed by irradiation with an external NIR laser (Figure 17.1). This technique, which can be termed nanotube-enhanced PTT, has proven to be an effective treatment in a wide variety of human cancer models including cervical carcinoma (Kam et al. 2005), renal carcinoma (Torti et al. 2007; Burke et al. 2009), mouth carcinoma (Moon et al. 2009), prostate adenocarcinoma (Fisher et al. 2010), breast adenocarcinoma (Ding et al. 2011), ascitic carcinoma (Burlaka et al. 2010), and lymphoma (Chakravarty et al. 2008; Marches et al. 2009) *in vitro* as summarized in Table 17.3. Efficacy also has been demonstrated *in vivo* in numerous syngeneic mouse models of cancer (Huang et al. 2010; Robinson et al. 2010) and in human xenografts grown in mice (reviewed by Iancu and Mocan 2011 and summarized in Table 17.3).

The clinical model for the use of nanomaterials as heat transduction agents is based on laser-induced thermotherapy (LITT) (O'Neal et al. 2004), a photothermal ablation technique in which an NIR laser is used to directly heat a target tissue, such as a tumor, above the thermal ablation temperature threshold of approximately 55°C (Kangasniemi et al. 2004; O'Neal et al. 2004; Nikfarjam et al. 2005). As a result, protein denaturation, membrane lysis, and coagulative necrosis occur, leading to cell

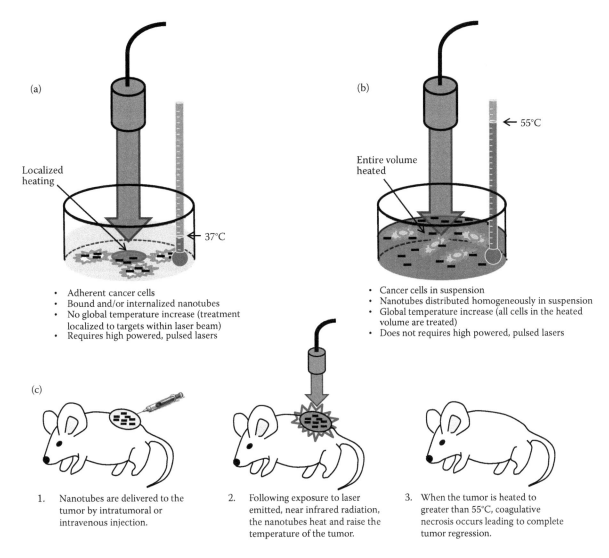

(a)

Localized heating

← 37°C

- Adherent cancer cells
- Bound and/or internalized nanotubes
- No global temperature increase (treatment localized to targets within laser beam)
- Requires high powered, pulsed lasers

(b)

← 55°C

Entire volume heated

- Cancer cells in suspension
- Nanotubes distributed homogeneously in suspension
- Global temperature increase (all cells in the heated volume are treated)
- Does not requires high powered, pulsed lasers

(c)

1. Nanotubes are delivered to the tumor by intratumoral or intravenous injection.

2. Following exposure to laser emitted, near infrared radiation, the nanotubes heat and raise the temperature of the tumor.

3. When the tumor is heated to greater than 55°C, coagulative necrosis occurs leading to complete tumor regression.

FIGURE 17.1 Schematic illustration nanotube enhanced photothermal therapy. (a) Selective photothermal heating of cancer cells. Adherent cancer cells readily take up carbon nanotubes following coincubation. Excess unbound nanotubes are washed away. Following exposure to laser emitted near-infrared radiation (NIR), cells that have taken up nanotubes rapidly heat and die. The amount of heat generated is insufficient to significantly raise the temperature of the surrounding media, making treatment localized to the cellular level. (b) Volumetric photothermal heating of cancer cells. Cells are dispersed in aqueous media containing well suspended nanotubes. After exposure to NIR, the entire volume heats significantly, causing cancer cell death. (c) Heating of tumors *in vivo*. Typical *in vivo* studies generally have taken a volumetric approach to therapeutic heating mediated by nanotube enhanced photothermal therapy. Nanotubes can be directly injected into the tumor or delivered via the tumor vasculature. After exposure to NIR via an external laser, the nanotube containing tumor rapidly heats up, causing coagulative necrosis and cancer cell death.

death. A major limitation of LITT has been an inability to consistently achieve thermoablative temperatures throughout the target lesion and to confine treatment exclusively to the tumor (Chen et al. 2005; Gnyawali et al. 2008). Therefore, to be of clinical benefit, CNTs must greatly improve the deposition of heat following NIR exposure.

Several studies have explored the potential of CNTs to improve LITT by enhancing deposition of heat following NIR exposure. Kam et al. (2005) first described the use of SWCNTs for PTT. After 60 s of exposure to 808-nm laser radiation, little heat was generated in the absence of SWCNTs. However, in the presence

of SWCNTs, the absorbance was robust, raising the temperature of the solution to more than 55°C (i.e., into the established thermoablative range). This temperature increase was sufficient to kill HeLa cells (a well-characterized cervical carcinoma cell line) that previously had internalized SWCNTs following incubation period of a few hours. Efficient conversion of tissue penetrating wavelengths of NIR into heat is fundamental to the application of this and other nanomaterials that seek to treat nonsuperficial cancerous lesions *in vivo*.

After this initial study, further testing of the heating of cancer cells with SWCNTs upon NIR irradiation was conducted *in vitro* and *in vivo*. Huang et al. (2010) observed that the quantity of heat generated and the efficacy of the therapy improved with both increasing SWCNT concentration and increased laser exposure (energy deposition). In the same study, both tumor reduction and a modest survival benefit were seen in a subcutaneous syngeneic murine squamous cell carcinomas model following intratumoral injection of 1 mg/mL of SWCNTs and irradiation with a low power (200 mW/cm²) NIR laser for 10 min. The treatment resulted in a maximum temperature increase in the tumor of 18°C as measured by IR thermometry, indicating that the thermal ablation threshold was reached. However, this treatment failed to achieve a durable cancer remission. Uneven tumor ablation and tumor recurrence were observed, and necrosis was seen in nearby normal tissue, indicating significant heat transfer away from the tumor site and into the surrounding non-tumor region.

By comparison, Moon et al. (2009) demonstrated that heat localization and therapeutic efficacy of SWCNT-mediated PTT could be greatly improved through the use of higher powered laser irradiation (3 W/cm²) for a shorter irradiation time (3 min). Following intratumoral injection of SWNCTs (120 μg/mL) into mice bearing subcutaneous xenografts of human mouth carcinoma cells and NIR irradiation, the tumors were completely destroyed, with no apparent toxicity, side effects, or tumor recurrence during several months of follow-up. Nearby normal tissue was spared, whereas tumor sections stained positively for TUNEL, suggesting apoptosis as the mode of cell death. However, even in the absence of SWCNTs the irradiation procedure itself resulted in significant burning of the target region. Ideally, the total energy needed to induce thermal ablation should be minimized to reduce off-target damage.

In this regard, MWCNTs may offer a significant potential advantage over SWCNTs: as previously noted, because of the unique structure of MWCNTs, they absorb NIR far more efficiently than SWCNTs, thus requiring only a fraction of the incident radiation or concentration SWCNTs needed to generate an equivalent increase in temperature (Torti et al. 2007; Burke et al. 2009). Analysis of the literature suggests that in most studies, the total energy needed to achieve thermal ablation *in vitro* with SWCNTs typically is on the order of 100 to several hundred Joules and the required SWCNT concentrations are greater than 100 μg/mL (see Tables 17.2 and 17.3). Using MWCNTs, thermal ablative temperatures could be achieved *in vitro* following exposure of 10 μg/mL of MWCNTs to as little as 4 J, resulting in the death of 99% of treated cancer cells (Biris et al. 2009). Although Ghosh et al. (2009) conducted detailed experiments to examine the influence of nanotube concentration, laser power, and duration of laser exposure on the heating of SWCNTs and MWNCTs, no systematic effort has been made to determine the ideal combination of CNT type, concentration, and NIR needed to minimize the required energy dose for therapeutic heat generation. Thus, treatment parameters still vary widely (see Tables 17.2 and 17.3).

For example, effective treatment of Erlich ascitic carcinoma cells *in vitro* was achieved following incubation of the cells with MWCNTs at a 100 μg/mL concentration and irradiation with an NIR laser (780–1400 nm range) at 3.5 W/cm² for 1.5–2 min, heating the culture media to between 50°C and 70°C (Burlaka et al. 2010). Fisher et al. treated renal adenocarcinoma cells (RENCA) and PC-3 human prostate cancer cell lines with 100 μg/mL MWCNTs and by 5 min exposure to NIR (15.3 W/cm² at 1064 nm). The resulting 42°C temperature increase in the culture media surrounding the cells was sufficient to kill (Fisher et al. 2010). In this latter study, cell internalization of MWCNTs was measured using fluorescence and transmission electron microscopy. Consistent with previous reports (Kostarelos et al. 2007), MWCNTs were taken up by cells and translocated to the nucleus. With increasing incubation duration, a greater number of MWNTs were observed in cellular vacuoles and nuclei. In Section 17.6, the role that cell binding and internalization of CNTs play in the efficacy of treatment will be discussed in more detail.

The number of MWCNTs needed per cell for effective thermal therapy has been studied (Torti et al. 2007). Human CRL1932 renal adenocarcinoma cells grown in a monolayer were incubated with increasing concentrations of nitrogen-doped MWCNTs corresponding to estimated MWCNT to cell ratios of 1:1, 100:1, and 1000:1. Following exposure to NIR (3 W/cm² of 1064 nm NIR with for 4 min), a 23°C temperature rise was observed in the culture media of cells exposed to MWCNTs at the 1000:1 ratio, with near-complete cell death. No significant heating differences of cell death was observed in cells exposed to MWCNTs at the 1:1 or 100:1 ratios as compared to control heated cells heated in the absence of MWCNTs.

In a follow-up study, Burke et al. (2009) demonstrated that cellular uptake of MWCNTs before NIR exposure was not necessary for *in vitro* cancer cell killing. Murine RENCA cells were homogeneously dispersed in phosphate buffered saline containing 100 μg/mL of MWCNTs (Burke et al. 2009). Following a brief exposure to NIR (3 W/cm² of 1064 nm NIR with for 30 s), 98% of cancer cells were killed, whereas 45 s of treatment killed 100% of the cells. *In vivo* studies demonstrated that intratumoral injection of increasing doses of MWCNTs (10, 50, or 100 μg) into RENCA tumors implanted in the flanks of nude mice significantly decreased tumor growth and increased survival in a dose-dependent manner following exposure to NIR (3 W/cm² of 1064 nm NIR with for 30 s). In the absence of MWCNTs, this laser treatment resulted in minimal superficial burning. Durable remission was observed in 80% of mice receiving the 100 μg MWCNT dose combined with NIR for the length of the 6-month study. Although MWCNTs remained at the injection site, no toxicity was detected. In contrast, all mice administered MWCNTs in the absence of NIR did not survive beyond 3 months, and no difference in tumor growth was observed between these animals and untreated controls.

This study presented a key demonstration of the capability of MWCNTs coupled with laser irradiation to enhance the *in vivo* treatment of tumors through more controlled thermal deposition leading to increased tumor injury. Furthermore, Burke et al. examined the induction of heat shock proteins (HSPs) 27, 70,

TABLE 17.2 Photothermal Cancer Therapy Using Near-Infrared Radiation to Heat Carbon Nanomaterials *In Vitro*

Nanomaterial	Functionalization	Cancer Model	Optimal Nanotube Conc.	NIR Parameters: Wavelength; Irradiance; Duration; Laser Mode	Radiant Exposure (Total Energy)[a]	Maximum Temp.[b]	Therapeutic Efficacy	Reference
SWCNH (nanohorn)	Acid oxidized; bovine serum albumen coated	5RP7 rat fibroblasts cells transformed by the c-Ha-ras oncogene	20 µg/mL	670 nm; 160 mW/cm²; 5 min; continuous	48 J/cm² (38 J)	>41°C	Nanohorns were incubated with cell monolayer for 24 h then irradiated. 68% of cells killed following treatment.	Zhang et al. (2008)
	Pluronic F127 coated	RENCA murine kidney cancer cells	85 µg/mL	1064 nm; 40 W/cm²; 1 min; continuous	2400 J/cm² (300 J)	up to 85°C	Nanohorns were incubated with cell monolayer for 24 h then irradiated.; >90% cell death following treatment.	Whitney et al. (2011)
SWCNT	Acid oxidized	BT474 breast cancer	100 µg/mL	800 nm; 200 mW/cm²; 1 min; continuous	12 J/cm² (Energy: N/A – spot size unknown)	minimal temp. increase	Nanotubes mixed with suspended cells then irradiated. No temperature increase was measured. Cell death attributed to photoacoustic shock wave rather than thermal effect. Cell death rate not quantified.	Panchapakesan et al. (2005)
	DNA coated	U251 human glioblastoma	10 µg/mL	808 nm; 2 W/cm²; 5 min; continuous	600 J/cm² (226 J)	up to 50°C	Nanotubes were mixed with suspended cells then irradiated. Near 100% cell death following treatment.	Markovic et al. (2011)
	Acid oxidized; PEG coated; antibody targeted	MCF-7 human breast cancer cells	Not reported	808 nm; 800 mW/cm²; 3 min; continuous	144 J/cm² (113 J)	55–60°C	Nanotubes were mixed with suspended cells and incubated for 30 min, then irradiated. Near 100% cell death following treatment. Highly selective for targeted population.	Ning et al. (2007)
	Acid oxidized; antibody targeted	SK-BR-3 human breast cancer cells	4 µg/mL	808 nm; 5 W/cm²; 3 min; continuous	900 J/cm² (137 J)	>38°C	Nanotubes were incubated for 24 h with cell monolayer, washed, then irradiated. Near 100% cell death following treatment with targeted SWCNTs. Highly selective for targeted population.	Xiao et al. (2009)
	Chitosan coated; folic acid targeted	Hep G2 human hepatocellular carcinoma cells	20 µg/mL	1064 nm; 200 mW/cm² average; 20 s; Q-switched, millisecond pulsed	4 J/cm² (28 J)	minimal temp. increase	Nanotubes were incubated for 3–5 h with cell monolayer then irradiated. >85% cell death following treatment. Cause of death attributed to photoacoustic shock wave rather than thermal effect. Highly selective for targeted population.	Kang et al. (2009)
	Phospholipid-PEG coated; antibody targeted	Daudi human Burkitt's lymphoma cells	90 µg/mL	808 nm; 5 W/cm²; 7 min; continuous	2100 J/cm² (Energy: N/A – spot size unknown)	temp. not reported	Nanotubes were mixed with suspended cells for 20 min, washed, then irradiated; >90% cell death following treatment with targeted SWCNTs. Highly selective for targeted population.	Chakravarty et al. (2008)
	DNA coated or phospholipid-PEG coated; folic acid targeted	HeLa human cervical carcinoma cells	25 µg/mL	808 nm; 1.4 W/cm²; 2 min; continuous	168 J/cm² (approx. 1187 J)	>70°C	Nanotubes were mixed with suspended cells, washed, then irradiated. Extensive cell death reported but not quantified following treatment. Highly selective for targeted population.	Kam et al. (2005)

(continued)

TABLE 17.2 Photothermal Cancer Therapy Using Near-Infrared Radiation to Heat Carbon Nanomaterials *In Vitro* (Continued)

Nanomaterial	Functionalization	Cancer Model	Optimal Nanotube Conc.	NIR Parameters: Wavelength; Irradiance; Duration; Laser Mode	Radiant Exposure (Total Energy)[a]	Maximum Temp.[b]	Therapeutic Efficacy	Reference
	Acid oxidized; antibody targeted	BT-474 human breast carcinoma cells	approx. 5–10 μg/mL	808 nm; 9.5 W/cm²; 9 min; continuous	5130 J/cm² (Energy: N/A – spot size unknown)	temp. not reported	Nanotubes were incubated with cell monolayer for 24 h, washed, then irradiated. Approx. 40% of targeted cells were killed. Highly selective for targeted population.	Marches et al. (2011)
	Phospholipid-PEG; folate targeted	EMT6 murine breast carcinoma cells	3.5 μg/mL	980 nm; 1 W/cm²; 2 min; continuous	120 J/cm² (Energy: N/A – spot size unknown)	up to 70°C	Nanotubes were incubated with cell monolayer for 2 h, washed, then irradiated. >85% cell death observed after treatment. Highly selective for targeted population.	Zhou et al. (2011)
	Chitosan coated; antibody targeted	Glioblastoma cells expressing high or low levels of CD133, isolated from patient samples	2.5 μg/well of 24 well plate	808 nm; 2 W/cm²; 5 min; continuous	600 J/cm² (678 J)	temp. not reported	Nanotubes were incubated 6 h with cell monolayer then washed away and irradiated; >95% cell death following treatment. Highly selective for targeted population.	Wang et al. (2011)
MWCNT	Nitrogen-doped; Pluronic F127 coated	CRL 1932 human kidney cancer cells	approx. 83 μg/mL	1064 nm; 3 W/cm²; 4 min; continuous	720 J/cm² (171 J)	46°C	Nanotubes were incubated for 24 h with cell monolayer then irradiated. >95% of cells dead after treatment.	Torti et al. (2007)
	Raw (as produced)	Erlich ascitic murine carcinoma cells	100 μg/mL	780–1400 nm; 3.5 W/cm²; 1.5 min; continuous	315 J/cm² (Energy: N/A– spot size unknown)	Up to 50°C	Nanotubes were mixed with suspended cells then irradiated. >95% cell death following treatment.	Burlaka et al. (2010)
	HCl cleaned	HeLa human cervical cancer cells	10 μg/mL	1064 nm; approx. 32 W/cm²; 4 s; continuous	128 J/cm² (4 J)	temp. in laser beam >75°C	Nanotubes were incubated for 48 h with cell monolayer then irradiated. >98% cell death following treatment.	Biris et al. (2009)
	Pluronic F127 coated	RENCA murine kidney carcinoma cells alone	100 μg/mL	1064 nm; 3 W/cm²; 45 s; continuous	135 J/cm² (26 J)	>53°C	Nanotubes were mixed with suspended cells then irradiated. Near 100% cell death following treatment.	Burke et al. (2009)
	Pluronic F127 coated	RENCA murine kidney carcinoma and PC3 human prostate carcinoma cells	100 μg/mL	1064 nm; 15.3 W/cm²; 5 min; continuous	4590 J/cm² (900 J)	66°C	Nanotubes were mixed with suspended cells then irradiated. Near 100% cell death following treatment. Note: Cells were maintained at 25°C before treatment. Heating curves indicate that 30–60 s NIR should be sufficient to kill cells maintained at 37°C.	Fisher et al. (2010)
	Oxidized, human serum albumen coated	HepG2 and CRL-4020 human hepatocellular carcinoma cells	1–50 μg/mL	808 nm; 64 W/cm²; 2 min; continuous	7680 J/cm² (approx. 240 J)	temp. not reported	Nanotubes were Incubated for 1 min to 24 h with cell monolayer then irradiated; >95% cell death at all conc. MWCNTs after 5 or longer incubation.	Iancu et al. (2011)
	Acid oxidized; antibody targeted	Human stNB-V1 neuroblastoma cells; PC12 Neuron-derived rat pheochromocytoma cells	5–10 μg/mL	808 nm; 0.6–6 W/cm² for 10 min followed by 6 W/cm² for 5 min; continuous	approx. 3000 J/cm² (Energy: N/A – spot size unknown)	temp. not reported	Nanotubes were incubated 6 h with cell monolayer then washed and irradiated; All treated cells died. Highly selective for targeted population.	Wang et al. (2011)

[a] Radiant exposure was calculated from the product of the NIR irradiance and the exposure time. For dosimetry calculations (total energy), the radiant exposure was multiplied by the surface area of the laser beam (laser spot size) incident upon the surface of the tumor or cancer cell target as reported in the cited reference.

[b] Maximum temperature indicates the maximum temperature achieved following laser irradiation as reported by the cited references. Depending on the technique used to measure temperature, this may be a local (cellular) temperature measurement or a macroscale (averaged over a volume of several milliliters) temperature measurement.

TABLE 17.3 Photothermal Cancer Therapy Using Near-Infrared Radiation to Heat Carbon Nanomaterials *In Vivo*

Nanomaterial	Nanotube Functionalization	Cancer Model	Nanotube Dose/ Route	NIR Parameters: Wavelength; Irradiance; Duration; Mode	Radiant Exposure (Total Energy)[a]	Maximum Temp.[b]	Therapeutic Efficacy	Ref.
SWCNH (nanohorn)	Acid oxidized; bound to BSA	RAS-transformed 5RP7 rat fibroblast cells implanted in nude mice	30 µg; intratumoral	670 nm; 160 mW/ cm²; treatment given in 5 locations per tumor, 3 min each day for 10 days; continuous	28.8J/cm² per treatment × 5 locations for 10 days; (5.7 J per site per treatment (cumulative dose approx. 280 J/tumor))		Significant tumor regression observed in nanotube treated mice following irradiation, although no complete regression observed.	Zhang et al. (2008)
SWCNT	PEG coated	SCCVII murine squamous carcinoma cells implanted in C₃H/ HeN mice	60–100 µg; intratumoral	785 nm; 200 mW/ cm²; 10 min; continuous	120 J/cm²	590 J; 48°C	Complete tumor regression for 5/8 nanotube treated mice for at least 45 days following irradiation.	Huang et al. (2010)
	Phospholipid-PEG coated	4T1 murine breast cancer cells implanted in BALB/c mice	70 µg; intravenous	808 nm; 600 mW/ cm²; 5 min; continuous	180 J/cm²	approx. 90 J (assuming 5 mm tumor diameter)	Complete tumor regression in 10/10 nanotube treated mice for at least 60 days following irradiation.	Robinson et al. (2010)
	Phospholipid-PEG coated	KB human epidermoid mouth carcinoma cells implanted in nude mice	12 µg; intratumoral	808 nm; 3.8 W/cm²; 3 min; continuous	684 J/cm²	435 J; Up to 70°C in solution – *in vivo* temp. not reported	Complete tumor regression in 4/4 nanotube treated mice for at least 25 days following irradiation.	Moon et al. (2009)
	PEG-grafted poly(maleic anhydride-octadecene) coated	4T1 murine breast carcinoma cells implanted in BALB/c mice	100 µg; intravenous	808 nm; 1 W/cm²; 5 min; continuous	300 J/cm²	N/A – spot size not reported	Tumor regression observed in 5/7 treated mice following irradiation. No tumor regrowth observed in 3/7 treated mice for at least 2 weeks following irradiation.	Liu et al. (2011)

(continued)

TABLE 17.3 Photothermal Cancer Therapy Using Near-Infrared Radiation to Heat Carbon Nanomaterials *In Vivo* (Continued)

Nanomaterial	Nanotube Functionalization	Cancer Model	Nanotube Dose/ Route	NIR Paramenters: Wavelength; Irradiance; Duration; Mode	Radiant Exposure (Total Energy)[a]	Maximum Temp.[b]	Therapeutic Efficacy	Ref.
	phospholipid-PEG coated; folate targeted	EMT6 murine breast carcinoma cells implanted in BALC/c mice	Dose estimated to be 20–25 μg based on typical mouse weight of 20–25 g; intratumoral	980 nm; 1 W/cm²; 5 min; continuous	300 J/cm²	Estimated to be 60 J based on 5 mm tumor diameter; Up to 65°C	Tumor growth was not evaluated. However, >85% tumor cell death observed in nanotube treated mice 3 h after irradiation.	Zhou (2009)
MWCNT	Pluronic F127 coated	MDA-MB-231 human breast cancer implanted in nude mice	100 μg; intratumoral	1064 nm; 3 W/cm²; 30 s; continuous	90 J/cm²	21.6 J; temp. not reported	Complete tumor regression in 4/4 nanotube treated mice for at least 30 days following irradiation.	Ding et al. (2011)
	DNA coated	PC3 human prostate cancer cells in nude mice	50 μg; intratumoral	1064 nm; 2.5 W/cm²; pulsed 5 s on, 3 s off for 14 repetitions (70 s exposure time); pulsed	175 J/cm²	Energy: N/A – spot size not reported	Tumor regression observed in 4/4 nanotube treated mice with indication of recurrence in some animals 3 weeks after irradiation.	Ghosh et al. (2009)
	Pluronic F127 coated	RENCA murine kidney carcinoma cells alone or implanted in nude mice	100 μg; intratumoral	1064 nm; 3 W/cm²; 30 s; continuous	90 J/cm²	18 J; >70°C	Complete tumor regression in 8/10 nanotube treated mice for at least 300 days following irradiation.	Burke et al. (2009)
	Chitosan coated; antibody targeted	Glioblastoma cells expressing high or low levels of CD133, isolated from patient samples, implanted in nude mice	10 μg; cells incubated with nanotubes *in vitro* then implanted in mice	808 nm; 2 W/cm²; 5 min; continuous	600 J/cm²	678 J; temp. not reported	Complete tumor regression in 3/3 nanotube treated mice without metastasis for at least 5 weeks following irradiation.	Wang et al. (2011)

[a] Radiant exposure was calculated from the product of the NIR irradiance and the exposure time as reported in the cited reference. For dosimetry calculations (total energy), the radiant exposure was multiplied by the surface area of the laser beam (laser spot size) incident upon the surface of the tumor or cancer cell target as reported in the cited reference.

[b] Maximum temperature indicates the maximum temperature achieved following laser irradiation as reported by the cited references. Depending on the technique used to measure temperature, this may be a surface temperature measurement or an internal temperature measurement.

and 90 as an indirect measure of heat generation in full-depth tissue sections taken from tumors of mice 16 h after NIR exposure. HSPs are induced by elevated temperatures (typically in excess of 43°C) and serve as endogenous cellular markers of thermal stress. They observed minimal expression for all HSPs in untreated tumors, but in tumors treated with NIR in the absence of MWCNTs, significant HSP expression was observed. Maximal HSP27, HSP70, and HSP90 expression was induced proximal to the incident laser (near the skin) and then gradually diminished with increasing depth. In contrast, in tumors treated with NIR plus MWCNTs, the temperature elevation was sufficient to induce coagulative necrosis in much of the tumor, thus preventing significant HSP induction. However, HSPs were seen at deeper tissue levels, near the interface between tumor and normal tissue. These results demonstrate that NIR irradiation combined with MWCNTs can be used to extend the depth of thermal therapy.

Similar observations were made *in vitro* by Fisher et al. (2010), again demonstrating the combination of MWNTs and NIR can dramatically decrease cell viability without inducing HSP expression—possibly indicating a necrotic rather than apoptotic cell death mechanism. A necrotic mode of cell death offers a significant advantage over many conventional therapies that rely on induction of pro-death signal transduction pathways (Gottesman 2002; Bergamaschi et al. 2003; Pommier et al. 2004), because it does not provide selective pressure to induce the evolution of treatment-resistant cancer cell clones. However, necrosis has not been universally observed as the mechanism of cell death following treatment with NIR and MWCNTs; Kratz (2010) reported a dramatic increase in apoptosis in Hep G2 human hepatocellular carcinoma cells following exposure to MWCNTs and NIR. To date, no extensive research has been conducted on factors that may influence the mechanism(s) of cell death induced by the combination of CNTs and NIR, and this is clearly an area where more knowledge is needed.

The thermal effects generated by CNTs may have benefits in addition to direct thermal ablation of cancer cells. For example, hyperthermia can increase the permeability of tumor vasculature. This can enhance the delivery of drugs into tumors, as well as synergistically enhance cytotoxicity when combined with chemotherapy or radiotherapy (Falk and Issels 2001). In this regard, the use of mild NIR irradiation of MWCNTs to rapidly heat cancer cells to temperatures below the thermal ablation threshold has been shown to increase the uptake of codelivered chemotherapeutic drugs and enhance cancer cell death both *in vitro* and *in vivo* in a murine ascites tumor model (Levi-Polyachenko et al. 2009). Similarly, it was also demonstrated that photothermal heating of SWNTs chemically conjugated with platinum-based chemotherapeutics was significantly more effective than either therapy alone (Feazell et al. 2007). Such strategies may allow increased cancer selectivity of chemotherapeutic agents or reduce the dose necessary for efficacy and thus reduce the toxicity of such treatments.

17.6 Tumor Selective Nanotube Binding and Uptake for Enhanced Anticancer PTT

Once CNTs come into contact with cells, they can easily pass through cell membranes (Kostarelos et al. 2007). The mechanism of CNT internalization is still not completely understood and is greatly influenced by the physicochemical properties of the tubes themselves (Raffa et al. 2010). It is believed that they can enter the cell both passively by diffusion across the lipid membrane and are transported actively by endocytosis or receptor-mediated endocytosis (Shi Kam et al. 2004; Cai et al. 2005; Chen et al. 2008). Internalization of CNTs by cells offers an opportunity for extremely confined heating effects. For example, Kam et al. demonstrated that brief NIR excitation (six pulses, each 10 s long, at 1.4 W/cm²) of SWCNTs taken up by endocytosis can trigger endosomal rupture with no apparent adverse toxicity (Kam et al. 2005; McDevitt et al. 2007; Welsher et al. 2008). Moreover, the combination of NIR and SWCNTs was shown to selectively release noncovalent molecular cargoes (DNA in this case) from nanotube carriers. In contrast to the nontoxic effect observed following brief, pulsed exposure to NIR, extensive cell death was observed after cells with internalized SWNTs were continuously exposed to NIR for 2 min of radiation under a 1.4 W/cm² power (Kam et al. 2005).

This result hinted that if SWNTs could be selectively internalized into cancer cells, NIR irradiation of the nanotubes could then selectively activate or trigger cell death without harming normal cells. One strategy to allow for more selective therapy would be to specifically (or actively) target CNTs to tumors. This can be accomplished by conjugation of peptides, proteins, or antibodies to the surface of CNTs, which has been shown to increase the specificity of tumor targeting (Kam et al. 2005; McDevitt et al. 2007; Welsher et al. 2008). The addition of targeting ligands involves modifying CNT surface chemistry (functionalizing), which could affect the optical absorbance and thermal properties of CNTs. Several studies have shown that under limited experimental conditions, the optical properties of SWCNTs were retained after adding targeting moieties (Chakravarty et al. 2008; Marches et al. 2009). To date, no detailed studies have been conducted to compare the photothermal behavior of such functionalized tubes to their nonfunctionalized precursors over a broad set of conditions.

An early approach to selective targeting adopted by several groups was to use folic acid (FA) as a targeting ligand to direct CNTs to FA receptors (FR), which are overexpressed on many tumors (Kamen and Smith 2004). In one such study conducted *in vitro*, SWCNTs (average length 150 nm) were conjugated to FA, and directed toward human cervical cancer cells with low FR expression, or HeLa cells overexpressing FR. The FR overexpressing cells internalized the targeted SWCNTs (as verified by fluorescent labeling), whereas normal cells with low FR expression did not take up the CNTs. After the cells were heated using 808-nm laser at 1.4 W/cm² continuously for 2 min, extensive cell death was observed for the FR overexpressing cells, whereas cells with low FR expression remained intact and exhibited normal proliferation behavior (Kam et al. 2005). Similar results were observed by Zhou (2009), and by Kang et al. (2009), who targeted FA conjugated SWCNTs to human hepatocellular carcinoma cells (HepG2) overexpressing FR. In the latter study, the effect of NIR combined with FA-targeted SWCNTs on FR overexpressing HepG2 cells was found to be dependent on the nanotube concentration, with cell death rising from 50% in cells treated with CNTs at a concentration of 2 µg/mL, to greater than 85% cell death at a concentrations of 20 µg/mL and higher. In contrast, similar treatment of HepG2 cells with low FR expression resulted in less than 10% cell death at SWCNT concentrations of 20 µg/mL and less. However, selectivity of this treatment was greatly reduced at higher nanotube concentrations; significant cell death (35–40%) was observed in HepG2 expressing low levels of FR following treatment with 50 µg/mL SWCNT dispersions, indicating that nonspecific binding and uptake of CNTs may reduce the selectivity of this treatment.

Although homogeneous upregulation of FR is found in up to 90% of some cancers including ovarian and brain cancers, other solid tumors, such as those found in the breast, have more variable folate receptor expression with only about 50% showing overexpression (Leamon and Reddy 2004). Therefore, folate is useful as a targeting modality in only a fraction of cancers. The versatility of CNTs as a platform allows for display of many types of ligands other than small molecules such as folate including antibodies with binding affinity for the breast cancer associated receptor Her2 (Marches et al. 2011; Xiao et al. 2009). In one such study, anti-Her2 conjugated SWCNTs were seen to bind to Her2 expressing, SK-BR-3 human breast adenocarcinoma cells, but remained at the surface without internalizing (Xiao et al. 2009). Little binding was observed in Her2 negative MCF-7 breast adenocarcinoma cells following exposure to anti-Her2 conjugated SWCNT. After washing away unbound CNTs and adding fresh media to the cells, treatment with an NIR laser at 5 W/cm^2 for 2 min killed 97% of SK-BR-3 cells. Strikingly, minimal cell death was observed using nontargeted SWCNTs combined with NIR, and the Her2 antibody alone had no effect. Similarly, MCF-7 cells, to which the anti-Her2 conjugated SWCNTs did not bind, were spared from thermal ablation upon identical treatment conditions.

Further refinement of this treatment strategy was achieved by Marches et al. (2011), who also conjugated SWCNTs to an anti-Her2 antibody. In this case, the anti-Her2 targeting moiety induced internalization of the SWCNTs following binding to Her2 expressing cancer cells. By tracking the relative cell binding and internalization of the anti-Her2 conjugated SWCNTs over time, Marches et al. determined that cells containing internalized CNTs were more sensitive to NIR-mediated thermal ablation than cells that bind to, but do not internalize, the CNTs. Moreover, in a mixed population of cells expressing or not expressing Her2, NIR-mediated cell damage was restricted primarily to Her2 expressing cells that bound and internalized the CNTs (Marches et al. 2011), demonstrating the possibility of

tailoring this type of therapy to differentiate between targets on a cell by cell basis.

One significant advantage of CNTs over other nanoparticles is that because of the large surface area/volume ratio, CNTs can effectively display more than one targeting ligand on their surface. This allows CNTs to be engineered to bind to multiple receptors overexpressed on cancer cells, a strategy that can be used to expand both the tropism and specificity of cancer-targeted CNTs. For example, antibodies targeting Her2 and another receptor overexpressed in breast cancers, insulin-like growth factor 1 receptor (IGF1R) were attached to SWCNTs (Ning et al. 2007). In cell culture studies, the binding of each targeting ligand did not affect the other, allowing specific binding to both MCF-7 cells (which express IGF1R but not Her2) and BT474 cells (a human breast ductal carcinoma cell line that expresses Her2 but not IGF1R). Following NIR irradiation, almost all of the cells with targeted SWCNTs were killed. NIR irradiation of cells exposed to nonspecifically targeted SWCNTs killed fewer cells (less than 50%). Significantly, the estimated power used to kill each cell was calculated to be approximately 200 nW (an order of magnitude less than other protocols) (Ning et al. 2007).

Perhaps the most striking example of the possibility of selective thermal cancer ablation mediated by CNTs was recently demonstrated through the use of SWCNTs designed to accumulate at an intracellular target. In 2010, Zhou et al. demonstrated that, following coating with a phospolipid modified to contain polyethylene glycol (PEG) in the head group, SWCNTs selectively accumulate intracellularly at the mitochondrial membrane. When exposed to NIR, these SWCNTs selectively destroy the target mitochondria, thereby inducing mitochondrial depolarization, cytochrome *c* release, and caspase 3 activation. *In vivo*, treatment with these modified SWCNTs suppressed tumor growth in a murine breast cancer model, resulting in complete tumor regression in some cases (Zhou et al. 2011).

Strategies to selectively target CNTs to cancer cells for thermal ablation therapy are not limited to the use of SWCNTs. Several studies have demonstrated that similar approaches to those described above for SWCNTs can be applied to MWCNTs, allowing for the greater heat transduction capability of MWCNTs to be used for targeted PTT. In one study, MWCNTs were conjugated to monoclonal antibody directed GD2, a carbohydrate antigen overexpressed in neuroblastomas (Wang et al. 2009). Binding and internalization was found to be specific for GD2 expressing neuroblastoma cells, whereas control rat neuroendocrine tumor cells that did not express GD2 did not take up these MWCNTs. Cells were heated using an 808-nm laser that gradually increased from 0.6 to 6 W/cm^2 over 10 min and was maintained at 6 W for 5 min more. This treatment caused necrosis in nearly all GD2+ cancer cells (as determined by calcein staining) but not in control cells. Only the cells within the laser zone were killed, and a clear boundary of living cells delineated the border between the treatment zone and cells that were not illuminated by the laser (Wang et al. 2009).

A second example of selectively targeting MWCNTs involved functionalizing the surface of MWCNTs with human serum albumin in order to target the albumin-binding Gp60 receptor expressed in hepatocellular carcinoma cells (Kratz 2010). Specific internalization was seen in Gp60 expressing HepG2 cells, but not in normal hepatocytes. Uptake was shown to be mediated by receptor binding, followed by caveolin-dependent endocytosis. Following NIR irradiation (808 nm; 2 W/cm^2 for 2 min) cell death directly correlated with increased concentrations of CNTs and increased CNT exposure time. There was a 5- to 6-fold increase in apoptosis of cancer cells compared to hepatocytes. Treatment with nonfunctionalized MWCNTs produced no significant differences in cell death between the cancer and non-cancer cell lines.

Targeted thermal ablation therapies based on CNTs are being tested for the treatment of cancers that are highly resistant to current therapies, including glioblastomas (Wang et al. 2011). In glioblastomas and other brain tumors, the CD133 receptor appears to be a cancer stem cell marker associated with malignancy, tumor recurrence, and poor survival. (Singh et al. 2004; Beier et al. 2008; Zeppernick et al. 2008). Cancer stem cells have been putatively identified as self-renewing therapy-resistant populations in many types of tumors (Jordan et al. 2006). In glioblastomas, CD133+ subpopulations are enriched following radiotherapy, are radio- and chemotherapy-resistant, and are responsible for restoring tumor cells after treatment (Singh et al. 2004; Bao et al. 2006; Lee et al. 2006). Treatment strategies based on targeting this subpopulation may be able to prevent the development of resistance to therapy.

Recently, Wang et al. (2011) used MWCNTs chemically conjugated to a monoclonal antibody directed against CD133 to target these cells. In cell culture experiments, they observed specific internalization of these targeted MWCNTs via endocytosis in glioblastoma cells that expressed CD133, but not in cells that did not. Importantly, these were not immortalized cells, but were cells that had been freshly isolated from patients. To test the selectivity of nanotube-enhanced PTT, mixed populations of both cell types were incubated with 2.5 mg of the MWCNTs for 6 h and then were irradiated with an 808-nm laser at 2 W/cm^2 for 5 min. Flow cytometry confirmed that CD133+ cells were killed, whereas CD133− cells were spared. These effects were further recapitulated by an *in vivo* xenograft model in which CD133 expressing glioblastoma cells were pretreated with targeted MWCNTs before injection into mice. The cells took up the MWCNTs, and xenograft growth was abolished and no metastases were detected after NIR exposure (Wang et al. 2011). This was a key demonstration of the potential of CNTs to treat glioblastomas and other currently untreatable cancers.

Cancer stem cells are particularly well described in breast cancer, where they have been shown to be highly resistant to standard chemotherapy and radiotherapy (Al-Hajj et al. 2003; Diehn and Clarke 2006). Recent work by Burke et al. (2012) describes the utility of MWCNT-mediated thermal ablation in treating this otherwise therapy-resistant cell population. Initially, bulk (non-stem) and stem breast cancer cells were treated with hyperthermia delivered by incubation in a circulating water bath to mimic conventional clinical hyperthermic therapy, and changes in cell viability were determined. By this method, breast cancer stem cells were significantly more resistant to hyperthermia than bulk breast cancer cells across the entire treatment range. In contrast, when both cell types were heated to the same final temperature using 50 µg/mL amidated MWCNTs and NIR laser radiation (1064 nm; 3 W/cm^2), stem and bulk breast cancer cells were equally sensitive to MWCNT-mediated thermal therapy. The authors demonstrated that this was due to the induction of rapid and robust necrotic cell death following MWCNT-mediated thermal therapy that was largely absent after conventional hyperthermia. Collectively, these findings demonstrate that nanotube-mediated hyperthermia is functionally distinct from hyperthermia delivered by other means, and may represent a significant therapeutic advance for the treatment of refractory, stem cell-driven cancers.

17.7 Systemic Delivery and Biocompatibility of CNTs for PTT

In order for CNT enhanced PTT to spare normal tissue from heating, it is important for the CNTs to be selectively taken up by tumor cells and not by normal cells. Ideally, this could be achieved following systemic administration of tumor targeted CNTs. Significant effort has been made to achieve selective targeting of nanoparticles to tumor sites based on both passive and active targeting (Brannon-Peppas and Blanchette 2004; You et al. 2006; Li and Huang 2008; Ruenraroengsak et al. 2010; Van Lehn et al. 2010; Yoo et al. 2010). Passive targeting refers to strategies that attempt to achieve tumor delivery without utilizing specific biological (ligand–receptor) interactions by correlating the physicochemical and surface characteristics of the nanoparticle with the pathophysiology and anatomy of the target site. Active delivery is inherently dependent on passive delivery to reach the tumor site, but also adds to nanoparticles the ability to associate or interact with specific biological moieties by attachment of ligands with an enhanced binding affinity for complementary cellular receptors, as discussed above.

For the past 20 years, passive strategies to selectively deliver nanoparticles to tumor sites have relied on the enhanced permeability and retention effect. The strategy proposes that tumor specificity of nanoparticles that remain in circulation for long periods can be achieved because of the nanoparticles' ability to extravasate through the leaky vasculature surrounding the tumor and enter the tumor site (Maeda et al. 2000; Liu et al. 2008). Long blood circulation frequently is achieved through coating nanoparticles with steric stabilizers such as PEG, which inhibit blood clearance by reducing the uptake of nanoparticles by macrophages and other components of the mononuclear phagocyte system. Several groups have demonstrated that PEG coating can greatly improve the tumor localization of CNTs following intravenous injection in mice (Cato et al. 2008; Liu et al. 2009; Bhirde et al. 2010).

Recently, this strategy for passively targeting nanotubes to tumors intravenously was shown to be effective for nanotube-enhanced PTT (Robinson et al. 2010). Short (140 nm), PEG-coated SWCNTs at a dose of 3.6 mg/kg were injected via the tail vein into mice bearing 4T1 murine breast tumors. Three days later, SWCNT accumulation at the tumor site was confirmed by detection of the inherent NIR photoluminescence of the SWCNTs (1100–1400 μm wavelength) using an optical imaging system. The tumors were heated for 5 min at 0.6 W/cm^2 with 808 nm laser, resulting in complete tumor ablation. All treated mice survived without recurrence for the duration of the 6-month study, and no toxicity was seen. This was an extremely significant step toward the development of a tumor-selective, systemically delivered photothermal ablation agent. It is important to note that in this experiment, the nanotubes were not modified to display a specific targeting ligand, nor were they internalized by the cancer cells themselves. More research is needed to determine if active targeting will offer an additional benefit *in vivo*.

Although these results indicate that CNTs offer great promise for targeted hyperthermia of cancer, the translation of CNTs from an interesting nanomaterial to an effective pharmaceutical product is still in its nascence. The toxicity, and therefore the ability to analyze the potential risk–benefit balance for these materials, will ultimately determine their long-term clinical fate. Unfortunately, the development of an accurate toxicity profile of CNTs is a complicated matter. Not only do structural characteristics, such as diameter and length, greatly influence biological and toxicological responses following injection of CNTs (Donaldson et al. 2006), but also changes in surface functionalization, which alter adsorption properties, electrostatic interactions, hydrophobicity/hydrophilicity, and influence the stability of CNT dispersions, also affect CNT toxicity (Dyke and Tour 2004; Lacerda et al. 2006). Finally, toxicity may also be due to by-products from CNT manufacturing, including residual catalysts such as Co, Fe, Ni, and Mo (Lacerda et al. 2006).

The toxicity of CNTs has been reviewed extensively elsewhere (Ai et al. 2011; Beg et al. 2011; Donaldson et al. 2011; Kaiser et al. 2011; Stella 2011; Uo et al. 2011; Zhang et al. 2011). Therefore, only a few key studies will be highlighted here. It should be noted that most toxicity studies have focused on environmental exposure following inhalation of pristine CNTs that have not been modified from their "as produced" state (Jia et al. 2005; Lam et al. 2006; Smart et al. 2006; Warheit 2006; Kolosnjaj et al. 2007; Wick et al. 2007; Kostarelos 2008; Koyama et al. 2009). For example, concerns have been raised regarding the possibility of very long (10–20 μm) nanotubes to induce an asbestos-like reaction (Schipper et al. 2008), or for inhalation of nanotubes to cause dose-dependent granulomatous pneumonia, oxidative stress, and acute inflammatory and cytokine responses, with fibrosis and decrease in pulmonary function (Shvedova et al. 2005), and the possibility that CNTs may elicit an immune or allergic response (Park et al. 2009; Ryman-Rasmussen et al. 2009).

CNTs are inherently hydrophobic, and the toxicity of the pristine CNTs may be largely due to their hydrophobicity (Sayes et al. 2006). CNTs used for biomedical applications must be modified in some way from their pristine, as-produced, condition in order to render them suitable for dispersion in aqueous environments (Burke et al. 2011). Typical modifications include acid oxidation of the CNT exterior to introduce carboxyl groups (reviewed by Tasis et al. 2003), "wrapping" CNTs in long-chain surfactants (reviewed by Nakashima and Fujigaya 2007), and as noted above, linking antibodies or other targeting moieties to the CNT surface both to aid in their dispersion and promote their accumulation in tumor tissue (Liu et al. 2007; McDevitt et al. 2007). Such chemical modifications can improve the overall toxicity profile of CNTs and enhance their body clearance (Dyke and Tour 2004; Lacerda et al. 2006; Sayes et al. 2006).

The use of CNTs as nanomedicines will necessitate bypassing the body's natural defenses, possibly through intravenous delivery, where alternative toxicities might be observed. Because of the ability of intravascularly delivered particulates to induce undesired thrombotic events (Radomski et al. 2005; Dobrovolskaia et al. 2008; Mayer et al. 2009; Semberova et al. 2009), demonstration that CNTs are compatible with blood is particularly critical. Recently, the role played by chemical functionalization of CNTs on blood toxicity following intravenous injection into mice was examined (Burke et al. 2011). It was found that pristine MWCNTs were substantially more thrombogenic than chemically functionalized MWCNT following intravenous injection in mice. At a dose of 250 μg, pristine MWCNT were acutely lethal, inducing blockage of the pulmonary vasculature. In contrast, an equivalent dose of covalently functionalized MWCNTs exerted little effect on coagulation *in vivo*, with their sole measurable effect being a transient depletion of platelets. Consistent with this, the majority of studies in which mice were injected with CNTs that have been chemically functionalized to improve aqueous dispersion have not shown any long-term or chronic toxicity (Lacerda et al. 2006; Singh et al. 2006; Liu et al. 2008; Yang et al. 2008; Deng et al. 2009; Qu et al. 2009). Although larger and longer-term studies must be undertaken before the toxicity profile of CNTs is fully understood, it appears that CNTs can be designed to be biocompatible and suitable for systemic delivery.

17.8 Perspectives on Translational Potential of CNT-Enhanced PTT

CNTs have the potential to play a key role in the next generation of photothermal agents. They offer a unique capacity for designing and tuning optical, thermal, and cancer-selective properties that are not possible with other types of therapeutic vectors. Clinical applications of CNT-enhanced photothermal ablation could not only provide rapid and localized heating in response to NIR, but also be compatible with noninvasive imaging to spatially define the margins of the target lesion, assess the distribution of injected CNTs within the tumor, ensure the that the appropriate thermal dose is achieved in the target area, and track the response of the treated area to therapy over time. This combined functionality allows CNTs to overcome many of the drawbacks of traditional thermotherapy and may allow for expanded clinical use of image-guided LITT and improved therapeutic outcomes for cancer patients following such treatment.

Conceptually, benefits to patients offered by CNTs could be safely achieved using minimally invasive methods. For example, CNTs could be infused directly into a tumor or the main blood supply of a tumor could be identified intraoperatively by ultrasound to allow for intra-arterial infusion of CNT dispersions followed by laser irradiation using a mini-laser guided by a videoscopic or other real-time imaging modality. As noted by Iancu and Mocan (2011), the potential benefits of such a treatment strategy include reduced postoperative pain, more rapid recovery and decreased hospitalization, fewer surgical or wound complications, and improved cosmetics. Additionally, the next era of thermal therapy could include not only the use of nanoparticles for ablation of tissues, but the evaluation of nanoparticles for codelivery of chemotherapeutic agents to cancer cells. Efficient strategies for selectively targeting CNTs to tumors following intravenous or arterial injection may further enhance treatment efficacy.

Further research will be required to critically assess the potential for toxicity as well as evaluate pharmacologic properties of newly designed targeted CNTs. Since these properties will depend on the precise particle under evaluation, extensive materials assessment will be a critical preamble to successful clinical development. The encouraging results obtained to date with a variety of CNTs suggest that we have only begun to plumb their therapeutic potential.

Acknowledgments

This study was supported in part by National Institutes of Health grants R01CA12842 and R01CA128428-02S1, National Institutes of Health training and career development grants T32CA079448 and K99CA154006, and grant W81XWH-10-1-0434 from the Department of Defense.

References

Ai, J., E. Biazar et al. 2011. Nanotoxicology and nanoparticle safety in biomedical designs. *International Journal of Nanomedicine* 6:1117–1127.

Ajayan, P. M., M. Terrones et al. 2002. Nanotubes in a flash— ignition and reconstruction. *Science* 296(5568):705.

Al-Hajj, M., M. S. Wicha et al. 2003. Prospective identification of tumorigenic breast cancer cells. *Proceedings of the National Academy of Sciences of the United States of America* 100(7):3983–3988.

Ananta, J. S., B. Godin et al. 2010. Geometrical confinement of gadolinium-based contrast agents in nanoporous particles enhances T(1) contrast. *Nature Nanotechnology* 5(11):815–821.

Avouris, P., M. Freitag et al. 2008. Carbon-nanotube photonics and optoelectronics. *Nature Photonics* 2(6):341–350.

Ayala, P., R. Arenal et al. 2010. The physical and chemical properties of heteronanotubes. *Reviews of Modern Physics* 82(2):1843–1885.

Bao, S., Q. Wu et al. 2006. Glioma stem cells promote radioresistance by preferential activation of the DNA damage response. *Nature* 444(7120):756–760.

Beg, S., M. Rizwan et al. 2011. Advancement in carbon nanotubes: Basics, biomedical applications and toxicity. *Journal of Pharmacy and Pharmacology* 63(2):141–163.

Beier, D., J. Wischhusen et al. 2008. CD133 Expression and cancer stem cells predict prognosis in high-grade oligodendroglial tumors. *Brain Pathology* 18(3):370–377.

Berber, S., Y. K. Kwon et al. 2000. Unusually high thermal conductivity of carbon nanotubes. *Physical Review Letters* 84(20):4613–4616.

Bergamaschi, D., M. Gasco et al. 2003. p53 polymorphism influences response in cancer chemotherapy via modulation of p73-dependent apoptosis. *Cancer Cell* 3(4):387–402.

Bhirde, A. A., S. Patel et al. 2010. Distribution and clearance of PEG-single-walled carbon nanotube cancer drug delivery vehicles in mice. *Nanomedicine* 5(10):1535–1546.

Biris, A. R., S. Ardelean et al. 2011. Studies on near infrared optical absorption, Raman scattering, and corresponding thermal properties of single- and double-walled carbon nanotubes for possible cancer targeting and laser-based ablation. *Carbon* 49(13):4403–4411.

Biris, A. S., D. Boldor et al. 2009. Nanophotothermolysis of multiple scattered cancer cells with carbon nanotubes guided by time-resolved infrared thermal imaging. *Journal of Biomedical Optics* 14(2):021007.

Brannon-Peppas, L., and J. O. Blanchette. 2004. Nanoparticle and targeted systems for cancer therapy. *Advanced Drug Delivery Reviews* 56(11):1649–1659.

Brigger, I., C. Dubernet et al. 2002. Nanoparticles in cancer therapy and diagnosis. *Advanced Drug Delivery Reviews* 54(5):631–651.

Burke, A., X. Ding et al. 2009. Long-term survival following a single treatment of kidney tumors with multiwalled carbon nanotubes and near-infrared radiation. *Proceedings of the National Academy of Sciences of the United States of America* 106(31):12897–12902.

Burke, A. R., R. N. Singh et al. 2011. Determinants of the thrombogenic potential of multiwalled carbon nanotubes. *Biomaterials* 32(26):5970–5978.

Burke, A. R., R. N. Singh et al. 2012. The resistance of breast cancer stem cells to conventional hyperthermia and their sensitivity to nanoparticle-mediated photothermal therapy. *Biomaterials* 33(10):2961–2970.

Burlaka, A., S. Lukin et al. 2010. Hyperthermic effect of multiwalled carbon nanotubes stimulated with near infrared irradiation for anticancer therapy: *In vitro* studies. *Experimental Oncology* 32(1):48–50.

Cai, D., J. M. Mataraza et al. 2005. Highly efficient molecular delivery into mammalian cells using carbon nanotube spearing. *Nature Methods* 2(6):449–454.

Cassell, A. M., J. A. Raymakers et al. 1999. Large scale CVD synthesis of single-walled carbon nanotubes. *The Journal of Physical Chemistry B* 103(31):6484–6492.

Cato, M. H., F. D'Annibale et al. 2008. Cell-type specific and cytoplasmic targeting of PEGylated carbon nanotube-based nanoassemblies. *Journal of Nanoscience and Nanotechnology* 8(5):2259–2269.

Chakravarty, P., R. Marches et al. 2008. Thermal ablation of tumor cells with antibody-functionalized single-walled carbon nanotubes. *Proceedings of the National Academy of Sciences of the United States of America* 105(25):8697–8702.

Chen, J., S. Chen et al. 2008. Functionalized single-walled carbon nanotubes as rationally designed vehicles for tumor-targeted drug delivery. *Journal of the American Chemical Society* 130(49):16778–16785.

Chen, W. R., M. Korbelik et al. 2005. Enhancement of laser cancer treatment by a chitosan-derived immunoadjuvant. *Photochemistry and Photobiology* 81(1):190–195.

Dai, H. 2002. Carbon nanotubes: Synthesis, integration, and properties. *Accounts of Chemical Research* 35(12):1035–1044.

Day, E. S., J. G. Morton et al. 2009. Nanoparticles for thermal cancer therapy. *Journal of Biomechanical Engineering-Transactions of the Asme* 131(7):074001.

Deng, X. Y., F. Wu et al. 2009. The splenic toxicity of water soluble multi-walled carbon nanotubes in mice. *Carbon* 47(6):1421–1428.

Diehn, M., and M. F. Clarke. 2006. Cancer stem cells and radiotherapy: New insights into tumor radioresistance. *Journal of the National Cancer Institute* 98(24):1755–1757.

Ding, X., R. Singh et al. 2011. Development of iron-containing multiwalled carbon nanotubes for MR-guided laser-induced thermotherapy. *Nanomedicine (London)* 6(8):1341–1352.

Dobrovolskaia, M. A., P. Aggarwal et al. 2008. Preclinical studies to understand nanoparticle interaction with the immune system and its potential effects on nanoparticle biodistribution. *Molecular Pharmaceutics* 5(4):487–495.

Doherty, S. P., D. B. Buchholz et al. 2006. Semi-continuous production of multiwalled carbon nanotubes using magnetic field assisted arc furnace. *Carbon* 44(8):1511–1517.

Donaldson, K., R. Aitken et al. 2006. Carbon nanotubes: A review of their properties in relation to pulmonary toxicology and workplace safety. *Toxicology Science* 92(1):5–22.

Donaldson, K., F. Murphy et al. 2011. Identifying the pulmonary hazard of high aspect ratio nanoparticles to enable their safety-by-design. *Nanomedicine (London)* 6(1):143–156.

Dresselhaus, M. S. 2004. Applied physics: Nanotube antennas. *Nature* 432(7020):959–960.

Dresselhaus, M. S., G. Dresselhaus et al. 1995. Physics of carbon nanotubes. *Carbon* 33(7):883–891.

Dumitrica, T., M. E. Garcia et al. 2004. Selective cap opening in carbon nanotubes driven by laser-induced coherent phonons. *Physical Review Letters* 92(11):117401.

Dyke, C. A., and J. M. Tour. 2004. Overcoming the insolubility of carbon nanotubes through high degrees of sidewall functionalization. *Chemistry* 10(4):812–817.

Ebbesen, T. W., and P. M. Ajayan. 1992. Large-scale synthesis of carbon nanotubes. *Nature* 358(6383):220–222.

Esfarjani, K., A. A. Farajian et al. 1999. Electronic and transport properties of N-P doped nanotubes. *Applied Physics Letters* 74(1):79–81.

Falk, M. H., and R. D. Issels. 2001. Hyperthermia in oncology. *International Journal of Hyperthermia* 17(1):1–18.

Feazell, R. P., N. Nakayama-Ratchford et al. 2007. Soluble single-walled carbon nanotubes as longboat delivery systems for platinum(IV) anticancer drug design. *Journal of the American Chemical Society* 129(27):8438–8439.

Fisher, J. W., S. Sarkar et al. 2010. Photothermal response of human and murine cancer cells to multiwalled carbon nanotubes after laser irradiation. *Cancer Research* 70(23):9855–9864.

Fujii, M., X. Zhang et al. 2005. Measuring the thermal conductivity of a single carbon nanotube. *Physics Review Letters* 95(6):065502.

Gannon, C. J., P. Cherukuri et al. 2007. Carbon nanotube-enhanced thermal destruction of cancer cells in a noninvasive radiofrequency field. *Cancer* 110(12):2654–2665.

Ge, Z. B., Y. J. Kang et al. 2005. Thermal transport in Au-core polymer–shell nanoparticles. *Nano Letters* 5(3):531–535.

Geohegan, D. B., A. A. Puretzky et al. 2007. In situ time-resolved measurements of carbon nanotube and nanohorn growth. *Physica Status Solidi (B)* 244(11):3944–3949.

Ghosh, S., S. Dutta et al. 2009. Increased heating efficiency and selective thermal ablation of malignant tissue with DNA-encased multiwalled carbon nanotubes. *ACS Nano* 3(9):2667–2673.

Gnyawali, S. C., Y. Chen et al. 2008. Temperature measurement on tissue surface during laser irradiation. *Medical & Biological Engineering & Computing* 46(2):159–168.

Gottesman, M. M. 2002. Mechanisms of cancer drug resistance. *Annual Review of Medicine* 53(1):615–627.

Hagen, A., G. Moos et al. 2004. Electronic structure and dynamics of optically excited single-wall carbon nanotubes. *Applied Physics A Materials Science & Processing* 78(8):1137–1145.

Hanson, G. W. 2005. Fundamental transmitting properties of carbon nanotube antennas. *Ieee Transactions on Antennas and Propagation* 53(11):3426–3435.

Hartland, G. V., M. Hu et al. 2004. Investigation of the properties of gold nanoparticles in aqueous solution at extremely high lattice temperatures. *Chemical Physics Letters* 391(4–6): 220–225.

Hartman, K. B., S. Laus et al. 2008. Gadonanotubes as ultrasensitive pH-smart probes for magnetic resonance imaging. *Nano Letters* 8(2):415–419.

Hata, K., D. N. Futaba et al. 2004. Water-assisted highly efficient synthesis of impurity-free single-walled carbon nanotubes. *Science* 306(5700):1362–1364.

Hone, J. 2000. Electrical and thermal transport properties of magnetically aligned single wall carbon nanotube films. *Applied Physics Letters* 77(5):666.

Hone, J., M. C. Llaguno et al. 2000. Electrical and thermal transport properties of magnetically aligned single walt carbon nanotube films. *Applied Physics Letters* 77(5):666–668.

Huang, N., H. Wang et al. 2010. Single-wall carbon nanotubes assisted photothermal cancer therapy: Animal study with a murine model of squamous cell carcinoma. *Lasers in Surgery and Medicine* 42(9):638–648.

Iancu, C., and L. Mocan. 2011. Advances in cancer therapy through the use of carbon nanotube-mediated targeted hyperthermia. *International Journal of Nanomedicine* 6:1675–1684.

Iancu, C., L. Mocan et al. 2011. Enhanced laser thermal ablation for the *in vitro* treatment of liver cancer by specific delivery of multiwalled carbon nanotubes functionalized with human serum albumin. *International Journal of Nanomedicine* 6:129–141.

Iijima, S. 1991. Helical microtubules of graphitic carbon. *Nature* 354(6348):56–58.

Imholt, T. J., C. A. Dyke et al. 2003. Nanotubes in microwave fields: Light emission, intense heat, outgassing, and reconstruction. *Chemistry of Materials* 15(21):3969–3970.

Javey, A., J. Guo et al. 2003. Ballistic carbon nanotube field-effect transistors. *Nature* 424(6949):654–657.

Jia, G., H. F. Wang et al. 2005. Cytotoxicity of carbon nanomaterials: Single-wall nanotube, multi-wall nanotube, and fullerene. *Environmental Science & Technology* 39(5): 1378–1383.

Jordan, C. T., M. L. Guzman et al. 2006. Cancer stem cells. *New England Journal of Medicine* 355(12):1253–1261.

Kaiser, J. P., M. Roesslein et al. 2011. Carbon nanotubes — curse or blessing. *Current Medicinal Chemistry* 18(14):2115–2128.

Kam, N. W., and H. Dai. 2005. Carbon nanotubes as intracellular protein transporters: Generality and biological functionality. *Journal of the American Chemical Society* 127(16):6021–6026.

Kam, N. W., M. O'Connell et al. 2005. Carbon nanotubes as multifunctional biological transporters and near-infrared agents for selective cancer cell destruction. *Proceedings of the National Academy of Sciences of the United States of America* 102(33):11600–11605.

Kamen, B. A., and A. K. Smith. 2004. A review of folate receptor alpha cycling and 5-methyltetrahydrofolate accumulation with an emphasis on cell models in vitro. *Advanced Drug Delivery Reviews* 56(8):1085–1097.

Kanemitsu, Y. 2011. Excitons in semiconducting carbon nanotubes: Diameter-dependent photoluminescence spectra. *Physical Chemistry Chemical Physics* 13(33):14879–14888.

Kang, B., D. Yu et al. 2009. Cancer-cell targeting and photoacoustic therapy using carbon nanotubes as bomb agents. *Small* 5(11):1292–1301.

Kangasniemi, M., R. J. McNichols et al. 2004. Thermal therapy of canine cerebral tumors using a 980 nm diode laser with MR temperature-sensitive imaging feedback. *Lasers in Surgery and Medicine* 35(1):41–50.

Keblinski, P., D. G. Cahill et al. 2006. Limits of localized heating by electromagnetically excited nanoparticles. *Journal of Applied Physics* 100(5):054305–054309.

Kempa, K. 2002. Gapless plasmons in carbon nanotubes and their interactions with phonons. *Physical Review* B66(19): 195406–195410.

Keren, K., R. S. Berman et al. 2003. DNA-templated carbon nanotube field-effect transistor. *Science* 302(5649):1380–1382.

Kim, K. Y. 2007. Nanotechnology platforms and physiological challenges for cancer therapeutics. *Nanomedicine* 3(2):103–110.

Kim, P., L. Shi et al. 2001. Thermal transport measurements of individual multiwalled nanotubes. *Physics Review Letters* 87(21):215502.

Klingeler, R., S. Hampel et al. 2008. Carbon nanotube based biomedical agents for heating, temperature sensoring and drug delivery. *International Journal of Hyperthermia* 24(6):496–505.

Kolosnjaj, J., H. Szwarc et al. 2007. Toxicity studies of carbon nanotubes. *Bio-Applications of Nanoparticles* 620:181–204.

Kong, L. B., S. Li et al. 2010. Electrically tunable dielectric materials and strategies to improve their performances. *Progress in Materials Science* 55(8):840–893.

Konig, K. 2000. Multiphoton microscopy in life sciences. *Journal of Microscopy* 200(Pt 2):83–104.

Kostarelos, K. 2008. The long and short of carbon nanotube toxicity. *Nature Biotechnology* 26(7):774–776.

Kostarelos, K., L. Lacerda et al. 2007. Cellular uptake of functionalized carbon nanotubes is independent of functional group and cell type. *Nature Nanotechnol* 2(2):108–113.

Kostarelos, K., A. Bianco et al. 2009. Promises, facts and challenges for carbon nanotubes in imaging and therapeutics. *Nature Nanotechnol* 4(10):627–633.

Koyama, S., Y. A. Kim et al. 2009. In vivo immunological toxicity in mice of carbon nanotubes with impurities. *Carbon* 47(5):1365–1372.

Kratz, F. 2010. Albumin, a versatile carrier in oncology. *International Journal of Clinical Pharmacology and Therapeutics* 48(7):453–455.

Lacerda, L., A. Bianco et al. 2006. Carbon nanotubes as nanomedicines: From toxicology to pharmacology. *Advanced Drug Delivery Reviews* 58(14):1460–1470.

Lam, C. W., J. T. James et al. 2006. A review of carbon nanotube toxicity and assessment of potential occupational and environmental health risks. *Critical Reviews in Toxicology* 36(3):189–217.

Leamon, C. P., and J. A. Reddy. 2004. Folate-targeted chemotherapy. *Advanced Drug Delivery Reviews* 56(8):1127–1141.

Lee, J., S. Kotliarova et al. 2006. Tumor stem cells derived from glioblastomas cultured in bFGF and EGF more closely mirror the phenotype and genotype of primary tumors than do serum-cultured cell lines. *Cancer Cell* 9(5):391–403.

Lehman, J. H., M. Terrones et al. 2011. Evaluating the characteristics of multiwall carbon nanotubes. *Carbon* 49(8): 2581–2602.

Levi-Polyachenko, N. H., E. J. Merkel et al. 2009. Rapid photothermal intracellular drug delivery using multiwalled carbon nanotubes. *Molecular Pharmaceutics* 6(4):1092–1099.

Li, S. D., and L. Huang. 2008. Pharmacokinetics and biodistribution of nanoparticles. *Molecular Pharmaceutics* 5(4):496–504.

Liu, C. H., and S. S. Fan. 2005. Effects of chemical modifications on the thermal conductivity of carbon nanotube composites. *Applied Physics Letters* 86(12):17541–17546.

Liu, Y. Q., and L. Gao. 2005. A study of the electrical properties of carbon nanotube-NiFe$_2$O$_4$ composites: Effect of the surface treatment of the carbon nanotubes. *Carbon* 43(1):47–52.

Liu, Z., W. Cai et al. 2007. In vivo biodistribution and highly efficient tumour targeting of carbon nanotubes in mice. *Nature Nanotechnology* 2(1):47–52.

Liu, Z., K. Chen et al. 2008. Drug delivery with carbon nanotubes for *in vivo* cancer treatment. *Cancer Research* 68(16):6652–6660.

Liu, Z., S. M. Tabakman et al. 2009. Preparation of carbon nanotube bioconjugates for biomedical applications. *Nature Protocols* 4(9):1372–1382.

Liu, Z. W., A. Bushmaker et al. 2011. Thermal emission spectra from individual suspended carbon nanotubes. *ACS Nano* 5(6):4634–4640.

Liu, X. W., H. Q. Tao et al. 2011. Optimization of surface chemistry on singled-walled carbon nanotubes for *in vivo* photothermal ablation of tumors. *Biomaterials* 32(1):144–151

Maeda, H., J. Wu et al. 2000. Tumor vascular permeability and the EPR effect in macromolecular therapeutics: A review. *Journal of Controlled Release* 65(1–2):271–284.

Marches, R., P. Chakravarty et al. 2009. Specific thermal ablation of tumor cells using single-walled carbon nanotubes targeted by covalently-coupled monoclonal antibodies. *International Journal of Cancer* 125(12):2970–2977.

Marches, R., C. Mikoryak et al. 2011. The importance of cellular internalization of antibody-targeted carbon nanotubes in the photothermal ablation of breast cancer cells. *Nanotechnology* 22(9):095101.

Markovic, Z. M., L. M. Harhaji-Trajkovic et al. 2011. In vitro comparison of the photothermal anticancer activity of graphene nanoparticles and carbon nanotubes. *Biomaterials* 32(4):1121–1129.

Mayer, A., M. Vadon et al. 2009. The role of nanoparticle size in hemocompatibility. *Toxicology* 258(2–3):139–147.

McDevitt, M. R., D. Chattopadhyay et al. 2007. Tumor targeting with antibody-functionalized, radiolabeled carbon nanotubes. *Journal of Nuclear Medicine* 48(7):1180–1189.

Meng, J., J. Meng et al. 2008. Carbon nanotubes conjugated to tumor lysate protein enhance the efficacy of an antitumor immunotherapy. *Small* 4(9):1364–1370.

Milne, W. I., K. B. K. Teo et al. 2004. Carbon nanotubes as field emission sources. *Journal of Materials Chemistry* 14(6):933–943.

Moon, H. K., S. H. Lee et al. 2009. In vivo near-infrared mediated tumor destruction by photothermal effect of carbon nanotubes. *ACS Nano* 3(11):3707–3713.

Nakashima, N., and T. Fujigaya. 2007. Fundamentals and applications of soluble carbon nanotubes. *Chemistry Letters* 36(6):692–697.

Nikfarjam, M., V. Muralidharan et al. 2005. Mechanisms of focal heat destruction of liver tumors. *Journal of Surgical Research* 127(2):208–223.

Ning, S., S. Lu et al. 2007. Integrated molecular targeting of IGF1R and HER2 surface receptors and destruction of breast cancer cells using single wall carbon nanotubes. *Nanotechnology* 18(31):315101.

Nobuhito, I. et al. 2007. Synthesis-condition dependence of carbon nanotube growth by alcohol catalytic chemical vapor deposition method. *Science and Technology of Advanced Materials* 8(4):292.

O'Neal, D. P., L. R. Hirsch et al. 2004. Photo-thermal tumor ablation in mice using near infrared-absorbing nanoparticles. *Cancer Letters* 209(2):171–176.

Osman, M. A., and D. Srivastava. 2001. Temperature dependence of the thermal conductivity of single-wall carbon nanotubes. *Nanotechnology* 12(1):21–24.

Panchapakesan, B., S. Lu et al. 2005. Single-wall carbon nanotube nanobomb agents for killing breast cancer cells. *NanoBioTechnology* 1(2):133–139.

Pantarotto, D., C. D. Partidos et al. 2003. Immunization with peptide-functionalized carbon nanotubes enhances virus-specific neutralizing antibody responses. *Chemistry & Biology* 10(10):961–966.

Pantarotto, D., R. Singh et al. 2004a. Functionalized carbon nanotubes for plasmid DNA gene delivery. *Angewandte Chemie. International Edition in English* 43(39):5242–5246.

Pantarotto, D., J. P. Briand et al. 2004b. Translocation of bioactive peptides across cell membranes by carbon nanotubes. *Chemical Communications (Cambridge)* 1:16–17.

Park, E. J., W. S. Cho et al. 2009. Pro-inflammatory and potential allergic responses resulting from B cell activation in mice treated with multi-walled carbon nanotubes by intratracheal instillation. *Toxicology* 259(3):113–121.

Patchkovskii, S., J. S. Tse et al. 2005. Graphene nanostructures as tunable storage media for molecular hydrogen. *Proceedings of the National Academy of Sciences of the United States of America* 102(30):10439–10444.

Patri, A. K., A. Myc et al. 2004. Synthesis and *in vitro* testing of J591 antibody–dendrimer conjugates for targeted prostate cancer therapy. *Bioconjugate Chemistry* 15(6):1174–1181.

Picou, L., C. McMann et al. 2010. Spatio-temporal thermal kinetics of *in situ* MWCNT heating in biological tissues under NIR laser irradiation. *Nanotechnology* 21(43):435101.

Plech, A., S. Kurbitz et al. 2003. Time-resolved X-ray diffraction on laser-excited metal nanoparticles. *Europhysics Letters* 61(6):762–768.

Pommier, Y., O. Sordet et al. 2004. Apoptosis defects and chemotherapy resistance: Molecular interaction maps and networks. *Oncogene* 23(16):2934–2949.

Pop, E., D. Mann et al. 2006. Thermal conductance of an individual single-wall carbon nanotube above room temperature. *Nano Letters* 6(1):96–100.

Puretzky, A. A., D. B. Geohegan et al. 2000. Dynamics of single-wall carbon nanotube synthesis by laser vaporization.

Applied Physics A: Materials Science & Processing 70(2): 153–160.

Pustovalov, V. K., and V. A. Babenko. 2004. Optical properties of gold nanoparticles at laser radiation wavelengths for laser applications in nanotechnology and medicine. *Laser Physics Letters* 1(10):516–520.

Qu, G. B., Y. H. Bai et al. 2009. The effect of multiwalled carbon nanotube agglomeration on their accumulation in and damage to organs in mice. *Carbon* 47(8):2060–2069.

Radomski, A., P. Jurasz et al. 2005. Nanoparticle-induced platelet aggregation and vascular thrombosis. *British Journal of Pharmacology* 146(6):882–893.

Raffa, V., G. Ciofani et al. 2010. Physicochemical properties affecting cellular uptake of carbon nanotubes. *Nanomedicine* 5(1):89–97.

Reulet, B., A. Y. Kasumov et al. 2000. Acoustoelectric effects in carbon nanotubes. *Physical Review Letters* 85(13):2829–2832.

Richard, C., B. T. Doan et al. 2008. Noncovalent functionalization of carbon nanotubes with amphiphilic Gd3+ chelates: Toward powerful T-1 and T-2 MRI contrast agents. *Nano Letters* 8(1):232–236.

Rieke, V., and K. B. Pauly. 2008. MR thermometry. *Journal of Magnetic Resonance Imaging* 27(2):376–390.

Robinson, J. T., K. Welsher et al. 2010. High performance *in vivo* near-IR (>1 mum) imaging and photothermal cancer therapy with carbon nanotubes. *Nano Research* 3(11):779–793.

Ruenraroengsak, P., J. M. Cook et al. 2010. Nanosystem drug targeting: Facing up to complex realities. *Journal of Controlled Release* 141(3):265–276.

Ryman-Rasmussen, J. P., E. W. Tewksbury et al. 2009. Inhaled multiwalled carbon nanotubes potentiate airway fibrosis in murine allergic asthma. *American Journal of Respiratory Cell and Molecular Biology* 40(3):349–358.

Saito, R., M. Hofmann et al. 2011. Raman spectroscopy of graphene and carbon nanotubes. *Advances in Physics* 60(3):413–550.

Saito, Y., K. Hata et al. 2002. Field emission of carbon nanotubes and its application as electron sources of ultra-high luminance light-source devices. *Physica B-Condensed Matter* 323(1–4):30–37.

Salvador-Morales, C., W. W. Gao et al. 2009. Multifunctional nanoparticles for prostate cancer therapy. *Expert Review of Anticancer Therapy* 9(2):211–221.

Sayes, C. M., F. Liang et al. 2006. Functionalization density dependence of single-walled carbon nanotubes cytotoxicity in vitro. *Toxicology Letters* 161(2):135–142.

Schipper, M. L., N. Nakayama-Ratchford et al. 2008. A pilot toxicology study of single-walled carbon nanotubes in a small sample of mice. *Nature Nanotechnology* 3(4):216–221.

Schonenberger, C., and L. Forro. 2000. Mulitwall carbon nanotubes. *Physics World* 13:37–41.

Selvi, B. R., D. Jagadeesan et al. 2008. Intrinsically fluorescent carbon nanospheres as a nuclear targeting vector: Delivery of membrane-impermeable molecule to modulate gene expression in vivo. *Nano Letters* 8(10):3182–3188.

Semberova, J., S. H. De Paoli Lacerda et al. 2009. Carbon nanotubes activate blood platelets by inducing extracellular Ca2+ influx sensitive to calcium entry inhibitors. *Nano Letters* 9(9):3312–3317.

Shi Kam, N. W., T. C. Jessop et al. 2004. Nanotube molecular transporters: Internalization of carbon nanotube–protein conjugates into Mammalian cells. *Journal of the American Chemical Society* 126(22):6850–6851.

Shvedova, A. A., E. R. Kisin et al. 2005. Unusual inflammatory and fibrogenic pulmonary responses to single-walled carbon nanotubes in mice. *American Journal of Physiology-Lung Cellular and Molecular Physiology* 289(5):L698–L708.

Singh, R., D. Pantarotto et al. 2005. Binding and condensation of plasmid DNA onto functionalized carbon nanotubes: Toward the construction of nanotube-based gene delivery vectors. *Journal of the American Chemical Society* 127(12): 4388–4396.

Singh, R., D. Pantarotto et al. 2006. Tissue biodistribution and blood clearance rates of intravenously administered carbon nanotube radiotracers. *Proceedings of the National Academy of Sciences of the United States of America* 103(9): 3357–3362.

Singh, S. K., C. Hawkins et al. 2004. Identification of human brain tumour initiating cells. *Nature* 432(7015):396–401.

Sitharaman, B., K. R. Kissell et al. 2005. Superparamagnetic gadonanotubes are high-performance MRI contrast agents. *Chemical Communications* (31):3915–3917.

Smart, S. K., A. I. Cassady et al. 2006. The biocompatibility of carbon nanotubes. *Carbon* 44(6):1034–1047.

Stella, G. M. 2011. Carbon nanotubes and pleural damage: Perspectives of nanosafety in the light of asbestos experience. *Biointerphases* 6(2):P1–P17.

Stoyanov, S. R., A. V. Titov et al. 2009. Transition metal and nitrogen doped carbon nanostructures. *Coordination Chemistry Reviews* 253(23–24):2852–2871.

Talanov, V. S., C. A. Regino et al. 2006. Dendrimer-based nanoprobe for dual modality magnetic resonance and fluorescence imaging. *Nano Letters* 6(7):1459–1463.

Tasis, D., N. Tagmatarchis et al. 2003. Soluble carbon nanotubes. *Chemistry-A European Journal* 9(17):4001–4008.

Terrones, M., A. G. Souza et al. 2008. Doped carbon nanotubes: Synthesis, characterization and applications. *Carbon Nanotubes* 111:531–566.

Torti, S. V., F. Byrne et al. 2007. Thermal ablation therapeutics based on CN(x) multi-walled nanotubes. *Int J Nanomedicine* 2(4):707–714.

Uo, M., T. Akasaka et al. 2011. Toxicity evaluations of various carbon nanomaterials. *Dental Materials Journal* 30(3): 245–263.

Van Lehn, R. C., C. E. Sing et al. 2010. Multidimensional targeting: Using physical and chemical forces in unison. *Current Pharmaceutical Biotechnology* 11(4):320–332.

Vazquez, E., and M. Prato. 2009. Carbon nanotubes and microwaves: Interactions, responses, and applications. *ACS Nano* 3(12):3819–3824.

Vitetta, E. S., R. Marches et al. 2011. The importance of cellular internalization of antibody-targeted carbon nanotubes in the photothermal ablation of breast cancer cells. *Nanotechnology* 22(9):095101–095110.

Wadhawan, A. 2003. Nanoparticle-assisted microwave absorption by single-wall carbon nanotubes. *Applied Physics Letters* 83(13):2683.

Wang, C. H., Y. J. Huang et al. 2009. In vitro photothermal destruction of neuroblastoma cells using carbon nanotubes conjugated with GD2 monoclonal antibody. *Nanotechnology* 20(31):315101.

Wang, C. H., S. H. Chiou et al. 2011. Photothermolysis of glioblastoma stem-like cells targeted by carbon nanotubes conjugated with CD133 monoclonal antibody. *Nanomedicine* 7(1):69–79.

Wang, X., J. Ren et al. 2008. Targeted RNA interference of cyclin A2 mediated by functionalized single-walled carbon nanotubes induces proliferation arrest and apoptosis in chronic myelogenous leukemia K562 cells. *ChemMedChem* 3(6):940–945.

Wang, Y. 2004. Receiving and transmitting light-like radio waves: Antenna effect in arrays of aligned carbon nanotubes. *Applied Physics Letters* 85(13):2607.

Warheit, D. B. 2006. What is currently known about the health risks related to carbon nanotube exposures? *Carbon* 44(6):1064–1069.

Weisman, R. B. 2003. Carbon nanotubes — Four degrees of separation. *Nature Materials* 2(9):569–570.

Weissleder, R. 2001. A clearer vision for *in vivo* imaging. *Nature Biotechnology* 19(4):316–317.

Welsher, K., Z. Liu et al. 2008. Selective probing and imaging of cells with single walled carbon nanotubes as near-infrared fluorescent molecules. *Nano Letters* 8(2):586–590.

Whitney, J. R., S. Sarkar et al. 2011. Single walled carbon nanohorns as photothermal cancer agents. *Lasers in Surgery and Medicine* 43(1):43–51.

Wick, P., P. Manser et al. 2007. The degree and kind of agglomeration affect carbon nanotube cytotoxicity. *Toxicology Letters* 168(2):121–131.

Wong, S. S., E. Joselevich et al. 1998. Covalently functionalized nanotubes as nanometre-sized probes in chemistry and biology. *Nature* 394(6688):52–55.

Xiao, Y., X. Gao et al. 2009. Anti-HER2 IgY antibody-functionalized single-walled carbon nanotubes for detection and selective destruction of breast cancer cells. *BMC Cancer* 9:351.

Yang, S. T., X. Wang et al. 2008. Long-term accumulation and low toxicity of single-walled carbon nanotubes in intravenously exposed mice. *Toxicology Letters* 181(3):182–189.

Ye, Z., W. D. Deering et al. 2006. Microwave absorption by an array of carbon nanotubes: A phenomenological model. *Physical Review B* 74(7):075425.

Yi, W., L. Lu et al. 1999. Linear specific heat of carbon nanotubes. *Physics Review B* 59(14):R9015–R9018.

Yin, L. C., H. M. Cheng et al. 2011. Fermi level dependent optical transition energy in metallic single-walled carbon nanotubes. *Carbon* 49(14):4774–4780.

Yoo, J. W., E. Chambers et al. 2010. Factors that control the circulation time of nanoparticles in blood: Challenges, solutions and future prospects. *Current Pharmaceutical Design* 16(21):2298–2307.

You, C. C., A. Verma et al. 2006. Engineering the nanoparticle-biomacromolecule interface. *Soft Matter* 2(3):190–204.

Yu, X., B. Munge et al. 2006. Carbon nanotube amplification strategies for highly sensitive immunodetection of cancer biomarkers. *Journal of the American Chemical Society* 128(34):11199–11205.

Zeppernick, F., R. Ahmadi et al. 2008. Stem cell marker CD133 affects clinical outcome in glioma patients. *Clinical Cancer Research* 14(1):123–129.

Zhang, J. F., J. C. Ge et al. 2010. In vitro and *in vivo* studies of single-walled carbon nanohorns with encapsulated metallofullerenes and exohedrally functionalized quantum dots. *Nano Letters* 10(8):2843–2848.

Zhang, M., T. Murakami et al. 2008. Fabrication of ZnPc/protein nanohorns for double photodynamic and hyperthermic cancer phototherapy. *Proceedings of the National Academy of Sciences of the United States of America* 105(39):14773–14778.

Zhang, W., Z. Zhang et al. 2011. The application of carbon nanotubes in target drug delivery systems for cancer therapies. *Nanoscale Res Lett* 6:555.

Zharov, V. P., E. N. Galitovskaya et al. 2005. Synergistic enhancement of selective nanophotothermolysis with gold nanoclusters: Potential for cancer therapy. *Lasers in Surgery and Medicine* 37(3):219–226.

Zhou, F. 2009. Cancer photothermal therapy in the near-infrared region by using single-walled carbon nanotubes. *Journal of Biomedical Optics* 14(2):021009.

Zhou, F. F., D. Xing et al. 2010. New insights of transmembranal mechanism and subcellular localization of noncovalently modified single-walled carbon nanotubes. *Nano Letters* 10(5):1677–1681.

Zhou, F. F., S. N. Wu et al. 2011. Mitochondria-targeting single-walled carbon nanotubes for cancer photothermal therapy. *Small* 7(19):2727–2735.

V

Future Outlook

18

Regulatory Issues for Clinical Translations

J. Donald Payne
Oncolix

18.1 Introduction

It is an understatement to describe medical product development as a highly complex process. Once research has established "proof of principle," great care should be taken to comply with the labyrinth of requirements of regulatory authorities to develop a clinically useful product. The written and verbal guidance of regulatory authorities is applicable to each step along the way.

It is helpful to remember three points for this process. First, it is helpful to start with "the end in mind"—what is the appropriate patient population, how will the product be packaged and administered to the patient, how will the physician measure effectiveness, etc. Second, the clinical translation pathway is not a straight line, and there are decision points and alterations along the way, and higher levels of precision and control are required at the end than at the beginning. Third, the *in vivo* behavior of the product, including its biodistribution and mechanism of action, will have a significant impact on the clinical translation requirements.

Additionally, regulatory oversight of nanotechnology-based products is evolving. As the properties, and/or potential safety risks, of this class become more evident, the regulatory requirements may change.

As discussed in this book, the radiation oncology application generally is composed of an energy source and a nanocomponent. The energy source may be either an existing therapy (e.g., ionizing radiation) for which the nanocomponent alters the effect (e.g., radiation dose enhancement or radioprotection). Alternatively, the outcome may be a new medical application resulting from the combination of the energy and nanocomponent (e.g., ablation through alternating magnetic fields and superparamagnetic iron oxide particles). The application may also require new methods of control, including modifications of existing software or new treatment planning methods. The approach to clinical translation must consider all these aspects.

As the medical use may be a modification of an existing therapy or an entirely new treatment, the regulatory issues arising from each require careful consideration. This chapter discusses some, but not necessarily all, of the regulatory issues important in clinical translation. It is impractical to discuss the issues pertinent to every market, so the discussion is limited to the United States and the European Union (EU).

18.2 Regulatory Classification of the Product in the United States

In the United States, medical products are generally classified as drugs, medical devices, or a combination product (both a drug and a device). The classification as either a device or drug will have a significant effect on the clinical translation activities, affecting the nature and scope of toxicity testing before human use as well as the clinical trial process. The mechanism of action of the therapy is important in determining the regulatory pathway, particularly if classification as a medical device is desired.

The U.S. Food and Drug Administration (FDA) is the final arbiter of the application of its own rules, but the principal distinction between a drug and medical device is included in Section 201 of 21 U.S.C. 321. A medical device "does *not* achieve any of its primary intended purposes *through chemical action* within or on the body of man or other animals and…is *not dependent upon being metabolized* for the achievement of any of its primary intended purposes" (emphasis added) (FDA-CDRH). Accordingly, if the therapy does not work through chemical action or its metabolism, it *may* be classified as a medical device. However, the FDA may make simplifying assumptions and classify similar therapies together as either a drug or device, so classification decisions should be made through discussions with FDA.

Nanotechnology-based applications in radiation oncology generally are a "system" involving an energy source (such as ionizing radiation, alternating magnetic fields, or near-infrared energy) as well as a nanomaterial. The nanomaterial may be an energy-transducing solid particle, a vector to deliver a sensitizing drug or a similar agent, or a combination of the two, and may be passively or actively targeted to a cell by a targeting molecule. In many cases, a therapeutic benefit occurs only when the energy source and particle are combined. The regulatory classification should be determined for the entire therapeutic "system" as well as the nanocomponent, and this may result in each component receiving a different classification.

The regulatory pathway of the energy source is often previously defined. The energy source for ablation therapies such as radiofrequency ablation, high-intensity focused ultrasound, or laser ablation have been developed under the medical device guidelines. In these therapies, the energy is directly deposited in tissue, creating a thermal effect. The mechanism of action is generally mechanical or energy transduction, not chemical or metabolic. Accordingly, lasers, magnetic field generators, and radiofrequency generators are generally classified as medical devices.

The total therapeutic system as well as the nanotechnology component, however, may require a *de novo* classification determination by the regulatory authority. If the system (energy source plus nanocomponent) does not act by chemical interaction with the body or through its metabolism, the system and the nanocomponent would technically meet the definition of a medical device.

As an example, consider photothermal ablation using gold-based nanoparticles. Certain particles used in photothermal ablation (e.g., nanoshells) have a gold exterior and are considered inert. The mechanism of action in photothermal therapy is the transduction of energy from a near-infrared laser into heat to thermally ablate the target cells. This mechanism of action does not involve chemical or metabolic activity by the particle. Accordingly, this product has been classified by the FDA as a medical device, and a human trial of this photothermal ablation system (gold nanoshell and near-infrared laser) as a medical device has been allowed by the FDA (ClinicalTrials.gov; Hirsch et al. 2003; O'Neal et al. 2004; Schwartz et al. 2009).

It should be noted that the nanomaterial in this clinical trial does not have an active targeting component and relies on the enhanced permeability and retention (EPR) effect for tumor accumulation (Maeda 2001). The addition of targeting molecules, such as an antibody or peptide with an affinity for a tumor cell surface molecule, complicates the regulatory classification.

The FDA has established an Office of Combination Products (OCP) to determine the classification of such products. In addition, OCP will determine which group at FDA (drug, device or biologics group) will be the lead center for regulation of the product (FDA-OCP).

If the mechanism of action of the nanocomponent is by chemical or metabolic action, it is likely that this component will be classified as a drug. Hypothetically, a radiation sensitizing particle that acts to increase oxygenation at a tumor site (say, by release of a chemical that modifies hemoglobin) may be designated by the FDA as a drug or a combination product.

The classification of a product as a medical device can only be determined with certainty through discussions with the FDA. One company has announced its intent to commence a clinical trial for a radiation-enhancing nanoparticle composed of HfO_2 directly injected into the tumor. This high Z-particle serves to enhance the effect of ionizing radiation in a manner described in previous chapters. Although the details of regulatory discussions were not publicly available, the company announced that the FDA will classify the particle as a drug, but that the particle will be classified as a medical device in the EU (http://www.nanobiotix.com/news/).

If the system (including the energy source) and the nanocomponent are principally regulated by FDA as a medical device, an Investigational Device Exemption (IDE) filing may be required to conduct a clinical trial. If the system and nanocomponent are principally regulated as a drug, an Investigational New Drug (IND) filing may be required to conduct a clinical trial. Each process has different requirements, each with an evidence-based approach to evaluating potential toxicity effects.

18.3 Regulatory Classification of the Product in the EU

In the EU, nanotechnology-based products are generally classified either as medicinal products (i.e., drugs) or as medical devices. As in the United States, the mechanism of action of the therapy is important in determining the regulatory pathway, particularly if classification as a medical device is desired. Council Directive 93/42/EEC of 14 June 1993 states that a medical device "*does not* achieve its principal intended action in or on the human body by pharmacological, immunological or metabolic means, but…may be assisted in its function by such means" (emphasis added) (EU-Directive_93/42_EEC). Although the EU's definition of a medical device is similar to that of the United States, the same product can be classified as a drug in one jurisdiction and a medical device in the other. In the EU, the entities performing the classification and regulation of medical devices are with private, licensed "Notified Bodies" and not a government agency. However, the decisions of the Notified Body can be overturned by a government agency, so a governmental determination may be wise in the event of ambiguity.

Products that are *not* clearly medical devices or medicinal products (drugs) are referred to as "borderline," and this category also includes products that are clearly a combination of a drug and device. (See MEDDEV 2.1/3 rev.3 "Borderline products, drug-delivery products and medical devices incorporating, as integral part, an ancillary medicinal substance or an ancillary human blood derivative." December 2009.) As with U.S. combination products, the EU also requires that in making a determination of classification as a medical device or medicinal product, "particular account shall be taken of the principal mode of action of the product" (EC-Directorate F-Unit F3). A formal request for determination of the classification of the product as a device or medicinal product may be requested from one of the member states of the EU.

One nanoparticle-based thermal ablation product has been approved for marketing in the EU as a medical device. MagForce Nanotechnologies AG received approval in 2010 to market its Nano Cancer® therapy, which uses an alternating magnetic field to generate heat from magnetic nanoparticles injected into brain tumors. These magnetic particles are generally considered to be inert, and the mechanism of action is by transduction of the alternating magnetic field into heat, a physical effect rather than chemical or metabolic. MagForce received separate device approvals for its alternating magnetic field generator, for the planning software used in treatment, and for the magnetic particles (http://www.magforce.de/en/home.html).

Similar to the United States, the use of a nanoparticle to deliver a drug with a pharmacological, immunological, or metabolic effect likely would be classified as a drug.

18.4 Preclinical Toxicity Testing of Medical Devices

The safety of the medical device must be established in animal models before conducting a clinical trial. The importance of the accurate classification of the product as a drug or medical device becomes readily apparent as one considers the range of pharmacology and toxicity testing required to demonstrate safety. The absence of chemical action or metabolization narrows the scope of testing required for a medical device; in contrast, the chemical effect or metabolism of a drug necessitates more extensive preclinical testing. First, we discuss the testing of medical devices.

The United States and the EU have substantially harmonized the required toxicity testing of medical devices, with both following the provisions of ISO10993 "Biological Evaluation of Medical Devices." However, as a notable exception, the U.S. FDA also requires additional tests for products with certain characteristics (FDA-CDRH).

These standards provide a risk-based approach to testing, with the scope of testing dependent on the length of the body's exposure to the medical device and the physiological systems exposed. The regulations define contact with the body as either limited (less than 24 h), prolonged (24 h to 30 days), or permanent (>30 days). The longer the exposure of the body to the medical device, the more extensive the preclinical testing required.

Two factors will generally have a significant effect on the testing required for nanoparticles classified as devices. The first is the period of exposure of the body to the nanoparticle, which is a function of the rate of degradation of the particle, if any, and route of clearance from the body (urine, feces). The second major factor is the number and extent of organ systems exposed to the nanoparticle. Exposure is generally related to route of administration (systemic, intratumoral injection) and the degree of migration or extravasation (from the blood or lymphatic systems) of the particle.

Nanoparticle-based medical devices will generally require more extensive testing than simple medical devices. Preliminary preclinical work should address certain questions in order to select the appropriate toxicology tests. Although not an exhaustive list, the following questions should be addressed:

- Do particles delivered through the blood stream or injected into tumors extravasate into normal tissue? Most nanoparticles rely on the EPR effect for delivery, and are too large for normal tissue extravasation. However, if the particles do extravasate, an evaluation of the effect on major organ systems (central nervous system, respiratory system, etc.) should be conducted. Recent literature would suggest that many nanoparticles do not extravasate, reducing the need for major organ pharmacology testing (O'Neal et al. 2004; James et al. 2007; Goodrich et al. 2010).

- Are the particles cleared from the blood through the reticuloendothelial system or urine? Many nanoparticles are too large for urine clearance, so the principal mechanism of clearance may be the liver and spleen, with macrophages playing the predominant, if not exclusive, role. If there is evidence of uptake by normal hepatic cells in addition to macrophages, additional concerns may be raised regarding liver toxicity.

- Over what time period are particles cleared from the body? The longer the retention period, the greater the need for chronic toxicity testing before entering human trials. Recent evidence indicates certain particles may be retained over long periods and are essentially "permanent" implants. If the particles are retained in the spleen, it may be appropriate to address the effect, if any, on immune function. If the particles are retained in the liver, a chronic toxicity (6 to 12 months) study in animals may be necessary to evaluate potential toxicity.

- Do the particles degrade and, if so, what are the degradation products? Some nanoparticles are composed of materials that either naturally decompose or are digested by cells. Other nanoparticles, principally the metallic-based particles, are not degraded but are subject to leaching. Upon exposure to an energy source (x-rays, magnetic fields, lasers), do the particles degrade or generate byproducts? The potential toxicity of the decomposition, degradation or leaching materials should be evaluated and may require separate testing.

If the nanomedical radiation oncology application involves an energy source, the preclinical testing program should also evaluate safety issues associated with the combination of the energy source and nanomaterial. For example, for thermal applications, separate testing of the laser, alternating magnetic field, or radiofrequency generator may have been conducted. However, the effect when combined with the nanomedicine should be evaluated. The energy dosimetry may require adjustment if the nanomedicine alters known safety levels. Additionally, the nanomedicine may create unwanted bystander effects, such as ablation of cells in clearance organs (e.g., liver or spleen) if these are within the field of energy application. The following table attempts to demonstrate the possible combinations and the potential testing required under each. This matrix is not definitive, and any testing program should be established by discussion with the appropriate regulatory agency.

| Energy Source | Nanocomponent Effect | |
	Acts Synergistically	Establishes a New Therapy
Preexisting device, no change in use	Testing may be limited to nanocomponent and combination effects	All components must be tested for the new indication
Preexisting device, new software required	Testing may be limited to nanocomponent, the new software, and combination effects	All components must be tested for the new indication
Preexisting device, new medical application	All components must be tested for the new indication	All components must be tested for the new indication
New device	All components must be tested for the new indication	All components must be tested for the new indication

Based on the preliminary data, the appropriate toxicity tests should be discussed with the regulatory body (the FDA or the selected Notified Body in the EU). These toxicity tests must be performed under appropriate quality assurance standards, or "good laboratory practices" (GLP). As a result, these tests are often performed by external vendors with established quality assurance systems.

These tests must be conducted using material that meets the minimum level of reproducibility, purity stability, and sterility as described under "Manufacturing Issues" below.

A national laboratory, the Nanotechnology Characterization Laboratory (NCL), has been established to assist in the clinical translation of nanomaterials. The NCL accepts material from researchers and commercial enterprises for testing, and conducts a rigorous series of tests involving physical characterization and toxicity. The NCL also publishes on its website various protocols related to acceptable test methods that may be used by others (http://ncl.cancer.gov/).

18.5 Preclinical Toxicity Testing of Drugs

As with medical devices, the FDA and the EU have substantially harmonized the preclinical testing required for drugs (or medicinal products). An international cooperative group, the International Conference on Harmonisation of Technical Requirements for Registration of Pharmaceuticals for Human Use (ICH), has published a series of guidelines for preclinical testing for drugs (ICH-Safety; FDA-S9 March_2010). These guidelines have generally been adopted by the FDA and published as Guidance documents.

Of particular importance to radiation oncology is Guideline S9 "Nonclinical Evaluation for AntiCancer Pharmaceuticals," dated October 29, 2009. This Guidance was adopted by the FDA in March 2010. Because of the life-threatening nature of metastatic disease, Guidance S9 generally limits the degree of nonclinical (*in vitro* and preclinical) testing required for drugs intended to treat patients with late-stage or advanced disease. Although this limitation may not be applicable to all nanomedicine radiation oncology applications, it may be pertinent to the initial clinical trials that are often restricted to late-stage patients. In certain cases, the Guidance suggests that certain studies can be deferred until after Phase I testing, and in other cases the studies may be avoided entirely. However, specific evidence or concerns noted in preliminary preclinical testing should override these limitations (FDA_95-1; ICH_S9 29Oct2009).

Key components of this Guidance document that may speed clinical testing in late stage cancer patients include the following:

- Stand-alone safety pharmacology studies need not be conducted before human trials if vital organ function is assessed in the general toxicology studies. These assessments would include the effect on cardiovascular, respiratory and central nervous systems.
- Comprehensive absorption, distribution, metabolism, and excretion studies are not necessary before human testing, and pharmacokinetic data may be limited to those factors which would facilitate dosing studies [peak plasma levels, AUC (area under the curve), and half-life].
- Reproductive toxicity and genotoxicty studies may not be necessary before clinical trials (ICH_S9 29Oct2009; FDA-S9 March_2010).

Accordingly, toxicity studies in two species, with appropriate toxicokinetic profiles and clinical evaluations, may be the minimum requirement to initiate human trials in late-stage cancer patients. To the extent that the ultimate patient population is expected to be earlier stage cancer patients, additional testing in animals may be conducted in parallel to the initial human testing.

As with medical device preclinical testing, all of these studies should be conducted under GLP.

18.6 Manufacturing Issues

Medical devices must be manufactured under a quality assurance program. The FDA has adopted Quality System Regulations (QSR), the medical device equivalent to current Good Manufacturing Practices (cGMP). Drugs and medicinal products must be manufactured under cGMP. For simplicity herein, we will refer to both QSR and cGMP as cGMP.

It is important to note that cGMP for investigational products is generally a continuum, with greater quality assurance required for marketed products than at the initial stage of preclinical testing. However, a major problem with many preclinical programs is the failure to adequately control and document the production of material used in the GLP toxicity testing. At a minimum, the material (drug or device) used in the GLP studies to support a clinical trial must be manufactured under controlled conditions, be produced under written records meeting specified criteria, and be tested for sterility and purity. Additionally, the material manufactured for the initial clinical trial must be substantially identical and manufactured under the same controls as the material used in the GLP preclinical studies.

Careful consideration should be given to manufacturing issues before commencing preclinical testing. The presence of endotoxins, process residuals, or impurities could adversely affect the outcome of toxicity tests. Additionally, if the formulation or packaging do not support a sufficient stability for potential use in a clinical trial, subsequent alterations to the product or the manufacturing process may require new GLP preclinical studies.

18.7 Clinical Testing of the Product

In the United States and the EU, a clinical trial of a medical device or drug will generally require regulatory clearance. In the United States, this clearance is granted by the FDA after the filing of an IDE or IND. In the EU, a Clinical Trial Application for a medicinal product must be filed and approved by one of the member states. Separately, an institutional review board (called an Ethics Committee in the EU) must approve the human study.

The regulatory filing requirements in both the United States and the EU include the requisite preclinical studies to establish toxicity levels, an assessment of the risks and benefits for human testing, and a written protocol describing the clinical study.

In developing the clinical protocol, key factors to consider include:

- The patient population in which the therapy will be tested
- The study endpoints and how success will be measured
- The risks of patient noncompliance or slow recruitment

18.8 Nanotechnology Environment

To date, the regulatory authorities in the United States and the EU have indicated that the existing framework for drug and device approvals is sufficient to evaluate any special risks related to nanomedical products. However, these authorities continually reevaluate this framework. The FDA has conducted workshops related to manufacturing and preclinical testing issues (FDA_Workshop 2010). The FDA recently issued a draft Guidance document, "Considering Whether an FDA-Regulated Product Involves the Application of Nanotechnology" as part of its ongoing effort to reassess the adequacy of its regulatory oversight. It is unknown whether these efforts will lead to specific regulatory requirements for nanomedical applications in general, or for radiation oncology in particular (FDA_Nano 2010).

References

ClinicalTrials.gov. Pilot Study of AuroLase(tm) Therapy in Refractory and/or Recurrent Tumors of the Head and Neck. Accessed November 28, 2011. http://clinicaltrials.gov/ct2/show/NCT00848042?term=aurolase&rank=1.

EC-Directorate F-Unit F3 MEDDEV 2. 1/3 rev 3: Guidance document—Borderline products, drug-delivery products and medical devices incorporating, as an integral part, an ancillary medicinal substance or an ancillary human blood derivative.

EU_Directive_93/42_EEC COUNCIL DIRECTIVE 93/42/EEC of 14 June 1993 concerning medical devices. C. o. t. E. Communities.

FDA-CDRH. 05/03/2009 Blue Book Memo 95-1 Use of International Standard ISO-10993, 'Biological Evaluation of Medical Devices Part 1: Evaluation and Testing.' Accessed November 28, 2011. http://www.fda.gov/MedicalDevices/DeviceRegulationandGuidance/GuidanceDocuments/ucm080735.htm.

FDA-CDRH. 03/01/2010. Is The Product A Medical Device? Accessed November 28, 2011. http://www.fda.gov/medicaldevices/deviceregulationandguidance/overview/classifyyourdevice/ucm051512.htm.

FDA-OCP. 07/15/2009. Office of Combination Products: Classification and Jurisdictional Information. Accessed November 28, 2011. http://www.fda.gov/CombinationProducts/JurisdictionalInformation/default.htm.

FDA-S9. March_2010. Guidance for Industry S9 Nonclinical Evaluation for Anticancer Pharmaceuticals.

FDA_95-1. 05/03/2009. Blue Book Memo 95-1 Use of International Standard ISO-10993, 'Biological Evaluation of Medical Devices Part 1: Evaluation and Testing.' Accessed November 28, 2011. http://www.fda.gov/MedicalDevices/DeviceRegulationandGuidance/GuidanceDocuments/ucm080735.htm.

FDA_Nano. 2010, 06/14/2011. Draft Guidance "Considering Whether an FDA-Regulated Product Involves the Application of Nanotechnology." Accessed November 28, 2011. http://www.fda.gov/RegulatoryInformation/Guidances/ucm257698.htm#_ftn1.

FDA_Workshop. 2010. Public Workshop—Medical Devices and Nanotechnology: Manufacturing, Characterization, and Biocompatibility Considerations, September 23, 2010 from http://www.fda.gov/MedicalDevices/NewsEvents/WorkshopsConferences/ucm222591.htm.

Goodrich, G. P., L. Bao et al. 2010. Photo thermal therapy in a murine colon cancer model using near-infrared absorbing gold nanorods. *Journal of Biomedical Optics* 15(1):018001-1.

Hirsch, L. R., R. J. Stafford et al. 2003. Nanoshell-mediated near-infrared thermal therapy of tumors under magnetic resonance guidance. *Proceedings of the National Academy of Sciences of the United States of America* 100(23):13549–13554.

ICH-Safety. Safety Guidelines. November 28, 2011. http://www.ich .org/products/guidelines/safety/article/safety-guidelines.html.

ICH_S9. 29Oct2009. Guideline S9: Nonclinical Evaluation for AntiCancer Pharmaceuticals.

James, W. D., L. R. Hirsch et al. 2007. Application of INAA to the build-up and clearance of gold nanoshells in clinical studies in mice. *Journal of Radioanalytical and Nuclear Chemistry* 271(2):455–459.

Maeda, H. 2001. The enhanced permeability and retention (EPR) effect in tumor vasculature: The key role of tumor-selective macromolecular drug targeting. *Advances in Enzyme Regulation* 41:189–207.

O'Neal, D. P., L. R. Hirsch et al. 2004. Photo-thermal tumor ablation in mice using near infrared-absorbing nanoparticles. *Cancer Letters* 209(2):171–176.

Schwartz, J. A., A. M. Shetty et al. 2009. Feasibility study of particle-assisted laser ablation of brain tumors in orthotopic canine model. *Cancer Research* 69(4):1659–1667.

19

Clinical Translations of Nanotechnology: Present and Future Outlook

Dev K. Chatterjee
M. D. Anderson Cancer Center

Sang Hyun Cho
Georgia Institute of Technology

Sunil Krishnan
M. D. Anderson Cancer Center

19.1 Introduction

In the past few years, there has been considerable interest in the applications of nanotechnology to radiation oncology, primarily focusing on the use of metal nanostructures as a means of enhancing the effect of radiation in a multitude of ways. In this book, we have explored this rapidly expanding field, and detailed the many different areas that nanotechnology is growing to overlap the discipline of radiation oncology. It seems probable that some, if not most, of these research efforts will develop further in the near future to impact clinical care. In this final chapter, we aim to look back on the state-of-the-art of these growing fields in order to summarize and harmonize the many offshoots of technology for a more holistic view of the whole field, and also to look forward and try to imagine what the future will hold for this multidisciplinary marriage between the technology family and the medicine family. This chapter is aimed at professionals from both sides of the divide—for the scientist as well as the clinician—wishing to keep abreast of latest developments. In keeping with this aim, details of research methodologies or nanoparticle structures have been generally omitted—the interested reader is always welcome to review the relevant chapter for more detailed information.

Today, radiation oncologists face several challenges every time they go to treat patients with localized tumors. Accurately planning and delivering a prescribed dose is challenging enough; however, the task is made several times more complicated by the inherent resistance of a fraction of cancer cells to radiation therapy (RT) and the narrow therapeutic window for enhancing the efficacy of treatment. Since normal tissue also suffers significant comorbidity from radiation exposure, methods to improve tumor targeting simultaneously with overcoming inherent radioresistance are urgently needed. Several clinical methods have gained traction—including intensity modulated RT, stereotactic RT, image-guided RT and the use of charged particle therapy. Although these have distinct advantages, there is ample room for further improvement—especially those beyond the traditional approaches. One of the most promising of these is the use of nanotechnology.

Very broadly, nanotechnology—in the present context—is almost synonymous with nanoparticles, an umbrella term that refers to particles whose greatest dimension is about 100 nm or less. These can be of a variety of shapes—from spheres to thin rods; in a variety of materials—organic and/or inorganic; can be solid, hollow, or with a core–shell structure; and usually possess specialized coatings. The small size gives these particles a large interaction cross section because of the relatively large ratio of surface area to volume and a large volume of distribution, while still allowing more significant latitude in design and composition than would be possible for single molecules. A very important aspect of this *in vivo* distribution is the spontaneous accumulation of nanoparticles in solid tumors (Maeda et al. 2003) because of enhanced permeation through leaky neoangiogenic blood vessels in tumors, and prolonged retention therein due to the dearth of draining lymphatics and sieve such as disordered extracellular matrix (Dvorak et al. 1988). Apart from this fortunate passive accumulation, nanoparticles can be more specifically tumor-targeted using peptides or other suitable ligands, and can possess unusual physical properties not observed in their bulk metal counterparts.

We envision a number of points of intersection between nanotechnology and radiation oncology. In all of these envisioned synergisms, a common theme is that of delivering greater radiation dose to the tumor while sparing normal tissues. Since the interaction between ionizing radiation and a cell can be conceptualized as either a physical phenomenon where atoms are ionized or a biological phenomenon where ionized free radicals mediate secondary molecular biologic events, we will stratify potential means of radiosensitization of cancers along these broad definitions (Table 19.1). On the one hand, one can use nanoparticles to increase the sensitivity of cancer cells to radiation by increasing tumor temperature (thermoradiotherapy). This approach modulates the physiology of tumor vascular blood flow, the biology of processing molecular damage from radiation, and the intrinsic sensitivity of cells to radiation. Another means to increase the sensitivity of tumor cells is to deliver radiation sensitizers specifically to cancer cells. These two methods can be said to cause *biological dose enhancement*, where effective radiation dose is increased by modification of the cellular response downstream of the radiation. A different method of enhancing radiation effect for a given radiation dose is by trapping more of the radiation within the tumor, or *physical dose enhancement*. This can be accomplished by using nanoparticles made of high atomic number elements, typically gold. Interaction of radiation of suitable energy (or wavelengths) with gold releases a shower of secondary electrons that multiply the cellular damage. Details of these methods have been described in earlier chapters in this book. There have also been reports of other ingenious uses of nanotechnology to aid RT, particularly through the delivery of radionuclides directly or activation of nonradioactive nanoparticles to create radionuclides. In this chapter, we will briefly discuss the state-of-the-art of these methods and project our vision for future work that could directly impact the practice of radiotherapy.

TABLE 19.1 Nanotechnology Applications in Radiation Oncology

- Biologic sensitization (by sensitizing tumor cells to radiation)
  - Nanoparticle-mediated thermoradiotherapy
  - Nanoparticle-mediated delivery of radiosensitizing drugs to tumor
- Physical sensitization (by "trapping" ionizing radiation specifically in the tumor)
  - Delivery of high atomic number elements to the tumor
  - Delivery of radioactive isotopes to tumors
  - Nanoparticle-mediated neutron capture therapy

19.2 Nanoparticle-Mediated Thermoradiotherapy

Localized and selective heating of cancer cells can directly result in tumor ablation, whereas mild temperature hyperthermia increases tumor sensitivity to radiation (thermoradiotherapy). Hyperthermia has been recognized as a useful complement to more conventional anticancer therapies; nonetheless, the means of achieving hyperthermia in a controlled, consistent fashion has remained elusive. Most conventional methods of generating hyperthermia, whether external or interstitial, are at least minimally invasive and lack the means to monitor temperature during thermal therapy. Consequently, the reporting of such treatments is based on point measurements of temperature within the tumor, a measurement that is subject to significant sampling errors depending on the location of the measurement probe. Spatially, the temperature increase within tumors is nonuniform (with "cold spots" or "heat-sinks" along vasculature) and temporally changing, whereas there is minimal control over temperature patterns since real-time monitoring and adaptive feedback control are rarely used. Metal nanoparticles, particularly gold and iron, offer an alternative approach to tumor heating. When noble metals—such as gold—are reduced to nanometer dimensions of the same order of magnitude as light waves, they develop unique optical properties of strong scattering and absorption of specific wavelengths of light. This happens because the free electrons oscillate resonantly when excited with light of proper wavelength, a phenomenon known as localized surface plasmon resonance. The resonant energy is either dissipated as radiation (Mie scattering) or converted to heat (absorption). In these nanoscale size regimes, intravenously administered metallic nanoparticles accumulate passively within tumors by leaking through the larger fenestrations and pores in vascular endothelial linings of tumor blood vessels which are, by default, chaotic, immature, and not fully formed. This preferential sequestration of nanoparticles within tumors compared to normal tissues is often termed the "enhanced permeability and retention" (EPR) effect. In the case of gold nanoparticles (GNPs), a laser light focused on the tumor laden with nanoparticles heats up the nanoparticles, which promptly transfer this heat to adjacent tumor tissues because of their high thermal conductivity (Jain et al. 2008). Other nanoparticles have also been tried for thermotherapy—ferromagnetic nanoparticles (iron, iron oxide, or core–shell mixtures of these) can also be heated up by the application of an external alternating magnetic field (Kim et al. 2008). Nanoparticle thermotherapy (NPTT) is often championed as a stand-alone anticancer approach, but the uneven distribution of the nanoparticles in the tumor and the difficulty of achieving ablative temperatures in the tumor without collateral damage to adjacent normal tissues dampen enthusiasm for this approach. However, the underlying principle of NPTT is more ideally suited for exploitation as an adjunct to RT when mild-temperature hyperthermia (as opposed to thermoablation) within tumors is conceived using nanoparticles (Diagaradjane et al. 2008). When gold nanoshell-laden tumors are illuminated with a near-infrared (NIR) laser, even a mild increase in temperature results in an initial increase in tumor perfusion and a consequent reduction in the extent of hypoxia within the core of the tumor (Figure 19.1). This reduction in hypoxia facilitates a greater response of tumors to subsequent radiation via a well-known direct correlation between oxygenation of tumors and radiosensitivity. Furthermore, unique to this form of hyperthermia, combining hyperthermia with radiation results in vascular disruption and extensive tumor necrosis, possibly due to the perivascular localization of these relatively large nanoparticles (150 nm diameter) that do not traverse deep into tumor parenchyma. Furthermore, the temperature rise can be measured noninvasively using magnetic resonance thermal imaging. This novel integrated antihypoxic and localized vascular disrupting therapy can potentially be combined with other conventional antitumor therapies and opens up the field of NPTT. In another indication of the novelty of this NPTT approach when combined

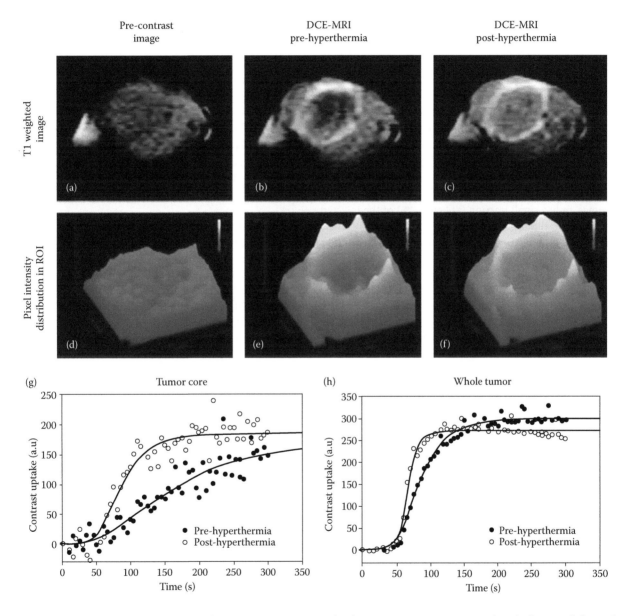

FIGURE 19.1 Gold nanoshell mediated hyperthermia in tumors. T1-weighted magnetic resonance imaging (MRI) of tumors laden with gold nanoshells (a) before infusion of intravenous contrast, (b) after infusion of intravenous contrast (dynamic contrast enhanced, DCE-MRI), and (c) DCE-MRI images immediately after hyperthermia. The corresponding 3D pixel intensity distribution profiles are presented as pseudocolor rendering in (d–f). Enhanced contrast (bright tumor center) observed in post-hyperthermia DCE-MRI when compared to pre-hyperthermia shows increased perfusion after gold nanoshell-mediated hyperthermia. Pre- and post-hyperthermia contrast uptake estimated from the ROI encompassing the tumor core and whole tumor is illustrated in (g) and (h), respectively. (Reprinted with permission from Diagaradjane, P. et al., *Nano Lett.*, 8(5), 1492–1500, 2008.)

with radiation, Atkinson et al. (2010) have demonstrated more recently that nanoparticle-induced hyperthermia may be specifically more damaging to cancer stem cells, resulting in more effective eradication of tumors. This result provides further impetus in taking the combination of NPTT and radiotherapy to the clinic. Several clinical trials currently underway using GNPs for plasmonic thermal destruction of atherosclerotic plaques or as delivery vehicles for oral insulin (clinicaltrials.gov) provide evidence of their clinical acceptability. Two early phase clinical trials specifically for cancer applications are also underway (personal communications, Glenn Goodrich, Nanospectra Biosciences Inc.). One of the advantages of taking GNPs through the gamut of preclinical and clinical testing is that, as of now, these particles are categorized by the federal drug agencies as "devices" and not "drugs"; this considerably reduces the time and cost of clinical translation.

The major challenge facing GNP-mediated hyperthermia is seeking appropriate clinical scenarios for advantageous application. NIR light, despite its deep tissue penetrating properties, cannot successfully heat up nanoparticles a few centimeters under the unbroken skin. Coupled to this is the fact that despite targeting via active ligands and PEGylation for prolonged circulation times, only a small fraction (as low as 1%) of the injected dose reaches the tumor. When combined, these factors make adequate heating of deep-seated tumors a difficult challenge to surmount.

Therefore, GNPs should be used for hyperthermia under appropriate clinical situations where they can be reasonably efficient for heat generation. Some of these situations would be where the tumor is sufficiently superficial (e.g., appropriate for head and neck tumors, skin tumors, post-mastectomy chest wall) or within tissues with low attenuation coefficients for NIR light (breast), where the tumor is inside a cavity that can be reached with a laser probe (e.g., colon cancers with a modified colonoscope), or where the tumor bed has been exposed intraoperatively. Simultaneously, further efforts are continuing to delineate the effect of size, shape, and surface composition on circulation times and tumor uptake, to maximize the accumulation of particles within tumors.

A method of increasing the penetration of electromagnetic radiation to heat up nanoparticles inside the body is to replace GNPs with iron oxide nanoparticles and NIR with alternating magnetic field. Alternating magnetic fields (AMF) are very safe for humans and only heats up the magnetic nanoparticles through a combination of hysteresis and Neel relaxation. These physical phenomena intrinsic to ferromagnetic nanoparticles cause an increase in temperature in the nanoparticles, and this temperature is dissipated into the tissues. Since the required concentration of ferromagnetic nanoparticles to generate a sufficiently large specific absorption rate in an alternating magnetic field is not typically achievable via systemic administration of nanoparticles, this technique is not suitable for thermoablation but can be used for hyperthermia to accentuate the effects of radiotherapy and chemotherapy. Even so, in its current incarnation, the typical method of getting these particles into the tumor

is via direct interstitial injection. Although this is not elegant from a distribution standpoint, one advantage of these particles is that they can be imaged by magnetic resonance imaging (when concentrations are sufficiently high in tissue), and deficiencies in distribution can be fixed by directing the interstitial injection to such areas of shortage. This method of tumor heating using water-soluble 15-nm particles with a magnetic core and a silane coating has been successfully commercialized by Magforce AG, a German company based in Berlin. The success of this method has gained considerable recognition in recent time, with the patented NanoTherm therapy (launched in 2011) having attracted over US\$30 million in equity financing and obtained European regulatory approval. At least three clinical trials—for glioblastoma, prostate cancer, and pancreatic cancer—are in progress using this therapy. Among these, the glioblastoma trials have already successfully completed Phase I (feasibility) and Phase II (efficacy) en route to obtaining European Union regulatory approval. Entry to the U.S market is still in its infancy—the Food and Drug Administration clearance mechanisms are likely to involve independent approvals of the nanoparticle, the alternating magnetic field generator, and the software used for adaptive treatment planning.

19.3 Nanoparticle-Mediated Delivery of Radiosensitizing Drugs to Tumor

Several chemicals are currently known to possess radiosensitization properties, and many new ones are being reported daily (Che et al. 2010; Kim et al. 2010a; Morgan et al. 2010; You et al. 2010). Often, these drugs have tumor selectivity by virtue of preferential sensitization of hypoxic cells to radiation (normal tissues are less prone to hypoxia than tumor cores that have outgrown their blood supply), preferential cytotoxicity to rapidly proliferating cells (as with bioreductive drugs that tumors incorporate more of), or differential radiation damage repair capabilities (less repair within tumors than normal tissues). Alternatively, synergy between drugs and radiation may be spatial where radiation counters localized disease, whereas drugs address metastatic disease primarily. In all of these scenarios, greater penetration of drugs into the hostile tumor microenvironment (beyond physical and biological barriers posed by the interstitial matrix of collagen, glycosaminoglycans, and proteoglycans that harbors the most aggressive and radioresistant cells that thrive in low pO_2, low pH environments) is highly desirable. Nanoparticles could serve as convenient chaperones that relay radiosensitizers to tumors more efficiently than radiosensitizers alone, especially when the radiosensitizer in question is genetic material (siRNA, shRNA, miRNA, DNA, etc). In fact, nanoparticles are the major nonviral means of delivery of genetic material to cells (Bondi and Craparo 2010). In a report, poly(D,L-lactide-co-glycolide) (PLGA) nanoparticles containing ataxia telangiectasia–mutated antisense oligonucleotides were found to be taken up by mouse squamous carcinoma cells, and sensitized them both *in vitro* and *in vivo* to irradiation (Zou et al. 2009). In another study from the same university, antisense epidermal growth factor receptor (EGFR) gene-containing nanoparticles were tested on the same cell line both *in vitro* and *in vivo* and was also found to enhance cell death through radiation (Ping et al. 2010).

Nanoparticle-mediated delivery has been demonstrated to improve the action of conventional hypoxic radiosensitizers such as paclitaxel and etanidazole (Jin et al. 2007). PLGA nanoparticles between 80 and 150 nm were used as the delivery vehicle of choice. It was shown that released drug effectively sensitized hypoxic tumor cells to radiation. The radiosensitization by nanoparticles containing both paclitaxel and etanidazole appeared to be significantly better than that of single drug-loaded nanoparticles. Similarly, PEG–PLA polymer micelles have been used as vehicles for delivery of a bioreductive drug, beta-lapachone, to prostate cancers that overexpress NAD(P) H:quinone oxidoreductase-1 enzyme, a member of the cytoplasmic 2-electron reductase family (Blanco et al. 2007). Such radiosensitization by nanoparticle-drug formulations can be further enhanced when the nanoparticle payload can be specifically engineered to home onto tumor cells as demonstrated in head and neck cancer cells and xenografts by folate decoration of nanoparticles with a core of PLGA laden with docetaxel and a capsule of lecithin and PEG (Werner et al. 2011). A novel twist to this strategy of nanoparticle-mediated delivery of cytotoxic radiosensitizers is to identify short peptides that bind specifically to irradiated tumor cells, but not unirradiated tumor cells or normal cells, and to decorate lipid-based nanoparticles laden with chemotherapeutic drug with these peptides. This identification of novel peptides uses a method called "phage display," where the bacteriophage virus is used to display a library of peptides on its surface to screen for peptides that can discriminate between irradiated and unirradiated cells (Lowery et al. 2011). Adopting an entirely different approach, investigators have used nanoparticles to deliver therapeutic doses of nitric oxide (a known radiosensitizer) gas to tumor tissues via photochemical methodologies utilizing transition metal complexes that are nitric oxide precursors. The advantage of the photochemical strategy is that it allows for precise control of the timing, location, and dosage for the targeted delivery of a bioactive agent (Ostrowski and Ford 2009). Along similar lines, external spatiotemporal control of drug release within tumors can be accomplished using thermosensitive liposome laden with drug that are ruptured by hyperthermia—both the hyperthermia and the site-specific drug release sensitize the tumor to radiation.

19.4 Nanoparticle-Mediated Radiation Dose Enhancement/Radiosensitization

There are many similarities between physical and biological dose enhancement of radiation on cancer cells. In both methods, nanoparticles are loaded selectively into cancer cells, and they increase the probability of the cell killing when exposed to radiation. The difference lies in the mechanism of sensitization:

whereas in the former method, the nanoparticle interacts with the cells mainly through interactions with the incident radiation, in the latter the interaction is directly with the cells.

Generally, all other factors being equal, for radiosensitive tumors, increasing the dose of external beam radiation therapy (EBRT) results in better tumor control (Peeters et al. 2006; Pollack et al. 2002; Zietman et al. 2005). However, radiation toxicity at higher doses limits the maximal effective dose to about 60–65 Gy even with hyperfractionated dosing schedules. Nanotechnology can help increase the efficacy of the radiation treatment without increasing the radiation dosage. This can be accomplished either by making the cells more radiosensitive (biological dose enhancement) or by "trapping" the ionizing radiation in a more tumor-specific way as it passes through the tumor (physical dose enhancement).

19.4.1 Overview of Basic Principles

Ionizing radiation causes damage to tumor cells either by directly breaking the DNA strands or by secondary electrons released through interactions between radiation and tissue elements (e.g., photoelectric absorption, Compton effect, and pair production, for photons), which create short-lived highly reactive oxygen species (ROS), which in turn cause secondary damage to vital cellular structures such as the DNA. In general, ionizing radiation deposits its energy without any discrimination between the normal and tumor tissues, although some tumor specificity of radiation can be achieved by various geometrical targeting approaches utilized for modern RT. "Trapping" a greater amount of the energy in the tumor will not only increase the radiation damage to tumor cells, but also reduce the damage to normal cells (Figure 19.2). In theory, this can effectively be accomplished by preloading the tumor with nanoparticles composed of elements capable of absorbing more radiation energy than tissue elements, for example, having larger photoelectric absorption cross sections for the gamma/x-ray photons. When irradiated by suitable energy (or wavelength) of radiation, these highly absorbing elements produce lots of low-energy (approximately on the order of keV or lower) secondary electrons (e.g., photoelectrons and Auger/Coster–Kronig electrons), most of which deposit their energy locally within the tumor. According to recent computational studies (Cho et al. 2009; Jones et al. 2010), the secondary electron fluence within the tumor irradiated by low energy (<~100 keV) photon sources could be increased by as much as 2 orders of magnitude if the tumor was loaded with GNPs at low concentrations (approximately on the order of 0.1 wt.%). The predicted level of increase in the secondary electron fluence would in turn lead to 2 orders-of-magnitude increase in electron energy deposition around GNPs within the tumor. Typically, the production of these secondary electrons happens as a chain reaction (e.g., photoelectric absorption followed by "Auger cascade") and is strongly dependent on the atomic number (Z) of elements (e.g., photoelectric absorption $\propto Z^3$). These characteristics offer some distinct advantages over other conceivable approaches of using high-Z elements to increase the secondary

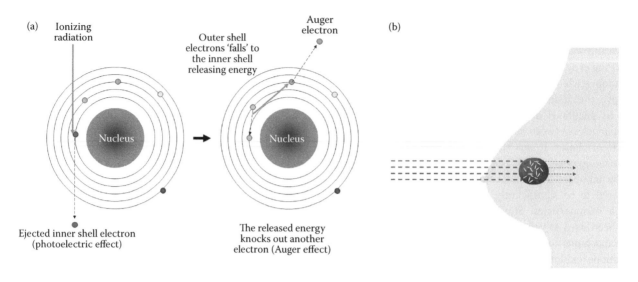

FIGURE 19.2   (See color insert.) Physical dose enhancement using nanoparticles. (a) Auger effect of high atomic number atoms exposed to radiation. Possible fluorescence (or characteristic) x-ray emission is not illustrated here. (b) Radiation can be "trapped" in tumors loaded with nanoparticles containing such atoms. Red sphere denotes the breast tumor loaded with the yellow gold nanorods.

electron production (e.g., pair production $\propto Z$) for physical dose enhancement.

Over the years, there have been numerous attempts to capitalize on the idea of using high-Z elements to enhance tumor dose (or radiosensitize the tumor). Researchers have investigated various forms of high-Z media including bromine (Furusawa et al. 1991), iodine (Kada et al. 1970; Briggs et al. 2011; Santos Mello et al. 1983; Nath et al. 1990; Norman et al. 1997), gadolinium (Goorley et al. 2004; Robar et al. 2002), platinum (Le Sech et al. 2000, 2001), and gold (Herold et al. 2000). In recent years, considerable research has focused on the development of approaches using various metallic nanoparticles (e.g., platinum and gold) (Chithrani et al. 2010; Kobayashi et al. 2002; Hainfeld et al. 2004; Kim et al. 2010b; Polf et al. 2011; Jain et al. 2011; Chang et al. 2008; Roa et al. 2009), taking advantage of the high tumor specificity of nanoparticles under passive/active targeting scenarios and the high-Z nature of the base metal. As explained in greater detail throughout the preceding chapters, GNP-based approaches seem to be the most promising among various methods because of the higher Z number of gold than other metals considered and more favorable toxicity profile of gold for human applications, for example, in comparison with platinum. Moreover, bioconjugated GNPs for active targeting of tumor cells (Fry and Pitcher 1990) offer a unique option to not only induce physical dose enhancement but also control the location of such enhancement to occur within the tumor, thereby modulating the overall radiation response of tumors under a given irradiation scenario. This possibility has significant implications for the clinical implementation of the so-called GNP-aided radiation therapy (GNRT) (Cho et al. 2009). More discussion on this aspect will be presented in the next subsection.

In general, various radiobiological outcomes seen in previous *in vitro/in vivo* studies with GNPs (e.g., 20% or more radiosensitization) can be reasonably accounted for, at least qualitatively,

by physical considerations. For example, the remarkable outcome seen in the first successful animal study by Hainfeld et al. (2004) (e.g., l-year survival rate of 86% vs. 20% for mice irradiated with/without GNP injection, respectively) could be attributed to significant increase in the photo-/Auger electron fluence within the tumor (including blood vessels) loaded with high-Z GNPs during kilovoltage x-ray irradiation. The increased secondary electron fluence possibly resulted in greater physical damage to tumor cells and more prominently to endothelial cells lining the tumor blood vessels because of the higher blood gold content at the time of kilovoltage x-ray irradiation (Cho et al. 2009; Hainfeld et al. 2004). Additionally, *in vitro* models have suggested that GNP-mediated radiosensitization depends on a number of factors such as the GNP size and concentration/internalization, the cell type, and the radiation type and energy. The physical explanation for many of these determinants of radiosensitization become somewhat obvious when estimates of physical dose enhancement are derived on a nano-/cellular scale by more elaborate physical models (See Chapter 10 for details).

Although the physical picture of GNP-mediated radiosensitization can be obtained somewhat intuitively, exact molecular mechanisms for this intriguing phenomenon are not well known and have been the subject of active investigation by many research groups in recent years. Many of these studies have focused on finding evidence of increased DNA double-strand breaks (DSB) due to the presence of GNPs within the cell culture (i.e., around or inside the cells) during the irradiation. By estimating the number of radiation-induced foci such as γ-H2AX and 53BP1, one study (Chithrani et al. 2010) found a positive correlation between the increase in the DSB and the number of GNPs internalized within the cells. Based on the counting of 53BP1 foci within GNP-treated cells after irradiation, however, another study (Jain et al. 2011) reported no increase in the DSB formation. These contradictory findings could be, in part, attributed to the difference in experimental conditions. For example, the efficiency of cellular uptake (or internalization) of GNPs is known to depend on the particle size. Consequently, by applying the above-mentioned physical considerations on a nano-/cellular scale alone (e.g., distance between GNPs and DNA, the range of secondary electrons from GNPs), one would expect a different outcome between the experiments with 50-nm-diameter (Chithrani et al. 2010) and 1.9-nm-diameter GNPs (Jain et al. 2011), even under the same experimental conditions. Although this example shows the potential of a physical model being applicable to a biological problem, there are purely biological issues for which any physical modeling effort becomes less meaningful. For instance, one study indicated that irradiated GNPs induced activation of the CDK kinases leading to acceleration in the G0/G1 phase and arrest of cells in the G2/M phase, accompanied by increased expression of cyclin B1 and cyclin E (Roa et al. 2009). Additionally, according to a recent study (Butterworth et al. 2010), 1.9-nm-diameter GNPs, which have been used in many recent studies, seem to interact directly with the cells, causing significant cytotoxicity, apoptosis, and increased oxidative stress. All of these exemplify some unique challenges in elucidating precise molecular mechanisms associated with GNP-mediated radiosensitization. Therefore, it will require considerable amount of research effort to properly identify all possible pathways associated with GNP-mediated radiosensitization.

19.4.2 Outlook for Clinical Implementation

Despite the abundant phenomenological evidence for tumor dose enhancement (radiosensitization) using high-Z media and kilovoltage x-rays in particular, the actual clinical implementation of this idea has been seriously questioned especially for EBRT treatments of deep-seated human tumors because of insufficient penetration of kilovoltage x-rays into condensed media such as human tissue. Here, we provide an overview of potential ways to translate this novel approach to clinical practice within the context of GNRT (Table 19.2). We expect these strategies will also be applicable to other similar approaches using high-Z metallic nanoparticles. Assuming eventual regulatory clearance of GNPs for human applications, the implementation of GNRT needs to be in general based on some careful consideration of key parameters such as tumor type and location, tumor targeting strategy (e.g., active vs. passive), delivery

TABLE 19.2 Clinical Implementation Scenarios of GNRT

- GNRT as contrast-enhanced radiation therapy
  - Intravenous or Intratumoral injection of untargeted GNPs
  - Irradiation immediately (on the order of minutes) after GNP injection
  - Most effective for superficial tumors
- GNRT as cellular-targeted radiation therapy
  - Intravenous injection of targeted or untargeted GNPs
  - Irradiation long (on the order of hours or day) after GNP injection
  - Possibly effective for both superficial and deep-seated tumors

method of GNPs (e.g., intratumoral vs. intravenous), timing of irradiation after the injection of GNPs (e.g., <5 min vs. 24 h), and radiation type (e.g., photons vs. electrons) and energy (e.g., kilovoltage vs. megavoltage). These parameters could be put in perspective in terms of the two classifications of GNRT (discussed in the following subsections).

19.4.2.1 GNRT as Contrast-Enhanced RT

Owing to their higher Z number and better tumor specificity, GNPs can be more effective in enhancing the tumor dose than conventional high-Z contrast media when applied as agents for the so-called contrast-enhanced radiation therapy (CERT) (Herold et al. 2000; Garnica-Garza 2009; McMahon et al. 2008; Robar 2006; Verhaegen et al. 2005). In fact, the aforementioned animal study by Hainfeld et al. (2004) was a successful demonstration of CERT, because the infusion of GNPs was immediately (within 2 min) followed by 250 kVp x-ray irradiation and the tumor gold content was largely an index of the vascularity of the tumor with GNPs serving merely as contrast agents at such early time points. The CERT approach assures a high overall tumor gold content (on the order of 1 wt.%, although heterogeneously distributed), possibly similar to that achievable via intratumoral injection of GNPs, and both experimental and computational data clearly demonstrate the potential for potent radiosensitization. Assuming the nonuniform dose distribution with kilovoltage sources can be handled by more refined beam delivery techniques such as a rotational delivery, there is still the critical challenge posed by the generation of potentially catastrophic collateral damage to blood vessels within normal tissues, especially those present along the beam path during EBRT treatment. A well-known problem with kilovoltage RT is that of dose enhancement to the bone/skull—using conventional or low-energy enhanced megavoltage photon beams could circumvent this problem, but the level of dose enhancement would also become less impressive (Cho 2005). Despite some positive outlook based on purely computational studies (Garnica-Garza 2009; Verhaegen et al. 2005), therefore, it is virtually impossible to conclude that the utility of GNRT via contrast-enhanced kilovoltage RT extends far beyond the treatment of superficial tumors even after careful consideration of collateral damage to normal tissue blood vessels. Obviously, GNRT of such tumors could also be implemented with electron beams. Meanwhile, although predicted to be less effective, GNRT implementation using low-energy enhanced megavoltage photon beams might still be beneficial in some clinical scenarios (see Figure 19.3 for an illustration of a hypothetical case) and is worth investigating further, especially since such beams from the flattening-filter-free mode of linear accelerators are routinely available these days. Finally, it is readily apparent that GNRT might actually be feasible with much less trouble, compared to that mentioned above, if it is implemented via brachytherapy. In particular, as argued before (Cho et al. 2009), GNRT implementations using low-energy high dose rate brachytherapy sources such as 50 kVp x-ray and [169]Yb sources look promising and warrant further investigation for eventual clinical translation.

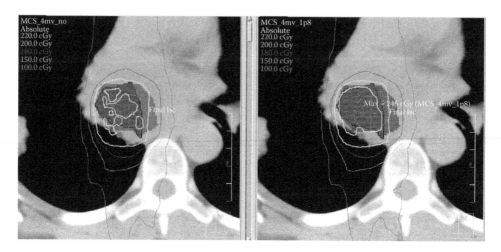

FIGURE 19.3   (See color insert.) Monte Carlo–calculated dose distributions for a lung tumor using a Cu target–produced unflattened 4 MV photon beams: a tumor without (left panel) and with (right panel) GNPs (assuming uniform distribution at 1.8% by weight). Dose enhancement up to 10% within the GNP-loaded tumor (yellow isodose line) can be seen in right panel, whereas other isodose lines remain virtually unchanged, despite softer photon energy spectra and larger buildup doses, compared to conventional megavoltage beams at the same nominal photon energy.

19.4.2.2 GNRT as Cellular-Targeted Radiation Therapy

In contrast to the above approach, one may consider irradiating the tumor after a longer interval (say, 24 h) following the intravenous injection of unconjugated GNPs (i.e., passive targeting). As shown in a previous study (Chang et al. 2008), the majority of GNPs at such a time point are expected to have already been cleared from the blood compartment, whereas extravasated GNPs (via the EPR effect) are expected to still remain within the tumor. Thus, regardless of radiation type, an EBRT implementation of GNRT might be less likely to cause unwanted collateral damage to normal tissues, although the efficacy of GNRT would become questionable because of the significantly lower overall tumor gold content compared to any CERT-type implementation. However, the local gold concentration within the tumor, such as that in the perivascular space where untargeted GNPs tend to be confined according to some studies (Diagaradjane et al. 2008), could still be significant enough to induce meaningful microscopic dose enhancement followed by increased radiation damage to nearby critical structures such as tumor vasculature. In fact, significant radiation sensitization effect seen in a previous *in vivo* study with an electron beam (Chang et al. 2008) could be attributed to such a disruption of tumor vasculature.

Taking this paradigm one step further, GNPs could be conjugated to antibodies or peptides directed against tumor or tumor vascular antigens to enable more tumor-specific delivery of GNRT. This active targeting strategy could not only increase the tumor gold content (Fry and Pitcher 1990; Qian et al. 2008) but also locate GNPs at the cell surface or even inside the cell, potentially creating more serious damage (e.g., DSB) to the cellular nucleus or DNA, the main target of radiation damage, via short-range secondary electrons emanating from GNPs. Consequently, active targeting would help improve the efficiency of GNP-mediated radiosensitization and thereby significantly

reduce the amount of GNPs required to induce significant radiosensitization effect during GNRT treatments. In fact, our unpublished data with gold nanorods conjugated with an anti-EGFR antibody strongly support this hypothesis. More studies are necessary to assess the efficacy of active targeting for GNRT with various radiation sources capable of generating secondary electrons from GNPs (e.g., photons, electrons, protons). Upon successful completion of such investigations, GNRT will become more widely applicable with less stringent constraints.

19.5 Internal Irradiation of Tumors Using Radioactive Nanoparticles

Internal irradiation of tumors can be achieved either using naturally radioactive isotopes that are ferried to the tumor or injected into the tumor, or using an external trigger to activate a nonradioactive nanoparticle to generate a radioactive isotope. The concept of ferrying radionuclides to tumors via systemic administration is most elegantly achieved when metabolic pathways ensure tumor-specific accumulation of the radioisotope (as with radioactive iodine and the thyroid gland). Alternatively, radiolabeled antibodies (radioimmunotherapy) can be used to home radioisotopes onto tumors. Localized implantation of radioisotopes in tumors is a strategy used in interventional radiology where intra-arterial instillation of radioactive microparticles (beads or resins) leads to embolization within tumor vessels preferentially—this strategy works best for liver tumors since the tumor derives most of its blood supply from the hepatic artery (that is cannulated and infused with the microparticles), whereas the normal liver receives most of its blood supply from the portal vein. The strategies described so far have largely achieved tumor-directed RT without the use of nanoparticles. Introduction of nanoparticles carrying payloads of radionuclides and targeted to tumors, offers the opportunity to combine the advantages of minimally invasive means of introduction into the body and highly efficient means of concentrating radioactivity within tumors both by passive accumulation and targeting ligands.

Whereas conventional radioimmunotherapy uses monoclonal antibodies that are labeled with a single radioactive atom, various forms of nanoparticles—including liposomes, microparticles, nanoparticles, micelles, dendrimers, and hydrogels—coated with antibodies deliver more radiation dose per recognition event (Hamoudeh et al. 2008a). A simulation of this situation showed that when treated with a 5-nm-diameter antibody-linked nanoparticle composed of the beta-emitting radionuclide $^{90}Y_2O_3$, tumor cells can reach up to 50 Gy (Bouchat et al. 2007). However, just as with GNPs, irregularity of vasculature and the presence of a necrotic core were shown to have a noticeable influence on the deposited dose. The delivery of radionuclides in nanoparticles (and microparticles) is now a maturing field and there have been several excellent reviews detailing these advances (Hamoudeh et al. 2008a; Sofou 2008; Williams et al. 2008) (Figure 19.4a). Among these recent advances, capitalizing on the unique peptide structure of apoferritin with its vacant core, radioactive yttrium encapsulated within a protein shell has been described (Wu et al. 2008). The 8-nm-diameter nanoparticle was functionalized with biotin to allow further decoration of the shell with tumor-targeted antibodies. Similarly, there are reports of biodegradable and biocompatible lipid–polymer hybrid nanoparticles loaded with a combination of chemotherapeutic drugs such as docetaxel and radionuclides such as ^{111}In and ^{90}Y that have shown good therapeutic efficacy in prostate cancer models (Wang et al. 2010). Direct injection of radioactive gold into prostate cancer xenografts has also yielded excellent tumor control probabilities as outlined in Chapter 12.

19.6 Neutron Capture Therapy Using Nanoparticles

Neutron capture therapy uses nonradioactive isotopes that accumulate preferentially within tumors upon systemic administration and an external trigger via neutron bombardment of these nonradioactive isotopes to generate a localized nuclear reaction within the tumor that results in tumor-specific killing (Figure 19.4b). The use of an external trigger is a unique feature of this therapy that offers the ability to control the spatial and temporal release of radioactivity. The classical example of such a treatment approach is to use a tumor-seeking chemical containing boron-10 and an epithermal neutron to initiate a reaction characterized by

$$^{10}B + n \rightarrow [^{11}B] \rightarrow ^{7}Li + \alpha + 2.31 \text{ MeV}$$

where n is a thermal neutron striking the boron atom after collisions with nitrogen and hydrogen along the way have slowed it down from an epithermal neutron to a low-energy thermal neutron and $[^{11}B]$ is an excited intermediate state that instantly disintegrates to a high-energy alpha particle and a high-energy stable ^{7}Li nucleus. The alpha particle and the lithium nucleus have short path lengths of <10 μm, roughly the diameter of a typical cell, where they cause a dense cluster of ionizations accounting for the high linear energy transfer (LET) within a small area. Clearly, such a technique of localized tumor irradiation yields optimum therapeutic benefits when the boron has cleared from the vascular compartment and concentrated densely and homogeneously within tumors without significant accumulation within normal tissues, and the neutron beam does not cause significant toxicity via interactions with normal tissues along its path. Despite the elegance of the concept of a localized detonation of a radioactive device within tumors via remote control, clinical implementation of boron neutron capture therapy (BNCT) has been largely stymied by the lack of uniform and tumor-specific accumulation of nontoxic boron-containing compounds, the lack of significant tissue damage beyond the boron-containing tumor cell, the poor penetration of the incident neutron beam, the collateral damage from high LET protons generated by interaction of neutrons with nitrogen

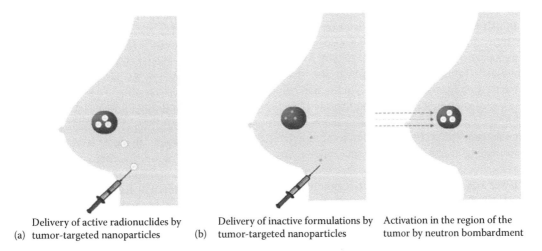

(a) Delivery of active radionuclides by tumor-targeted nanoparticles

(b) Delivery of inactive formulations by tumor-targeted nanoparticles

Activation in the region of the tumor by neutron bombardment

FIGURE 19.4 Nanoparticles for delivery of radiation through radionuclides by (a) delivery of radioactive radionuclides or (b) by neutron capture therapy.

(neutron capture) and hydrogen (recoil protons) in normal tissues along the beam path, and the lack of dedicated neutron irradiation sources in oncologic centers (rather, these treatments have been performed in collaboration with nuclear reactor facilities). Accumulated clinical experience in the United States, Japan, and Europe has spanned a spectrum of cancers including gliomas, melanomas, and recurrent head and neck cancers with both thermal and epithermal neutrons generated largely from nuclear reactors rather than accelerators. Efforts are underway to develop more advanced irradiators and more tumor-specific boron-containing compounds that can be tested in randomized clinical trials.

From the foregoing narrative on BNCT, it is evident that nanoparticulate formulations of boron would, in theory, offer the advantages of passive or active tumor targeting and greater boron density within each such targeted tumor cell, resulting in greater neutron dose capture (in essence, a larger tumor cell absorption cross section for neutrons) than single radioisotope molecules. In a proof-of-principle experiment, boron nanoparticles were formulated and incubated *in vitro* with tumor cells, which were then implanted in mice to create tumors preloaded with the nanoparticles (Petersen et al. 2008). Neutron activation of these tumors resulted in a significant improvement in survival. Similarly, a 100-nm liposomal boron delivery system for neutron capture therapy also demonstrated preferential tumor accumulation (Nakamura et al. 2009).

In addition to the use of nanoparticulate formulations for enhanced BNCT, other candidate neutron absorbers in nanoparticle embodiments have also been proposed. One study reported the use of dirhenium decacarbonyl $[Re_2(CO)_{10}]$ encapsulated in poly-L-lactide (PLLA) nanoparticles as a neutron-activatable radiopharmaceutical (Hamoudeh et al. 2008a). These nanoparticles had 23% w/w rhenium (compared to radionuclides described above, which are loaded in the region of 10%) in a stable, spherical form. Neutron irradiation resulted in random PLLA chain scission and consequent agglomeration of particles, but they were readily redispersible in solution for *in vivo* administration. The same group likewise reported similar results from radioactive holmium salt also encapsulated in PLLA nanoparticles (Hamoudeh et al. 2008b). Another interesting proposal is to use gadolinium (which also shares the property of activation by neutron capture) in the nanoparticulate form as a means of dual imaging and therapy (Arrais et al. 2008). Chitosan nanoparticles loaded with the magnetic resonance contrast agent Gd-DTPA were found to accumulate better in human malignant fibrosis histiocytoma cells *in vitro* than free Gd-DTPA, but this accumulation resulted in lower T1 reduction and lower magnetic resonance contrast enhancement (Fujimoto et al. 2009). It remains to be seen whether nanoparticle formulations will increase accumulation of neutron-absorbing nonradioactive isotopes within tumors and thereby increase tumor dose without increasing normal tissue dose to realize the promise of enhanced therapeutic gain with neutron capture therapy.

19.7 Other Methods of Enhancing Radiotherapy Using Nanoparticles

Evolving strategies to enhance radiation response of tumors that are described in preceding chapters of this book include the use of quantum dots to combine photodynamic therapy with RT and the use of nanoparticulate radioprotectors of normal tissues to widen the therapeutic window for RT.

The photodynamic therapy approach relies on the accumulation of a photosensitizer preferentially within tumors and its activation by illumination to generate singlet oxygen species that are tumor-confined and cytotoxic. These singlet oxygen species can synergize with RT to generate more DNA strand breaks and greater cell kill. In the nanoparticulate embodiment of this concept, the photosensitizer is linked to a nanoparticle that can be activated by external radiation to generate light which, in turn, activates the photosensitizer to generate tumor-specific singlet oxygen species that further enhance the antitumor effects of radiation (Chen and Zhang 2006). The unique advantage of such an approach is the ability to perform *in situ* photodynamic therapy in internal nonluminal organs via deep-penetrating ionizing radiation (rather than being confined to tissues that are superficial or accessible by endoscopes), spatial and temporal approximation of the photodynamic therapy with RT, and the ability to tightly conform the localization of nanoparticles and the photodynamic/radiation therapy to the geographical contours of the tumor. Progress in advancing such a concept has been described in Chapter 11.

Analogous to the delivery of radiosensitizing drugs to tumor tissues, selective delivery of radioprotective drugs to healthy tissues is an effective way to reduce morbidity associated with RT. This is especially relevant for radiosensitive tissues, such as the gut mucosa or bone marrow, especially in therapeutic scenarios where they face inevitable radiation exposure. The trouble with using nanoparticles to selectively deliver radioprotectors to normal tissues passively is that, by definition, nanoparticles tend to extravasate more from tumor vasculature into tumors than from normal vasculature into normal tissues. Consequently, radioprotection using normal endothelium penetrating nanoparticles may be best used in scenarios where radiation is not being used simultaneously to treat tumors (e.g., to mitigate radiation syndromes from accidental radiation exposure of healthy individuals) or where selective accumulation in normal tissues can be achieved by active targeting.

An example of the first principle is the fabrication of an orally bioavailable formulation of amifostine, the only clinically used radioprotective agent, that acts largely via scavenging of ROS. Greater alkaline phosphatase activity in normal tissues accounts for the greater conversion of amifostine to its active metabolite *N*-(2-mercaptoethyl)1,3-diaminopropane (WR-1065) in normal tissues than in tumor. Oral administration of a PLGA nanoparticle containing WR-1065 was shown to significantly protect mice from whole body irradiation as evidenced by decreased bone marrow and intestinal toxicity and better survival than control groups (Pamujula et al. 2008).

An intriguing nanoparticle-mediated radioprotection concept involves the use of cerium oxide (CeO_2) nanoparticles that protect normal tissue from radiation-induced damage by scavenging ROS that mediate the majority of radiation-induced cellular injury. In an *in vivo* murine model, the CeO_2 nanoparticles were well tolerated and prevented the onset of radiation-induced pneumonitis when delivered to live animals exposed to high doses of radiation (Colon et al. 2009). Although nanoceria are credited with the unique attribute of autocatalysis and regeneration that allows cyclical free radical scavenging, there is no inherent selectivity for radioprotection of normal tissues as opposed to tumors. Along similar lines, fullerene nanoparticles have shown promise as ROS scavengers that possess radioprotective properties *in vitro* (Bogdanovich et al. 2008) and *in vivo* (Daroczi et al. 2006).

Capitalizing on the observation that melanin, a naturally occurring pigment, possesses radioprotective properties, melanin-coated silica nanoparticles were synthesized and administered intravenously to melanoma-bearing nude mice (Schweitzer et al. 2010). Nanoparticles concentrated in bone marrow within 3 h of administration and subsequent radioimmunotherapy with [188]Re-labeled 6D2 melanin-binding antibody was associated with lower hematologic toxicity without any observable tumor protection.

19.8 Caveats and Outlook for Clinical Translation

Unlike classical pharmaceutical sensitizers of RT, nanoparticle-mediated sensitization of tumors to RT offers the possibility and the promise of being able to quantify and visualize their accumulation within tumors. In turn, this might serve as the basis for image-guided therapy and quantitative dosimetry where physical and biological consequences could be modeled and predicted before treatment. As noted in previous chapters, rapid and recent advances in noninvasive qualitative and quantitative imaging techniques have opened up the possibility that nanoparticle-mediated radiosensitization could distinguish itself from other forms of radiosensitization by being quantifiable and predictable *a priori* (before administration of RT).

The unique physicochemical properties of nano-scale formulations of bulk metals have opened up numerous opportunities to incorporate the use of nanotechnology within biomedicine. By the same token, these interactions between nanomaterials and biological systems have the potential to pose unique challenges. The most self-evident of these is the need to ensure their immediate and long-term safety and tolerability in humans. Some of the specific biocompatibility concerns are outlined in the chapter on toxicity of nanoparticles (Chapter 5), and the governmental oversight that regulates the transition from bench to bedside are addressed in Chapter 18. In general, gold and ferromagnetic nanoparticles have the advantage of some prior use in bulk form, at least, in clinical settings with the expectation that they may be relatively safe even in nanoparticulate form. Nevertheless, clinical advancement will require meticulous testing and documenting of safety and tolerability of the nanoparticulate incarnation with or without additional functionalities added.

The next concern with nanoparticle use in biomedical applications is the issue of nonspecific uptake and clearance by the body. There is ample evidence in the literature that nanoparticles are frequently opsonized by plasma proteins (primarily complement factors) and this aids their clearance via the reticuloendothelial system (liver, spleen, lymph nodes, etc). Ways to circumvent this nonspecific uptake and increase circulatory time of nanoparticles include (1) reducing the size of nanoparticles (particles in a size regime of <5–6 nm tend to be cleared more rapidly through the kidneys than larger particles); (2) changing the surface charge of particles (as a broad generalization it may be fair to state that positively charged particles are rapidly cleared from the bloodstream by opsonization; a small negative surface charge, due to coulombic interactions, keeps nanoparticles in suspension without clumping; a neutral charge minimizes chemical interactions and maintains long circulatory times); (3) surface modification of nanoparticles with biomolecules to render the nanoparticles stealth properties that enhance evasion from macrophages in the liver and spleen (usually via coatings of polyethylene glycol, dextran, etc.); and (4) inhibition of macrophage activity in the liver (using Kupffer cell inactivators). From the standpoint of efficient transport across the endothelial lining of blood vessels, a preponderance of data suggests that elongated nanoparticles extravasate more easily from blood vessels than spherical nanoparticles because of their propensity to travel along the periphery of an advancing column of blood (whereas spherical particles travel within the center of this column of blood with a parabolic leading edge) and their larger surface for contact with endothelial cells lining the vessel wall. Not surprisingly, therefore, size, shape, charge, and surface functionalization of nanoparticles contribute to immense variability in biodistribution and pharmacokinetics—attributes that need to be tested for each nanoparticle and for every surface functionalization of the nanoparticle.

Lastly, clinical applicability is limited by the physical constraints of external energy sources that trigger activation of intratumoral nanoparticles. For instance, for plasmonic nanoparticles that are activated by NIR light, the limited depth of penetration of NIR light narrows the therapeutic applicability to superficial tumors such as skin tumors, endoscopically accessible tumors, post-mastectomy inflammatory breast cancers, cancers treated with interstitially implanted brachytherapy sources, possibly cancers residing in tissues with a lower linear attenuation coefficient for NIR light such as breast tissues, and scenarios where tumor bed irradiation is performed intraoperatively. Similarly, alternating magnetic fields might be capable of remotely activating ferromagnetic nanoparticles that have accumulated in large enough quantities in deep-seated tumors but focusing the AMF fields on just the tumor and not the entire body/limb/trunk remains a technical challenge. Issues related to BNCT and neutron irradiation have also been alluded to earlier. In essence, each type of nanoparticle and energy source comes with its unique operating constraints that make it more or less suitable for specific oncologic applications. A greater understanding of these operating constraints, many of which have been elaborated upon in this book, will hopefully serve as a firm foundation for advancing nanoparticle-based strategies to enhance the efficacy of clinical RT.

19.9 Conclusions

Although nanoparticulate formulations have been around for centuries, the discovery of novel properties of matter when confined to nano-scale dimensions coupled with recent advances in their characterization and visualization on the nano-scale has led to a surge in interest in nanotechnology. Among biomedical applications, spontaneous entrapment of circulating nanoparticles within tumors through their extravasation from leaky tumor vasculature offers the greatest promise of revolutionizing the imaging and therapy of cancers. This platform for passive accumulation of nanoparticles in tumors serves as a springboard for such varied strategic paradigms as active targeting where biomolecules decorating the surface of nanoparticles dock to tumor cells and ferry the nanoparticulate payload to tumors more efficiently, release of payloads occurs in the tumor microenvironment or upon external trigger, biomarkers are sensed and monitored *in vivo*, dual imaging and therapeutic constructs permit image-guided therapy, and other such applications. Undoubtedly, the versatility of design and function of nanoparticles can be exploited in a number of ways to enhance almost any therapeutic procedure. RT is no exception, and nanoparticles can help in several ways as outlined in the preceding paragraphs and elaborated on in the preceding chapters. The broad spectrum of such exploitable interactions between nanoparticles and radiation includes photoelectric radiation dose enhancement by high atomic number particles, localized heating by plasmon resonant or magnetically activatable particles, ferrying of radiosensitizing agents to tumors or radioprotective agents to normal tissues, nanoparticulate radioactive isotopes, neutron activatable nonradioactive isotopes, and nanoparticulate means of combining photodynamic therapy and radiation. Research at this interface between radiation oncology and nanotechnology is still in its infancy and, to date, these initiatives have been largely confined to proof-of-principle experiments and modeling. Nevertheless, the intense excitement related to these early forays has fueled the expectation that one or more of these techniques may be deployed in the clinic in the next few years to realize tangible therapeutic gains for RT. In parallel, a greater understanding of the nano-scale physical and biological underpinnings of nanoparticle–radiation interactions might shed light on more efficient mechanisms to push this emerging frontier to greater heights.

References

Arrais, A. et al. 2008. Carbon coated microshells containing nanosized Gd(III) oxidic phases for multiple bio-medical applications. *Chemical Commununications (Cambridge)* 45: 5936–5938.

Atkinson, R. L. et al. 2010. Thermal enhancement with optically activated gold nanoshells sensitizes breast cancer stem cells to radiation therapy. *Science Translational Medicine* 2(55):55ra79.

Blanco, E. et al. 2007. Beta-lapachone-containing PEG-PLA polymer micelles as novel nanotherapeutics against NQO1-overexpressing tumor cells. *Journal of Controlled Release* 122(3):365–374.

Bogdanovic, V. et al. 2008. Fullerenol C60(OH)24 effects on antioxidative enzymes activity in irradiated human erythroleukemia cell line. *Journal of Radiation Research* (Tokyo) 49(3):321–327.

Bondi, M. L., and E. F. Craparo. 2010. Solid lipid nanoparticles for applications in gene therapy: A review of the state of the art. *Expert Opinion on Drug Delivery* 7(1):7–18.

Bouchat, V. et al. 2007. Radioimmunotherapy with radioactive nanoparticles: First results of dosimetry for vascularized and necrosed solid tumors. *Medical Physics* 34(11): 4504–4513.

Briggs, B. et al. 2011. Photosensitization by iodinated DNA minor groove binding ligands: Evaluation of DNA double-strand break induction and repair. *Journal of Photochemistry and Photobiology B* 103(2):145–152.

Butterworth, K. T. et al. 2010. Evaluation of cytotoxicity and radiation enhancement using 1.9 nm gold particles: Potential application for cancer therapy. *Nanotechnology* 21(29):295101.

Chang, M. Y. et al. 2008. Increased apoptotic potential and dose-enhancing effect of gold nanoparticles in combination with single-dose clinical electron beams on tumor-bearing mice. *Cancer Science* 99(7):1479–1484.

Che, S. M. et al. 2010. Cyclooxygenase-2 inhibitor NS398 enhances radiosensitivity of radioresistant esophageal cancer cells by inhibiting AKT activation and inducing apoptosis. *Cancer Investigation* 28(7):679–688.

Chen, W., and J. Zhang. 2006. Using nanoparticles to enable simultaneous radiation and photodynamic therapies for cancer treatment. *Journal of Nanoscience and Nanotechnology* 6(4):1159–1166.

Chithrani, D. B. et al. 2010. Gold nanoparticles as radiation sensitizers in cancer therapy. *Radiation Research* 173(6):719–728.

Cho, S. H. 2005. Estimation of tumour dose enhancement due to gold nanoparticles during typical radiation treatments: A preliminary Monte Carlo study. *Physics in Medicine and Biology* 50(15):N163–N173.

Cho, S. H., B. L. Jones, and S. Krishnan. 2009. The dosimetric feasibility of gold nanoparticle-aided radiation therapy (GNRT) via brachytherapy using low-energy gamma-/x-ray sources. *Physics in Medicine and Biology* 54(16):4889–4905.

Colon, J. et al. 2009. Protection from radiation-induced pneumonitis using cerium oxide nanoparticles. *Nanomedicine* 5(2):225–231.

Daroczi, B. et al. 2006. In vivo radioprotection by the fullerene nanoparticle DF-1 as assessed in a zebrafish model. *Clinical Cancer Research* 12(23):7086–7091.

Diagaradjane, P. et al. 2008. Modulation of in vivo tumor radiation response via gold nanoshell-mediated vascular-focused hyperthermia: Characterizing an integrated antihypoxic and localized vascular disrupting targeting strategy. *Nano Letters* 8(5):1492–1500.

Dvorak, H. F. et al. 1988. Identification and characterization of the blood vessels of solid tumors that are leaky to circulating macromolecules. *American Journal of Pathology* 133(1):95–109.

Fry, D. E., and D. E. Pitcher. 1990. Antibiotic pharmacokinetics in surgery. *Archives of Surgery* 125(11):1490–1492.

Fujimoto, T. et al. 2009. Accumulation of MRI contrast agents in malignant fibrous histiocytoma for gadolinium neutron capture therapy. *Applied Radiation and Isotopes* 67(7–8 Suppl):S355–S358.

Furusawa, Y., H. Maezawa, and K. Suzuki. 1991. Enhanced killing effect on 5-bromodeoxyuridine labelled bacteriophage T1 by monoenergetic synchrotron X-ray at the energy of bromine K-shell absorption edge. *Journal of Radiation Research (Tokyo)* 32(1):1–12.

Garnica-Garza, H. M. 2009. Contrast-enhanced radiotherapy: Feasibility and characteristics of the physical absorbed dose distribution for deep-seated tumors. *Physics in Medicine and Biology* 54(18):5411–5425.

Goorley, T., R. Zamenhof, and H. Nikjoo. 2004. Calculated DNA damage from gadolinium Auger electrons and relation to dose distributions in a head phantom. *International Journal of Radiation Biology* 80(11–12):933–940.

Hainfeld, J. F., D. N. Slatkin, and H. M. Smilowitz. 2004. The use of gold nanoparticles to enhance radiotherapy in mice. *Physics in Medicine and Biology* 49(18):N309–N315.

Hamoudeh, M. et al. 2008a. Holmium-loaded PLLA nanoparticles for intratumoral radiotherapy via the TMT technique: Preparation, characterization, and stability evaluation after neutron irradiation. *Drug Development and Industrial Pharmacy* 34(8):796–806.

Hamoudeh, M. et al. 2008b. Radionuclides delivery systems for nuclear imaging and radiotherapy of cancer. *Advanced Drug Delivery Reviews* 60(12):1329–1346.

Herold, D. M. et al. 2000. Gold microspheres: A selective technique for producing biologically effective dose enhancement. *International Journal of Radiation Biology* 76(10):1357–1364.

Jain, P. K. et al. 2008. Noble metals on the nanoscale: Optical and photothermal properties and some applications in imaging, sensing, biology, and medicine. *Accounts of Chemical Research* 41(12):1578–1586.

Jain, S. et al. 2011. Cell-specific radiosensitization by gold nanoparticles at megavoltage radiation energies. *International Journal of Radiation Oncology, Biology, Physics* 79(2): 531–539.

Jin, C. et al. 2007. Radiosensitization of paclitaxel, etanidazole and paclitaxel+etanidazole nanoparticles on hypoxic human tumor cells in vitro. *Biomaterials* 28(25):3724–3730.

Jones, B. L., S. Krishnan, and S. H. Cho. 2010. Estimation of microscopic dose enhancement factor around gold nanoparticles by Monte Carlo calculations. *Medical Physics* 37(7):3809–3816.

Kada, T., T. Noguti, and M. Namiki. 1970. Radio-sensitization with iodine compounds. I. Examination of damage in deoxyribonucleic acid with *Bacillus subtilis* transformation system by irradiation in the presence of potassium iodide. *International Journal of Radiation Biology & Related Studies in Physics, Chemistry & Medicine* 17(5):407–418.

Kim, I. A. et al. 2010a. HDAC inhibitor-mediated radiosensitization in human carcinoma cells: A general phenomenon? *Journal of Radiation Research (Tokyo)* 51(3):257–263.

Kim, J. K. et al. 2010b. Therapeutic application of metallic nanoparticles combined with particle-induced x-ray emission effect. *Nanotechnology* 21(42):425102.

Kim, J. et al. 2008. Photothermal response of superparamagnetic iron oxide nanoparticles. *Lasers in Surgery and Medicine* 40(6):415–421.

Kobayashi, K. et al. 2002. Enhancement of X-ray-induced breaks in DNA bound to molecules containing platinum: A possible application to hadrontherapy. *Radiation Research* 157(1):32–37.

Le Sech, C. et al. 2000. Strand break induction by photoabsorption in DNA-bound molecules. *Radiation Research* 153(4):454–458.

Le Sech, C. et al. 2001. Enhanced strand break induction of DNA by resonant metal-innershell photoabsorption. *Canadian Journal of Physiology and Pharmacology* 79(2): 196–200.

Lowery, A. et al. 2011. Tumor-targeted delivery of liposome-encapsulated doxorubicin by use of a peptide that selectively binds to irradiated tumors. *Journal of Controlled Release* 150(1):117–124.

Maeda, H. et al. 2003. Vascular permeability enhancement in solid tumor: Various factors, mechanisms involved and its implications. *International Immunopharmacology* 3(3):319–328.

McMahon, S. J. et al. 2008. Radiotherapy in the presence of contrast agents: A general figure of merit and its application to gold nanoparticles. *Physics in Medicine and Biology* 53(20):5635–5651.

Morgan, M. A. et al. 2010. Mechanism of radiosensitization by the Chk1/2 inhibitor AZD7762 involves abrogation of the G2 checkpoint and inhibition of homologous recombinational DNA repair. *Cancer Research* 70(12):4972–4981.

Nakamura, H. et al. 2009. Development of boron nanocapsules for neutron capture therapy. *Applied Radiation and Isotopes* 67(7–8 Suppl):S84–S87.

Nath, R., P. Bongiorni, and S. Rockwell. 1990. Iododeoxyuridine radiosensitization by low- and high-energy photons for brachytherapy dose rates. *Radiation Research* 124(3): 249–258.

Norman, A. et al. 1997. X-ray phototherapy for canine brain masses. *Radiation Oncology Investigations* 5(1):8–14.

Ostrowski, A. D., and P. C. Ford. 2009. Metal complexes as photochemical nitric oxide precursors: Potential applications in the treatment of tumors. *Dalton Transactions* 48:10660–10669.

Pamujula, S. et al. 2008. Radioprotection in mice following oral administration of WR-1065/PLGA nanoparticles. *International Journal of Radiation Biology* 84(11):900–908.

Peeters, S. T. et al. 2006. Dose-response in radiotherapy for localized prostate cancer: Results of the Dutch multicenter randomized phase III trial comparing 68 Gy of radiotherapy with 78 Gy. *Journal of Clinical Oncology* 24(13):1990–1996.

Petersen, M. S. et al. 2008. Boron nanoparticles inhibit tumour growth by boron neutron capture therapy in the murine B16-OVA model. *Anticancer Research* 28(2A):571–576.

Ping, Y. et al. 2010. Inhibition of the EGFR with nanoparticles encapsulating antisense oligonucleotides of the EGFR enhances radiosensitivity in SCCVII cells. *Medical Oncology* 27:715–721.

Polf, J. C. et al. 2011. Enhanced relative biological effectiveness of proton radiotherapy in tumor cells with internalized gold nanoparticles. *Applied Physics Letters*. 98(19):193702.

Pollack, A. et al. 2002. Prostate cancer radiation dose response: Results of the M. D. Anderson phase III randomized trial. *International Journal of Radiation Oncology, Biology, Physics* 53(5):1097–1105.

Qian, X. et al. 2008. in vivo tumor targeting and spectroscopic detection with surface-enhanced Raman nanoparticle tags. *Nature Biotechnology* 26(1):83–90.

Roa, W. et al. 2009. Gold nanoparticle sensitize radiotherapy of prostate cancer cells by regulation of the cell cycle. *Nanotechnology* 20(37):375101.

Robar, J. L. 2006. Generation and modelling of megavoltage photon beams for contrast-enhanced radiation therapy. *Physics in Medicine and Biology* 51(21):5487–5504.

Robar, J. L., S. A. Riccio, and M. A. Martin. 2002. Tumour dose enhancement using modified megavoltage photon beams and contrast media. *Physics in Medicine and Biology* 47(14):2433–2449.

Santos Mello, R. et al. 1983. Radiation dose enhancement in tumors with iodine. *Medical Physics* 10(1):75–78.

Schweitzer, A. D. et al. 2010. Melanin-covered nanoparticles for protection of bone marrow during radiation therapy of cancer. *International Journal of Radiation Oncology, Biology, Physics* 78(5):1494–1502.

Sofou, S. 2008. Radionuclide carriers for targeting of cancer. *International Journal of Nanomedicine* 3(2):181–199.

Verhaegen, F. et al. 2005. Dosimetric and microdosimetric study of contrast-enhanced radiotherapy with kilovolt x-rays. *Physics in Medicine and Biology* 50(15):3555–3569.

Wang, A. Z. et al. 2010. ChemoRad nanoparticles: A novel multifunctional nanoparticle platform for targeted delivery of concurrent chemoradiation. *Nanomedicine (London)* 5(3):361–368.

Werner, M. E. et al. 2011. Folate-targeted polymeric nanoparticle formulation of docetaxel is an effective molecularly targeted radiosensitizer with efficacy dependent on the timing of radiotherapy. *ACS Nano* 5(11):8990–8998.

Williams, L. E., G. L. DeNardo, and R. F. Meredith. 2008. Targeted radionuclide therapy. *Medical Physics* 35(7):3062–3068.

Wu, H. et al. 2008. Apoferritin-templated yttrium phosphate nanoparticle conjugates for radioimmunotherapy of cancers. *Journal of Nanoscience and Nanotechnology* 8(5):2316–2322.

You, Z. Y. et al. 2010. The radiosensitization effects of Endostar on human lung squamous cancer cells H-520. *Cancer Cell International* 10(1):17.

Zietman, A. L. et al. 2005. Comparison of conventional-dose vs high-dose conformal radiation therapy in clinically localized adenocarcinoma of the prostate: A randomized controlled trial. *Jama* 294(10):1233–1239.

Zou, J. et al. 2009. Inhibition of ataxia-telangiectasia mutated by antisense oligonucleotide nanoparticles induces radiosensitization of head and neck squamous-cell carcinoma in mice. *Cancer Biotherapy & Radiopharmaceuticals* 24(3):339–346.

Index

Page numbers followed by f and t indicate figures and tables, respectively.